More praise for *Cosmology's Century*

"A century of big ideas and powerful instruments has led us to the current model of our universe, with its inflationary beginning, cosmic structure built by the gravity of dark-matter particles, and accelerated expansion caused by dark energy. *Cosmology's Century* is a firsthand account of that remarkable period by Jim Peebles, who led this grand adventure with his manifold contributions and broad influence. A must-read for any serious student of cosmology."

—Michael S. Turner, Kavli Foundation and University of Chicago

"Jim Peebles has surely contributed more to the history of our understanding of the large-scale structure and evolution of the universe than anyone else still in a position to write about it; so written about it he has, and magnificently!"

—Virginia Trimble, Former President, International Astronomical Union, Division of Galaxies and the Universe

"An inspiring history of cosmic ideas."

—Joseph Silk, author of *The Infinite Cosmos: Questions from the Frontiers of Cosmology*

"Peebles offers a broad and deep description of cosmology, presenting the history of the field as well as many of the side turns, dead ends, and wrong paths that researchers explored along the way. I really enjoyed reading this book."

—David W. Hogg, New York University

COSMOLOGY'S CENTURY

Cosmology's Century

AN INSIDE HISTORY OF OUR MODERN UNDERSTANDING OF THE UNIVERSE

P. J. E. Peebles

PRINCETON UNIVERSITY PRESS
PRINCETON & OXFORD

Published by Princeton University Press
41 William Street, Princeton, New Jersey 08540
6 Oxford Street, Woodstock, Oxfordshire OX20 1TR

press.princeton.edu

ISBN 978-0-691-19602-2
ISBN (e-book) 978-0-691-20166-5

British Library Cataloging-in-Publication Data is available

Editorial: Jessica Yao and Arthur Werneck
Production Editorial: Brigitte Pelner
Jacket/Cover Design: Chris Ferrante
Production: Jacqueline Poirier
Publicity: Matthew Taylor (US), Katie Lewis (UK)
Copyeditor: Cyd Westmoreland

Jacket Image: Planck's Cosmic Microwave Background (CMB) Map, 2013, ESA and the Planck Collaboration

This book has been composed in Miller

Printed on acid-free paper ∞

Printed in the United States of America

10 9 8 7 6 5 4 3 2 1

To Alison, my best friend for six decades

CONTENTS

PREFACE AND ACKNOWLEDGMENTS

IT IS REMARKABLE that we can say with some confidence what the universe was like far away and in the remote past. The well-tested theory grew out of starting ideas from a century ago, in a reasonably simple way compared to other branches of natural science, and relatively few people were sifting through the clues to how to make progress. I have been one of that party for over half of the century since Einstein started us in about the right direction, and this book is my opportunity to put down what I understand to be what was happening and my impressions of why. It is generally acknowledged that in natural science, we take poor care of our history. I aim to present the story warts and all: the brilliant insights and lucky guesses, the roads not taken and the mistakes large and small, and the accumulation bits of evidence that at last began to fit together in a way that makes sense. The relative simplicity of the story makes it a good illustration of how natural science really is done.

I intend this book to combine an objective history of this subject with my own recollections, the latter indicated by use of the first person. I think the two are not seriously incompatible. We are not approaching a final theory (if there is such a thing), which means that assessments of where we are within our incomplete and approximate state of establishment of natural science call for judgments that cannot be objective. We must operate with subjective assessments of what we hope is reasonably objective evidence, while bearing in mind that some pieces of evidence are a lot more objective and informative than others.

I date the convergence of evidence to a credible theory of the large-scale nature and evolution of the universe to the years 1998 to 2003, and I end my account at this revolutionary half decade around the turn of the century. As I complete writing this book, in 2019, I enter occasional comments about what has happened since the revolution. It would be tedious to keep repeating "at the time of writing," so I leave this to be understood when I feel the context suggests it.

I begin this history with Einstein's introduction of the transformative general theory of relativity, which allows quantitative analyses of the nature of a universe without edges. What came before Einstein is important—it informed later thinking—but my comments are limited to the conceptual problems with an unbounded universe in Newtonian physics. I have taken the liberty of simplifying the story of what happened after Einstein by omitting paths that I do not think were useful (even as foils to more successful thinking) and are not likely to be missed. I apologize for and would appreciate being informed of inadvertent omissions of lines of research that arguably have socially redeeming value.

The post-revolution history of this subject is important, and it is interesting also to consider what things might be like in the distant future and what might have happened in the remote past, before the earliest stages of evolution about which we have useful evidence. But I do not discuss these considerations.

People now contributing to advances in research in cosmology should be aware of the history of their subject. I offer this book as a place to look up what happened years ago that helped set the community straight, more or less; it is complicated, of course. Cosmologists already know the technical aspects of this history, which are not all that difficult. I intend the explanations in this book to be understandable to an undergraduate who is thinking of majoring in physical science, and to a nonscientist who is fascinated by what has been learned about stars, galaxies, and the expanding universe and is willing to skip over the technicalities and pay attention to the descriptions. Details are useful to those inclined to examine them, so I have placed in footnotes those comments that I consider relevant but not essential to the broader picture described in the text. Footnotes also offer definitions of astronomers' sometimes curious conventions; they are best tracked down through the index. There are equations in the main text, because they are important to the story. Where the equations are dense, I intend the narrative to keep the big picture visible. There's nothing wrong with skimming over equations, to be sorted out later if they're found to be really needed for the reader's purpose. And I offer introductory and concluding sections in which I discuss the situation without equations.

I term this book a history, but it is written in the tradition of the physics I know and love, save only for my attempt to avoid our superficial creation stories. Discussions with professional historians have taught me that my approach certainly could be complemented by assessments in the traditions of historians and sociologists. I have written little about personalities, for example, or the evolving nature of support for research, and I do not mention much about means of communication, in earlier times by letters exchanged within the old boy network, now perhaps through blogs, of all things. Communications at conferences remain important, but conferences in this subject have become increasingly specialized as the reach of cosmology has grown, a troubling development in the eyes of many in my generation. But I must leave all this to real historians, philosophers, and sociologists of science, who I hope will understand that I transgress on their traditions because I operate in the traditions of a practicing physicist.

I offer thoughts about the nature and philosophy of the enterprise of natural science, informed by what happened in cosmology, in Section 1.1 and Chapter 10. We see in this history that scientists act as they do because they behave much like people in general, though they tend to be more compulsive about it. And with each advance in science, we see an addition to the evidence that there is an objective physical reality and that we are probing ever more deeply into its nature. There is nothing new in all this, but I think the examples to be drawn from the history of modern cosmology are particularly clear and informative, because the subject is relatively simple.

I have already presented portions of this story. What happened in George Gamow's research group in 1948 is considered in detail in Peebles (2014). The work in Bob Dicke's Gravity Research Group that was so important to the development of experimental gravity physics—and led to the recognition of the sea of thermal radiation remnant from the hot early universe—is reviewed in Peebles (2017). Recollections of research in the 1960s by those who were involved in the identification and interpretation of this fossil thermal radiation are in the book *Finding the Big Bang* by Peebles, Page, and Partridge (2009).

References to the research papers I consider important to the story are indicated by authors' names followed by the year and are listed in the References section at the end of the book. The list is dismayingly long, but it has to be: although this has been a relatively small science, its development took a lot of work. I have selected samples of the pioneering contributions and apologize to colleagues who have different opinions about this subjective matter. Page numbers following references indicate where the papers are cited in the text.

Some of the quotations in this book are taken from the literature, and the sources are so indicated. Where the quote is in French or German, I add my translation, sometimes condensed and aided by Google. This book offered an excellent opportunity to ask for recollections from those who have long memories of research in this subject. Quotes drawn from them for the purpose of this book are marked by the author's name and "personal communication." I have profited also from the advice of younger people, and from the wonders of the Internet. I am particularly thankful for NASA's Astrophysics Data System Bibliographic Services archive, a most useful tool for tracking down research papers from times past.

Figures that illustrate data can be influential, and the evolving nature of these figures is a part of the history. I am grateful to colleagues who gave me figures they made and own; their names are mentioned in the captions. The figures I made for this book, or made in times past but never published, have no references in the captions. Captions state sources of the many figures that have been taken from the literature, and the copyright holder can be traced through the reference to the publication. Copyright holders have a broad variety of prescriptions for statements of permission to reproduce, and their conditions for permission range from casual statements that reuse of figures is OK to payments required to reproduce two of the figures in this book taken from the publication of an otherwise respectable scholarly society. I take this confusion of permissions to be a consequence of the natural desire of publishers to keep some control over their content while the ease of taking figures from the literature for use in lectures can readily spill over into publications. I apologize for any permissions to reproduce I may have improperly stated or overlooked, and if notified will make amends in later printings.

The color plates that appear within chapter 9 are a sample of the actors in this history; I mean them to be reminders of the people behind those equations and measurements. I apologize to valued colleagues whose photos could have been included if space in this book and the energy to collect them had been more freely available. The text accompanying the photographs is my opportunity to comment on the stories behind the images, the analog in print of teachable moments.

My choice of units follows customs that tend to differ in different lines of research. In some parts of cosmology, the units usually are chosen so the velocity of light is unity. These equations look odd to me when the symbol c is entered, a matter of conditioning of course, but I follow tradition, which seems appropriate, since this is a history. In other places, Planck's constant $\hbar$ is unity, or Newton's constant G is unity. The old centimeter, gram, second units are being replaced by meters, kilograms, seconds. I suppose this is a sensible move, but the change is slow, and again I follow the history in staying with the former.

The index lists only a few of the pioneers of cosmology. This is a subjective choice, as is whatever else is deemed appropriate for an index. It would make no sense to place in the index the many appearances in the text of the word "redshift," so I index only the definition that appears early in the text. The word "inflation" appears a lot, too, and I enter the first significant commentary about the concept and later page numbers in which cosmological inflation is particularly relevant. But such algorithms are only of limited help with so many decisions.

This account may seem overly centered on the small town of Princeton in the small state of New Jersey. That is inevitable, in part because I have been a member of Princeton University since arriving here as a graduate student in 1958, but inevitable in even larger part because a good deal of the story happened here. My role in this story was aided by sabbatical leaves at the California Institute of Technology; the University of California, Berkeley; the Dominion Astrophysical Observatory in British Columbia; the University of Cambridge; and on two occasions, the Institute for Advanced Study in Princeton. I learned a lot at these places.

I have benefited from advice from and recollections of many colleagues: Neta Bahcall, John Barrow, Dick Bond, Steve Boughn, Michele Cappellari, Claude Carignan, Ray Carlberg, Rick Carlson, Robin Ciardullo, Don Clayton, Shaun Cole, Ramanath Cowsik, Marc Davis, Richard Dawid, Jaco de Swart, Jo Dunkley, John Ellis, Wyn Evans, Sandra Faber, Kent Ford, Ken Freeman, Carlos Frenk, Masataka Fukugita, Jim Gunn, David Hogg, Piet Hut, David

Kaiser, Steve Kent, Bob Kirshner, Al Kogut, Rocky Kolb, Andrey Kravtsov, Rich Kron, Malcolm Longair, Gary Mamon, John Mather, Adrian Melott, Liliane Moens, Richard Mushotzky, Kieth Olive, Jerry Ostriker, Lyman Page, Bruce Partridge, Will Percival, Saul Perlmutter, Mark Phillips, Joel Primack, Martin Rees, Adam Riess, Brian Schmidt, Jerry Sellwood, Joe Silk, David Spergel, Ed Spiegel, Paul Steinhardt, Matthais Steinmetz, Michael Strauss, Alex Szalay, Alar Toomre, Rien van de Weygaert, Hugo van Woerden, Steve Weinberg, Rainer Weiss, Cyd Westmoreland, Simon White, Ned Wright, Jessica Yao, and Matias Zaldarriaga. I surely have forgotten to mention some; my sincere apologies.

COSMOLOGY'S CENTURY

CHAPTER ONE

Introduction

THE STORY OF how cosmology grew is fairly simple, compared to what people have been doing in other branches of science, but still complicated enough that sorting it out requires a better plan than the common practice in science. Papers reporting research in cosmology and other parts of physics usually begin with an outline of what came before. Abandoned ideas and roads not taken are seldom mentioned, and there is the natural human tendency to follow patterns of attributions found in introductions in other recent papers. This builds evolving creation stories that efficiently set the current context for the research to be described. We tell these creation stories in the classroom for a quick introduction to what we are really interested in: the nature of the science. But the stories tend to be at best only vaguely related to what actually happened. Their gross incompleteness may not be a problem for ongoing research, except of course when good ideas have been overlooked or abandoned and lost. But the creation stories leave a woefully incomplete and inaccurate impression of how science is done.

To do better, we have to look further back in time, and we certainly have to consider the ideas that seemed interesting but were falsified or otherwise found not to be so interesting after all. A closer account of how cosmology grew presented in chronological order would be awkward, because different parts of what became the established theory were making progress at different rates following different methods and motivations until they started to come together. This account accordingly presents histories of six lines of research that were developing more or less separately. They are reviewed in Chapters 2 to 7. The advantage is a modest degree of continuity within each chapter. The disadvantage is the need to refer back and forth in time to what was happening in different lines of research. The arrangement is explained in more detail in Section 1.2 in this chapter, in the form of an outline and guide to the story to come. But first let us consider our traditions of research in the natural sciences, with particular attention to the operating conditions in cosmology.

1.1 The Science and Philosophy of Cosmology

The starting assumption for cosmology, as in all branches of natural science, is that nature operates by kinds of logic and rules that we can discover by careful examination of what is observed, informed by past experience of what has worked. The results are impressive; I urge any who might disagree to consider the rich fundamental physics employed in the construction and operation of their cellphones. But despite the many demonstrations of its power, physics, along with all the rest of natural science, is incomplete. Maybe discoveries to come will make the physical basis for science complete, revealing the final rules by which nature operates. Or maybe it's successive approximations all the way down.

The standard and accepted methods of science must be adapted to what can be done, of course. In physical cosmology and extragalactic astronomy, we can look but never touch. In cosmology, we cannot run the experiment again; we must instead resort to what can be inferred from fossils of times past. We find some fossils relatively nearby, as in the rocks on Earth and the stars in our galaxy and others, all of which have their own creation stories. Our past light cone offers us views of times past, because radiation detected here has been approaching us at the speed of light: the greater the distance of an object, the earlier in the evolution of the universe it is observed. Our light cone integrated through human history captures an exceedingly thin slice of what has been happening, but it reveals the way things were over a long range of time in a large universe that offers a lot to see and to seek to interpret.

The research path to where we are now in cosmology is marked by debates on open questions, as is usual in natural science. But the issues in cosmology have been defended and criticized with considerably more vigor than might have been expected from the modest weight of the evidence at the time. This was in part because observations that might settle questions in cosmology have tended to seem just out of reach or perhaps just barely possible. And I think an important factor has been the tendency to take a personal interest in the nature of our world. Is the universe really evolving, or might it be in a steady state? If evolving, how might it all end, in a big crunch or a big freeze? And where did it all come from? Such debates are quieter now, because we at last have a theory that passes an abundance of tests, but they continue.

Research in cosmology in the twentieth century usually was done in small groups, often an individual working alone or maybe with a colleague or a student or two. In the twenty-first century, ongoing research in cosmology grew richer and called for larger groups to develop special-purpose equipment for data acquisition, which in turn called for groups of comparable size to reduce the data and interpret it. Big Science has become important to this subject: We have to get used to gathering data in vast amounts, analyzing these data, and employing massive numerical simulations that help bridge the gap between theory and observation. But Big Science best takes aim at well-motivated and sharply defined questions. The main considerations in this book are about how small groups working on seemingly independent lines of research found their results coming together in a cosmology that looked good enough to call for the demanding tests afforded by Big Science. I date this revolutionary convergence to a credible theory to the half decade from 1998 to 2003.

Research certainly continued to be active and productive after the revolution; the difference is that the community had agreed on a paradigm, in Kuhn's (1962) terms. (This is what the majority was thinking, of course; not all agreed.) An example of the adherence to the normal science of cosmology is the study of how the galaxies formed and evolved, which builds theories of galaxy formation on the standard and accepted theory of the evolution of the universe. Normal scientific research of this sort may uncover anomalies that point to a still better underlying theory. This is a point of particular interest in cosmology, because the theory is at the same time well and persuasively tested and particularly incomplete.

Our present normal science of cosmology includes an excellent case for the presence of dark matter that interacts weakly if at all with ordinary matter. There are tight constraints on the properties of dark matter, but no clear evidence exists of detection of this substance other than the inference from the effects of its gravitational attraction. Some argue that dark matter will remain only hypothetical until there is more evidence of it than that: maybe detection in the laboratory, maybe indications of what it is doing to galaxies apart from holding them together. Others argue that the case for dark matter already is so tight that it is abundantly clear that the dark matter really exists. The same applies to Einstein's cosmological constant, Λ. It has gained a new name: dark energy. But that is a poor disguise for a fudge factor that we accept because it serves to unify theory and observations so well. There are other fudge factors, hypotheses to allow the theory to save the phenomena, in the present standard science of cosmology and in all the other branches of natural science. Research in the sciences continues to improve tests of our theories that, whether intended or not, may lead to better theories that inspire new tests. And they might on occasion replace fudge factors with unified theories in paradigms that bring parts of this enterprise closer together. It happens.

The physical cosmology that is the subject of this history is an empirical science, that is, it is based on and tested by what can be observed or measured by detectors, such as microscopes and telescopes and people. But we must pay attention to the role of theory, and intuition, and what Richard Dawid (2013 and 2017) terms "nonempirical theory assessment." The prime example in this history is that during most of the past century of research in cosmology, the community majority implicitly accepted Einstein's general theory of relativity. Few pointed out that this is an enormous extrapolation from the few meager tests of general relativity that we had in the 1960s. By the 1990s, as research in cosmology was starting to converge on a well-tested theory, there were demanding checks of the predictions of general relativity on scales ranging from the laboratory to the solar system, probing out to length scales of about 10^{13} cm. But the application to cosmology on the scale of the Hubble length, about 10^{28} cm, extrapolates from the precision tests by some fifteen orders of magnitude in length scale. This was not often mentioned, in my experience, and when mentioned, it tended to make some scientists a little uneasy, at least temporarily. In the first decades of the twenty-first century, the parts of general relativity that are relevant to the standard cosmology have passed an abundance of demanding tests. In short, the theory Einstein built on laboratory experiments was seriously tested only by the orbit of the planet Mercury. (The test of the prediction of the gravitational deflection of light by the mass of the sun, led by the people pictured in Plate III, was heavily cried up but in retrospect, their evidence seems marginal.) We find that this theory successfully extrapolates to applications on the immense scales of the observable universe. It is a remarkable result.

General relativity is an elegant extension of electromagnetism in flat spacetime; it has been said that it is a theory waiting to be found (though that is easier to say in hindsight). The faith in its extrapolation exemplifies the powerful influence and very real successes of nonempirical theory assessment. Of course, influential nonempirical assessments can mislead: Consider that in the 1930s through the 1990s, few objected to the assertions by respected experts that Einstein's cosmological constant, Λ, surely may be discarded. The evidence now is that Λ, under its new name—dark energy—is an essential part of our well-tested cosmology.

The practice of nonempirical assessments is sometimes termed "post-empiricism," but I have not found this term in Dawid's writing. Dawid (in a personal communication, 2018) states instead that

> non-empirical assessment as I understand it crucially depends on the ongoing collection of empirical data elsewhere in the research field and on the continued search for empirical confirmation of the theory under scrutiny. In a "post-empirical" phase where no substantially new data comes in any more, non-empirical assessment would get increasingly questionable and eventually would come to a halt as well.

This is consistent with what I understand to be normal practice in the physical sciences. That is, I have in mind the kind of nonempirical assessments we have been practicing all along without thinking much about it.

I take account of three other kinds of assessments: personal; community, though some may disagree; and pragmatic. The first two speak for themselves. I take examples of the third from cosmology. The usual practice has been to analyze data and observations in terms of general relativity. This surely has been due in part to the beauty of the theory, and in part to respect for Albert Einstein's magnificent intuition. But it was important also that the use of a common theory allowed comparisons of conclusions from independent analyses of the same or different data on a common fundamental ground. I do not imagine much thought has been given to this point, but I believe the implicitly pragmatic approach in cosmology (and I suppose in other branches of natural science) has helped reduce the chaos of multiple theories.

The pragmatic approach to science, if carried too far, could waste time and resources by directing research along a path as it grows increasingly clear that something is wrong. And even if the popular and pragmatically chosen path proves to be leading us in a useful direction, it can be important to have well-defended alternatives to standard ideas to motivate careful evaluations of approved ideas and observations. It may reveal corrections large or small that point toward a more profitable path. For example, a stimulating proposal in the mid-twentieth century was that textbook physics may have to be adjusted to include continual spontaneous creation of matter. The brave souls who argued for this steady-state cosmology were not always gently treated, but from what I saw, they gave as good as they got in debates over the relative merits of the general relativity and steady-state world views, arguments that were more intense than warranted by the evidence for or against either side. The idea of continual creation in the universe as it is now is no longer seriously considered in cosmology, but it had a healthy effect. New ideas can inspire defense and attacks that stimulate research, while a pragmatic defense of the old ways may help keep research from degenerating into confusion.

An important example of an implicitly pragmatic assessment is the general acceptance of Einstein's proposal that the universe is homogeneous in the average over local irregularities. Prior to the 1960s, there was scant evidence of this. Maps of distributions of the galaxies across the sky suggested instead that the galaxies are moving away from one another into space that is asymptotically empty or close to it, as in a fractal galaxy distribution. But whether by accident or design, this quite pertinent thought was put aside for the most part, and the main debate kept more sharply focused on the concepts of evolution or else a steady state of a nearly homogeneous universe. The first serious evidence for homogeneity came a half century after Einstein, from research for other purposes in the 1960s, as will be discussed in Chapter 2. Whether by good luck or good taste, the community was not much distracted by the elegant but wrong idea of a fractal universe.

It is not always easy to see why some issues receive much more attention than others; I suppose such things are to be considered eventualities. We do have reasonably clear standards for rejecting an apparently interesting idea. For example, the steady-state cosmology introduced in 1948 is elegant, but its predictions clearly violate the later accumulation of empirical tests. I do not know of a clear prescription for a move in the other direction, namely, the promotion of a working model to a standard theory. We might use the term "community opinion" to describe such decisions.

In 1990, general relativity usually was taken to be the appropriate basis for the study of the large-scale nature of the universe, but as argued above, it was an implicitly pragmatic assessment that the theory was serving well as a working basis for research. In 2003, after the revolution, the cosmological tests gave weight to the community opinion that the universe actually is well described by general relativity applied to the set of assumptions in what became known as the ΛCDM cosmological model. The introduction of these assumptions, including Einstein's cosmological constant Λ and the hypothetical cold dark matter, is reviewed in Section 8.2. Some disagreed, to be sure, but to most the accumulation of evidence (reviewed in Chapter 9) had become tight enough to have emboldened talk of what "really happened" far away and in the remote past, based on the ΛCDM theory. The notion of reality is complicated, so a more secure statement would be that whatever happened—and we assume something did happen—left traces that closely resemble those predicted by ΛCDM. And the traces are abundant and well enough cross-checked that the community opinion, including mine, is that this theory almost certainly is a useful though incomplete approximation to what actually happened.

1.2 An Overview

I have sorted this history of cosmology into lines of research that operated more or less independently of one another through stretches of time in the twentieth century. I consider the developments in each of the lines of research roughly in chronological order, but because different lines of research were at best only loosely coordinated, there have to be references back and forth in time as different lines of research started to interact. This outline is meant to explain how I have arranged the presentation of the research and how it all fits together, at least roughly, apart from the wrong turns taken.

I begin in Chapter 2 with considerations of Albert Einstein's (1917) proposal, from pure thought, that a philosophically sensible universe is homogeneous and isotropic: no preferred center or direction, no observable edges to the universe as we see it around us. That of course is apart from the minor irregularities of matter concentrated in people and planets and stars. Einstein's homogeneity is essential to the thought that we might be able to find a theory of the universe as a whole rather than of one or another of its parts. It was an inspired intuitive vision or maybe just a lucky guess; Einstein certainly had no observational evidence that suggested it. The history of how Einstein's thought was received and tested exemplifies the interplay in science between theory and practice, sometimes reinforcing each other; sometimes in serious tension; and, as in this case, sometimes aided by unexpected developments. Because I have not found a full discussion elsewhere, I consider in some detail the development of the evidence that supports what became known as Einstein's cosmological principle.

Einstein's general theory of relativity predicts that a close-to-homogeneous universe has to expand or contract. Expansion was indicated by astronomers' observations that starlight from galaxies of stars is shifted to the red, as if Doppler shifted, because the galaxies are moving away from us. Chapter 3 reviews the importance of the discovery that the Doppler shift, or redshift, is larger for galaxies that are farther away. This is the expected behavior if the universe is expanding in a nearly homogeneous way. The big bang cosmology discussed in Sections 3.1 and 3.2 uses the general theory of relativity to describe the evolution of a near-homogeneous expanding universe.

We should pause here to note that the name, "big bang," is inappropriate, because a bang connotes an event in spacetime. Unlike a familiar bang, this cosmology has nothing to do with a special position or time. The theory is instead a description of cosmic evolution of a universe that is homogeneous on average, and it attempts to follow cosmic evolution to the present from the earliest time of formation of fossils that can be observed and interpreted. That has come to include the epoch of light-element formation, when the temperature of the universe was some nine orders of magnitude larger than it is now. This is a spectacular extrapolation back in time, but not to a bang, and not to a singular start of things: We must assume that something different happened before the singularity. Simon Mitton (2005) concludes that Fred Hoyle coined the term "big bang" for a lecture on BBC radio in March 1949. It was meant as a pejorative; Hoyle favored the steady-state picture. Though unfortunate, the name "big bang" is commonly accepted. I have not encountered a better term, and the pragmatic assessment is that it is to be used in this book.

It was important that there were testable alternatives to the big bang picture; these alternatives inspired the search for tests. The leading idea, the steady-state model, is discussed in Section 3.3. It will be termed the "1948 steady-state model" to distinguish it from variants introduced later. In contrast to the prominence of the steady-state alternative to the big bang model through the mid-1960s, the leading alternative to Einstein's idea of homogeneity—a fractal distribution of matter—only became widely discussed after we at last had reasonably clear evidence of homogeneity (Section 2.6).

Hermann Bondi's (1952, 1960) book *Cosmology* in two editions, gives a valuable picture of thinking at the time. Which if either of the big bang or 1948 steady-state models, or perhaps some other model then still being considered, is the most reasonable and sensible, and on what grounds, empirical or nonempirical? Helge Kragh (1996) presents a historian's perspective of this mainstream research in cosmology up to the 1960s. Sections 3.4–3.7 augment these sources with my thoughts about the similarities and differences of assessments of the two cosmologies. I take it that in the 1950s and early 1960s, nonempirical issues account for the lack of popularity of the steady-state model in many quarters, despite its greater predictive power for observers.

The weaker predictive power of the big bang model may help account for the abundance of nonemipirical assessments discussed in Section 3.5.

The greatest effort devoted to the empirical study of the big bang cosmological model in the years around 1990 was the measurement of the mean mass density. Sections 3.6.3 and 3.6.4 review the considerable variety of these probes, and Section 3.6.5 offers an overview of what was learned. The motivation for this large effort was in part to see whether the mass density is large enough that its gravity will cause the expansion to stop and the universe to collapse, and the results were important for the empirical establishment of cosmology. But I think in large part the motivation became simply that this is a fascinating problem whose resolution is difficult but maybe not quite impossible.

The topic of Chapter 4 is the informative fossils left from a time when the universe was very different from now, dense and hot enough to produce the light elements and the sea of thermal radiation that nearly uniformly fills space. Since it was (and is) exceedingly difficult to imagine how the light elements and the radiation with its thermal spectrum could have originated in the universe as it is now, these fossils were a valuable addition to the evidence that our universe is evolving, not in a steady state. The book *Finding the Big Bang* (Peebles, Page, and Partrige 2009) recalls how these fossils were recognized in the mid-1960s, with recollections from those involved of how the recognition led to the research that produced the first good evidence that our universe really did evolve from a hot early state at about the rate of expansion predicted by general relativity. The tangled story of how Gamow and colleagues anticipated these fossils a decade before they were recognized is presented in the paper, "Discovery of the Hot Big Bang: What Happened in 1948" (Peebles 2014). Section 4.2 presents a shorter version of the main points. The sea of thermal radiation has become known as the cosmic microwave background, or CMB. The later developments leading to its central place in the revolution that established the ΛCDM cosmology are reviewed in Chapter 9. This theory of the expanding universe assumes the general theory of relativity applied to a close-to-homogeneous universe (Chapter 2), the presence of Einstein's cosmological constant Λ (Section 3.5), dark matter (Chapter 7), and particular choices of initial conditions (Section 5.2.6).

It was natural to explore how the very evident departures from Einstein's homogeneity—stars in galaxies in groups and clusters of galaxies—might have formed in an expanding universe. In the established cosmology, cosmic structure formed by the gravitational instability of the relativistic expanding universe. The early confusion about the physical meaning of this instability is an important part of the history. These considerations are reviewed in Chapter 5, along with assessments of early scenarios of how cosmic structure might have formed. The importance of these considerations for the convergence to the standard cosmology is a recurring topic throughout the rest of this book.

The subject of Chapter 6 is the astronomers' discoveries of apparent anomalies in the measurements of masses of galaxies and concentrations of galaxies. Other accounts of the exploration of these phenomena are in Courteau *et al.* (2014) and de Swart, Bertone, and van Dongen (2017). Fritz Zwicky was the first to recognize the phenomenon: He saw that the galaxies in the rich Coma Cluster of galaxies seem to be moving relative to one another too rapidly to be held together by the gravitational attraction of the mass seen in the stars in the galaxies in the cluster. One way to put it is that the mass required to hold this concentration of galaxies together by gravity seemed to be missing, always assuming the gravitational inverse square law of gravity (in the nonrelativistic Newtonian limit of general relativity). It was later seen that mass also seemed to be missing from the outer parts of spiral galaxies, based on the measurements discussed in Section 6.3 of circular motions of stars and gas in the discs of spiral galaxies. Much the same conclusion came from the studies described in Section 6.4 of how galaxies with prominent discs acquired their elegant spiral patterns. By the mid-1970s, it had become clear that understanding this is much easier if the seen mass is gravitationally held in near-circular motion in the disc with the help of the gravitational attraction of less-luminous matter that is more securely stabilized by more nearly random orientations of the orbits.

These observations pointed to a key idea for the establishment of cosmology: the existence of "dark matter," the new name for what was variously known as "missing," "hidden," or "subluminal" mass. The idea came almost entirely out of pursuits in astronomy, not cosmology, and for this purpose, the subluminal component need not be very exotic: low-mass stars would do, though they would have to be present in surprising abundance relative to counts of the more luminous observed stars. But in the 1970s, another key idea for cosmology was growing out of particle physicists' growing interest in the possible forms of nonbaryonic matter. Gas and plasma, people, planets, and normal stars are all forms of what is termed "baryonic matter." Most of the mass of baryonic matter is in atomic nuclei; the accompanying electrons are termed "leptons," but they are also counted in the mass of baryonic matter. The neutrinos are leptons that we now know have small but nonzero rest masses. Thus they act as nonbaryonic dark matter that contributes to the masses of galaxies, but in the standard cosmology, this contribution is much smaller than the total indicated by the astronomical evidence. We need a new kind of nonbaryonic matter.

The thought that the astronomers' subluminal matter is the particle physicists' nonbaryonic matter and the cosmologists' dark matter was and remains a conjecture at the time of writing. The only empirical evidence of the new nonbaryonic dark matter is the effect of its gravity. It has been a productive idea, however, that passes demanding checks. The particle physicists' considerations of nonbaryonic matter reviewed in Chapter 7 takes into account the condition that if this nonbaryonic matter were produced in the hot early stages of expansion of the universe, then its remnant mass density must not exceed that allowed by the relativistic big bang cosmological model (again, assuming the relativistic theory). But it is notable that cosmologists took over the notion of nonbaryonic dark matter before the particle physics community had taken much interest in the astronomers' evidence of the presence of subluminal matter.

The nonbaryonic dark matter most broadly discussed in the 1980s came in two varieties, cold and hot. The latter would be one of the known class of neutrinos with rest mass of a few tens of electron volts (Sections 5.2.7 and 7.1). The initially hot (meaning rapidly streaming) neutrinos in the early universe would have smoothed the mass distribution, and that smoothing would have tended to cause the first generation of structure to be massive systems that must have fragmented to form galaxies. The spurious indication in 1980 of a laboratory detection of a neutrino mass appropriate for the hot dark matter picture certainly enhanced interest in the indicated formation of galaxies by fragmentation. This model was considered but had to be rejected: the observations show hierarchical growth of structure, from smaller to larger mass distributions.

The prototype for the nonbaryonic matter that is an essential component of the established cosmology was introduced by particle physicists in 1977. The idea occurred to five groups who published in the space of 2 months. These papers do not exhibit much interest in the astronomers' subluminal mass phenomena, but the considerations certainly were relevant to subluminal matter. Was this a curious coincidence or an idea that somehow was "in the air?" This is considered a little further in Sections 7.2.1 and 10.4.

Sections 8.1 and 8.2 review why in the early 1980s cosmologists co-opted the astronomers' subluminal mass and the particle physicists' nonbaryonic matter in what became known as the standard cold dark matter, or sCDM, cosmological model. The letter "s" might be taken to mean that the model was designed to be simple (as it was) but it instead signified "standard," not because it was established but because it came first. It was meant to distinguish this version from the many variants to be considered in Section 8.4. A large part of the cosmology community soon adopted variants of the sCDM model as bases for exploration of how galaxies might have formed in the observed patterns of their space distribution and motions (Section 8.3), and for analyses of the effect of galaxy formation on the angular distribution of the sea of thermal radiation. This widespread adoption was arguably overenthusiastic, because it was easy to devise other models, less simple to be sure, that fit what we knew at the time. And it was complicated by the nonempirical feeling that space sections surely are flat. In general relativity that could be because the mass density is large enough to produce flat space sections, or because Einstein's cosmological constant, Λ, makes it so. The nonempirical reasons for preferring flat space sections, preferably without resorting to Λ, are discussed in Section 3.5. These reasons were influential and long-lasting enough to have played a significant role in the confusion of variants and alternatives to the sCDM idea considered in the 1990s.

The reduction of confusion in the years 1998–2003 was great enough to be termed a revolution. It was driven by the two great experimental advances discussed in Chapter 9. The first is the measurement of the relation between the redshift of the spectrum of an object and its brightness in the sky, given its luminosity: the cosmological redshift–magnitude relation. Its detection had been a goal for cosmology since the 1930s; it was at last accomplished by two independent groups at the turn of the century (Section 9.1). The second is the detailed mapping of the angular distribution of the CMB radiation. Work on this began in the mid-1960s, and coincidently also produced demanding constraints on cosmological models at the turn of the century. These results from the two sets of measurements, together with what was already known, made a tight case for the presence of Einstein's cosmological constant Λ and the nonbaryonic CDM in the relativistic hot big bang ΛCDM theory. It was a dramatic development.

It was proper to have asked whether the introduction of two very significant hypothetical components, CDM and Λ, along with all the other assumptions that go into the choice of a cosmological model, might only amount to adjusting the theory to fit the measurements. That line of debate did not become very prominent, because the ΛCDM cosmology that fit the two critical measurements brought together so many other lines of evidence in a tight network of empirical tests. This is the topic of Section 9.3.

By the year 2003, the community had at last settled on a respectably well-supported theory of the large-scale nature of the universe. Skeptics remained, as is appropriate, for this theory is an immense extension of the reach of established physics. Indeed, the 2003 theory has been modified to fit later measurements, but these changes amount to fine adjustments of parameters, not challenges to the basic framework of the theory. It is the nature of science to advance by successive approximations, and it would not be at all surprising to find that there is a still better theory than ΛCDM. But we have excellent reason to expect that a better theory will describe a universe that behaves much like ΛCDM, because ΛCDM passes an abundance of empirical tests that probe the universe in so many different ways.

I cannot think of any lesson to be drawn from this story of how cosmology has extended the boundaries of established science that cannot be drawn from other branches of natural science. This is no surprise, because cosmology operates by the methods of natural science. But I think there are lessons to be drawn with greater clarity in the relatively uncluttered historical development of this subject. My offerings are given in Chapter 10.

CHAPTER TWO

The Homogeneous Universe

MODERN COSMOLOGY GREW out of Albert Einstein's search for how his general theory of relativity might apply to the large-scale nature of the universe. Einstein's (1917) thought was that a philosophically reasonable universe is the same everywhere and in all directions, apart from minor irregularities, such as the observed concentrations of matter in planets and stars. This is a distinct departure from the tradition of research in natural science, which is to select for examination a level in a hierarchy of structure. It may be the examination of molecules; the atoms in molecules; the nuclei in atoms; the nucleons in nuclei; or the quarks and gluons in nucleons. One can examine structure on larger scales: the vast complexity of interactions of atoms and molecules in condensed matter, chemistry, and on up to biophysics; or the natures of planets around stars, stars in galaxies, or galaxies in groups and clusters and superclusters of galaxies. Einstein's thought was that this hierarchy of structures ends in something new to modern science: large-scale homogeneity. (Although not stated explicitly at first, the thought includes large-scale isotropy. That is, the universe is assumed to be invariant under rotations as well as translations.)

Einstein's homogeneity assumption allows us to consider and test the possibility of a theory of the universe as a whole, rather than a theory of a particular level in a hierarchy. If the universe is homogeneous in the large-scale average, then observations from our position may inform the theory of what the universe is like when observed from any other place. But we need evidence that this approximation is useful.

2.1 Einstein's Cosmological Principle

Einstein's (1917) original argument for the picture of large-scale homogeneity is difficult to assess. He argued against the idea that the material content of the universe might be confined to a single concentration, an island universe in otherwise empty space. If this were so, and the escape velocity were finite, then stars would evaporate, escaping the island universe. This behavior would be contrary to his implicit assumption that the universe is in a stationary state. If the escape velocity were arbitrarily large, then statistical relaxation would produce the occasional star moving with arbitrarily large speed. This might be taken to be contrary to the observation that the velocities of nearby stars are much smaller than the velocity of light. Both points would make some sense if the universe were not evolving and the stars had had time to approach statistical equilibrium. Einstein does not seem to have paused to consider that if energy is conserved, then the stars must eventually stop shining. And if stars nevertheless shine forever, then his homogeneous universe would be full of starlight. This is Olbers' paradox, and is certainly an unacceptable situation.

The argument that may be closer to what Einstein was thinking in 1917 is stated in *The Meaning of Relativity*, the publication of his lectures at Princeton University in 1921 (Einstein 1923). He pointed out that his general relativity allows solutions in which there is a single mass concentration outside of which spacetime is empty and asymptotically flat, or as Einstein put it, quasi-Euclidean. Motions of matter in this mass concentration would have the usual properties of acceleration, such as the flattening of a gravitationally bound rotating galaxy. But in a nonrelativistic mass concentration, this rotation would be relative to empty spacetime. Thus Einstein (1923, 109) wrote: "If the universe were quasi-Euclidean, then Mach was wholly wrong in his thought that inertia, as well as gravitation, depends upon a kind of mutual action between bodies."

A similar sentiment, expressed in Einstein (1917), is that (in an English translation): "In a consistent theory of relativity there can be no inertia *relative to "space,"* but only an inertia of masses *relative to one another.*" He went on to point out that in his general relativity, a single particle of mass in otherwise flat spacetime would have inertia, contrary to his stated view of relativity.

Einstein (1923, 110) argued that it is "probable that Mach was on the right road" in the relativity of inertia, and cited three examples:

1. The inertia of a body must increase when ponderable masses are piled up in its neighborhood.
2. A body must experience an accelerating force when neighboring masses are accelerated, and, in fact, the force must be in the same direction as the acceleration.
3. A rotating hollow body must generate inside of itself a "Coriolis field," which deflects moving bodies in the sense of the rotation, and a radial centrifugal field as well.

With all respect to Einstein's genius, we must observe that the first example, if meant as a local measurement, may follow from Mach's principle, but it is not true in general relativity. This theory predicts that an observer confined to a space small enough that tidal fields may be neglected sees the same universal local physics, including the usual properties of inertia, whatever the environment. An operational meaning of the second example seems to be equivalent to the third. This is the Lense-Thirring effect: An inertial frame of reference near a rotating massive body rotates relative to distant matter as if the inertial frame were dragged by the rotation of the massive body. The effect has since been observationally checked.

The prediction in general relativity is in line with the thought that acceleration, like motion, surely is meaningful only relative to what the rest of the universe is doing. This certainly seems to be the direction of Ernst Mach's thinking (as expressed in his book, *Die Mechanik in Ihrer Entwicklung Historisch-Kritisch Dargestellt*, and on pages 283–285 in the English translation in Mach 1960). And we must consider that Einstein's reading of what he termed "Mach's principle" led him to an idea that is now clearly established: The observable universe is very close to homogeneous. Debate continues on whether Einstein was right about this for the right reason.

To make acceleration relative within general relativity, Einstein had to remove the possibility of a quasi-Euclidean universe. He did so by proposing as a sort of boundary condition that the universe is homogeneous: It has no preferred center and no edges. Space is to be pictured as nearly uniformly filled everywhere with matter and radiation.

The paper by Willem de Sitter (1917a, 3) gives some indication of Einstein's thinking:

> The most desirable and the simplest value for the $g_{\mu\nu}$ at infinity is evidently zero. Einstein has not succeeded in finding such a set of boundary values[1] and therefore makes the hypothesis that the universe is not infinite, but spherical: then no boundary conditions are needed, and the difficulty disappears.... The idea to make the four-dimensional world spherical in order to avoid the necessity of assigning boundary-conditions, was suggested several months ago by Prof. Ehrenfest, in a conversation with the writer. It was, however, at that time not further developed.

(I cannot follow the comments in de Sitter's footnote.) Spherical space, closed as is the surface of a sphere, has no boundary on which we must assign conditions, and it can be assumed to be close to homogeneous. We see that the bold and eventually successful idea of homogeneity grew out of some mix of philosophy and intuition, supplemented by interactions with colleagues and perhaps aided by some measure of wishful thinking. It certainly was not based on any empirical evidence.

Edward Arthur Milne recognized the power of homogeneity in formulating a cosmology, and he named the assumption "Einstein's cosmological principle." Milne (1933) showed that, independent of general relativity, this principle with

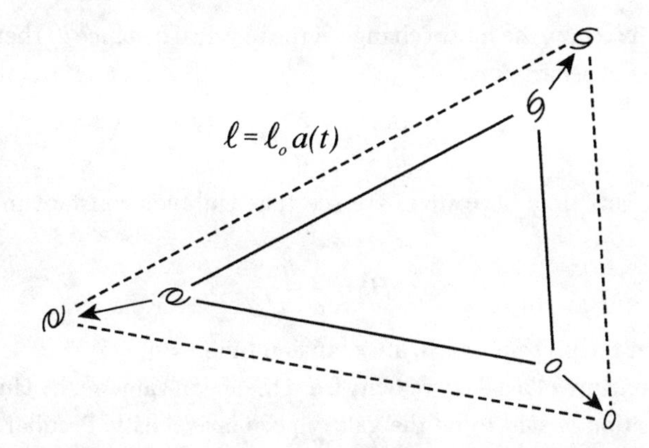

FIGURE 2.1. Homogeneous and isotropic expansion (Peebles 1980).

standard local physics accounts for a central feature of cosmology: the relation between the recession velocity v of a galaxy and its distance r,

$$v = cz = H_0 r, \tag{2.1}$$

where H_0 is the constant of proportionality. To see this, write the velocities of the galaxies as the vector relation $\vec{v} = H_0 \vec{r}$. Then an observer on galaxy a sees galaxy b moving away at velocity

$$\vec{v}_b - \vec{v}_a = H_0(\vec{r}_b - \vec{r}_a). \tag{2.2}$$

This shows that all observers see this same pattern of recession of the other galaxies, as required by homogeneity.

The expansion rate H_0 is known as Hubble's constant. The subscript is meant to indicate that H_0 is a measure of the present rate of expansion of the universe; in an evolving cosmology, the expansion rate is a function of time. Equation (2.1) is known as the redshift-distance relation, where the redshift z is defined in equation (2.1) for recession speeds well below the speed of light.

The redshift-distance relation is commonly termed Hubble's law. A vote by members of the International Astronomical Union would rename it the Hubble-Lemaître law, in recognition of Lemaître's prediction (discussed in Section 3.1). Others could have been named, too. Vesto Melvin Slipher's redshift measurements and Henrietta Leavitt's Cepheid period-luminosity relation were essential to Hubble's (1929) redshift-distance plot, and Milton Humason's redshift measurements in the 1930s were essential to establishing a clear and tight demonstration of the effect.

For another way to understand Milne's point, consider the three galaxies at the vertices of the triangle in Figure 2.1. If the galaxies are moving away from one another in a homogeneous and isotropic way, the angles of the triangle are unchanged while the length ℓ_i of each side increases by the same factor, $\ell_i \propto a(t)$. This has to be true of any triangle. That is, $a(t)$ is a universal expansion factor. With $l \propto a(t)$, the rate of change of the physical distance $l(t)$ between any two galaxies at separation l is

$$\frac{dl}{dt} = v = \frac{\dot{a}}{a}\ell(t). \tag{2.3}$$

The dot means time derivative. We see that Hubble's constant in equation (2.1) is

$$H_0 = \frac{1}{a}\frac{da}{dt}, \tag{2.4}$$

evaluated at the present epoch, at expansion time $t = t_0$.

The departure of a galaxy velocity from the mean value set by Hubble's law at that position is said to be the galaxy peculiar velocity. Peculiar velocities usually may be attributed to the gravitational pull of the growing clustering of mass in galaxies and concentrations of galaxies, but nongravitational forces produced by explosions may be important, too.

At nonrelativistic recession speeds, the cosmological redshift is defined as $z = v/c$, where c is the speed of light. This is a first-order Doppler shift. The distance at which Hubble's relation between distance and recession velocity extrapolates to the speed of light, $r_{\rm H} = cH_0^{-1} \sim 10^{28}$ cm, is the Hubble length. Consideration of the relativistic correction to equation (2.3) for galaxies at this great distance begins in Section 3.2.

2.2 Early Evidence of Inhomogeneity

In the 1930s, the cosmological principle passed an important empirical check: The prediction from homogeneity of the redshift-distance relation in equation (2.1) was shown to fit the tight tests discussed in Section 2.3. But homogeneity was not suggested by maps of the galaxy distribution. Charlier (1922) presented a map of the distribution across the sky of the known nebulae. Among the objects in Charlier's map are clusters of stars in our galaxy, and regions where starlight is reflected by clouds of dust, but most are extragalactic nebulae, that is, other galaxies of stars. Charlier pointed out that the map brings to mind hierarchical clustering: galaxies appear in clumps that are present in clumps of clumps, and so on, perhaps to indefinitely large scales. This was later named a "fractal universe."

A decade later, Harlow Shapley and Adelaide Ames at the Harvard College Observatory presented a catalog of the 1,249 known galaxies brighter than $m = 13$ (a measure of the brightness in the sky). Their maps of the angular positions in the two hemispheres of our galaxy are shown in Figure 2.2 (Shapley and Ames 1932). The left-hand panel shows the galaxies in the North hemisphere of our galaxy, the right-hand panel those in the South galactic hemisphere. The near absence of galaxies near the plane of our Milky Way galaxy is due to absorption of light by interstellar dust lying close to the plane

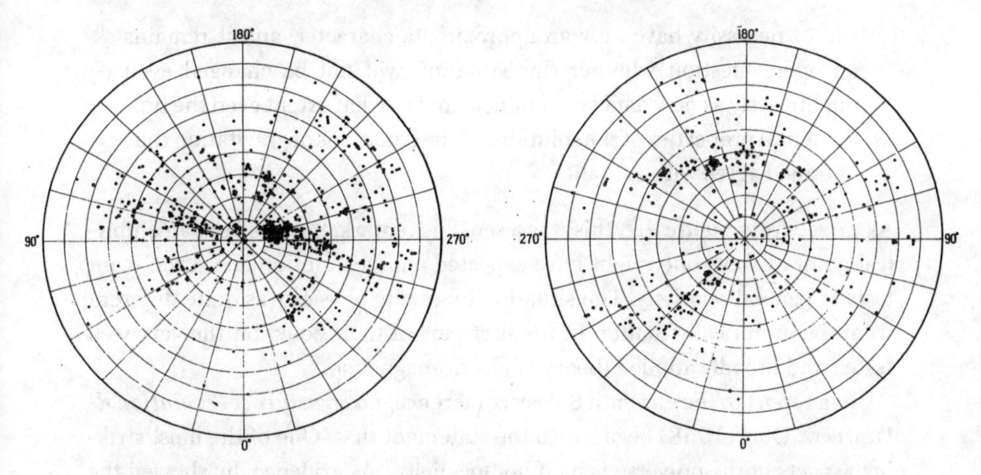

FIGURE 2.2. The Shapley and Ames (1932) map of galaxies brighter than apparent magnitude 13. Courtesy of the John G. Wolbach Library, Harvard College Library.

of our galaxy. The sky is clearer above and below the plane. The northern hemisphere on the left in the figure shows the many galaxies in the prominent concentration in and around the Virgo Cluster of galaxies. (The cluster is named for its position in the sky, near the stellar constellation Virgo.) De Vaucouleurs (1953 and 1958a) named the Virgo Cluster and the broad concentration of galaxies around it the Local Supercluster. This distinctly inhomogeneous distribution of the nearby galaxies is well established.

Willem de Sitter (1917a,b) presented discussions of Einstein's thoughts about the structure of the universe. Since de Sitter was a knowledgeable astronomer, he could have told Einstein about the nebulae, the thought that most are extragalactic, and the evidence that these extragalactic nebulae are not at all close to uniformly distributed. But I have not seen any indication that Einstein considered this observation and if so, whether it affected his thinking.

The possibilities in 1917 were that obscuration by dust is quite patchy even well away from the plane of our galaxy, or else that the observed distribution of galaxies does not at all resemble the homogeneity of the cosmological principle. Not much had changed by the 1950s except that the dust option was ruled out. The situation was recognized in the influential and informative book, *The Classical Theory of Fields* (Landau and Lifshitz 1951, the English translation of the 1948 Russian edition). It presents an admirable exposition of the special and general theories of relativity, but there is little mention of data in this book or in the others in their series on theoretical physics. A rare exception is the comment about Einstein's homogeneity assumption in Landau and Lifshitz (1951, 332):

> Although the astronomical data available at the present time give a basis for the assumption of uniformity of this density, this assumption

> can of necessity have only an approximate character, and it remains an open question whether this situation will not be changed even qualitatively as new data are obtained, and to what extent even the fundamental properties of the solutions of the equations of gravitation thus obtained agree with actuality.

As we see from Figure 2.2, this was a sensible remark, though from an empirical point of view, one might have expected another caution about the scant tests of general relativity. The situation in gravity physics was quite different from the empirical situation in the first part of their book, on the very well tested and broadly applied theory of electromagnetism.

In a report to the eleventh Solvay conference, *La structure et l'évolution de l'univers,* Oort (1958) began with the statement that "One of the most striking aspects of the universe is its inhomogeneity." As evidence, he showed the Shapley and Ames (1932) map in Figure 2.2. He could have added that Abell's (1958) catalog of the more-distant rich clusters of galaxies shows them scattered across the sky in a clumpy fashion, as in superclusters of clusters. But the distribution of clusters in Abell's map (1958, Figure 7) does look distinctly less clumpy than the distribution of the much closer galaxies in the Shapley-Ames map.

2.3 *Early Evidence of Homogeneity: Isotropy*

There remained the possibility that the galaxies are uniformly distributed in the average over larger volumes than Shapley and Ames had sampled. Hubble (1926 and 1934) introduced a test, the variation of the counts of faint galaxies as a function of position across the sky. Away from the areas obscured by interstellar dust close to the plane of the Milky Way, Hubble (1934) typically found about 100 galaxies per square degree (reduced to standard observing conditions) to a limiting redshift he estimated to be about $z = 0.1$. This is deep, 10 percent of the speed of light, and is about ten times the distance sampled in the Shapley-Ames map. Hubble's counts at low galactic latitudes, plotted as the lower strings of data in Figure 2.3, are smaller than at high latitudes and show a systematic variation across the sky. Both are effects of obscuration by dust in variable amounts along lines of sight near the directions of the plane of the Milky Way. The upper strings of data are counts at 40–50 degrees above the plane, plotted as filled circles in the north galactic hemisphere and open circles in the south. The counts are similar in the two hemispheres and do not show a systematic tendency to vary with position across the sky. Hubble (1934, 62) concluded that

> On the grand scale, however, the tendency to cluster averages out. The counts with large reflectors conform rather closely with the theory of

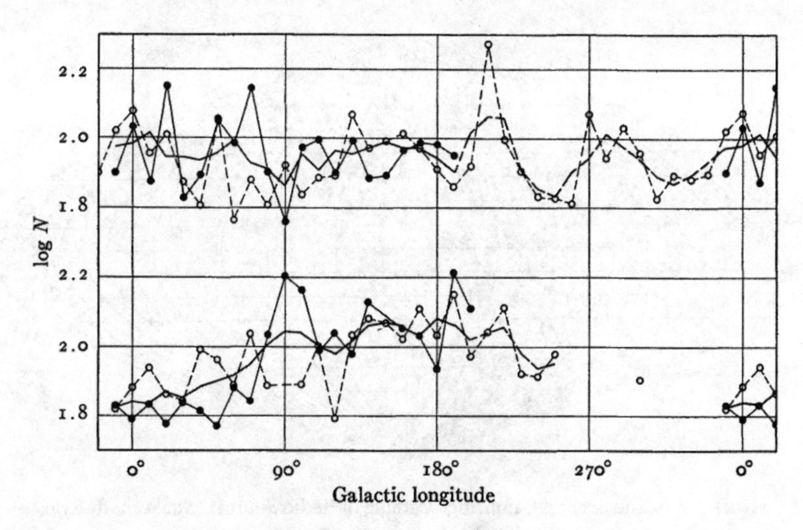

FIGURE 2.3. Hubble's (1934) counts of galaxies at high galactic latitudes in the upper curves, and at low latitudes in the lower curves. © AAS. Reproduced with permission.

> sampling for a homogeneous population. Statistically uniform distribution of nebulae appears to be a general characteristic of the observable region as a whole.

Bok's (1934, 8) considerations led him to the opposite conclusion:

> Different lines of evidence all indicate that the available material points to the existence of a widespread non-uniformity in the distribution of external galaxies, and that this tendency toward clustering is probably one of the chief characteristics of the part of the Universe within the reach of modern telescopes.

Bok was at the Harvard College Observatory, and he emphasized the clumpy distribution of galaxies in the Harvard Shapley-Ames map that came out of this observatory. He referred to Hubble (1934) but did not mention Hubble's Figure 4, which is reproduced here in Figure 2.3. Hubble took it to be an indication of approach to uniformity; Bok does not seem to have been convinced.

Hubble's interpretation seems to be the more reasonable to me, and I count it as the first indication that in the average over large enough volumes, the galaxy distribution approaches isotropy. That is easier to see now, of course. And it is easier to see that if we may take it that our position among the galaxies is not special, then the indication from this figure, though certainly preliminary, was that the galaxy distribution approaches homogeneity on large scales.

Another line of evidence opened in the 1950s with the ability to probe the universe at radio wavelengths and soon after that by X-ray and microwave detectors. Figure 2.4 shows the distribution of radio source positions across

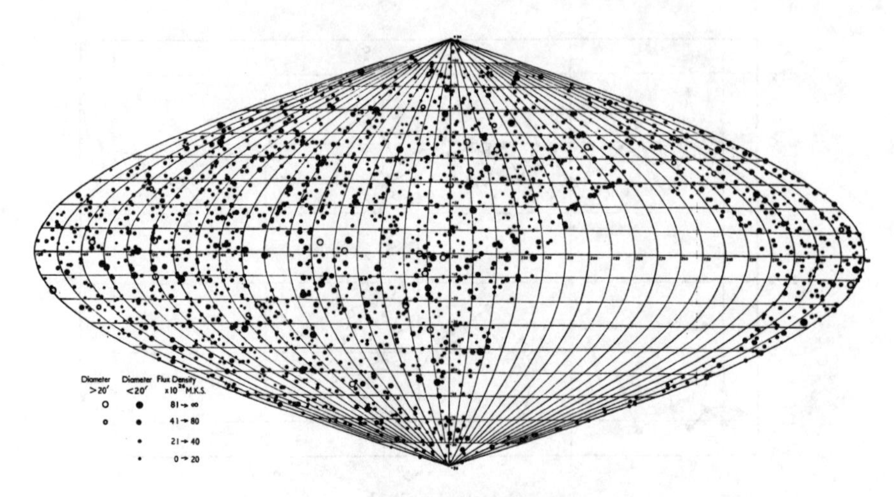

FIGURE 2.4. The Second Cambridge Catalog of Radio Sources (Shakeshaft, Ryle, Baldwin, et al. 1955).

the part of the sky surveyed in the *Second Cambridge Catalog of Radio Sources, 2C,* by Shakeshaft et al. (1955).[1] The sources were suspected then and are now known to be in galaxies. The catalog lists 1,936 sources at wavelength 3.7 meters (82 MHz). A few are close to the plane of the Milky Way and likely are in our galaxy. Others are spurious detections of sources in sidelobes, and some real sources are missing. The brightest extragalactic radio source in the sky, Cygnus A, is on the equator in this map and a quarter of the way in from the left-hand side. It is so bright in the radio that it obscures sources close to it in the map, accounting for the empty region around this object. (The large empty region to the lower right was not observed, because it always is below the horizon at the telescope.)

Optical identifications and redshift measurements of a few of these sources had suggested that many have redshifts large enough that the observations might show a detectible departure of the count of sources as a function of the radio flux density from what would be expected in the flat spacetime of special relativity. This is the cosmological test to be discussed in Section 3.4. Its application here was frustrated by spurious source detections and omissions. This systematic error has a less serious effect on the angular distribution of sources, however, and we see that the constant-area map of sources in Figure 2.4 does look about as expected in a homogeneous universe: no indication in any direction that the observations encounter an edge to the distribution of these objects.

We are in seas of X-ray and microwave radiation. The latter, later termed the "cosmic microwave background" (CMB), is the subject of Chapter 4. A 6-minute rocket flight gave the first evidence of the former, a sea of X-rays

1. This is the figure between pages 148 and 149 in Shakeshaft et al. (1955).

(Giacconi et al. 1962). The flight allowed little time for measurement of the X-ray angular distribution, but Gould's (1967) review indicated that this radiation does not vary across the sky by more than about 10 percent. Schwartz (1970) used the year-long scan of the sky by the OSO-III X-ray satellite (at 7.6 to 38 KeV, and angular resolution $\sim 10°$) to bound the X-ray anisotropy to 4 percent.

Recognition of the other component, the sea of microwave radiation, was presented by Penzias and Wilson (1965). By the end of the 1960s, Wilson and Penzias (1967) and Partridge and Wilkinson (1967) had found that this radiation is isotropic to better than 0.2 percent.

The isotropy observed at optical, radio, microwave, and X-ray wavelengths seriously constrains ideas about the large-scale nature of our universe. The maps in Figures 2.3 and 2.4 require that, if the space distribution of galaxies is not close to homogeneous, then it is at least close to spherically symmetric about our position. That would seem to be a curious arrangement of matter, and it would be curious, too, that there are enormous numbers of other galaxies that would seem to have equally suitable homes for observers such as us. It is difficult to imagine we would be so special as to be close to the center of symmetry. The easier interpretation is that our universe is close to homogeneous, meaning observers on other galaxies also would see isotropic distributions of sources.

Another picture to be considered is that the X-ray and microwave radiation backgrounds are isotropic because the universe contains a homogeneous sea of radiation that has nothing to do with the galaxy distribution. This might work if spacetime is static. But if we accept the evidence from galaxy redshifts that the galaxies are moving apart, then most would have to be moving through a uniform sea of radiation. The Doppler effect would cause an observer moving through the radiation to find that the radiation is brighter than average in the direction the galaxy is moving and dimmer in the opposite direction. The radiation we observe is close to isotropic, so again we would have to be in an exceedingly special galaxy, one of the few that are moving slowly through the radiation. Most galaxies would be moving through the sea more rapidly, the most distant at near-relativistic speeds. Why should we be in this special situation?

Another situation we might consider accepts that the radiation is uniformly distributed in a curved spacetime that describes homogeneous and isotropic space sections, consistent with the cosmological principle, but that the galaxies are distributed in a clumpy fashion even on arbitrarily large scales, as in a fractal distribution. This picture might have been defended in the 1960s by supposing that the X-ray background did not come from the galaxies and that we could ignore the gravitational disturbance to spacetime caused by the mass in the regions occupied by the galaxies. Issues of this sort were discussed by Wolfe and Burbidge (1970) and Peebles (1971a). The conclusion is that it is difficult to imagine a model for clumping of matter on scales approaching the Hubble length that fits all the observations of isotropy. And there is more.

2.4 Early Evidence of Homogeneity: Counts and Redshifts

In his book, *Cosmology*, Bondi (1952, 14, 15) took note of the clumpy distribution of the galaxies but did not express much concern about it. On page 48 he states that Hubble and Humason's (1931) redshift-distance measurements and Hubble's (1936) galaxy counts, which Bondi had discussed earlier in the book, "strongly point towards the correctness of the hypothesis" of Einstein's cosmological principle: large-scale homogeneity. Consistent with this, Oort (1958), despite his emphasis on inhomogeneity in maps of galaxy positions, was willing to estimate the mean mass density of a homogeneous universe on the basis of Hubble's deep counts. Let us consider now this evidence for the cosmological principle.

Milne demonstrated that the cosmological principle predicts Hubble's law, the linear relation between galaxy distance and recession speed, v (or redshift, z) in equation (2.1). Hubble's (1929) announcement of evidence for the linear relation did not present a very tight case, but Hubble and Humason (1931) improved that. They found the relation is a good approximation for the most luminous galaxies observed out to redshift $z \sim 0.06$, and Hubble (1936) reported an even deeper check, to $z \sim 0.1$. (Hubble 1929 had used his measurements of distances of individual galaxies with Slipher's 1917 redshift measurements. The later tests with Humason assume the inverse square law to get ratios of distances to galaxies. This allows a check of linearity of the redshift-distance relation that leaves the constant of proportionality, H_0, undetermined.)

These redshift-distance tests reached recession velocities approaching 10 percent of the velocity of light, impressively deep probes of the universe, and they encountered no serious challenge to homogeneous expansion. Hubble and Humason did not mention it, but a valuable by-product of their test of the redshift-distance relation is this check of the cosmological principle. But we should bear in mind that although Hubble's law is characteristic of homogeneity, it does not require it. If a violent explosion caused the galaxies to fly apart into initially empty space, then in time, the more rapidly moving galaxies would be farther away. This velocity sorting would produce Hubble's law in a clumpy universe, of course assuming the gravitational interactions of the clumps may be ignored.[2] Sections 3.6.3 to 3.6.5 discuss the later use of

2. To be explicit, a collection of particles moving apart radially with constant speeds v_i and initial distances r_i are after time t at distances $d_i = r_i + v_i t$. At arbitrarily large t, this is arbitrarily close to $v_i = d_i/t$.

departures from Hubble's law to probe the gravitational effect of departures from a mass distribution that is homogeneous in the mean.

The other probe Bondi mentioned is the count of galaxies as a function of their brightness in the sky. If the space distribution of galaxies is homogeneous on average over the volumes sampled, and if we can neglect the relativistic corrections that are important at great distances, then the count of galaxies brighter than the received energy flux density f varies with f as

$$N(>f) \propto f^{-3/2}. \tag{2.5}$$

To see this, consider the inverse square law: A galaxy with luminosity L produces starlight energy flux density $f = L/(4\pi r^2)$ at distance r (neglecting obscuration and relativistic corrections). The galaxies with luminosity L that are observed to be brighter in the sky than f thus are observed at distances $r < \sqrt{L/(4\pi f)}$. In a homogeneous distribution, the count of galaxies with luminosity L that are brighter in the sky than f is proportional to the volume within distance r, which is proportional to $r^3 \propto f^{-3/2}$. This power-law scaling applies to galaxies in each class of luminosity, so it applies to the counts summed over galaxies of all luminosities, resulting in equation (2.5).

The astronomers' measure of energy flux density f is the apparent magnitude[3]

$$m = -2.5 \log_{10} f + \text{a constant}. \tag{2.6}$$

The count-magnitude relation for a statistically homogeneous distribution of galaxies is then

$$\log_{10} N(<m) = 0.6m + \text{another constant}. \tag{2.7}$$

The simple but valuable relation in Equations (2.5) and (2.7) was first applied to star counts. It revealed the limited extent of our Milky Way galaxy of stars by the departure from this relation.

Hubble (1926, 366) compared the relation to counts of galaxies and concluded that

> The agreement between the observed and computed log N over a range of more than 8 mag. is consistent with the double assumption

3. The astronomers' measure of intrinsic luminosity, L, is the absolute magnitude M defined by

$$M = -2.5 \log_{10} L + \text{another constant}, \quad m - M = 5 \log_{10} d/(10 \text{ pc}).$$

The distance is d, and the distance modulus is $m - M$, normalized to $m = M$ at 10 parsecs distance, where 1 parsec is roughly 3 light years. Measurements of apparent and absolute magnitudes specify the window of wavelengths in which the radiant energy is measured, the correction for obscuration by the atmosphere and dust in the Milky Way and the source, and the shift of wavelengths if the distance is large. But that calculation should be left to more able hands.

of uniform luminosity and uniform distribution or, more generally, indicates that the density function is independent of the distance.

This early recognition of evidence of homogeneity from galaxy counts is impressive, but the case was based on heterogeneous samples. The more systematic compilation of counts in Hubble (1936, 186) reaches impressively large distances, to recession velocities of about 40 percent of the speed of light (in the estimate by Peebles 1971a, 37). These counts increase with decreasing energy flux density f a little less rapidly than the $f^{-3/2}$ law. It could mean that the universe at great distances is slightly less dense than nearby, or that Hubble had a modest systematic error in his apparent magnitude scale, or perhaps that he had detected the relativistic correction. But we can conclude that the counts did not offer any indication that Hubble's observations of distant galaxies were reaching an edge to the realm of the galaxies.

The $f^{-3/2}$ law for counts, and the redshift-magnitude relation, assume space between the galaxies is fully transparent. Zwicky (1929) asked whether light passing through the great distances of intergalactic space might suffer a friction of some sort that causes photons to lose energy, a concept that became known as "tired light." With Einstein's expression for the energy of a photon, $\varepsilon = h\nu$, the tired light picture would indicate that photon wavelengths increase as they travel great distances and lose their energy ε. Might this have produced the redshifts of the galaxies? And might the friction also make free space slightly opaque? Hubble and Tolman (1935) proposed a test of the first, and implicitly the second, from the variation of galaxy surface brightnesses[4] with redshift, modeled as

$$i \propto (1+z)^{-r}. \tag{2.8}$$

In the standard theory, redshifts are the result of the expansion of the universe. This produces index $r = 4$ by the considerations discussed in Section 4.1: one power of $(1+z)$ comes from the loss of energy of each photon as it is redshifted, one power from the decrease in the rate of reception of photons, and two from the Doppler aberration of solid angle. And if free space in the expanding universe were not fully transparent, it would make $r > 4$. In a static tired-light universe, only the first effect operates, meaning $r = 1$, assuming space is transparent.

This elegant surface-brightness test is unaffected by space curvature, but its application is complicated by the difficulty of modeling the evolution of galaxy intrinsic surface brightnesses as stellar populations evolve. But we have a demanding test from another direction: the sea of microwave radiation

4. Radiation surface brightness is the net flux of energy integrated over frequency per unit area, time, and solid angle. In a static situation, the surface brightness along the path of a light ray is constant. This is Liouville's theorem applied to light modeled as a gas of photons. A Doppler shift in flat or curved spacetime produces the index $r = 4$ in equation (2.8).

discussed in Chapter 4. The close-to-thermal spectrum shown in Figure 4.7 shows that surface brightness evolution closely agrees with the Doppler effect with $r = 4$. Scattering by free space, not absorption, would not disturb the thermal spectrum of the sea of radiation, but it would tend to smooth the radiation anisotropy. The tests for this effect reviewed in Chapter 9 indicate that as much as a few tens of percent of the radiation from high redshift, back to the dark ages, may have been Thomson scattered by free electrons in intergalactic plasma (as first indicated in Spergel et al. 2003). In the standard big bang model, this means galaxy counts increase with decreasing flux density f less rapidly than expected in equation (2.5), apart from the effects of galaxy evolution, and redshifts increase with increasing apparent magnitude less rapidly than expected from equation (2.1). But the effects of this Thomson scattering are small at redshifts less than unity.

Let us note also that relativistic corrections are important for the modern deeper and more precise observations discussed in Section 9.1, but not for what could be done in the 1930s. The impressive thing is that there were observations in the 1930s that probed to galaxies distant enough that their redshifts indicate they are moving away from us at near the speed of light, and the observations did not encounter an edge to the realm of the galaxies.

2.5 *The Universe as a Stationary Random Process*

Following Jerzy Neyman (1962), a more formal statement of Einstein's cosmological principle is that the universe is assumed to be a realization of a stationary (statistically homogeneous and isotropic) random (stochastic) process. The stationary condition means that expectation values are independent of position and direction; only relative positions matter. The concept can only be empirically useful if the realization of the process that is our observable universe offers a close to fair sample, from which we can find estimates of the statistical measures of the galaxy distribution that usefully approximate these measures in the idealized, infinitely realizable process. The test is the check of reproducibility of statistical estimates that sample different parts of the sky at different ranges of distance.

Jerzy Neyman and Elisabeth Scott at the University of California, Berkeley, introduced a pioneering program of statistical analyses of the galaxy space distribution. They fitted counts of galaxies in cells in the sky to a model of galaxies in clusters of ν members, where the random number ν may assume the value unity, and the clusters may be in superclusters containing random numbers of clusters. They constrained model parameters by second moments of the counts. This program was motivated at least in part by Donald Shane's observational program at the nearby Lick Observatory. He was leading the cataloging of the million or so brightest galaxies in the sky (Neyman, Scott, and Shane 1954; Shane and Wirtanen 1954 and 1967). The Neyman et al. program clarified the philosophy of statical analyses of extragalactic objects, and it foreshadowed the halo occupation distribution program that became a useful tool for analyses of galaxy distributions in the twenty-first century (e.g., Berlind and Weinberg 2002). But their program is not well suited to probe for a large-scale approach to homogeneity. That was achieved by measurements of galaxy N-point position correlation functions.

Limber (1953 and 1954), Rubin (1954), and Totsuji and Kihara (1969), introduced the use of the two-point statistical measure of the galaxy distribution. All used or mentioned the ongoing Lick counts. Limber's (1954, 656) description of how he estimated the two-point angular correlation of galaxy counts in cells is worth recording here:

> The number of nebulae per square degree for each degree along such a parallel was recorded separately on each of two strips [of paper]. In order to obtain $\overline{NN}_\phi$ for this parallel, one strip was displaced ϕ degrees with respect to the other, and then the values on the two strips which were adjacent after the displacement were multiplied together, and the mean value of these products was obtained.

The quantity $\overline{NN}_\phi$, after normalization and subtraction of shot noise, is an estimate of the angular two-point correlation function at separation ϕ; it has come to be a widely used statistic in extragalactic astronomy. But fuller use of the data from the Lick survey and other catalogs awaited the advances in computation in the 1970s that replaced Limber's labor-intensive method. I took advantage of this in the program of statistical analysis with colleagues at Princeton University. The results are summarized in Peebles (1980).

The N-point correlation functions represent the structure of the universe by a distribution of point-like particles: perhaps galaxies, perhaps mass elements. The probability that a particle is found in the volume element dV is

$$dP = n\, dV. \tag{2.9}$$

This defines the mean particle number density, n. In the assumed stationary process, n is independent of position. The probability that a particle is found in the volume element dV at distance r from a particle is

$$dP = n(1 + \xi(r))\, dV. \tag{2.10}$$

This defines the reduced two-point correlation function, $\xi(r)$ (where "reduced" simply means removal of the first term in parentheses and removal of the factor n). Under the assumption of statistical homogeneity and isotropy, this two-point statistic can only be a function of the separation r of the two points. If the realization we observe has presented us with a good approximation to a fair sample, then the estimate of $\xi(r)$ from the observations is a good approximation to the function in the idealized random process. The higher-order reduced N-point functions, $N > 2$, are similarly defined, as discussed at length in Peebles (1980).

The mean (expectation value or ensemble average) number of particles within distance r of a particle is, from equation (2.10),

$$\langle N(<r)\rangle = n\,V + n \int_0^r 4\pi r^2\,dr\,\xi(r), \tag{2.11}$$

where V is the volume within distance r (and distances are small compared to the Hubble length, so we can think in term of flat space).

In a stationary random Poisson process, each particle position is assigned independently of where the other particles are. In this case, $\xi = 0$, and the mean number of neighbors is the usual product of the number density n with the volume V within radius r. To avoid confusion, note that in a stationary Poisson process, the volume V randomly placed contains nV particles on average, but the volume placed on a randomly chosen particle biases the count of particles contained to $nV + 1$ particles on average.

The second term in equation (2.11) is the mean number of neighbors in excess of the Poisson distribution. It can be negative, if particles tend to avoid one another. If the two-point function is positive and a power law, which is close to what is observed for the galaxies, we have

$$\xi(r) = (r_0/r)^\gamma, \quad \langle N(<r)\rangle = n\,V + \frac{4\pi n}{3-\gamma} r_0^\gamma r^{3-\gamma}. \tag{2.12}$$

Another way to write the second expression is the fractional difference between the mean number of neighbors within distance r of a particle and the mean number $N = nV$ expected if positions were uncorrelated,

$$\frac{\delta N}{N} = \frac{\langle N(<r)\rangle - nV}{nV} = \frac{3}{3-\gamma}\left(\frac{r_0}{r}\right)^\gamma. \tag{2.13}$$

The parameter r_0 is a measure of the clustering length in this power-law model. At $r \ll r_0$, the positions are distinctly clustered: A typical particle has many more neighbors than expected if positions were unrelated. At $r \gg r_0$, the mean departure from an uncorrelated Poisson distribution is a small fractional perturbation to the count.

Measuring $\xi(r)$ and the higher order functions requires finding a way around the relatively large uncertainties of galaxy distance measurements. The approach is indirect: Infer $\xi(r)$ from estimates of the angular two-point correlation correlation function $w_d(\theta)$ in maps of angular positions of galaxies that have distance estimates d in some chosen range of values. The errors in the galaxy distance estimates are assumed to be uncorrelated, though one can devise corrections for that. The probability distribution of distance errors is supposed to be reasonably well understood. And we have the powerful assumption that the distribution is statistically isotropic.

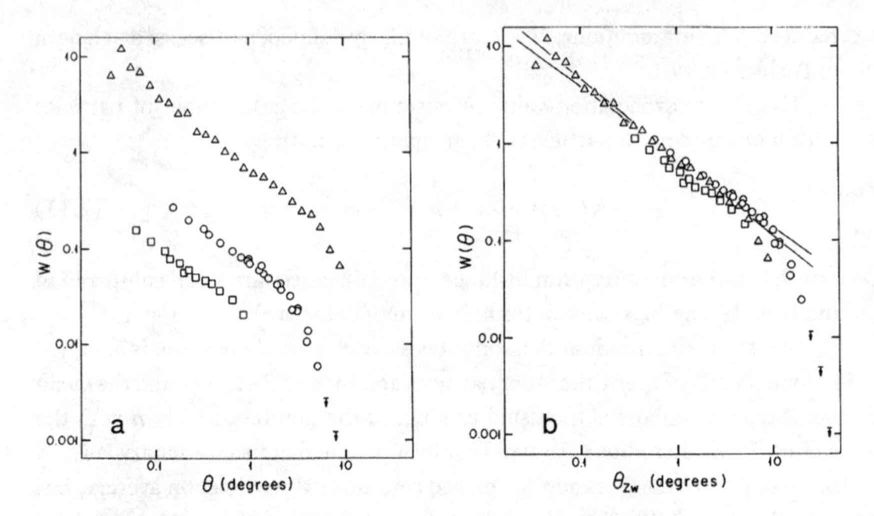

FIGURE 2.5. Scaling test of the galaxy two-point angular correlation functions in Panel (a) for the Zwicky (triangles), Shane-Wirtanen (circles), and Jagellonian (squares) samples. Panel (b) is the result of applying the scaling relation in equation (2.15) to the angular functions (Groth and Peebles 1977). © AAS. Reproduced with permission.

Following the definition of the spatial two-point function in equation (2.10), the angular two-point function $w_d(\theta)$ in a sample of galaxies within nominal distance limit d is defined by the probability of finding a galaxy in the element $d\Omega$ of solid angle at angular distance θ from a galaxy in the sample,

$$dP = \mathcal{N}(1 + w_d(\theta))\, d\Omega, \qquad (2.14)$$

where $\mathcal{N}$ is the mean number of galaxies per unit solid angle (per steradian if angles are measured in radians).

If we have a fair sample, and if the spatial function $\xi(r)$ decreases more rapidly than r^{-1} at large separations, then the angular function at small angular separations, $\theta \ll 1$ radian, scales with the sample depth d as (Peebles 1973a, eq. [69])

$$w_d(\theta) = d^{-1} W(\theta d). \qquad (2.15)$$

The function $W(x)$ at fixed $x = \theta d$ is independent of the depth d. (Relativistic corrections are applied in Groth and Peebles 1977, §V). To understand how d enters equation (2.15), consider that the angular function evaluated at fixed θd probes structure at a fixed linear scale. As d is increased, the angular correlation function averages over increasing numbers of realizations at linear scale θd along the line of sight, which averages out the clustering seen in the angular distribution.

Figure 2.5, from Groth and Peebles (1977), shows a quantitative test for fair samples of a statistically homogeneous galaxy distribution: a spatially stationary random process. The angular two-point functions in Panel (a) are estimates from catalogs of galaxy angular positions to three limiting apparent magnitudes. The triangles are derived from the nearest sample, the Zwicky et al. (1961–1968) *Catalogue of Galaxies and Clusters of Galaxies*. The circles are based on the deeper Lick catalog. Shane and Wirtanen (1954, 1967) published the Lick counts summed in 1-square-degree cells. Seldner et al. (1977) reduced the original counts in 10×10 arcmin cells to standard conditions to obtain the correlation function plotted as circles. The still-deeper Rudnicki et al. (1973) catalog of galaxies in the Jagellonian field yields the two-point function plotted as squares.

The angular correlation functions in Panel (a) in Figure 2.5 decrease with increasing sample depth, in the direction expected if deeper samples better average out the fluctuations in a stationary random process. The quantitative check in Panel (b) is the result of applying the scaling relation in equation (2.15), where the ratios of limiting depths d of the samples are taken to be the cube roots of the ratios of the mean galaxy counts per steradian. The reasonably close agreement of scaled functions makes a reasonably good case that the spatial function is reliably determined. That is, the evidence from this scaling test is that the correlation function estimates have not been seriously distorted by systematic errors, such as variable obscuration across the observed parts of the sky. And for our purpose, the point is that the consistency of the scaled functions is evidence in support of the assumption that we have observations of a fair sample of a stationary process.

On smaller scales in Figure 2.5, the angular two-point function is well approximated by a power law. This translates to the power-law spatial function $\xi(r)$ in equation (2.12). Groth and Peebles (1977) found $\gamma = 1.77$ with $r_0 = 4.7h^{-1}$ Mpc. The Harvard-Smithsonian Center for Astrophysics (CfA) redshift survey (discussed in Section 3.6.4) had better control of redshifts; it increased the length scale to (Davis and Peebles 1983a)

$$\gamma = 1.77, \quad r_0 = 5.4 \pm 0.3h^{-1}\ \mathrm{Mpc}, \text{ at } 10\ \mathrm{kpc} \lesssim hr \lesssim 3\ \mathrm{Mpc}. \tag{2.16}$$

Hubble's constant is written as $H_0 = 100h$ km s^{-1} Mpc^{-1} (as in equation (3.15)).[5]

Totsuji and Kihara (1969) seem to have been the first to find the power-law form in equation (2.16). Their estimate of the index is $\gamma = 1.8$, impressively close to what followed from later, better data.

The mean number of galaxies within distance r of a galaxy in excess of the mean found in a randomly placed sphere of this radius is the integral over $\xi(r)$ in equation (2.11). A related measure is the mean-square fluctuation in the

5. The length units in this book are megaparsecs and kiloparsecs, where 1 Mpc $= 10^3$ kpc $= 10^6$ parsecs, or about 3 million light years, as mentioned in footnote 3 in Section 2.4.

count N of galaxies in a randomly placed sphere of radius r. For the power-law correlation function, this latter statistic is[6]

$$\left(\frac{\delta N}{N}\right)^2 = \frac{\langle (N - \langle N \rangle)^2 \rangle}{\langle N \rangle} = J_2 \left(\frac{r_0}{r}\right)^\gamma, \quad J_2 = 1.82 \text{ for } \gamma = 1.77. \tag{2.17}$$

A measure of the transition from nonlinear clustering on small scales to small departures from homogeneity on large scales is the sphere radius r_{cl} at which the galaxy counts fluctuate from the average by the root-mean-square fractional amount unity:

$$\frac{\delta N}{N} = 1 \text{ at clustering length } r_{cl} = 7.6h^{-1} \text{ Mpc}. \tag{2.18}$$

The small value of this characteristic clustering length compared to the Hubble length, $H_0 r_{cl}/c \sim 0.003$, is an indication that the observable universe presents us with many different patches of clustering that allow many probes of the galaxy distribution that may be expected to yield fair and secure statistical measures of this random process. Patterns are seen in galaxy maps on considerably larger scales than r_{cl}, but they are small fractional fluctuations in the counts of galaxies averaged over larger scales.

The scaled two-point angular correlation function in Panel (b) in Figure 2.5 breaks below the power law at large separation. Because the angular function is a convolution of the spatial function over the range of distances sampled, the spatial function $\xi(r)$ rises slightly above the power law at $r \sim$ 10 Mpc and then falls below it. This is demonstrated by Soneira and Peebles (1978, Fig. 6). The break from a power law is shown with greater precision in Efstathiou, Sutherland, and Maddox (1990) and Zehavi et al. (2011, Fig. B22).

These statistical measures apply to the common large galaxies, such as the Milky Way, that contribute the bulk of the cosmic mean luminosity density. Their characteristic luminosity is written as L^*. The more numerous galaxies with $L \ll L^*$ have close to the same clustering parameters as $L \sim L^*$ galaxies. The rare giants with $L \sim 10L^*$ are more strongly clustered. For example, Masjedi et al. (2006) find that the Luminous Red Galaxy (LRG) sample from the Sloan Digital Sky Survey (SDSS) has clustering length about twice that of $L \lesssim L^*$ galaxies. This is consistent with the tendency of the most massive galaxies to appear preferentially in the most massive clusters of galaxies, because the cluster positions are more strongly clustered than are common $L \sim L^*$ galaxies (Peebles and Hauser 1974, eq. [47]; Bahcall and Soneira 1983). Kaiser (1984) made the excellent point that massive concentrations of galaxies are

6. This ignores the shot-noise term in equation (60.3) and uses the analytic expression for J_2 in equation (59.3) in Peebles (1980).

expected to be more strongly correlated than are the common $L \sim L^*$ galaxies in a positively correlated Gaussian random process, as observed. This is discussed in Section 3.5.3.

2.6 *A Fractal Universe*

Bondi (1952, 14, 15) mentioned another kind of statistical homogeneity: a clustering hierarchy, or what later became known as a fractal galaxy distribution. For example, imagine that particles, perhaps galaxies, are placed in clusters, clusters are placed in second-order clusters, second-order clusters are in third-order clusters, and so on, perhaps continuing to indefinitely large scales. In a scale-invariant clustering hierarchy, or fractal, the mean number of galaxies (or the mean amount of mass) within distance r of a particle (or mass element) is the limit of equation (2.12) as $n \to 0$ and $r_0 \to \infty$. This amounts to

$$\langle M(<r) \rangle \propto r^{3-\gamma} = r^D, \quad D \equiv 3 - \gamma. \tag{2.19}$$

In astronomers' units, this is

$$\log \langle M(<m) \rangle = 0.2Dm + \text{constant}. \tag{2.20}$$

The distribution is said to have fractal dimension D, with $0 < D < 3$ in three dimensions. If the distribution is spatially homogeneous, then $D = 3$, as usual. If $D < 3$, the distribution may be homogeneous in another sense: that is, each element of mass finds itself in statistically the same hierarchy of clusters within clusters and on up. But if $D < 3$, the mean mass density averaged over arbitrarily large scales is arbitrarily close to zero.

The Newtonian gravitational potential energy of a mass M concentrated within radius r is on the order of $U \sim GM/r$. In a fractal mass distribution with dimension D, the potential energy on the scale r thus varies as $U \propto r^{D-1}$. A pure scale-invariant fractal in three dimensions with $D = 1$ thus has gravitational potential that diverges only as the logarithm of the length scale on arbitrarily small and large scales. If kinetic energy scales like potential energy, then velocities would would be safely below the velocity of light over a broad range of scales. This could be an elegantly arranged universe, but it is not ours.

The galaxy distribution on scales less than about 10 Mpc approximates a fractal with dimension $D = 1.23$ (equation (2.16)). The three- and four-point correlation functions also agree with a simple fractal hierarchical clustering pattern with this dimension (Groth and Peebles 1977; Fry and Peebles 1978). In a gravitationally bound and stable clustering pattern, the relative velocity dispersion of particles scales as the square root of the mean gravitational potential difference at their separation. Since the small-scale galaxy distribution has D slightly greater than unity, one might expect the galaxy relative velocity dispersion to increase slowly with increasing length scale. This is observed. But the departure downward from the power law form on a scale of about 20 Mpc is a well-established departure from scale invariance.

Pietronero, Gabrielli, and Sylos Labini (2002) make the interesting point that the ratios of depths of the catalogs in Figure 2.5 are scaled from the mean angular densities as $d \propto \mathcal{N}^{1/3}$, which assumes large-scale homogeneity. This argument is circular if we are seeking to check that the galaxy density has a nonzero mean. The circularity is mitigated by the fact that the galaxy distribution is sampled on length scales large compared to the clustering length in equation (2.18). And we have now an independent check: the weak lensing distortion of background galaxy images caused by the masses concentrated around foreground galaxies yields the galaxy-mass cross-correlation function $\xi_{g\rho}(r)$. Sheldon et al. (2004) find that $\xi_{g\rho}(r)$ is a good approximation to a power law at the range of separations $0.04 \lesssim r \lesssim 12$ Mpc, with $\gamma = 1.79 \pm 0.06$ and $r_0 = (5.4 \pm 0.7)h^{-1}$ Mpc. Within the uncertainties, these values agree with the parameters in equation (2.16) based on the scaling test for the galaxy-galaxy function.

In his books, *Les objets fractals*, Benoît Mandelbrot (1975 and 1989) reviews earlier discussions of clustering hierarchies and names them "fractals." He presents ample examples of fascinating fractal patterns, including mathematical constructions and pragmatic considerations, such as the measurement of the length of the coastline of Brittany, which behaves as a fractal because the length depends on the spatial resolution at which it is measured. Mandelbrot's stroke of genius forced attention on many interesting and practical applications of fractals. Perhaps it was inevitable that he should consider the idea that the galaxy distribution is fractal.

Others were thinking along similar lines. I noted in Section 2.2 Charlier's (1922) argument that the galaxies seem to be arranged in a hierarchal clustering pattern. Indeed, this is now well established at the distances Charlier could observe. Carpenter (1938) argued that the galaxy distribution fits $\langle M(<r)\rangle \propto r^D$ with dimension $D = 1.5$ in Mandelbrot's notation. Along with Oort (1958) and Abell (1958), de Vaucouleurs (1970) pointed out that maps of the galaxy spatial distribution probing out to the greatest distances that could be reliably surveyed offered no hint of convergence to homogeneity. Oort was willing to consider that the universe approaches homogeneity on still larger scales, on the basis of Hubble's deep galaxy counts, but de Vaucouleurs proposed that Carpenter's scaling relation extends to much smaller and much larger scales, in a "universal density-radius relation" with fractal dimension he put at $D = 1.3$.

Hubble's (1936) deep galaxy counts shown in Figure 16 in his book, *The Realm of the Nebulae*, fit fractal dimension $D = 2.6$. Reconciling this with large-scale homogeneity, $D = 3$, and neglecting relativistic corrections requires the postulate that a systematic error exists in Hubble's distance scale. Reconciliation with de Vaucouleurs' $D = 1.3$ requires a larger systematic error in the other direction. Gérard de Vaucouleurs certainly considered this point: He told me that he examined Hubble's photographic plates for the deep galaxy counts but could not check the magnitude calibration, because the plates had faded. The systematic error needed to reconcile Hubble's counts with de Vaucouleurs' fractal dimension certainly was worth considering, but to be considered also is the line of evidence from the clustering scaling test in Figure 2.5. De Vaucouleurs' $D = 1.3$ is quite close to the measured correlation function power law on smaller scales, $D = 1.23$, but the measurements reviewed on page 30 indicate a break downward from the power law at separation ~ 20 Mpc: a break from scale invariance.

A qualitative point to be considered is the appearance of maps of angular positions of galaxies within a given distance, d, for different values of d. In a scale-invariant fractal with dimension $D < 3$, the number of particles in the map increases with increasing distance d, but the fractional fluctuations $\delta N/N$ of particle numbers across the sky are statistically independent of d. This follows from the scale invariance of the fractal distribution: If the fluctuations $\delta N/N$ grew smaller with increasing distance d, it would define a characteristic distance $d_{\rm nl}$ at which $\delta N/N$ at a chosen angular scale decreases through unity to small, linear, fractional fluctuations across the sky. But a scale-invariant fractal does not have a characteristic length $d_{\rm nl}$. For early examples of this test, compare the Shapley-Ames map of the angular positions of relatively nearby galaxies in Figure 2.2 (which shows the large number of galaxies near the north relative to the south galactic poles) to Hubble's (1934) deep counts of galaxies across the sky at higher galactic latitudes in the two hemispheres in Figure 2.3, and to the angular positions of distant radio galaxies in Figure 2.4. We see convergence toward isotropy that is contrary to a scale-invariant fractal behavior. In addition, if the galaxy distribution were fractal, the reduced correlation functions in Panel (a) in Figure 2.5 would not decrease with the increasing depths of the three samples.

At the Bern Conference on the *Jubilee of Relativity Theory* (Jubilee for the 1905 special theory; 40 years for the general theory), Oskar Klein (1956) discussed yet another world picture: Perhaps the galaxies are drifting apart into empty flat space after an explosion of a local concentration of matter. Klein considered that the total mass M_0 in the galaxies, and the radius R_0 in which matter was concentrated prior to the explosion, might satisfy $GM_0 \sim R_0c^2$. This is a relativistic concentration at about the Schwarzschild radius. It could be in accordance with the near-relativistic expansion indicated by Hubble and Humason's observations of galaxy redshifts that are not far below unity. Klein had some good arguments. Velocity sorting would put more rapidly moving galaxies farther away, approaching Hubble's relation $v = H_0r$. Explosions are familiar; why not consider a particularly large one that scattered the galaxies?

Flat spacetime is familiar; why bother with the notion of spacetime curvature? The clumpy and irregular galaxy distribution that might be expected from an explosion is familiar; why bother with the homogeneity that was not seen in galaxy maps at the time? Klein's commonsense model was viable then and might have been expected to have attracted broader interest than it did. Klein (1966) continued the argument for this picture, in a paper with the title "Instead of Cosmology." Hannes Alfvén (1965) added to Klein's picture a matter-antimatter universe, whose expansion was driven by the radiation pressure derived from annihilation of much of the matter and antimatter. But by this time, the observations had seriously challenged the notion that the observable universe has an edge.

2.7 Concluding Remarks

In the 1950s, there was some observational evidence for the cosmological principle, large-scale homogeneity, from Hubble's demonstration of the approximate isotropy of distant galaxy positions shown in Figure 2.3 and from the linearity of the Hubble and Humason redshift-distance relation. A nearly exact homogeneity and isotropy would offer the great convenience of an analytic solution to Einstein's field equation and Milne's elegant derivation of Hubble's law. But an alternative picture, a fractal universe, had the support of one of the best observational astronomers in the years around 1970, Gérard de Vaucouleurs, and it had the support of Benoît Mandelbrot's elegant examples of fractals in mathematics and physics. The fractal picture for the galaxy distribution rightly attracted attention and inspired debates and research. But a more vigorous promotion of the idea earlier, in the 1950s, could have been more productive, because by the 1970s, the totality of evidence reviewed in this chapter had made it clear that the fractal picture is not promising. The evidence in the 1970s instead favored the cosmological principle, which is best put as the assumption that our universe is a realization of a stationary random process. This assumption is implicit in the demanding cosmological tests reviewed in Chapter 9, and it continues to pass checks based on consistency.

Arguments from elegance and observation can instruct or mislead; we see examples of both in the history of the cosmological principle. We also see that, at least on occasion, confusion of this sort can be resolved. Cosmology in the 1970s could operate on the reasonably secure assumption of large-scale homogeneity, and the evidence of it has continued to grow more secure.

An abstract stationary random process has no edge. That may be true of our universe, or there may be edges where the universe is different from what we see. It would have to be far enough away not to have significantly disturbed the thermal radiation propagating to us from close to the maximum distance that can be observed, because this radiation has been closely mapped and checked against theory, which includes the cosmological principle, as part of the tests. Maybe a deeper cosmology will predict which it is: edges at some great distance or none at all. Let us leave to future generations the debate over what it might mean to conclude that a prediction of this sort is to be considered established but not testable in principle.

CHAPTER THREE

Cosmological Models

TWO SPATIALLY HOMOGENEOUS world pictures captured most of the attention in cosmology from the late 1940s through the mid-1960s: an evolving universe and a universe in a statistically steady state. The evolving model describes expansion according to general relativity from an exceedingly dense early condition often termed the big bang. (The discussion on page 7 explains why this usage is inappropriate but lasting and so adopted here.)

In the big bang model, a straightforward extrapolation of its evolution back in time ends at a singularity: a manifest failure of standard general relativity. The usual hopeful thought is that there is a better theory that eliminates the singularity; the cosmological inflation picture to be discussed in Section 3.5.2 is the most common line of discussion in this direction. But to be considered in this chapter is the study of cosmic evolution after inflation, or whatever happened in the very early universe: how we arrived at the big bang model, developed nonempirical assessments of it, and made a start on empirical tests. The following chapters review the lines of research that finally established the hot big bang model at the turn of the century, though that has left open the question of what happened before the big bang.

In the alternative world picture, the continual creation of matter keeps the near-homogeneously expanding universe in a steady state (in the average over a span of time that was left to be discussed). It lacked Einstein's endorsement, but skillful proponents kept the picture visible in England though generally less so at other research centers. The steady-state cosmology is much more predictive than the big bang, which might have been expected to have added more than it did to general interest in the model.

3.1 Discovery of the Relativistic Expanding Universe

To set notation: the line element in the standard form for the relativistic theory of an expanding homogeneous universe will be written in the standard form

$$ds^2 = dt^2 - a(t)^2 \left[\frac{dr^2}{1 - r^2 R^{-2}} + r^2 \left(d\theta^2 + \sin^2 \theta d\phi^2 \right) \right]. \tag{3.1}$$

The spatial coordinates r, θ, ϕ are comoving (that is, fixed to the mean streaming motion of the material contents of the universe), and the factor R^{-2} is a constant (which, despite the notation, can be positive or negative). This expression is usually termed the Robertson-Walker form, after the recognition by Robertson (1929) and Walker (1935) that it follows from spatial homogeneity and isotropy in a spacetime described by a line element. It applies to the steady-state cosmology as well as the relativistic picture.

Comoving observers, at fixed r, θ, ϕ, keep the proper physical time t in equation (3.1). The expansion factor $a(t)$ in equation (3.1) appears in the discussion of the linear redshift-distance relation in Figure 2.1 and equation (2.3). Two events at the same world time t, at coordinate distance r from an observer who sees that the events are separated by the small angle $\delta\theta$, are at physical separation $\delta l = a(t) r \delta\theta$. If the constant R^{-2} is small enough to be neglected, the physical distance from the origin to coordinate radius r at given world time t is $l = a(t) r$, and the rate of change of the physical distance is

$$v = \frac{dl}{dt} = Hl, \quad H = \frac{1}{a}\frac{da}{dt}. \tag{3.2}$$

This is Hubble's law in equation (2.1), with H evaluated at the present epoch, and ignoring space curvature and the relativistic correction for observation along the past light cone instead at fixed t.

In general relativity with Einstein's cosmological constant Λ, the expansion parameter $a(t)$ satisfies the Friedman-Lemaître equations:

$$\left(\frac{1}{a}\frac{da}{dt} \right)^2 = \frac{8}{3}\pi G \rho(t) - \frac{1}{a^2 R^2} + \Lambda, \quad \frac{1}{a}\frac{d^2 a}{dt^2} = -\frac{4}{3}\pi G (\rho + 3p) + \Lambda, \tag{3.3}$$

with the expression for the local conservation of energy being

$$\frac{d\rho}{dt} = -\frac{3}{a}\frac{da}{dt}(\rho + p). \tag{3.4}$$

The mean mass density, including the mass equivalent in radiation energy, is $\rho(t)$, and the pressure is p (with units chosen so the speed of light is unity). To understand the energy equation, recall that the energy $\varepsilon = \rho V$ in a container of volume V changes when the volume changes at the rate $d\varepsilon/dt = -p\, dV/dt$ when the pressure is p. The energy equation follows by setting $\varepsilon = 4\pi\rho a^3/3$ and working out the time derivative.

The factor R^{-2} in the first expression in equation (3.3) may be considered a constant of integration in the sense that, with the energy equation (3.4), the time derivative of $(da/dt)^2$ in the first expression is the second expression. But in general relativity, R^{-2} also defines the geometry of space at fixed t. If R^{-2} in equation (3.1) is positive, the spatial geometry is closed—the analog of the surface of a sphere; if negative, it is the shape of a saddle. If $R^{-2}=0$, the spacetime is said to be cosmologically flat, even though spacetime may be curved.

In the first step to modern cosmology, Einstein (1917) found the static solution to his field equation in general relativity for a universe that is homogeneous and isotropic, consistent with his thinking at the time about Mach's principle. Perhaps the condition that it is static seemed perfectly reasonable at the time. To get this solution, he had to modify his original relativistic field equation by adding what became known as the cosmological constant, Λ. Then, if pressure can be neglected, the conditions $da/dt=0=d^2a/dt^2$ in equation (3.3) require that the mass density ρ and the parameter R^{-2} representing space curvature in equation (3.1) satisfy

$$\Lambda=\frac{4}{3}\pi G\rho=\frac{1}{3a^2R^2}. \tag{3.5}$$

Here aR is the physical radius of curvature of space.

Eddington (1923, 166) noticed that equation (3.5) represents a curious situation: A dynamical variable, the mass density, must agree with a constant of nature, Λ. He wrote that "the question at once arises, by what mechanism can the value of λ [now Λ] be adjusted to correspond with M [a measure of the mean mass density]?" One might also wonder what happens if the mass is rearranged so this condition is locally violated. These questions offered an early hint that Einstein's static model is unstable. This is discussed in Chapter 5.

It takes nothing from Einstein's genius to note that a static universe does not make physical sense in conventional thinking: Recall the Olbers problem discussed in Section 2.1. The Russian Alexander Friedman showed the way out, by generalizing Einstein's solution to a homogeneous expanding or contracting model universe (Friedman 1922, 1924). Einstein's first judgments—that Friedman did not have a correct solution to the field equation in general relativity, and then that the solution is correct but unphysical—have been well reviewed, as by Goenner (2001) and Longair (2006). The Olbers problem with an unlimited buildup of starlight is removed in an expanding universe, because we can assume that the stars have limited lifetimes, in accordance with local energy conservation. And there is extra help from the expansion of the universe, which dilutes the energy density of the starlight. But I have seen no evidence that Friedman recognized he had solved Olbers' problem.

It is unfortunate that Friedman died before recognizing a possible connection between theory and observation: the observed tendency of galaxy spectra to be shifted to the red in proportion to their distance, as would be expected in a homogeneously expanding universe.

The Hungarian mathematical physicist Kornel Lanczos (1923) introduced a coordinate labeling of de Sitter's (1917b) solution for empty homogeneous and isotropic spacetime with a positive cosmological constant. In the notation of equation (3.1), Lanczos' form (in his eq. [32]) is

$$ds^2 = dt^2 - \cosh^2\left(\sqrt{\Lambda}t\right)\left[\frac{dr^2}{1-\Lambda r^2} + r^2\left(d\theta^2 + \sin^2\theta d\phi^2\right)\right]. \qquad (3.6)$$

This is Friedman's solution for closed-space sections with a positive cosmological constant in the limit of vanishing mass density. Lanczos did not take explicit note of the possible relation to the astronomical evidence, however. The Belgian Georges Henri Joseph Édouard Lemaître (1925, 192) also reported this solution for vanishing mass density, and he pointed out that the solution "gives a possible interpretation of the mean receding motion of spiral nebulae."

The American physicist Howard Percy Robertson (1928) reported another coordinate labeling of de Sitter's spacetime, this one cosmologically flat, with $R^{-2} = 0$ and expansion parameter $a \propto e^{\sqrt{\Lambda}t}$. He also pointed to the possible relation to the astronomers' redshift phenomenon.

We see from equations (3.2) and (3.3) that in de Sitter's solution, where pressure and mass density vanish, the physical distance from the origin of a radially moving test particle satisfies

$$\frac{d^2 l}{dt^2} = \Lambda l. \qquad (3.7)$$

In the solution $l \propto \cosh\sqrt{\Lambda}t$, the particle falls toward the origin and then moves away, a behavior known in the 1920s as de Sitter scattering. In this solution, and in the solution $l \propto e^{\sqrt{\Lambda}t}$, the late-time behavior is the same: The velocity of recession of a particle becomes proportional to its distance from the origin independent of initial conditions. The same applies to velocity sorting of particles moving away from an explosion: When mass and Λ can be neglected, the end result is that more rapidly moving particles are farther away.

Friedman (1922 and 1924) found the evolving matter-filled solution to Einstein's field equation. Considering Einstein's skepticism in the early 1920s, it is interesting to see in Einstein (1931, 236) his positive attitude in the statement that

> Es ist von verschiedenen Forschern versucht worden, den neuen Tatsachen durch einen sphärischen Raum gerecht zu werden, dessen Radius P zeitlich veränderlich ist. Als Erster und unbeeinflußt durch Beobachtungstatsachen hat A. FRIEDMAN[1] diesen Weg eingeschlagen, auf dessen rechnerische Resultate ich die folgenden Bemerkungen stütze. Dieser geht demgemäß von einem Linienelement von der Form ... Bemerkenswert ist vor allem, daß die allgemeine Relativitätstheorie HUBBELS neuen Tatsachen ungezwungener (nämlich ohne λ-Glied) gerecht werden zu können scheint als dem nun empirisch in die Ferne gerückten Postulat von der quasi-statischen Natur des Raumes.

My condensed translation, aided by Google, is:

> Different researchers have attempted to do justice to the new facts by considering a spherical space whose radius is a function of time. A. Friedman was the first to have embarked on this path, unaffected by the observational facts. It is remarkable that the general theory of relativity seems to be able to cope with Hubble's new facts more easily (and without the λ component) than with the postulate of the quasi-static nature of space.

We see that Einstein agrees that his general relativity can do justice to the new facts, and without the Λ term (which he wrote as λ). Why the mention that Friedman was not influenced by observational facts? It can be taken to mean that the redshift-distance relation implicit in Friedman's solution is a prediction of what was later observed by Hubble, whom Einstein mentioned, and others.

Lemaître (1927) generalized his 1925 coordinate labeling to the solution for a matter-filled homogeneous spacetime.[1] Friedman found it first, but the evidence is that Lemaître's discovery presented in 1927 was made independently. Consistent with that is a footnote in Lemaître (1929) that thanks Einstein for telling him about the important work by Friedman. The papers Lemaître (1931a and 1950) also refer to Friedman's prior discovery.

For the purpose of this book, the important advance in Lemaître's 1927 paper is the demonstration that in the expanding matter-filled model, spatial homogeneity allows the redshift of a galaxy to remain proportional to its distance as the universe expands. The essential distinction is that earlier discussions ignore mass, so the linear redshift-distance relation does not require Einstein's homogeneity: Velocity sorting would do. The linear redshift-distance relation in a universe with matter follows from homogeneity; it does not require general relativity (as one sees from equation (2.2) on page 15). But all this is much easier to see now, of course.

Lemaître (1927 and 1931a) referred to earlier discussions of a possible linear relation between galaxy redshifts and distances in empty de Sitter spacetime by Lanczos (1922), Weyl (1923), and Lundmark (1924). The German mathematical physicist Hermann Weyl knew of the observations that galaxy spectra tend to be shifted toward the red, and he proposed that this is because matter has moved apart, based on what causality would suggest must be pictured as matter moving on geodesics diverging from a common origin in the asymptotic past. These would be orbits with $l \propto e^{\sqrt{\Lambda}t}$. We may consider this an early example of Klein's (1956) explosion picture, but aided by de Sitter scattering.

1. I am grateful to John Peacock for the argument that Friedman's solution was the more general: Lemaître considered only closed space sections, while Friedman (1922 and 1924) presented the solutions for the closed, open, and cosmologically flat cases.

Lemaître (1927 and 1931a) did not claim that there is empirical evidence for the linear relation redshift-distance. His impression of the observational situation is suggested by the comment in a footnote in Lemaître (1927, 56) that

> Certains auteurs ont cherché à mettre en évidence la relation entre v et r et n'ont obtenu qu'une très faible corrélation entre ces deux grandeurs. L'erreur dans la détermination des distances individuelles est du même ordre de grandeur que l'intervalle que couvrent les observations et la vitesse propre des nébuleuses (en toute direction) est grande (300 Km./sec. d'après Strömberg), il semble donc que ces résultats négatifs ne sont ni pour ni contre l'interpretation relativistique de l'effet Doppler. Tout ce que l'imprécision des observations permet de faire est de supposer v proportionnel à r et d'essayer d'éviter une erreur systématique dans la détermination du rapport v/r. Cf. LUNDMARK. The determination of the curvature of space time in de Sitter's world M. N., vol. 84, p. 747, 1924.

In brief,

> Attempts to find the relation between v and r have shown at best a weak correlation. Since the distance errors are comparable to the range of redshifts (300 km s^{-1} according to Strömberg), all the observations allow is to suppose v is proportional to r and to try to avoid systematic error in the determination of v/r.

His reference is to Lundmark (1924). The discussion of the redshift-distance relation in this paper may not have inspired confidence. Lundmark had a reasonable estimate of the distance to the nearest large galaxy, the Andromeda Nebula M 31, from Öpik's (1922) ingenious interpretation of its velocity of rotation.[2] But this galaxy is in the Local Group, and it has a negative redshift. Lundmark had only rough estimates of distances to a few other nearby galaxies from apparent magnitudes of variable stars compared to novae in the Milky Way. He supplemented this with redshifts and distances of globular star clusters, but they are much closer and are surely parts of the Milky Way galaxy. Lemaître (1927) did not refer to (and maybe did not notice) the later, more encouraging results reported in Lundmark (1925). This paper takes note of reasonably good distances to M 31 and its companions from Hubble's

2. In outline, let $r = \theta D$ be the radius of M 31 at the observed angular size θ and the wanted distance D, let v be the speed of rotation of M 31, and let $v_{\odot}$ be the speed of motion of Earth around the Sun with mass $M_{\odot}$ at distance $r_{\odot}$. Then the mass M of M 31 satisfies $M/M_{\odot} \approx (\theta D/r_{\odot})(v/v_{\odot})^2$, in Newtonian mechanics. The observed energy flux density from M 31 is $f \approx L/D^2$, where L is its luminosity. The combination yields the distance D in terms of the observables θ and f and the mass-to-light ratio M/L, which might be expected to be similar to that of the stars in the Milky Way if M 31 is another galaxy of stars. Öpik found $D = 450$ kpc, impressively close to the modern value, $D = 780$ kpc.

(1925) observations of Cepheid variable stars with Leavitt's relation between the Cepheid period and luminosity. Lundmark (1925, 867) asserts that

> A rather definite correlation is shown between apparent dimensions and radial velocity, in the sense that the smaller and presumably more distant spirals have the higher space-velocity.

This is in line with the cosmological redshift-distance relation, but Lundmark (1925) does not explain the supporting evidence.

Given this state of affairs, it does not seem surprising that Lemaître (1927) concluded in his footnote that the data are not good enough to check the redshift-distance relation. It agrees with Lemaître's (1950, 2) later recollection: "Naturellement, avant la découverte et l'étude des amas de nébuleuses, il ne pouvait être question d'établir la loi de HUBBLE; mais seulement d'en calculer le coefficient." That is: "Naturally, before the discovery and study of many nebulae, it could not be a question of establishing Hubble's law; but only calculating the coefficient."

Lemaître's (1927) computation of the coefficient mentioned in his footnote assumes the redshift-distance relation is linear and that the distance errors are as often high as they are low, so the errors tend to cancel in the mean. Then a good estimate of the constant of proportionality, H_0, is the mean of the recession velocities divided by the mean of the distance estimates. Lemaître (1927) had a list of 42 galaxies with measured redshifts and apparent magnitudes, almost entirely from Slipher. For distances, he used Hubble's (1926) relation between galaxy apparent magnitude m and distance r in parsecs: $\log r = 0.2m + 4.04$.[3] Hubble based this on distances of six plus a possible seventh galaxy from Leavitt's (1912) relation between Cepheid variable star periods and luminosities. This is not a bad first approximation. Although the frequency distribution of galaxy luminosities is broad, the quite sharp upper cutoff on galaxy luminosities with the smaller volume of space sampled by less-luminous ones means galaxies selected by apparent magnitude have a modest dispersion in distances. And Lemaître's value of H_0 is not far from Hubble's (1929) estimate 2 years later, as will be discussed.

Lemaître had visited the University of Cambridge, where, according to the Plumian Professor of Astronomy and Experimental Philosophy, Arthur Stanley Eddington, Lemaître[4] "has been attending my lectures and pursuing his studies and investigations to my entire satisfaction during the academic

3. I am grateful to Stephen Kent for pointing this out to me. Lemaître's use of Hubble's (1926) galaxy magnitude-distance relation is discussed by Luminet (2013).

4. I am grateful to Liliane Moens for providing me with this and other information from the Archives Georges Lemaître, Université catholique de Louvain, Louvain-la-Neuve, Belgium.

year 1923–24." Lemaître then moved to the Massachusetts Institute of Technology in Cambridge, Massachusetts, where he published his 1925 paper in the short-lived MIT *Journal of Mathematics and Physics*. A publication in one of the journals influential physicists and astronomers read might have attracted attention to Lemaître's interesting result. On returning to Belgium, he published his 1927 paper in *Annales de la Société Scientifique de Bruxelles*. Again, this was not a journal leading figures followed. After several letters, Lemaître managed to draw Eddington's attention to the 1927 paper. Evidence of the response is to be seen in Eddington's postcard to de Sitter (from the Archives Georges Lemaître and reproduced in Plate III):

> Lemaitre's address is 40 rue de Namur, Louvain. A research student McVittie and I had been worrying at the problem and made considerable progress; so it was a blow to us to find it done much more completely by Lemaitre (a blow softened, as far as I am concerned, by the fact that Lemaitre was a student of mine.)
>
> By the way it was the report of your remarks & mine at the R.A.S. which caused Lemaitre to write to me about it.

George McVittie (1967, 295) recalled that

> Curiously enough this work [by Lemaître] at first attracted little attention. Nearly three years later, I was a research student of Eddington's and he had suggested that I work on the redshift problem. I well remember the day when Eddington, rather shamefacedly, showed me a letter from Lemaitre which reminded Eddington of the solution to the problem which Lemaitre had already given. Eddington confessed that, though he had seen Lemaitre's paper in 1927, he had completely forgotten about it until that moment. The oversight was quickly remedied by Eddington's letter to Nature of 1930 June 7, in which he drew attention to Lemaitre's brilliant work of three years before.

Eddington also generously arranged to have Lemaître's 1927 paper translated into English and published in a journal influential figures did and do follow, *Monthly Notices of the Royal Astronomical Society*.

Lemaître (1931a) set the stage for conspiracy theorists by not including in the English translation the footnote in his 1927 paper that helps understand how he arrived at a value of what became known as Hubble's constant. His value is close to what Robertson published the next year and Hubble the year after that in his announcement of evidence of the linear redshift-distance relation. The three early estimates are

$$
\begin{aligned}
H_0 &\approx 630 \text{ km s}^{-1}\text{ Mpc}^{-1} \quad \text{Lemaître (1927)},\\
H_0 &\approx 460 \text{ km s}^{-1}\text{ Mpc}^{-1} \quad \text{Robertson (1928)},\\
H_0 &\approx 500 \text{ km s}^{-1}\text{ Mpc}^{-1} \quad \text{Hubble (1929)}.
\end{aligned} \tag{3.8}
$$

The calculation of H_0 in the first line of equation (3.8) is displayed in equation (24) in Lemaître (1927) and discussed in the footnote. Mario Livio's (2011) admirable detective work reveals correspondence that clearly demonstrates the removal of the 1927 footnote from the 1931 English translation was Lemaître's decision. Robertson's (1928) estimate in the second line is based on Slipher's redshift measurements and Hubble's Cepheid variable distances of six galaxies. Robertson did not offer an assessment of the evidence that the relation may be linear.

Hubble (1929) attended to the observations, not the theory. He had distances for six or seven galaxies from observations of Cepheid variable stars, 13 distances for galaxies on the assumption that the most luminous stars have a common intrinsic luminosity, and several galaxies in the Virgo cluster at a not very well understood distance. But Lemaître, Robertson, and Hubble had essentially the same information, which is why their estimates of H_0 in equation (3.8) are similar. My impression is that Lemaître (1927) was unduly pessimistic about evidence for the linear relation, Robertson just trusted the theory, and Hubble (1929) may have been overly optimistic. The redshift-distance plot in Figure 1 in Hubble (1929) seems suggestive of the linear relation but hardly convincing. But Hubble was on the right track, as he and Humason showed in the 1930s.

Why did Lemaître remove the footnote? He states there that there had been no convincing claim of the linear relation. The evidence reviewed by van den Bergh (2011) is that Lemaître and Hubble were aware of earlier indications of a relation between galaxy redshifts and distances, and we see that Hubble referred to Lundmark's (1925) not very encouraging discussion of the evidence. But the situation changed in 1929, when Hubble explicitly announced evidence of the linear relation. The straightforward guess is that Lemaître removed the footnote because Hubble had made it obsolete. But in any case, we can be quite sure the footnote had not been removed in a conspiracy to obscure a prior empirical discovery that Lemaître did not even claim.

Let us pause to notice that Hubble's measurements of galaxy distances were based on Henrietta Leavitt's discovery of the relation between the luminosities and periods of Cepheid variable stars. She reported the key result in Leavitt (1912, 2):

> there is a simple relation between the brightness of the variables and their periods. The logarithm of the period increases by about 0.48 for each increase of one magnitude in brightness.

Hubble's 1929 paper largely relied on Slipher's (1917) redshift measurements. Slipher was one of the excellent astronomers Percival Lowell brought to his observatory constructed to check the possibility of canals on Mars made by an advanced civilization. And in the 1930s, Milton Humason played an important role with Hubble in their considerable extension of distances of galaxies with measured redshifts (as in Hubble and Humason 1931) that greatly improved the case for the linear redshift-distance relation. All this is not meant to depreciate Hubble's contributions—to my mind, he had just the right instincts for the observations that would advance extragalactic astronomy at that time—but to note that Hubble had help in arriving at Hubble's law.

3.2 The Relativistic Big Bang Cosmology

Lemaître's (1927) solution traces the expansion of the universe back in time to Einstein's (1917) static world model. This demonstrates the instability of Einstein's model: A slight homogeneous disturbance to the static situation sets the universe expanding (in Lemaître's solution) or else collapsing. Lemaître (1931b, 706) turned to the idea that the expansion traces back to a dense state. Perhaps, as he wrote, "the world has begun with a single quantum." Lemaître (1931c, 706) termed this "l'atome primitif"; it is now known as the big bang.

McCrea and Milne (1934) showed that if the pressure p and cosmological constant Λ can be ignored, then the Friedman-Lemaître equations (3.3) follow from Newtonian physics. To see this, consider that in a homogeneous universe with mass density ρ, a sphere with radius $a(t)$ much less than the radius of curvature of space contains mass $M = 4\pi\rho a^3/3$. In Newtonian mechanics, the gravitational acceleration of the radius of this sphere is determined by the mass it contains, $d^2a/dt^2 = -GM/a^2$, independent of the spherically distributed mass outside $a(t)$. This is the first of equations (3.3) when p and Λ vanish. The second equation is the integral of the first, where the constant of integration is R^{-2}. This expresses the Newtonian conservation of kinetic plus potential energies. These results follow because general relativity in the limit of $\Lambda = 0$ and for small velocities, and applied to regions small compared to spacetime curvature, is Newtonian mechanics. In particular, the McCrea and Milne argument assumes the flat spacetime of Newtonian mechanics, but that is an arbitrarily good approximation to the relativistic model if we choose a small enough sphere radius. Another theory with this same limiting behavior would do as well for the purpose of this chapter, of course.

We see from the second expression in the Friedman-Lemaître equations (3.3) that the source of gravity tending to slow the expansion is $\rho + 3p$, which is to say that in general relativity, pressure acts as active gravitational mass density. Consistency of the Friedman-Lemaître equations requires it. A Newtonian analog of sorts for the dual role of the constant R^{-2}—conserved kinetic plus potential energy in equation (3.3) and the measure of the curvature of space sections in equation (3.1)—is that the Newtonian energy of the matter in a sphere of comoving radius r is $U = -r^2/(2R^2)$, which is the departure from flat spacetime in equation (3.1).

Lemaître (1934, 12) introduced the thought that Λ may define the vacuum energy density. If Λ is nonzero then, he wrote,

> Everything happens as though the energy *in vacuo* would be different from zero. In order that absolute motion, i.e., motion relative to vacuum, may not be detected, we must associate a pressure $p = -\rho c^2$ to the density of energy ρc^2 of vacuum. This is essentially the meaning of the cosmical constant λ which corresponds to a negative density of vacuum ρ_0 according to
>
> $$\rho_0 = \frac{\lambda c^2}{4\pi G} \cong 10^{-27}\text{gr./cm.}^3 \tag{3.9}$$

His estimate of the vacuum energy density is too big, because he used Hubble's overestimate of Hubble's constant. The point was made, however: Λ sets the zero of energy, as far as gravity is concerned, and Λ acts as mass with homogeneous energy density and pressure (taking $c = 1$ as usual):

$$\rho_\Lambda = \frac{3\Lambda}{8\pi G}, \quad p_\Lambda = -\rho_\Lambda. \tag{3.10}$$

The sum of this effective pressure and mass density vanishes, as required to keep ρ_Λ constant in the energy equation (3.4) in an expanding universe. As we have seen, consistency of the two relativistic Friedman-Lemaître equations requires that pressure contributes to the active gravitational mass density in the amount $\rho_\Lambda + 3p_\Lambda = -2\rho_\Lambda$. And we can admire Lemaître's recognition that this vacuum energy density is not changed by a velocity transformation. One way to put it is that the relativistic stress-energy tensor $T^{\mu\nu}$ of a fluid at rest with energy density ρ and pressure p is diagonal with components ρ, p, p, p. With $p_\Lambda = -\rho_\Lambda$, the stress-energy $T_\Lambda^{\mu\nu}$ is proportional to the Minkowski metric tensor and therefore is unchanged by a Lorentz transformation. That is, Λ does not define a preferred frame of motion. (But the idea discussed in Section 3.5.1 that the value of Λ may be evolving, decreasing to its "natural" value, zero, does introduce the preferred motion in which Λ has no spatial gradient.)

The effective mass density ρ_Λ has come to be termed "dark energy." The term first appeared in Huterer and Turner (1999). But the effective negative pressure is not to be associated with the negative pressure of a fluid, which is an unstable situation unless p is exactly the negative of ρ.

The cosmological redshift is defined in an expanding (or contracting) homogeneous and isotropic metric spacetime as follows. Let λ be the physical wavelength of a freely propagating photon, and more generally the de Broglie wavelength of a freely moving particle. This wavelength is to be measured by a comoving observer, one moving with the mean streaming flow of the material. The wavelength is stretched in proportion to the expansion parameter: $\lambda \propto a(t)$.[5] This means the wavelength λ_{em} of a spectral feature measured by an observer at the comoving source at the time t_{em} of emission is related to the wavelength λ_0 of the feature measured by a comoving observer who detects the radiation at the present time t_0 by

$$1+z \equiv \frac{\lambda_0}{\lambda_{em}} = \frac{a(t_0)}{a(t_{em})}. \tag{3.11}$$

This defines the cosmological redshift z.

If the time t_{em} is close to the time t_0 of detection, then the first-order expansion of equation (3.11), at $z \ll 1$, is

$$v = cz \simeq \frac{1}{a}\frac{da}{dt} \times c(t_0 - t_{em}) = H_0 d, \tag{3.12}$$

where $d = c(t_0 - t_{em})$ is the physical distance between source and observer and H_0 is Hubble's constant (and I have put back the speed of light). This relation also follows by the arguments leading to equation (2.1).

The redshift gives the rate of change of the physical separation of source and observer in the following sense. Imagine two observers, each comoving with the local expansion of the universe, and separated by coordinate distance x. This is physical distance $r = a(t_0)x$ on the hypersurface of fixed world time t_0. If r is much less than the Hubble length cH_0^{-1}, and the two observers are connected by a string, then the string would have to be payed out at the rate $H_0 r$. In another limit, let the separation be larger than the Hubble length. Here the instantaneous physical separation at time t_0 is the sum of measurements by a dense row of observers, each of whom finds the physical distance to the next observer farther along the line connecting source and observer, and all evaluated at time t_0. The sum is $r = a(t_0)x$. The rate of change of this distance is $H_0 r$. It can exceed the velocity of light, but no one can see it.

To reduce the chance of confusion, let us pause to note that in the standard theory, the physical distances between the galaxies are increasing in the mean; the galaxies are physically moving apart. It is sometimes said that space is expanding, but I do not know how to give meaning to the expression, and it can be misleading. In particular, the galaxies themselves are not changing in size, apart from the effects of accretion of mass and the mass loss produced by the deaths of stars and such processes. A positive value of Einstein's cosmological constant Λ causes galaxies to be slightly more compact than would

5. To see this, imagine space is periodic with period length L. Fourier waves have to be continuous, so the allowed wavelengths are $\lambda = L/n$ for integer n. If the universe is homogeneously expanding, then $\lambda \propto a(t)$. If the mode frequency is large compared to the rate of change of the mode wavelength, $\dot{a}/a$, then adiabaticity says the field in the mode stays in the mode. And all this follows in the limit of arbitrarily large L.

be indicated by their mass alone. Apart from that, in general relativity the expansion of the universe has no effect on a galaxy or its contents, including us.

The Friedman-Lemaître equation (3.3) for the evolving rate of expansion of the universe is often written as

$$\frac{1}{a(t)}\frac{da(t)}{dt} = H_0\left[\Omega_r(1+z)^4 + \Omega_m(1+z)^3 + \Omega_k(1+z)^2 + \Omega_\Lambda\right]^{1/2}, \qquad (3.13)$$

$$\Omega_r + \Omega_m + \Omega_k + \Omega_\Lambda = 1.$$

In this convenient approximation,[6] the matter pressure is assumed to be small compared to its energy density ($p_m \ll \rho_m c^2$). The present rate of expansion of the universe is Hubble's constant (equations (2.1) and (3.12)):

$$H_0 = \frac{1}{a}\frac{da}{dt}, \quad \text{at} \quad t = t_0. \qquad (3.14)$$

The left side of equation (3.13) is the Hubble parameter evaluated at time t, and $1+z = a(t_0)/a(t)$ is the expansion factor from t to t_0. Hubble's constant is traditionally written as

$$H_0 = 100h \text{ km s}^{-1} \text{ Mpc}^{-1}. \qquad (3.15)$$

The use of the dimensionless Hubble parameter h offers a way to indicate the sensitivity to the extragalactic distance scale, which until relatively recently was uncertain by a factor of about two. By the end of the revolution, the measurements had converged to $h = 0.72 \pm 0.05$ (Freedman et al. 2001; Spergel et al. 2003).

The dimensionless cosmological parameters Ω_i are the fractional contributions to the square of the present expansion rate, $z = 0$, in equation (3.3). In equation (3.13), the mass is assumed to be well described as seas of relativistic and pressureless matter. The former is assumed to have the equation of state $p_r = \rho_r c^2/3$ of radiation, so it follows from equation (3.4) that the mass density of this component varies as $\rho_r \propto a(t)^{-4} \propto (1+z)^4$. The mass density in low-pressure matter varies as $\rho \propto (1+z)^3$. The third term in equation (3.13) is the contribution to the expansion rate by the space curvature term in equation (3.3); it varies as $(1+z)^2$. The last term represents Einstein's cosmological constant.

The dimensionless parameters Ω_i in equation (3.13) determine how the universe began and how it will end, and the time scale for what happens is set by the value of the Hubble parameter h in equation (3.15), provided of course that the theory is an adequate approximation to reality. One of the themes of

6. This assumes that the contents of the universe can be approximated as two components: pressureless matter and radiation that may include a sea of relativistic particles. Analyses of more complicated situations—as happens during the annihilation of the sea of thermal electron-positron pairs as the early universe expands and cools—must return to equation (3.3).

this book is the evolution of assessments of the values of these parameters, empirical and nonempirical, along with the emergence of evidence that this relativistic theory actually is a useful approximation to what happened.

A variant of the relativistic big bang cosmology was inspired by the large value of the ratio of the electrostatic to gravitational forces of attraction of an electron to a proton,

$$\frac{e^2}{Gm_p m_e} \sim 10^{40}. \tag{3.16}$$

Both forces are inverse square laws, so this dimensionless number is independent of the separation of the electron and proton. Eddington (1936) pointed out that this is a curiously large number to appear in a fundamental theory. Dirac (1938, 201) proposed a principle:

> Any two of the very large dimensionless numbers occurring in nature are connected by a simple mathematical relation, in which the coefficients are of the order of magnitude unity.

The ratio of the present characteristic cosmic expansion time, $t_c = H_0^{-1}$, to the atomic time unit, $e^2/m_e c^3$, where m_e is the electron mass, also is large, on the general order of equation (3.16). This suggested to Dirac that gravity is so much weaker than electromagnetism because the strength of gravity has been decreasing, roughly in inverse proportion to the age of the universe, thus preserving the similar values of these two ratios. In units chosen so the electron charge, the velocity of light, and masses are constant (and assuming m_e/m_p is constant; it's not a very small number) Newton's gravitational "constant" G would vary inversely with time, as

$$\frac{1}{G}\frac{dG}{dt} \sim -10^{-10}\,\text{year}^{-1}. \tag{3.17}$$

Pascual Jordan (1948) wrote down a field theory of this effect, Brans and Dicke (1961) enlarged on it, and Dicke played the leading role in probing for evidence of the effect. Peebles (2017) describes how Dicke led the effort to place arrays of optical corner reflectors on the Moon for precision timing of reflected laser pulses. That could be used to explore whether G might be decreasing. This program has shown that the rate of change of the strength of gravity is no more than two orders of magnitude smaller than equation (3.17) (Williams, Turyshev, and Boggs 2012). Dicke's program of experimental probes of gravity physics was important to the growth of cosmology, as also reviewed in Peebles (2017). And there still is willingness to suspect that the dimensionless parameters of physics are evolving, despite the absence of substantial evidence at the time of writing.

Among cosmologists, recent thinking about evolution of the parameters of physics is dominated by the curious value of Einstein's cosmological constant in the cosmology established at the turn of the century. Its value defines a characteristic mass density that may be compared to the characteristic Planck density defined by Planck's constant, Newton's constant, and the velocity of light:

$$\rho_\Lambda = \frac{\Lambda}{G} \sim 10^{-30} \text{ g cm}^{-3}, \quad \rho_{\text{Planck}} = \frac{c^5}{G^2 \hbar} \sim 10^{94} \text{ g cm}^{-3}. \tag{3.18}$$

The expression for ρ_{Planck} follows from dimensional analysis; it also is an order-of-magnitude estimate of the sum of zero-point energies of the quantum fields with wavelengths down to the Planck length. Dirac's principle helped inspire the thought by Ratra and Peebles (1988) that Λ is so small because it has been evolving to what might be its natural value, zero, for a long time. The notion that Λ may be evolving is considered further in Section 3.5.1.

3.3 The Steady-State Cosmology

Bondi and Gold (1948) proposed a generalization of Einstein's cosmological principle to their perfect cosmological principle: that the universe is a stationary random process in time as well as space. In an expanding universe, this requires that matter is continually created, gravity collecting the matter to form new galaxies that replace those moving apart as they age, preserving the steady state. Hoyle (1948) introduced an adjustment of general relativity to describe this continual creation of matter. And Bondi, Gold, and Hoyle led a spirited campaign in support of their steady-state cosmology.

The 1948 version implicitly assumed that the statistical fluctuations in time average out in the mean over the characteristic expansion time H_0^{-1}. Later versions allowed episodes of creation of matter and radiation, between which the universe may have evolved much like the relativistic model.

In the original 1948 steady-state cosmology, the expansion rate $\dot{a}/a$ is constant, so we see from equation (2.3) that the expansion parameter must evolve as

$$a(t) \propto e^{Ht}, \tag{3.19}$$

with H now a constant. Homogeneity means the line element can be written in the Robertson-Walker form in equation (3.1). The physical radius of curvature of sections of constant world time is $|a(t)R|$. Since $a(t)$ is a function of time, this physical radius can only be independent of time if $R \to \infty$, or $R^{-2} = 0$. The spacetime geometry thus is the same as the cosmologically flat relativistic model with negligible mass density: $\Omega_{\text{m}} = 0$, $\Omega_{\text{k}} = 0$, $\Omega_\Lambda = 1$.

Apart from the continual creation of matter, the steady-state model assumes standard local physics. Thus the cosmological redshift is defined by the expansion factor $a(t)$ in equation (3.11). The frequency distribution of ages τ of the galaxies works out to the 1948 steady-state prediction

$$\frac{dP}{d\tau} = 3He^{3H\tau}, \quad \langle \tau \rangle = \frac{1}{3tH}. \tag{3.20}$$

The count of galaxies as a function of observed redshift varies as (Bondi and Gold 1948, eqs. [1]–[6])

$$\frac{dN}{dz} \propto \frac{z^2}{(1+z)^3}, \tag{3.21}$$

and the observed bolometric (integrated over wavelengths) energy flux density f from a source with luminosity L observed at redshift z is

$$f = \frac{H^2 L}{4\pi c^2 z^2 (1+z)^2}. \tag{3.22}$$

3.4 *Empirical Assessments of the Steady-State Cosmology*

The great virtue of the 1948 steady-state model is that the background cosmology has only one parameter, Hubble's constant H, and the model postulates that the nature of the contents of the universe is not changing with time. Distant objects are observed as they were in the past (because of the light travel time), but in this model, they are statistically the same as nearby. This offers fixed targets for such tests as the distribution of ages of the galaxies in our neighborhood, the counts of radio sources as a function of energy flux density, and the relation between galaxy redshifts and distances. In the 1950s and early 1960s, this model offered a greater stimulus to the modest empirical advances in cosmology than the relativistic model.

In the year the steady-state model was introduced, Stebbins and Whitford (1948) announced evidence that the colors of early-type galaxies[7] are redder at larger redshifts. This challenged the steady-state postulate that the galaxy population does not evolve. Whitford (1954, 601) acknowledged the careful questioning of evidence of the effect by Bondi, Gold, and Sciama (1954, 601), reviewed the considerable advances in the observations since 1948, and concluded, "The foregoing discussion is not intended to indicate that the author considers the color-excess effect established beyond all reasonable doubt."

In the addendum to the second edition of his book, *Cosmology*, Bondi (1960) stated that the Stebbins-Whitford effect "was disproved." He may have had in mind Code's (1959) review of the situation and the conclusion that,

7. Early-type galaxies contain relatively little mass in gas and plasma and relatively few young stars. They may be termed lenticular if flattened, elliptical if less flat, cD if exceptionally large and luminous, and Luminous Red Galaxies if selected by color and for large luminosity. Late-type galaxies, including spirals such as our Milky Way, contain greater mass fractions of plasma and gas that support larger star formation rates. In the 1920s, it was suspected that early-type galaxies evolve into late-type, hence the names, but that proves to be far too simple. Hubble (1926, 326) stated that the terms early and late "express a progression from simple to complex forms" of galaxies.

with a more secure template of spectra of nearby ellipticals, the Stebbins-Whitford "color effect is virtually eliminated."

The observations that some astronomical sources of radio radiation are located in other galaxies, and that most radio sources are uniformly distributed across the sky, suggested that most are extragalactic. Some with known distances were seen to be luminous enough to be detectible at cosmologically interesting distances. This meant the count $N(<S)$ of sources as a function of the observed radio energy flux density S offered an interesting test. In a statistically homogeneous spatial distribution of radio sources, the counts at low redshifts vary as $N(<S) \propto S^{-3/2}$ (equation (2.5) with f replaced by the radio astronomers' S). The steady-state prediction of the form of $N(<S)$ when the observations reach redshifts approaching unity depends on the frequency distribution of source luminosities as a function of wavelength, which was not well known. But in the 1948 steady-state cosmology, the count of sources of given intrinsic luminosity increases with increasing redshift only as $N \propto \log z$ at $z \gg 1$, so the source counts to fainter flux densities S would expected to increase about as $N(>S) \sim \log S^{-1}$ (equations (3.21) and (3.22)). Ryle (1955), a leader in the development of radio astronomy, concluded that the source counts increase with decreasing flux density more rapidly than that. This would suggest radio sources were more luminous in the past, a contradiction of the 1948 steady-state model.

Debate over the interpretation of this piece of evidence was vigorous, at least in part because Martin Ryle, a steady-state skeptic, and Fred Hoyle, an advocate, were both at the University of Cambridge. The measurement itself was controversial, because the limited angular resolution and sidelobes of the radio telescope array caused systematic errors in the source flux density estimates. Peter Scheuer, who was Ryle's research student, found a statistical measure that helped correct for faint sources lost or artificially enhanced, with results that again indicated the growth of $N(<S)$ with decreasing S is faster than expected in the relativistic model without evolution or than predicted in the steady-state model (Ryle and Scheuer 1955). In his book, *The Cosmic Century*, Malcolm Longair recalls this episode, and how radio source counts eventually became a convincing challenge to the 1948 steady-state cosmology (Longair 2006, Sec. 12.3).

Sandage (1961a) considered how the great 200-inch Hale telescope at Palomar Mountain Range in southern California might test cosmological models. He concluded that a useful probe might be the galaxy redshift-magnitude relation, z-m (where the galaxy redshift z is defined in equation (3.11), and the galaxy apparent magnitude m is defined in equation (2.6)). The steady-state model makes a definite prediction of the z-m relation for galaxies with a given intrinsic luminosity or absolute magnitude (as defined in footnote 3 on page 23). One would of course observe many galaxies to take account of differences in luminosities. But in the steady-state cosmology, one could take it that observations at different redshifts sample statistically the same population. The situation is much more complicated in an evolving universe, because the galaxy population surely would evolve too, and with it the galaxy luminosities.

At the time of Sandage's (1961a) study of cosmological tests, the measured redshifts were well below unity, so one could focus on the first-order correction to the linear redshift-distance relation for the departure from a static Minkowskian spacetime. The usual measure was the deceleration parameter (sometimes termed the "acceleration parameter"):

$$\begin{aligned} q_0 &= \frac{a(t_0)\ddot{a}(t_0)}{\dot{a}(t_0)^2} = -1, \quad \text{in the 1948 steady-state model,} \\ &= \frac{\Omega_m}{2} - \Omega_\Lambda, \qquad \text{in the Friedman-Lemaître model.} \end{aligned} \tag{3.23}$$

The time derivatives are evaluated at the present time, t_0, and the expression for the Friedman-Lemaître model assumes pressure can be ignored. The value of q_0 in the steady-state cosmology follows from (equation (3.19)). It is the same as in a relativistic Friedman-Lemaître model dominated by Λ; both have the spacetime geometry of de Sitter's solution for a static empty universe with a cosmological constant.

The Humason, Mayall, and Sandage (1956, 151) assessment of measurements of the deceleration parameter led them to conclude that

> The foregoing analysis therefore suggests that any reasonable estimates of the errors in the measured magnitudes and in the values of $\dot{M}$ and $\dot{K}$ [these are the rates of evolution of galaxy luminosities and colors] require that $\ddot{R}_0$ [where $\ddot{R}_0 = \ddot{a}$ in equation (3.23)] be negative and that the expansion is decelerating. This result cannot be considered as established, however, until accurate values of K [the color term needed to correct the apparent magnitude for the effect of the shift of the spectrum] are available from Whitford's current work and until an adequate theory is worked out to explain the Stebbins-Whitford effect.

Baum's (1957) observations indicated $q_0 = 0.5 \pm 1$; Sandage (1961a) suggested that this is the best available measurement. It would seem from these papers that this line of observations could be approaching the ability to distinguish between the steady-state cosmology and relativistic models with $\Lambda = 0$. The *z-m* measurements did reach this critical point, but that happened four decades later (Section 9.1). The 1948 steady-state model was seriously challenged earlier, by recognition of the presence of the sea of thermal radiation to be discussed in Chapter 4, and before that by a less-celebrated but serious test, the distribution of galaxy ages, as follows.

When the steady-state model was introduced, the usually accepted value of Hubble's constant was an overestimate by a factor of seven. That set the characteristic expansion time to $T \equiv H^{-1} \simeq 2 \times 10^9$ y, a serious underestimate. Bondi (1952) concluded from a survey of the evidence from radioactive decay ages and stellar evolution ages that the oldest observed objects are significantly older than this value of H^{-1}. In the relativistic model, that would call for Lemaître's (1931d) proposal that a positive cosmological constant can cause the expansion rate to have been smaller in the past, making the expansion time longer than H^{-1}. Bondi and Gold (1948, 264) pointed to another way out: In the 1948 steady-state cosmology, galaxies have the distribution of ages shown in equation (3.20), and some are older than $T = H^{-1}$. They wrote that

> there is no reason to suppose that a particular nebula (such as our Milky Way) is of some age rather than another. In our theory there is therefore no difficulty whatever in taking the age of our galaxy to be anything indicated by local observations (such as say $5 - 8 \times 10^9$ years), although T is much shorter.

This remark is correct but incomplete. Gamow (1954a) pointed out that the color of a star population may be expected to evolve as the population ages and the more massive and bluer stars die out. This means that the broad distribution of ages in equation (3.20) would predict a broad distribution of colors of nearby galaxies. Gamow (1954a, 200) wrote that

> Dr. Baade informs me that the observations show that color indices of the elliptical galaxies in our neighborhood are remarkably constant, within 0.05 mag., which speaks against the possibility of mixed-age population predicted by the theory of the steady-state universe.

We have from equation (3.20) that the distribution of the ages t of the galaxies present at any time is $P(<t) = 1 - e^{-3Ht}$, so

$$10 \text{ percent are younger than } 0.035H^{-1}; \quad 10 \text{ percent are older than } 0.77H^{-1}. \tag{3.24}$$

This is a spread of a factor of 20. It seems reasonable to expect that such a wide range of ages would have produced noticeable differences among the stellar spectra of nearby galaxies, contrary to what was and is observed.

The correction to the extragalactic distance scale, and to $T = H^{-1}$, is reviewed in Section 3.6.1. The correction eased constraints on the relativistic model, and it was important also for the steady-state picture, because the old scale made the predicted mean age of the galaxies, $T/3$, seriously short compared to the age of the ordinary-looking Milky Way galaxy. This is illustrated by Bondi's (1960, 165) comment in the addendum to the second edition of *Cosmology*:

> It is not easy to appreciate now the extent to which for more than fifteen years all work in cosmology was affected and indeed oppressed by the

short value of T (1.8×10^9 years) so confidently claimed to have been established observationally.

Bondi was right, but he might have added another complaint: that theorists had not challenged the observers' estimate of T, as he and others had done for the Stebbins-Whitford effect.

The adjustment of the extragalactic distance scale relieved the problem of the mean age of galaxies in the 1948 steady-state cosmology, but Gamow's (1954a) point stood: The model predicts an apparently unreasonable spread of ages of local galaxies. Although Gamow offered a serious challenge, it is not mentioned in either edition of Bondi's (1952 and 1960) generally careful assessments of the situation. The NASA Astrophysics Data System lists no citations of Gamow's (1954a) paper prior to 1980. By this time, the 1948 steady-state cosmology was seriously challenged from other directions: the counts of quasars[8] as a function of redshift; the observations that the properties of galaxies differ at high and low redshift; the tightening web of evidence that helium and deuterium are remnants from cosmic evolution; and most direct and convincing, the thermal spectrum of the sea of microwave radiation (Section 3.4, Figure 4.7). Early steps in the exploration of the properties of this radiation were the first widely recognized challenge to a steady state. Gamow's challenge was earlier, but community assessments of evidence can be capricious.

It is worth considering that if the z-m measurements in the 1950s had succeeded, they would have shown that the deceleration parameter is consistent with the steady-state prediction, $q_0 = -1$, within reasonable allowance for measurement error. In the 1950s, this would have been a serious argument for the steady-state cosmology. But the value of q_0 was established much later, which simplifies the history.

The steady-state philosophy may be adjusted to the postulate of episodic creation of close-to-homogeneous distributions of matter and radiation dense and hot enough enough to have forced relaxation to statistical equilibrium, with formation of helium and deuterium after creation as the universe expanded and cooled. The evolution between episodes of creation might be well described by the relativistic model. This would be the hot big bang cosmology with a different creation story from the cosmological inflation picture for the very early universe (see Section 3.5.2). Hoyle and Narlikar (1966) presented an early vision of a quasi-steady-state cosmology. Later thinking about this alternative to the relativistic model is presented in Hoyle, Burbidge, and Narlikar (1993 and 2000). These later arguments had little discernible effect on developments in empirical cosmology from the 1970s on. But the 1948 steady-state cosmology was an important early stimulus for observations, as they began to lead us to the established evolving cosmology.

8. Quasars were known as quasi-stellar objects; they now are identified as active galactic nuclei (AGNs).

3.5 Nonempirical Assessments of the Big Bang Model

Sandage's (1961a) paper on possible probes of cosmology using the great telescope on Palomar has an appropriately conservative title: "The Ability of the 200-inch Telescope to Discriminate between Selected World Models." He could hope to falsify the steady-state model. But an empirical challenge to the relativistic big bang model is much more difficult, because this cosmology is far less predictive. Where the former has just one free parameter, H, the latter has three: the mean mass density and Λ along with the present expansion rate, H_0. And the evolving world model allows the galaxies to evolve, giving considerable freedom to adjust ideas about what the galaxy population has been doing to fit what is observed. It does not seem surprising then that nonempirical assessments were influential in early studies of the relativistic big bang cosmology. The history of thinking to be considered includes the notable influence of the cosmological inflation picture beginning in the early 1980s (see Section 3.5.2).

3.5.1 EARLY THINKING

The thought that the cosmological constant, and maybe also space curvature, are not likely to be interesting or significant in a satisfactory cosmological model traces back to Einstein and de Sitter (1932) (Their photograph is in Plate III, panel (c)). They pointed out that mass certainly is present and ought to be taken into account in a cosmological model, but nonzero values of Λ and space curvature are not needed to construct a relativistic model that might be compared to the observations available at the time. Thus they argued that it is sensible to begin with the simplest acceptable version of equation (3.13), in which

$$\Omega_{\rm m} = 1, \quad \Omega_{\rm r} = \Omega_k = \Omega_\Lambda = 0, \quad a \propto t^{2/3}, \quad H_0^2 = \frac{8}{3}\pi G \rho_0. \qquad (3.25)$$

This became known as the Einstein–de Sitter cosmological model.

Einstein and de Sitter (1932, 214) remarked that

> The curvature is, however, essentially determinable, and an increase in the precision of the data derived from observations will enable us in the future to fix its sign and to determine its value.

They did not mention that Λ also may be measured, maybe because Einstein had come to regret introducing this term. For example, Pais (1982, 288) quotes Einstein in a 1923 letter to Weyl:

> According to De Sitter two material points that are sufficiently far apart, continue to be accelerated and move apart. If there is no quasistatic world, then away with the cosmological term.

Examples of thinking about the Einstein–de Sitter model are in Robertson (1955): This model is of “some passing interest;” and Bondi: (1960, in the addendum to the second edition) for “its outstanding simplicity.” Meanwhile, other respected authorities were arguing that Λ has no place in physics. In the second edition of *The Meaning of Relativity*, Einstein (1945, 111) states that Λ “constitutes a complication of the theory, which seriously reduces its logical simplicity.”

Pauli (1958, 220), in the Supplementary Notes to the English translation of his book, *Theory of Relativity*, writes that Einstein “*completely rejected the cosmological term* as superfluous and no longer justified. I fully accept this new standpoint of Einstein’s.”

Landau and Lifshitz (1951, 338) state that “Nowhere in our equations do we consider the so-called cosmological constant, since at the present time it has finally become clear that there is no basis whatsoever for such a change in the equations of attraction.”

And let us also take note of Einstein’s 1947 comment about Λ in a letter to Lemaître (from the Archives Georges Lemaître, see footnote 4 on page 42): “Since I have introduced this term I had always a bad conscience ... I cannot help to feel it strongly and I am unable to believe that such an ugly thing should be realized in nature.”

John Archibald Wheeler’s intuitive feeling about the cosmological parameters was conditioned by the thought that acceleration could be meaningful only relative to the rest of matter, as Ernst Mach and Albert Einstein had argued (see Section 2.1). Wheeler’s thinking, expressed in Misner, Thorne, and Wheeler (1973, §21.12), was that this calls for closed space sections. It would mean that the parameter R^{-2} is positive in equations (3.1) and (3.3) (so $\Omega_k < 0$ in equation (3.13)). A universe that is quite close to Einstein–de Sitter, but with $R^{-2} > 0$, would do for this line of thinking.

Before there were convincing measurements of the parameters Ω_k and Ω_Λ (which represent space curvature and Einstein’s cosmological constant, respectively), arguments from numerical coincidences suggested that these two parameters likely are zero or exceedingly close to it. Consider first the Friedman-Lemaître equation (3.3) for the acceleration of the rate of expansion of the universe written in terms of the dimensionless cosmological parameters in equation (3.13):

$$\frac{1}{a}\frac{d^2a}{dt^2} = H_0^2\left[-\Omega_r(1+z)^4 - \frac{1}{2}\Omega_m(1+z)^3 + \Omega_\Lambda\right]. \tag{3.26}$$

The gravitational attraction of the mass tends to slow the rate of expansion, while a positive cosmological constant Λ tends to speed it up. The evidence is that Ω_r is small. If so, it is a good approximation to write the redshift at which the acceleration d^2a/dt^2 vanishes as

$$z_e = (2\Omega_\Lambda/\Omega_m)^{1/3} - 1 = 0.67 \text{ at } \Omega_m = 0.3,\ \Omega_\Lambda = 0.7. \tag{3.27}$$

The numerical value uses recent measures of the parameters, but the cube root makes their values scarcely matter. If Λ is positive, and it makes a significant contribution to the present expansion rate, then we flourish shortly after the special epoch at which the deceleration of the expanding universe changed to acceleration. With Hubbles's constant $H_0 \approx 70$ km s^{-1} Mpc^{-1}, the acceleration vanished about 6 Gyr ago, just when the solar system was starting to form. Why this coincidence? In the Einstein–de Sitter model with $\Lambda = 0$ and no space curvature, the mass density parameter is $\Omega_m = 1$ no matter when in the course of expansion of the universe it was measured. In the second edition of *Cosmology*, Bondi (1960) remarks that this is an arguably more reasonable-looking situation.

For another way to put this coincidences argument, consider equation (3.13) rewritten as

$$\left(\frac{\dot{a}}{a}\right)^2 = A(t)\left[1 + \frac{\Omega_m}{\Omega_r}\frac{a}{a_0} + \frac{\Omega_k}{\Omega_r}\left(\frac{a}{a_0}\right)^2 + \frac{\Omega_\Lambda}{\Omega_r}\left(\frac{a}{a_0}\right)^4\right], \tag{3.28}$$

with $A(t) = \Omega_r H_0^2 (a_0/a(t))^4$. The parameter $\Omega_r \sim 10^{-4}$ is the fractional contribution to the present expansion rate by the thermal radiation and accompanying neutrinos (taken here to be massless) in the hot big bang cosmology. To be specific, let us suppose that the matter density parameter is $\Omega_m = 0.1$, a reasonable approximation to what is seen around us, and let us consider the situation in the hot big bang cosmology when light elements were starting to form, at redshift $z = a_0/a \sim 10^{10}$ (see Chapter 4). If there is no cosmological constant, then with $\Omega_m = 0.1$, the curvature term is $\Omega_k = 0.9$, and the contribution of the curvature term in equation (3.28) to the expansion rate at the time of nucleosynthesis is

$$F_k = \frac{\Omega_k}{\Omega_r}\left(\frac{a}{a_0}\right)^2 \sim 10^{-16}. \tag{3.29}$$

If, on the other hand, there is no space curvature, and $\Omega_m = 0.1$, then the cosmological constant contributes $\Omega_\Lambda = 0.9$ to the present rate of expansion, and its fractional contribution to the expansion rate at $z = 10^{10}$ is

$$F_\Lambda = \frac{\Omega_\Lambda}{\Omega_r}\left(\frac{a}{a_0}\right)^4 \sim 10^{-36}. \tag{3.30}$$

In either of these cases, we flourish at an exceedingly special epoch, just as the curvature or Λ term, which would have been so small at $z \sim 10^{10}$, had become the dominant contribution to the expansion rate. This curious coincidence of timing is avoided if both space curvature and Λ are negligibly small now. Then the universe would have evolved from radiation dominated at $z \gtrsim 10^4$ to matter dominated, and after that the mass density would have remained very close to the Einstein–de Sitter value, $\Omega_m = 1$, the density parameter we would find whenever we happened to come on the scene and measure it.

This coincidence argument may have occurred to many who did not bother to publish, because it was not at all clear what to make of it. I remember its discussions in the early 1960s in meetings of Dicke's Gravity Research Group. It was published by Dicke (1970, 62), McCrea (1971, 151), and Dicke and Peebles (1979, 506–507). A counterargument, the anthropic principle, is to be considered later in this section.

Lemaître's (1934) thought that Einstein's cosmological constant Λ defines the vacuum energy density in equation (3.10) is discussed on page 46. The vexed problem with this in quantum physics is reviewed by Rugh and Zinkernagel (2002) and Peebles and Ratra (2003), as follows. In the 1930s, there was clear empirical evidence that the quantum zero-point energy is real and must be taken into account for consistency of quantum theory predictions with measurements of binding energies of molecules. The electromagnetic field is described by the same quantum mechanics; its zero-point energy surely is just as real. In standard and established physics, energy is equivalent to gravitating mass. But the sum of zero-point energies of modes of the electromagnetic field up to X-ray frequencies amounts to an absurdly large mass density for a relativistic cosmological model. If local Lorentz covariance is valid, this translates to an absurdly large value of Λ.

Wolfgang Pauli presents an interesting take on the problem in his 1933 *Handbuch der Physik* article on quantum mechanics. He states that (in the English translation in Rugh and Zinkernagel, 2002, 5) "it is more consistent from scratch to exclude a zero-point energy for each degree of freedom as this energy, evidently from experience, does not interact with the gravitational field." Pauli (1933) refers here to the zero-point energy of the electromagnetic field. But the zero-point energy of matter certainly does interact with the gravitational field: This is the well-tested equivalence of energy and gravitational mass. Since the zero-point energies of matter and radiation are the same quantum physics, this solution is not a comfortable one.

A hopeful thought is that some symmetry to be discovered forces the vacuum energy density to its only "reasonable" value, $\Lambda = 0$. We see this tradition in the paper "The End of Cold Dark Matter?" by Davis et al. (1992, 492). They present a discussion of the problem of reconciling the argument from cosmological inflation (Section 3.5.2) that space sections should be flat with the evidence from dynamics that there is not enough mass to make them flat without postulating a nonzero cosmological constant. But the authors note that

> From the point of view of a particle physicist, the value of Λ needed to work these miracles is extraordinarily small, 10^{120} times smaller than its 'natural' value.[62] Such fine tuning seems sufficiently unattractive that most cosmologists regard this solution as a long shot, preferring to think that some unknown symmetry principle requires the cosmological constant to be exactly zero.

The reference is to Weinberg's (1989) discussion of the problem with Λ. Bertschinger (1993, 308) expresses the opinion that

> One can go to the extreme of replacing some of the cold dark matter by a substance that is totally inert and unresponsive to gravity: vacuum energy density, otherwise known as a cosmological constant. ... Although the needed vacuum energy density is unnatural from the viewpoint of fundamental physics, this model appears astrophysically tenable at present and may lead to acceptable large-scale structure.

Another idea is that Λ is observed to be so small because it has been evolving to what might be its natural value, zero, for a long time. The earliest expression of this thought I have found is by Dolgov (1983), who considered a variant of the Brans and Dicke (1961) theory in which both Λ and the strength of the gravitational interaction are evolving to zero. Reuter and Wetterich (1987) explored the difficult problem of setting up field equations in which the value of an effective Λ evolves to zero. Peebles and Ratra (1988) showed how the potential of a scalar field can be chosen so its energy density rolls toward zero in the manner of a decreasing value of Λ, assuming that the quantum vacuum energy density is forced to vanish. Caldwell, Davé, and Steinhardt (1998) gave the name "quintessence" to this evolving effective Λ. Early thoughts about cosmological tests of the idea are presented in Ratra and Peebles (1988).

As the evidence for the presence of Λ grew (see Section 8.4), the pressure to learn to live with it despite the deep theoretical problem with the quantum vacuum energy density helped direct attention to yet another line of thought that became known as the anthropic principle. The idea traces back to Dicke's (1961) point that it is not surprising that the universe has been expanding for a long time, roughly 10^{10} years. It takes about that long for several generations of stars to produce and disperse the heavier elements we are made of, then for Earth to form and cool, and then for life to evolve into something that became interested in measuring the age of the universe. Also, we must be in a galaxy of at least modest mass that is capable of gravitationally capturing the debris from stars at the ends of their lifetimes for recycling to produce the heavy elements we need. These are consistency conditions, which of course assume nature operates in a consistent way.

It may be argued on similar but perhaps less secure grounds that consistency requires that we flourish in a universe that has been matter dominated, at least until fairly recently, so Ω_m is not far below unity. The gravitational assembly of mass concentrations (such as galaxies) requires matter-dominated expansion if structure grew out of small primeval departures from homogeneity (Chapter 5). Rees (1984) expresses this line of thinking in an argument that the space curvature parameter $|\Omega_k|$ likely is well below unity, as follows. He remarks that the condition that $|\Omega_k|$ does not make an unreasonably large contribution to the Friedman-Lemaître equation (3.13) translates to the condition that the radius of curvature in the Robertson-Walker line element is some 30 orders of magnitude larger than the Planck length, $(G\hbar/c^3)^{1/2} \sim 10^{-33}$ cm, expanded to the present from the time when the Hubble length would have been equal to the Planck length. If the ratio $\mathcal{R}$ of these lengths were much larger than this bound, it would mean space curvature is close to Einstein–de Sitter. But if $\mathcal{R}$ were much smaller, it would have meant that the universe collapsed or entered free expansion before the Milky Way galaxy could have formed. Rees (1984, 339) suggests that "One reaction to this requirement might be an anthropic one: the universe would not have offered a hospitable environment for galaxies and stars if $\mathcal{R}$ were not so huge."

Brandon Carter expanded on Dicke's (1961) argument from consistency and Hugh Everett's (1957) many-worlds picture of results of measurements in quantum physics, leading Carter (1974, 295, 296) to consider a

> "world ensemble" . . . of universes characterised by all conceivable combinations of initial conditions and fundamental constants . . . [and] Subject to the further condition that it is possible to define some sort of fundamental a priori probability measure on the ensemble, it would be possible to make an even more general kind of prediction based on the demonstration that a feature under consideration occurred in 'most' members of the cognizable subset [that allows the] existence of any organism describable as an observer.

Weinberg (1987) applied this line of thought to the value of the cosmological constant. If Λ were large and negative, it would have stopped the expansion of the universe and caused it to collapse before we could have come into existence; if large and positive, it would have made the rate of expansion too rapid to have allowed gravity to form the galaxy our existence requires. With Carter, Weinberg considered an ensemble of universes each with its own physics and with observers (such as us) present only in the tiny subset with our physics or close to it, and with a value of the vacuum energy density Λ close enough to zero to have allowed galaxies to form to provide a suitable environment. The natural value of Λ from the point of view of quantum physics is enormous: Weinberg (1989) put it at "118 decimal places" times what cosmology allows. (One way to arrive at this number is the dimensional analysis on page 50.) Weinberg proposed that universes that have larger values of Λ may be closer to natural and hence more abundant. In this case, we might expect to find ourselves in a universe with the absolute magnitude of Λ about as large as is compatible with our existence. This would be at least roughly comparable to the value of Λ indicated by the observations. The argument applies equally well to Rees' spacetime curvature radius: the natural value might be $\mathcal{R} \sim 1$; larger values might be less common in an ensemble of universes, and we might expect to find ourselves in a universe with a value of $\mathcal{R}$ that is about as small as is compatible with our existence. This, too, would be at least roughly consistent with the bounds we had then.

Some accept considerations of the multiverse and anthropic principles as reasonable and natural. Others object that applications have the flavor of just-so stories that we may hope will be replaced by deeper physics that accounts for such things as the bounds on Λ and $\mathcal{R}$ in a more direct, theory-based way.

The term "multiverse" to describe situations contemplated in anthropic arguments came into general use with the introduction of the cosmological inflation picture to be discussed next. Steinhardt (1983, 262) put it that such ideas suggest that "When a new bubble is formed, it regenerates a new universe which can never contact our own. New Universes are regenerated forever." Vilenkin (1983) and Linde (1986) presented similar thoughts. The title of Linde's influential paper is to be noted: "Eternally Existing Self-Reproducing Chaotic Inflationary Universe."

Let us consider some aspects of these ideas.

3.5.2 COSMOLOGICAL INFLATION

The most influential argument for the Einstein–de Sitter model with no cosmological constant and no space curvature, and for the multiverse contemplated in anthropic arguments, was inspired by the concept of cosmological inflation that grew out of a series of papers in the early 1980s.[9] The idea was rapidly accepted as an elegant answer to a demanding question: What was the expanding universe doing before its evolution could have been described by the classical Friedman-Lemaître solution? The answer in inflation is that the dominant term in the Friedman-Lemaître equation (3.3) in the very early universe was an analog of Λ, but it was much larger and was rolling toward smaller values. This hypothetical component might be the potential plus kinetic energies of a scalar "inflaton" field, and the field potential could be arranged so as to have caused near-exponential expansion during what may be termed the "inflation era." Ideas about what happened before the inflation era are not relevant for this history.

Inflation would have ended when the energy in the Λ-like component decayed and produced most of the entropy of the present universe, most of which ended up as the present sea of microwave radiation. The enormous expansion during inflation was taken to have to have stretched away density

9. The pioneering papers include Starobinsky (1980); Guth (1981); Mukhanov and Chibisov (1981); Sato (1981); Albrecht and Steinhardt (1982); Guth and Pi (1982); Hawking (1982); Linde (1982); and Bardeen, Steinhardt, and Turner (1983).

gradients, accounting for the nearly homogeneous universe we observe. This stretching would have forced the radius of curvature of the space we can observe to an enormously large value, amply satisfying Rees's condition of a large value for $\mathcal{R}$. That is, the thinking was that inflation requires the space curvature term Ω_k in equation (3.13) to be negligibly small. In the 1980s, the nonempirical arguments against Λ reviewed in Section 3.5.1 were persuasive too, and the general community opinion was that Λ ought to be negligibly small. This leaves the Einstein–de Sitter model. But we have learned to live with Λ, as will be discussed.

Any baryons present in the early stages of inflation would have been thoroughly dispersed, so baryons would have to have been produced out of the entropy from the decaying inflaton field present toward the end of inflation. Interest in the idea of baryogenesis was growing for other reasons (discussed by Sakharov (1967), who set the conditions under which it might happen, and Yoshimura (1978) and Dimopoulos and Susskind (1978), who were exploring its implementation) just as it was wanted for inflation. After the energy that drove inflation had decayed to the entropy that produced the baryons, the universe would have evolved according to the usual Friedman-Lemaître hot big bang cosmological model.

With the introduction of the inflation concept came the recognition that in standard quantum physics, the homogeneity produced by the great and rapid expansion during the inflation era would have been broken by quantum field fluctuations squeezed to classical spacetime curvature fluctuations that would be close to Gaussian and scale invariant. The first follows because the fluctuations are that of a nearly free field, the second because the curvature fluctuation amplitude is determined by the expansion rate, $\dot{a}/a$, which changes only slowly if the expansion is close to exponential. (The meaning of scale-invariant spacetime curvature fluctuations is discussed in Section 5.2.6.)

There were discussions of variants. Gott (1982) considered a phase transition in de Sitter's solution that would produce a spacetime with negative curvature of space sections: an open model. Kamionkowski et al. (1994) and Ratra and Peebles (1995) worked out the properties of an "open inflation" model. This was motivated by the evidence summarized in Peebles (1986) that the mean mass density is significantly less than the Einstein–de Sitter value. If so, then general relativity requires either open space sections, $\Omega_k > 0$, or a positive value of Einstein's Λ. The introduction of inflation in the early 1980s caused the general community preference to be for flat space sections.

How the inflation concept may be made more complete is debated but is not relevant to this story. The important historical point is the community assessment that inflation requires flat space sections. By the turn of the century, this was reinforced by an abundance of evidence that Ω_k is indeed close to zero and that Λ makes it so.

The rapid growth of influence of the inflation idea is illustrated by Frank Wilczek's comments in his summary of the June 1982 workshop on *The Very Early Universe* (Wilczek 1983, 475):

> By my count seventeen of the thirty-six lectures at the workshop were mainly about the idea of an inflationary Universe, and many of the other talks were heavily influenced by this idea. This count attests to the enormously seductive appeal of the idea. ... The most dramatic qualitative prediction of the inflation idea is that the Universe should be essentially flat—with well-known consequences for the mass density ($\Omega = 1$).

Bardeen (1986, 275), in his paper in the proceedings of the May 1984 conference *Inner Space/Outer Space*, wrote that

> A cold dark matter dominated inflationary universe is theoretically attractive for its predictive power. Barring extraordinary fine tuning of the present vacuum energy density (the cosmological constant), the cosmological density parameter Ω should be very close to one.

In the summary for the June 1985 conference on *Dark Matter in the Universe* Gunn (1987, 541) asked

> What then, are we to do if our prejudices demand that Ω be unity? In this connection it should be noted that not all inflationary scenarios demand simultaneous flatness and homogeneity, and that at least one primordial inflationary model, that of Gott, produces negatively curved homogeneous models naturally.

In the same conference proceedings, Davis (1987, 108) presented a review of the mass density measurements. He pointed out that the straightforward reading of the evidence is that $\Omega_m < 1$, but this assumes galaxies trace mass, which was questioned. Davis concludes with the statement that "there is no existing data that contradicts the notion that galaxies are a biased mass tracer and that we live in an $\Omega = 1$ universe." Kaiser (1986, 262) mentions his bias idea at the May 1984 conference: "the possibility of strong segregation of luminous galaxies from mass ... is a very attractive feature for those who find the idea that $\Omega \neq 1$ unpalatable."

The thought behind the last two statements is that the mass density may be large enough to make $\Omega_m = 1$ but that this large mass density is not readily observed, because most of the mass is not where the galaxies are. Implementation of this thought is reviewed in Section 3.5.3.

In Alan Guth's (1991, 1) introduction to the December 1990 conference *Observational Tests of Cosmological Inflation*, he reported that beginning two years after the introduction of inflation in the early 1980s, research journals were averaging 150 papers a year on inflation, which

Table 3.1. What Will Ω Turn Out to Be?

Ω	Votes
$1.001 < \Omega$	2
$0.999 < \Omega \leq 1.001$	28
$0.05 < \Omega \leq 0.999$	29
$\Omega \leq 0.05$	2
Don't know	71
Don't care	0

> shows the first (and most naive) fundamental argument for inflation ... the number of articles per year related to inflation, as tabulated from the SPIRES database at the Stanford Linear Accelerator Center. It is mainly a particle physics database, so some of the more astrophysical papers on the subject are probably not represented. In any case, one sees that the inflationary model has stirred up a lot of interest.

Guth went on to review the promise of cosmological inflation that so attracted the attention of particle physicists, cosmologists, and astronomers. This 1991 review is an update of his assessment in the introductory lecture at the December 1982 *Texas Symposium on Relativistic Astrophysics* (Guth 1984). The basic idea quickly found its influential place in cosmology and has changed little after that.

Not all agreed with this line of thinking, of course; Gunn mentioned Gott's open model. At the June 1985 conference, the participants were polled on the question "What will Ω turn out to be?" It is unfortunate that the poll failed to state the meaning of Ω. At the time, the condition from the concept of inflation that space sections are flat was usually taken to mean that the mass density in matter is equal to the Einstein–de Sitter value: $\Omega = \Omega_m = 1$. But if Einstein's cosmological constant Λ is not zero, and is set to $\Omega_\Lambda = 1 - \Omega_m$ (equation (3.13)), it would allow a low mass density, $\Omega_m < 1$, with flat space sections. This way to reconcile flat space sections with the evidence that the mean mass density is well below unity was being discussed (e.g., Peebles 1984b; Rees 1984; Turner, Steigman, and Krauss 1984; Kofman and Starobinsky 1985). But the more typical feeling in the 1980s is illustrated by Bardeen's (1986) comment that it would require "extraordinary fine tuning."

The poll results are shown in Table 3.1. About half of those voting expressed the sensible opinion that we just don't know. About a quarter expected that Ω is significantly less than unity. Maybe some of them had in mind the thought that the mass density in matter is low, as the observations suggested, but that Λ keeps space sections flat. But it is difficult to find indications of that thinking from the evidence I have seen and from my own recollections. I conclude that most who expected that Ω is well below unity simply were willing to live with curved space sections despite the elegance of the argument from inflation. Among those who reckoned that Ω will prove to be very close to unity, some may have had in mind resorting to the cosmological constant to make space sections flat, but again it seems more likely, according to my recollection, that most if not all who voted for $\Omega = 1$ had in mind $\Omega_m = 1$. And the quotations from conferences presented above show that this is what influential figures had in mind.

If $\Omega_m = 1$, it calls for serious measures: Show that galaxies are biased tracers of mass, or maybe abandon the relativistic cosmology. From the mid-1980s to the late 1990s, the most common approach was to accept general relativity and flat space sections and operate on the postulate that $\Omega_m = 1$ and galaxies are not fair tracers of mass. The first papers that explicitly argued for adding Λ to the Cold Dark Matter model (the CDM model discussed in Section 8.2) appeared in 1984, at about the same time as the comments just reviewed about what inflation may imply, but this thought was not well received until the mid-1990s.

3.5.3 BIASING

The long history of measurements of the mean mass density to be reviewed in Section 3.6.4 generally points to values of Ω_m that are significantly less than unity. The example noted above is in Davis's (1987) review at the conference, *Dark Matter in the Universe.* But the nonempirical considerations reviewed in Sections 3.5.1 and 3.5.2 made the Einstein–de Sitter model, with $\Omega_m = 1$, seem particularly reasonable. It was natural then to consider the possibility that the density estimates are biased low because most of the mass is outside the concentrations of galaxies. Recall that the gravitational attraction binding a star to a galaxy, or binding a galaxy to a cluster of galaxies, is unaffected by mass outside the concentration of objects of interest, apart from the usually minor effect of tidal fields. Thus if galaxies were more tightly clustered than mass, then mass density estimates that assume galaxies trace mass would underestimate the density; the galaxies would be biased mass tracers.

Rees (1985, 81p) offered a clear assessment of processes that might be expected to have produced this bias. His conclusion is that

> The effects discussed here are not merely *ad hoc* contrivances to shore up the philosophically attractive $\Omega = 1$ model against apparently conflicting evidence. It would be astonishing if none were important—if there were *no* large-scale environmental effects that influenced galaxy formation, and if light did indeed trace mass on all scales > 1 Mpc.

We see the influence of ideas: one nonempirical—inflation calls for matter density parameter $\Omega_m = 1$—another a combination of nonempirical and empirical—astrophysical processes surely have biased the space distribution of galaxies relative to mass—and a third empirical—isn't it curious that it is so hard to find evidence of this large mass density?

The biasing picture was inspired by Kaiser's (1984) explanation of why the positions of clusters of galaxies are more strongly correlated than are galaxy positions. It starts with the idea that galaxies formed out of the gravitational growth of small departures from a homogeneous mass distribution in the very early universe. Galaxies would have grown out of the occasional more prominent primeval upward mass fluctuations, and clusters, being much more rare, would have grown out of even more rare upward fluctuations, on a larger mass scale. If these initial departures from homogeneity were a random Gaussian process with a positive autocorrelation function on length scales of interest, then clusters would have a longer correlation length than would galaxies. Put another way, at a given separation, cluster positions would be expected to have a larger reduced two-point correlation function than galaxy positions, which is observed. The interesting idea this suggests is that galaxies would be expected to have a larger correlation function than that of the mass distribution.

Kaiser's point attracted prompt attention. In the May 1984 conference *Inner Space/Outer Space*, Bardeen (1986) and Kaiser (1986) argued for the key thought: Kaiser's effect suggests estimates of Ω_m are biased low, because they overestimate the degree of concentration of mass around galaxies. As Davis's (1987) comment recorded on page 64 illustrates, the idea was a welcome way to save the Einstein–de Sitter model. The thought remained influential until the revolution at turn of the century, when it at last became clear that galaxies trace mass well enough to have shown us that Ω_m really is well below unity. The case for this conclusion before the revolution is considered in Section 3.6.5.

Davis et al. (1985) introduced the consideration of biasing in numerical N-body simulations of cosmic structure formation. Carlos Frenk's recollection of this development is in Section 8.3. Some early numerical simulations of how galaxies formed supported the idea of biasing. Couchman and Carlberg (1992) found indications that the complexity of strongly nonlinear gravitational clustering may allow the appearance of low mass density in an Einstein–de Sitter universe. Katz, Hernquist, and Weinberg (1992) took account of the stresses of hydrodynamics as well as gravity. They found that their simulations of structure formation in a slice of an $\Omega_m = 1$ model universe (the standard CDM model discussed in Section 8.2) indicate the ratio of mass to stellar luminosity in their largest concentration of model galaxies, applied to the mean luminosity in the simulation, yields a mass density equivalent to $\Omega_m \sim 0.3$, maybe an illustration of biasing. But Cen and Ostriker (1992) modeled a broader slice of an $\Omega_m = 1$ model similar to that of Katz et al. The greater size of their simulation slice may account for its larger peculiar velocities. Cen and Ostriker (1992, L113) concluded that the $\Omega_m = 1$ model "can be ruled out on the basis of too large a predicted small scale velocity dispersion at greater than 95%

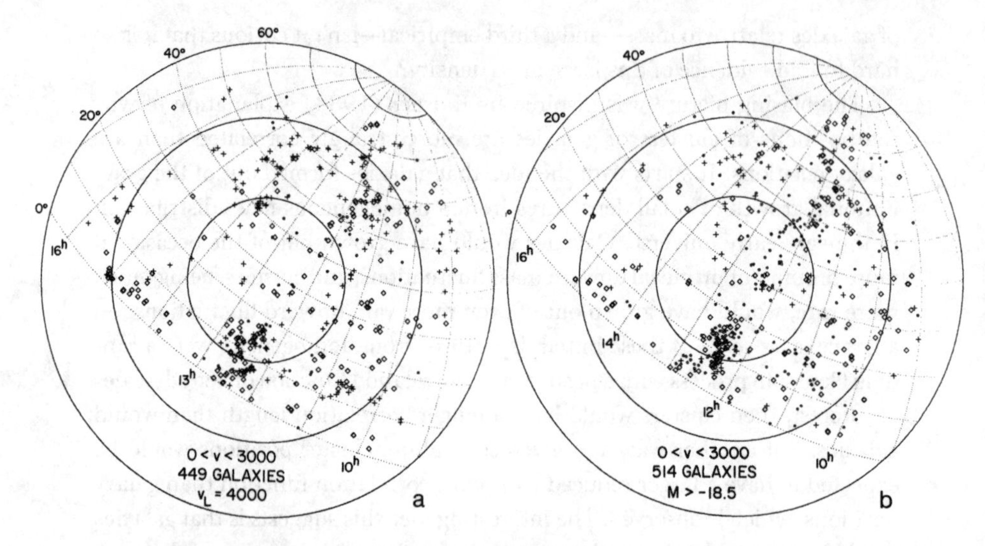

FIGURE 3.1. The Davis et al. (1982) magnitude-limited maps of the galaxies with redshifts $cz < 3{,}000$ km s^{-1}, showing the more luminous galaxies in Panel (a) and the less luminous ones in Panel (b). © AAS. Reproduced with permission.

confidence level; $\Delta v_{\rm rms,1D} = 715 \pm 135$ km s^{-1} compared with observations $\Delta v_{\rm rms,1D} = 340 \pm 40$ km s^{-1} on the $1h^{-1}$ Mpc scale." This agrees with the Davis and Peebles (1983a) conclusion from the semi-analytic approach discussed on page 89, and it argues against biasing.

Kaiser's (1984, 1986) argument was influential and had to be explored, but it was indirect enough to be debatable. Consider the situation at high redshift, when the seeds of galaxies large and small would have been slight departures from homogeneity, so the seeds might be expected to have had about the same distribution of properties everywhere. As the mass fluctuations grew, the young galaxies that happened to be in regions of higher ambient density might have grown more rapidly by accreting the more abundant surrounding matter. And a seed that would have grown into a large galaxy if it had been in a high density region might in a low density region show signs of a deprived youth: Perhaps instead of becoming a normal large galaxy, it would have grown into an irregular or dwarf galaxy. This seems to be a fairly reasonable prediction in the biasing scenario. But the prediction was already challenged by the evidence from the first adequate redshift survey (from the Center for Astrophysics program discussed in Section 3.6.4), as follows.

The rare very luminous galaxies with $L \sim 10L^*$ are observed to be more strongly clustered than normal large $L \sim L^*$ galaxies, consistent with the tendency of the $L \sim 10L^*$ galaxies to be in rich clusters of galaxies. But the distributions of normal large $L \sim L^*$ galaxies and the far more numerous ones with luminosities well below L^* have about the same space distribution. An early demonstration is shown in Figure 3.1. It compares the Davis et al. (1982)

maps of the space distributions of galaxies separately for the more luminous ones in Panel (a) and the less luminous ones in Panel (b). These samples are close to complete within the survey volume. The distributions of larger and smaller galaxies look quite similar to my eye, not what a simple picture of biasing might have predicted. The later quantitative demonstration of this phenomenon by Norberg et al. (2001) shows in their Figure 3 the clustering length defined in equation (2.16) as a function of the galaxy luminosity. One sees the large clustering length of the rare, most luminous galaxies and the near constant clustering length at $L \lesssim L^*$. An even tighter demonstration of this phenomenon is in Figure 7 in Zehavi et al. (2011).

The similar distributions of the $L \sim L^*$ galaxies that contribute most of the starlight and the much more numerous fainter galaxies does not naturally follow from Kaiser's biasing effect. Arguments along these lines were presented in Peebles (1986). But at the time, they were not very influential compared to the elegance of the Einstein–de Sitter cosmology.

3.6 *Empirical Assessments of the Big Bang Model*

Allan Sandage (1968, 93), in his 1967 Halley Lecture on Observational Cosmology, expressed his vision of the program of cosmological tests thus:

> It is remarkable that the theory gives *all* properties of the Friedmann models when *only two numbers are known—numbers which can be found from observations at the telescope.* These are (1) the present value of the Hubble expansion rate H_0, defined by the redshift-distance relationship $c\Delta\lambda/\lambda_0 = H_0 D$, and calibrated using nearby galaxies at distances D, and (2) the deceleration parameter q_0, related to the change of the expansion rate with time.

We must admit another parameter, Einstein's cosmological constant Λ, but this was not Sandage's point. His goal stated here and in Sandage (1961a) was to measure cosmological parameters. But these measurements would offer only a modest cosmological test: a check that its parameters can be adjusted to fit what is indicated by the measurements. A more demanding test requires more empirical constraints than free parameters. Perhaps that is what Sandage (1961b, 916) had in mind in the comment in his paper on cosmic times: "it seems important to follow all predictions of these specific models to see how well they do fit the observational data."

An informative example of the situation in the 1960s is the discussion of evidence of a sharp peak in the distribution of quasar redshifts at $z = 1.95$ by Petrosian, Salpeter, and Szekeres (1967) and Shklovsky (1967). They suggested that this redshift peak may be a signature of Lemaître's (1931d, 1934) hovering universe, in which there was a time in the past when the gravitational attraction of matter was nearly balanced by the repulsion effect of a positive Λ, so arranged that the expansion parameter was increasing only slowly, in a transient approximation to Einstein's static universe. Expansion from a big bang would have been slowed by the attraction of gravity before the hovering epoch and increased later by Λ. Lemaître had been interested in increasing the expansion time and setting an epoch for structure formation, for the reasons to be discussed in Section 3.6.1 and Chapter 5. Petrosian et al. (1967) and Shklovsky (1967) were interested in Lemaître's hovering model, because it could account for a peak in the distribution of redshifts at the redshift where the rate of expansion hovered near zero. Since Lemaître introduced central elements of relativistic cosmology, it was natural to pay attention to the hovering idea. O'Raifeartaigh et al. (2018) review later papers about it.

In the 1960s, the available constraints on the dimensionless parameters that represent the mass density and cosmological constant in the relativistic expansion rate equation (3.13) allowed adjustment to fit Lemaître's model for hovering at the redshift of the apparent peak in the quasar redshift distribution. Hovering calls for a special arrangement of the parameter values, which we tend to distrust, because it seems contrived; in fact, the hovering arrangement was later empirically ruled out. But special arrangements need not disqualify ideas: In the established ΛCDM cosmology we must live with the special arrangement that somehow caused the values of the mean mass density and the cosmological constant to be such that we flourish just when the dominant term in the rate of expansion of the universe is changing from the mass density to the cosmological constant. In short, the indication of a redshift peak did not challenge the relativistic cosmological model in the 1960s.

Before counting the redshift peak as evidence for Lemaître's hovering picture, we must consider two questions. First, is the observation reliable enough to be interesting? Second, is it a case of adjusting free parameters to fit the empirical constraints? That may be possible whether or not the theory is a useful approximation to reality. The first question might be addressed by better observations. As it happens, these observations replaced the sharp peak with a broader quasar redshift distribution that is plausibly attributed to cosmic evolution of the quasar phenomenon in galaxies. The second question would be addressed if we could check for consistency with other observations that independently require close to the same parameter values. Two early examples of this search for cross-checks are discussed here. The first, the comparison of the expansion time of the cosmological model and the time scales from astronomy and geology, is considered in Section 3.6.1. The second, the comparisons of the mass densities wanted for Lemaître's hovering model and for the Einstein–de Sitter model with the density inferred from a considerable variety of evidence, is reviewed in Section 3.6.4.

3.6.1 TIME SCALES

Eddington (1930) offered considerations of cosmic time scales in Lemaître's (1931a) original solution, the one that assumes the expansion of the universe traces back asymptotically to Einstein's static model. Eddington (1930, 677) wrote that

> We cannot calculate how long a period elapsed from the disturbance of Einstein equilibrium by the beginnings of evolutionary development until the deviation from equilibrium reached a serious amount; but from the time when the universe had reached, say, 1.5 times its initial radius to the present day, it is scarcely possible to allow more than 10^{10} years. If the sun has really existed as a star for 5 billion years, it is odd that it should have waited so long and then formed its system of planets just at the time the universe toppled into a state of dispersion.

This is a version of the coincidences argument discussed on page 59.

Eddington was thinking about Lemaître's original solution. Lemaître (1931b,c) replaced it with expansion from what he termed the primeval atom later became known as the big bang. Lemaître's (1933a) assessment of the time-scale situation took the radioactive decay age of the Earth and the characteristic cosmic expansion time to be

$$t_{\rm Earth} = 1.6\times 10^9 \text{ years: radioactive decay,} \qquad (3.31)$$
$$H_0^{-1} = 1.8\times 10^9 \text{ years: expansion time scale.}$$

Lemaître concluded that the Einstein and de Sitter (1932) model age, $t=2/(3H_0)$, would conflict with the radioactive decay age, and that the similarity of these two ages would indicate instead that the mass of the universe is not large enough to have appreciably slowed the expansion. Lemaître's (1933a, 85) reaction was that

> D'un point de vue purement esthétique, one peut peut-ètre le regretter. Ces solutions où l'universe se dilatait et se contractait successivement en se réduisant périodiquement á une masse atomique des dimensions du système solaire, avaient un charme poétique incontestable et faisaient penser au phénix de la légende.

Or in my translation:

> From a purely aesthetic point of view, one might regret it. Solutions in which the universe expands and contracts periodically to an atomic system with the dimensions of the solar system has an incontestable poetic charm that makes one think of the legendary phoenix.

We see an example of the longstanding interest in how the world will end: Is the mass of the universe large enough that the expansion will eventually stop and the universe collapse back to a big crunch?

Lemaître did not comment on a point that is interesting from the present perspective: the similar values of the two times in equation (3.31), obtained in very different ways, is either a curious coincidence or a hint that the big bang picture is at least a rough approximation to what happened. But that is in line with the scarce overt questioning of the relativistic theory among the few cosmologists active in the 1930s. We now know that the first line in equation (3.31) should be multiplied by a factor of three and the second by a factor of seven, which means these two quantities alone do not rule out a phoenix universe. But the broader point stands: The similarity of the quantities representing two different phenomena is notewrothy.

The expansion time of a relativistic universe with $\Lambda = 0$ is no greater than H_0^{-1}. The difference of H_0^{-1} from the estimate of Earth's age in equation (3.31) does seem uncomfortably short: Could the solar system have formed so soon after the big bang? An arguably attractive feature of adding a positive Λ to the cosmological model, and even the hovering epoch Petrosian, Salpeter, and Szekeres (1967) had considered, is that it would allow more time for stellar formation and evolution. Thus in a letter to Einstein (from the Archives Georges Lemaître named in footnote 4 on page 42) dated July 30, 1947, Lemaître wrote

> that the cosmical constant is necessary to get a time-scale of evolution which would definitely clear out from the dangerous limit imposed by the known duration of the geologic ages.

Einstein's reply dated September 26, 1947, is worth reporting:

> I can also understand that in the shortness of T_0 there exists a reason to try bold extrapolations and hypotheses to avoid contradiction with facts. It is true that the introduction of the λ term [now Λ] offers a possibility, it may even be that it is the right one.

Bart Bok (1946) reviewed a variety of evidence of cosmic time scales, including the times required for dynamical relaxation of systems of stars, a subject no longer pursued, as well as radioactive decay and stellar evolution ages. Bok concluded that (1946, 75)

> In our summary of the present status of the time-scale of the universe we have found mostly indications in favour of the short time-scale (3 to 5×10^9 years). ... The evidence derived from data on the universe of galaxies is not conclusive at the moment, and will probably remain indefinite until the questions about the expansion of the universe are settled. The field of stellar evolution is in a state of flux, but we find

> almost everywhere signs of youthful exuberance, notably among the supergiant stars.

Bok (1946, 69) did not express much unease about Hubble's short distance scale, or overmuch faith in the relativistic Friedman-Lemaître cosmological model, stating that

> In his own analysis of the existing dilemma Hubble has made it clear that there exists no special problem if the observed red-shifts are assumed to have their origin in some unknown principle of physics, rather than an actual expansion. If this interpretation should prevail, the observed red-shifts would not necessarily bear directly on the cosmic time-scale.

The error in the extragalactic distance scale that led Bok to question the idea that galaxy spectra are shifted to the red by the expansion of the universe, and led Lemaître to accept the big bang picture but argue that a positive Λ is needed, was corrected in two main steps. The first appears in the report of the 1952 Rome meeting of the International Astronomical Union. Walter Baade (1952, 397) is recorded to have reported that

> in the course of his work on the two stellar populations in M31, it had become more and more clear that either the zero-point of the classical Cepheids or the zero-point of the cluster variables must be in error. Data obtained recently—Sandage's colour-magnitude diagram of M 3—supported the view that the error lay with the zero-point of the classical Cepheids, not with the cluster variables. Moreover, the error must be such that our previous estimates of extragalactic distances—not distances within our own Galaxy—were too small by as much as a factor 2. ... Above all, Hubble's characteristic time scale for the Universe must now be increased from about 1.8×10^9 years to about 3.6×10^9 years.

The second step was Allan Sandage's (1958, 513) conclusion that

> The brightest stars are discussed as distance indicators for galaxies beyond the Local Group. It is probable that knots identified by Hubble as brightest stars in more distant resolved galaxies are really H II regions. From data in M100, the stars appear to be 1.8 mag. fainter than the knots. This correction, together with a correction of 2.3 mag. to Hubble's moduli for galaxies in the Local Group, suggests a total correction of about 4.1 mag. to the 1936 scale of distances. This gives $H \approx 75$ km/sec or $H^{-1} \approx 13 \times 10^9$ y, with a possible uncertainty of a factor of 2.

It became the tradition to write the value of Hubble's constant in terms of the dimensionless parameter h in equation (3.15). Hubble's (1929) first estimate was $h \simeq 5$. Hubble (1936) gave $h = 5.3$; Humason, Mayall, and Sandage

(1956) had $h = 1.8$; and Sandage (1958) gave $h = 0.75$. Later estimates include Sandage and Tammann (1984), $h = 0.50 \pm 0.07$; de Vaucouleurs and Corwin (1985), $h = 0.99 \pm 0.07$; Pierce and Tully (1988), $h = 0.85 \pm 0.10$; Jacoby, Ciardullo, and Ford (1990), $h = 0.87 \pm 0.13$; the Freedman et al. (2001) HST Key Project, $h = 0.72 \pm 0.08$; and Riess et al. (2018), $h = 0.7348 \pm 0.0166$. These numbers are based on measurements of redshifts and distances of relatively nearby galaxies. They are to be compared to the measurement $h = 0.674 \pm 0.005$ from the Planck Collaboration (2018) analysis of the observational constraints on the ΛCDM cosmological model assembled in Chapter 9. The difference from Riess et al. is statistically significant. At the time of writing, it is not known whether this difference is due to a subtle error in theory or observation, but the small size of the discrepancy suggests the correction will be a small adjustment in one or both, not a paradigm shift. Meanwhile, let us admire the progress in the difficult art of astronomical measurement of the extragalactic distance scale and Hubble's constant.

3.6.2 COSMOLOGICAL TESTS IN THE 1970s

In his 1969 Darwin Lecture, Martin Schwarzschild (1970, 14) reports that "Our present knowledge can, I believe, be properly summarized by giving an age for globular-cluster stars of 10 billion years, and I believe that this value is not likely to be off by more than 4 billion years in either direction."

Globular cluster stars are among the oldest in the Milky Way. Their ages can be compared to Dicke's (1969) estimate of the radioactive decay age of about 7×10^9 years for the isotopes of uranium, and Clayton's (1964) considerations of a broader variety of isotopes that suggested nucleosyntheses began about 13×10^9 years ago. These necessarily rough estimates are reasonably consistent with Schwarzschild's stellar evolution age, $10 \pm 4 \times 10^9$ years, and Sandage's (1958) expansion time scale, $H_0^{-1} \sim 13 \times 10^9$ years, give or take a factor of two. That is, there was a case to be made that stars could have formed and produced the heavy elements when the universe was appreciably younger, but after the universe had expanded from a state too dense for stars to have existed. In a similar way, the characteristic cosmological mass density, H_0^2/G, was known to be at least roughly in line with estimates of the cosmic mean mass density from counts and masses of galaxies to be discussed in Section 3.6.4.

A new probe of the mean mass density emerged in the 1960s with the recognition that the abundances of the isotopes of the light elements may have been set by thermonuclear reactions operating during the early stages of expansion of the hot big bang. This is discussed in Chapter 4. For the purpose of the present discussion, let us only note that the current value of the baryon mass density, ρ_{baryon}, sets the mass density in the early universe when nuclear reactions would have been building up the light elements. The greater the value of ρ_{baryon}, the greater the density during the early nuclear reactions and the closer to completion the reactions would be. Peebles (1966a,b) pointed out the particularly interesting value of D/H, the deuterium abundance relative to hydrogen. Wagoner, Fowler, and Hoyle (1967) presented the first careful comparison of the theory to the observations of light element abundances, with particular attention to deuterium. They translated that to a prediction of the present mean baryon mass density required to account for the light element abundances and compared it to Oort's (1958) estimate of the cosmic mean mass density in galaxies:

$$\begin{aligned} \rho_{\text{baryon}} &\simeq 2 \times 10^{-31}\ \text{g cm}^{-3} \qquad \text{Wagoner, Fowler, Hoyle;} \\ \rho_{\text{m}} &= 3 \text{ to } 7 \times 10^{-31} h^2\ \text{g cm}^{-3} \quad \text{Oort.} \end{aligned} \tag{3.32}$$

Wagoner, Fowler, and Hoyle (1967) did not claim that the rough consistency of these two densities is particularly significant. This was natural, because their deuterium abundance was quite uncertain. Maybe equally important, they also considered an alternative to the hot big bang cosmology, the Hoyle and Tayler (1964) proposal that helium and deuterium could have been produced in pre-galactic local explosions. But in hindsight, one sees the rough but very interesting consistency of the cosmic mean mass densities derived in different ways from two quite different phenomena. There is the complication of later evidence that only about a sixth of the matter is baryonic, while the rest—nonbaryonic dark matter—would not have taken part in nucleosynthesis. But at the empirical state of cosmology in the late 1960s, this is a reasonably minor detail. The important thing is that these considerations presented us with an approximate but meaningful cosmological test, along with the time scales in equation (3.31).

Gott et al. (1974) brought these tests together in Figure 3.2.[10] The horizontal axis is the matter density parameter Ω_{m} defined in equation (3.13). The vertical axis is Hubble's constant. The shaded regions are inconsistent with observations under the assumptions that Einstein's cosmological constant may be ignored but we can trust the relativistic cosmological model.

The two horizontal lines are conservative bounds on the value of Hubble's constant H_0 from estimates of the extragalactic distance scale. The vertical line to the far left is a lower bound on the matter density parameter Ω_{m} from counts of galaxies and dynamical estimates of galaxy masses. The vertical bound labeled "best Ω^*" is the mass density based on galaxy masses derived from the dynamics of motions of galaxies in groups and clusters of galaxies. This takes account of the nonbaryonic dark matter to be discussed in Chapter 6 that is largely concentrated around the outskirts of galaxies. The bounds in this figure assume $\Lambda = 0$, so the values of H_0 and Ω_{m} determine the time t_0 of expansion from high redshift (always assuming the relativistic cosmology).

10. Earlier examples along these lines, though with fewer observational constraints, are in Tomita and Hayashi (1963), who showed contours of the dimensionless product $H_0 t_0$ in the mass density–cosmological constant plane, and Tinsley (1967), who showed $H_0 t_0$ and the redshift-magnitude relation as functions of the mass density for the case $\Lambda = 0$.

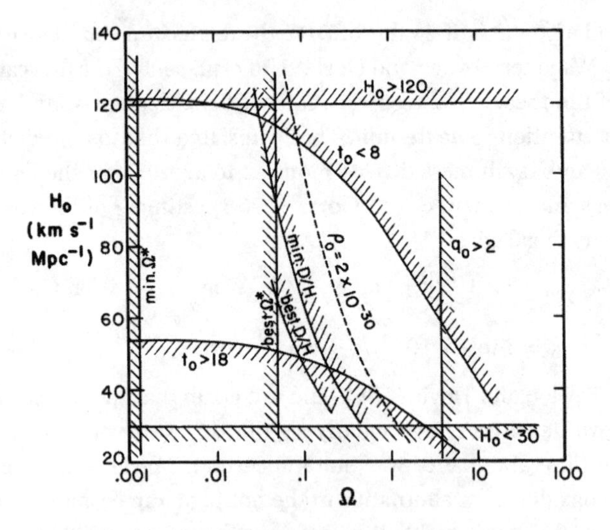

FIGURE 3.2. The Gott et al. (1974) cross-checks of cosmological parameters. © AAS. Reproduced with permission.

Their estimates of the bounds on t_0 from stellar evolution ages and radioactive decay ages are marked by the curves labeled t_0 (in units of 10^9 y). The vertical line to the right, labeled $q_0 > 2$, is the Gott et al. conservative upper bound on the deceleration parameter in equation (3.23).

Gott et al. estimated the baryon mass density from the hot big bang considerations mentioned above with an important new measurement of the abundance D/H of deuterium relative to hydrogen. The spectrometer on the Copernicus satellite detected in the spectrum of a hot star the ultraviolet Lyman series absorption lines produced by interstellar hydrogen and deuterium. The greater deuteron mass shifts its lines from the hydrogen absorption lines. Rogerson and York (1973) concluded from these data that the deuterium abundance relative to hydrogen in the interstellar medium is $D/H = 1.4 \pm 0.2 \times 10^{-5}$ by number. Gott et al. allowed that the primeval value of D/H may be as much as twice this value, because deuterium may have been destroyed in stars, or maybe half this value, if the satellite sampled regions where the abundance is unusually low for some reason. With Wagoner's (1973) improved nucleosynthesis computation, this suggested $\rho_{\rm baryon} = 4$ to 8×10^{-31} g cm^{-3}. The minimum and central values translate to the predicted values of $\Omega_{\rm baryon} h^2$ in the curves labeled D/H in Figure 3.2.[11]

11. Lyman Spitzer, Jr., who was based at Princeton University, was the principal investigator for the Copernicus spectrometer. Rogerson and York (1973) thanked Spitzer for pointing out the relevance of their deuterium measurement for cosmology. This work was centered at Princeton University, but I was not paying much attention then, having become fascinated by statistical analyses of the distributions in positions and peculiar motions of samples of extragalactic objects.

Gott et al. (1974) acknowledged two important caveats. First, the window of constraints on parameters is broader if Λ is allowed to be nonzero. Second, there may be matter that does not act like baryons. They referred to Cowsik and McClelland (1973), who with Szalay and Marx (1974) were considering the idea that neutrinos may have nonzero rest mass. This is reviewed in Section 7.1. It would mean that the neutrinos left over from the early universe, when neutrinos and radiation last were in thermal equilibrium, could make an appreciable contribution to the matter density. And Hawking (1971), Chapline (1975), and Carr (1975) were discussing yet another kind of mass that would not have taken part in big bang nucleosynthesis: small black holes that may have formed in the very early universe. These ideas would reduce the baryon density derived from D/H to a lower bound on the total.

A new application of the redshift-magnitude relation by Gunn and Oke (1975) suggested a new bound on the deceleration parameter, $q_0 \lesssim 0.33$. This would bring the bound from q_0 closer to the region allowed by D/H. But the authors cautioned that the measurement of q_0 is uncertain, not least because of the difficulty of correcting the measurement for evolution of galaxy luminosities. After comments on how the measurement might be improved, Gunn and Oke, (1975, 267) conclude that "When it is all done, the results will still be meaningless without good evolutionary corrections, of course; it is hoped that work on galaxy synthesis and evolution will progress to oblige in the next few years."

Sandage (1961b) and Tinsley (1967 and 1972) were quite aware of this challenge. The challenge remained part of community thinking at the turn of the century.

Despite all the caveats, Figure 3.2 represents an important advance: a systematic cross-check of constraints on parameters in more independent ways than there are parameters. All the constraints depend on two major assumptions: general relativity and the cosmological principle. There was accumulating evidence for the latter, large-scale homogeneity, in the 1970s (see Chapter 2). The former still was an enormous extrapolation from the empirical tests we had then, but it was being tested: If the relativistic cosmology is a useful approximation to reality, then constraints on the parameters based on different phenomena will be consistent. The next sections deal with a particularly rich check from the broad variety of probes of the mean mass density.

3.6.3 MASS DENSITY MEASUREMENTS: INTRODUCTION

Observed motions of matter that might be attributed to the acceleration of gravity offer measures of the mass that is gravitationally driving the motion, which can be interpreted in terms of the cosmic mean mass density. That translates to the value of the cosmological mass density parameter Ω_m in equation (3.13).

The value of the mean mass density, or later, Ω_m, has been a focus of attention from the beginning of modern cosmology, for reasons that have evolved with the science and society. Recall (on page 38) that Eddington (1923) drew attention to a curious situation in Einstein's (1917) static world model: that a dynamical variable, the mass density, is set equal to a combination of constants of nature (as in equation (3.5)). This is one of the problems with a static universe. Hubble (1926) applied Einstein's relation to his estimate of the mean mass density in galaxies to find the mass and size of the universe in Einstein's static model universe, but without comment about the problematic nature of this model. A decade later in his pioneering treatise on extragalactic astronomy, *The Realm of the Nebulae*, Hubble (1936) presented more detailed estimates of the mean mass density, but here he expressed no interest at all in what the result might mean, what it might say about an expanding world model. My impression is that Hubble meant to establish observable properties of the new realm of nature he was exploring, not to interpret the results in "the dreamy realms of speculation" (as he wrote on p. 202 in his book). With Hubble, Oort (1958) and van den Bergh (1961) presented estimates of the mean mass density but made no mention of how these measurements might test or constrain cosmological models.

The value of the mass density is important for tests of the relativistic cosmology. We see an early example in the paper by Einstein and de Sitter (1932). They proposed that de Sitter's earlier estimate of the mass density, largely following Hubble, might be consistent with Hubble's value of the expansion rate applied to equation (3.25) without the postulate of space curvature (and without a nonzero cosmological constant, though this is not stated). Though not mentioned, this rough degree of consistency meant that the relativistic model passed a test: The estimates of the cosmic mean mass density and expansion rate were not inconsistent with this model. The test is not at all tight, but it is meaningful. We see a later example in Figure 3.2. The Gott et al. (1974) paper does not mention this, but their figure shows that the relativistic model passes a still rough but real test.

Gott et al. (1974) did express an interest in what their collection of constraints might mean. The titles of their paper and their semi-popular account of what they found are "An Unbound Universe?" and "Will the Universe Expand Forever?" (Gott et al. 1976). Their answer was that yes, it seems likely that the expansion of the universe will continue into the indefinite future. The evidence reviewed in Section 4.6 that the mass density in baryons is well below Einstein–de Sitter was often noted in later considerations of the fit of light-element abundances to the prediction in the hot big bang model. Gott et al. did not pause to question whether their mass density estimate, if correct, really means the universe will expand forever: Is the theory to be trusted that far? And they did not mention the thoughts reviewed in Section 3.5 about what a philosophically attractive mass density would be. That issue became more widely discussed in the 1980s.

The issues of what we might expect the value of Ω_m to be, what its value might say about how the world will end, and what the value might mean for tests of cosmological models, all helped motivate the great program of measurements of the mass density in the 1980s through the 1990s. But as the program grew to become the most active and productive part of research in empirical cosmology in the 1990s, the goal became the measurement.

In the mid- to late 1990s, the evidence from the mass estimates that the value of Ω_m is less than Einstein–de Sitter grew more broadly influential among those searching for a cosmology that would fit the growing constraints reviewed in Chapter 8. That was followed by the revolution at the turn of the century, when the two other programs discussed in Chapter 9 also pointed to low Ω_m. But I shall argue that the mass density measurements discussed here and in the following two sections already made a good case for that conclusion.

The rich history and the considerable variety of methods of measurement employed in this program is illustrated by the sample of results listed in Table 3.2 and reviewed in Section 3.6.4. The increase of activity in the years around 1990 is suggested by the increasing density of entries listed by dates of publication, but it is somewhat understated by the less-dense sampling from 1990 on in what had grown to be an active productive science. The table ends when I take the revolution to have ended, in 2003.

The interpretation of these mass density probes was complicated by the question of how well the mass distribution may be modeled. If a considerable fraction of the mass were between the concentrations of galaxies, it would be missed, leaving Ω_m significantly underestimated. In Section 3.5.3, we saw how easy it was to imagine that the processes of galaxy formation may have made them poor tracers of the mass. And from the 1980s on, this line of thought was reinforced by the low values of the mass density estimates found under the assumption that galaxies are useful mass tracers, at about a third of the value in the elegant Einstein–de Sitter universe. The argument for the Einstein–de Sitter model is reasonable and sensible, in principle, and the idea that estimates of the mean mass density may be biased low accordingly was taken seriously in the 1980s through to the 1990s. Learning to live with a mass density less than Einstein–de Sitter was a difficult adjustment.

An early empirical argument that galaxies are likely to be useful mass tracers was reviewed in connection with Figure 3.1: If galaxy formation had been inhibited in low-density regions, one might have expected that the galaxies that are present in such regions would show signs of a deprived youth. But this is contrary to the evidence.

Another more broad-ranging argument developed during the course of the research summarized in Table 3.2. It revealed that, under the assumption when necessary that galaxies are useful mass tracers, the considerable variety of probes of the mass density on different length scales, ranging from ~ 0.3 Mpc to ~ 30 Mpc, at different degrees of density relative to the mean, and by a broad variety of considerations, are consistent with $\Omega_m \sim 0.3$. If

Table 3.2. Measures of the Cosmic Mean Mass Density

	Source	Ω_m	Comment
1	Hubble (1936)	0.002	galaxy counts and masses
2	Hubble (1936)	0.2	mass per galaxy in clusters
3	Oort (1958)	0.03	$j = 2.9 \times 10^8 h, M/L = 29h$
4	van den Bergh (1961)	0.024	$j = 2.7 \times 10^8 h, M/L = 25h$
5	Fall (1975)	0.01 to 0.05	Irvine-Layzer equation
6	Gott and Turner (1976)	0.08	$j = 0.9 \times 10^8 h, M/L = 240h$
7	Seldner and Peebles (1977)	0.69 ± 0.11	cluster $\xi_{c\rho}$ and ξ_{cg}
8	Peebles (1979)	0.4 ± 0.2	relative velocity dispersion
9	Yahil, Sandage, and Tammann (1980)	0.04 ± 0.02	Virgocentric flow
10	Davis et al. (1980)	0.4 ± 0.1	Virgocentric flow
11	Tonry and Davis (1981)	$0.5^{+0.3}_{-0.15}$	Virgocentric flow
12	Aaronson et al. (1982)	0.10 ± 0.03	Virgocentric flow
13	Davis and Peebles (1983a)	$0.2e^{\pm 0.4}$	relative velocity dispersion
14	Bean et al. (1983)	$0.14 \times 2^{\pm 1}$	relative velocity dispersion
15	Loh and Spillar (1986b)	$0.9^{+0.7}_{-0.5}$	redshift-magnitude relation[c]
16	Peebles (1986)	0.2 to 0.35	cluster $\xi_{c\rho}$ and ξ_{cg}
17	Yahil, Walker, and Rowan-Robinson (1986)	0.85 ± 0.16	motion of the Local Group
18	Strauss and Davis (1988)	0.4 to 0.9	motion of the Local Group
19	Blumenthal, Dekel, and Primack (1988)	~ 0.3	large-scale clustering[a]
20	Regős and Geller (1989)	$\lesssim 0.5$	clustercentric flow
21	Lynden-Bell, Lahav, and Burstein (1989)	~ 0.2	motion of the Local Group
22	Efstathiou, Sutherland, and Maddox (1990)	~ 0.3	large-scale clustering[a]
23	Bahcall and Cen (1992)	~ 0.25	rich clusters in CDM[a,c]
24	Strauss et al. (1992)	0.27 to 0.76	motion of the Local Group
25	Vogeley et al. (1992)	~ 0.3	large-scale clustering[a]
26	Briel, Henry, and Böhringer (1992)	0.14 ± 0.07	cluster baryon fraction[b,c]
27	White et al. (1993)	$\simeq 0.2$	cluster baryon fraction[b,c]
28	Dekel et al. (1993)	0.5 to 3	velocity and gravity fields
29	Fisher et al. (1994)	0.1 to 0.6	mean flow convergence
30	Hudson et al. (1995)	0.61 ± 0.18	velocity and gravity fields
31	Shaya, Peebles, and Tully (1995)	0.17 ± 0.10	Virgocentric flow
32	Davis, Nusser, and Willick (1996)	0.2 to 0.4	velocity and gravity fields
33	Bahcall, Fan, and Cen (1997)	0.34 ± 0.13	evolution of rich clusters[a,c]
34	Carlberg et al. (1997)	0.19 ± 0.06	cluster masses
35	Eke et al. (1998)	0.36 ± 0.25	evolution of rich clusters[a,c]
36	Willick and Strauss (1998)	0.31 ± 0.05	velocity and gravity fields
37	Schmoldt et al. (1999)	$0.43^{+0.29}_{-0.17}$	motion of the Local Group
38	Tadros et al. (1999)	$0.28^{+0.18}_{-0.14}$	mean flow convergence
39	Hamilton, Tegmark, and Padmanabhan (2000)	$0.23^{+0.13}_{-0.11}$	mean flow convergence
40	Percival et al. (2001)	0.29 ± 0.04	BAO[a,c]
41	Hawkins et al. (2003)	0.31 ± 0.09	mean flow convergence
42	Feldman et al. (2003)	$0.30^{+0.17}_{-.07}$	relative peculiar velocities[c]

[a] Assumes elements of the Cold Dark Matter (CDM) model.
[b] Assumes the Big Bang Nucleosynthesis (BBNs) theory.
[c] Insensitive to galaxy bias.

galaxies were seriously biased mass tracers, it would seem unlikely that galaxy formation conspired to produce a bias in the distribution of galaxies relative to mass that is independent of length scale over the considerable range sampled by the dynamical measures.[12] What is more, the measures in Table 3.2 with footnote c are less sensitive to the assumption that galaxies are useful mass tracers. They are of course sensitive to other considerations that may fail; this is true of all the measures. The empirical case for a reliable measurement of Ω_m rests on the consistency of results from a broad range of probes.

Assessment of this consistency of course requires charitably sensible allowance for systematic errors in difficult measurements. The weight of all these considerations is a personal judgment. I was persuaded that the mass density likely is less than Einstein–de Sitter by the mid-1980s (for the reasons reviewed in Peebles 1986). I remember the complaint of younger colleagues that I was only doing it to annoy. I was serious, but I must admit I enjoyed presenting this unpopular argument at conferences. A decade later, the weight of the evidence reviewed in Section 3.6.5 and presented later in Figure 3.5 had become great enough that the case for low mass density became a more common consideration in the search for a viable cosmological model, as reviewed in Chapter 8.

The review of the great program of mass density estimates in the next section must be lengthy, because the search for the value of Ω_m in the years around 1990 is a large part of the story of how modern empiricially-based cosmology grew. The program led to applications of a considerable variety of probes of the universe. This was important, because all the density measurements were challenging enough to call for cross-checks by other means. The methods that were developed in the program are of lasting physical interest, the history is an informative example of the great effort that can be inspired by an interesting and challenging problem, and the results of this program were influential in the thinking that led us to the cosmological model established at the end of the century.

In the years around 1990, it was common practice in this program to state results of dynamical mass density estimates in the combination $\beta = \Omega_m^{0.6}/b$ shown in equation (3.43), where the linear bias parameter b is defined in equation (3.42). It was a useful cautionary reminder of an uncertainty in the measurements. But I have entered the density measures in Table 3.2 that

12. At the larger observed length scales where the departures from homogeneity are small and would be expected to have been even smaller in the earlier universe when the galaxies formed (as discussed in Chapter 5), it is reasonable to expect that the differences between the distributions of galaxies and mass would have been small to begin with. If so, the growth of the clustering of mass and galaxies by gravity would have suppressed the differences by drawing together galaxies and mass alike. The effect is discussed in footnote 18 on page 94.

depend on the relative distributions of galaxies and mass on the assumption that galaxies are good mass tracers, $b \simeq 1$. This simplification distorts the history: The value of β (or its equivalent), not Ω_m, was the standard in reports of dynamical mass estimates in the years around 1990. But setting $b = 1$ has the practical advantage of reducing complications in an already lengthy review in the next section. More important, it makes it easier to see the development of the pattern of evidence that galaxies actually are useful mass tracers, with $b \simeq 1$. Lest the reader forget, in Section 3.6.4 I offer occasional reminders of this practice.

In preparation for the discussion in Section 3.6.4, note that the dynamical estimates of Ω_m are independent of the extragalactic distance scale that sets the value of Hubble's constant H_0. This is because the same relations among mass, acceleration, and velocity describe the expansion of the universe and the gravitational dynamics of systems of masses in the universe (apart from the effects of radiation in the early universe and the cosmological constant at low redshifts). It assumes we have the right gravity physics, of course, which is one of the things to be checked.

Entries in the history of mass estimates in Table 3.2 with footnote a depend on elements of the CDM cosmological model that became popular in the early 1980s (Chapter 8). Entries with footnote b combine dynamics with the theory of light-element formation in the early universe, mainly from the deuterium abundance D/H mentioned in the Section 3.6.2. Entries with footnotes a and b both depend on the value of H_0. To keep the comparisons of estimates reasonably straightforward, I use $h = 0.7$. Adjustments in the range usually considered, $0.5 \lesssim h \lesssim 1$, do not seriously change the picture. As noted above, the entries assigned footnote c are judged to be less sensitive to the assumption that galaxies trace mass, based on what was known in the years around 1990. These assignments are explained in the discussions in the next section.

3.6.4 MASS DENSITY MEASUREMENTS: HUBBLE TO THE REVOLUTION

Hubble's (1936) two mass densities in rows 1 and 2 in Table 3.2 are from his book, *The Realm of the Nebulae*. They are the products of his estimate of the number density of typical galaxies, from galaxy counts, with his two estimates of the typical mass of a galaxy. The mass follows from the condition that the gravitational acceleration $g \sim GM/r^2$ holding together an object of mass M and radius r balances the acceleration $g \sim v^2/r$ of matter moving at speeds $\sim v$ confined within distances r in the object. We have then $GM \sim v^2 r$. This is a rough approximation, but the evidence is that Hubble applied it in a sensible way. His first estimate used the mass in the luminous parts of galaxies. This is where most of the stars are, and where the stars make significant contributions to the total mass. Hubble's first mass density, $\Omega_{\text{stars}} \simeq 0.002$, thus may be compared to recent measurements of the mean mass density in stars, $\Omega_{\text{stars}} \sim 0.003$. (This measurement, by Cole et al. 2001, depends on the distance scale. As noted above, I set $h = 0.7$. Recall that Hubble's dynamical estimates of Ω_m in rows 1 and 2 are independent of h.) Hubble's second estimate used the mass per galaxy needed to hold a cluster of galaxies together. Since this takes reasonably fair account of the subluminal matter to be discussed in Chapter 6, Hubble's estimate, $\Omega_{\text{m}} \simeq 0.2$, may be compared to the value $\Omega_{\text{m}} = 0.315 \pm 0.007$ in the well-tested ΛCDM cosmology (Planck Collaboration 2018).

Both of Hubble's estimates are quite reasonable, through some combination of good luck and good management, though it is not easy to judge which of the two was more important. But Hubble certainly demonstrated an excellent sense of research in extragalactic astronomy in the 1930s despite his error in the distance scale. His larger estimate of the mass density takes account of mass between the galaxies, provided any bias in the galaxy distribution relative to the mass does not extend beyond the scale of clusters. I conclude that a judgment from the perspective of the 1990s or later is that his larger estimate is to be assigned real if modest significance in the weight of evidence for $\Omega_{\text{m}} \sim 0.3$.

Oort's (1958) dynamical mass density in row 3 is the product of his estimates of the mean luminosity density of starlight, j, from galaxy counts as function of luminosity, and the mean ratio of mass to stellar luminosity, M/L, in galaxies of two types, ellipticals and spirals, or early and late.[13] Oort chose M/L similar to that of Hubble for individual spiral galaxies, and a larger value, $M/L \sim 50$, for the ellipticals and other early types commonly found in clusters, because he knew that the mass per galaxy in clusters is large (assuming they are gravitationally bound, not flying apart, a consideration discussed in Chapter 6). His result is roughly the geometric mean of Hubble's two estimates. Van den Bergh's (1961) more detailed assessment of galaxy luminosities yielded the similar result in row 4 of Table 3.2.

Gott and Turner (1976) introduced the use of Schechter's (1976) expression for the galaxy luminosity function,

$$\frac{dn}{dL} = \phi^* \left(\frac{L}{L^*}\right)^{\alpha} e^{-L/L^*}, \tag{3.33}$$

13. The mass-to-light ratio M/L has units of solar masses per solar luminosity (where $M_\odot = 1.989 \times 10^{33}$ g, and $L_\odot = 3.83 \times 10^{33}$ erg s^{-1}). Since the spectrum of a star population may differ from that of the Sun, it is usual to state the wavelength band in which the luminosities of a star population and the Sun are compared. The luminosity L_B is measured in a standard wavelength band centered around $0.42\ \mu = 4200$ Å. Typical values are $M/L_B \sim 2 \pm 1$, depending on abundance of faint but numerous low-mass stars. The value of M/L scales with the value of Hubble's constant as $M/L \propto H_0$.

to find the mean luminosity density. The constant L^* is typical of the luminosities of the large nearby galaxies, including the Milky Way. Schechter's expression continues to be a good fit to much larger redshift surveys, as in the Norberg et al. (2002) analysis of the two-degree-field galaxy redshift survey (2dFGRS) of "more than 110,500 galaxies." They found

$$\alpha = -1.21 \pm 0.17, \quad j(b_J) = (1.82 \pm 0.17) \times 10^8 h \, L_\odot \, \text{Mpc}^{-3}. \tag{3.34}$$

This mean luminosity density is defined at the b_J pass band centered at wavelength 4500 Å = 450 nm. Since this is not far from the photographic magnitudes used in earlier entries in the table, we can compare $j(b_J)$ to the Oort and van den Bergh densities in rows 3 and 4. They are larger but reasonably close. Gott and Turner (1976) adopted $\alpha = -1$, which is close to the much later measurement in equation (3.34). Their adjustment of the parameters ϕ^* and L^* to fit their galaxy counts yielded the luminosity density used in row 6, which is rather lower than the established value.

The Oort and van den Bergh mass-to-light ratios are low because they do not take adequate account of the mass in subluminal matter in the far outskirts of the galaxies. Gott and Turner (1976) derived their estimate of M/L in row 6 from the relative motions of galaxies in their catalog of groups of galaxies, which takes better account of the subluminal matter. Their value of M/L agrees reasonably well with the Geller and Peebles (1973) estimate from the redshift space distortion effect (to be discussed beginning on page 87), and it is close to more recent measurements. Gott and Turner underestimated Ω_m largely because their estimate of the luminosity density is low, but their result is to be added to the weight of evidence, because their mass-to-light ratio is a reasonable measure of subluminous matter between the galaxies, as in groups and clusters.

Fall's (1975) approach is based on Irvine's[14] (1961) relation between the kinetic and potential energies of collisionless particles in an expanding universe:

$$\frac{d}{dt}(T+W) + \frac{1}{a}\frac{da}{dt}(2T+W) = 0, \; T = \frac{\langle v^2 \rangle}{2}, \; W = -\frac{G\rho_m}{2}\int d^3r \frac{\xi(r)}{r}. \tag{3.35}$$

The mean mass density is ρ_m. The last two expressions have been simplified by assuming a gas of equal-mass particles. The mean-square particle peculiar velocity relative to the Hubble flow is $\langle v^2 \rangle$. The particle two-point position correlation function (defined in equation (2.10)) is $\xi(r)$. Fall's bounds in row 5, $0.01 \lesssim \Omega_m \lesssim 0.05$, use his estimates $T \sim -W$ and $\langle v^2 \rangle^{1/2} \sim 300 \text{ km s}^{-1}$. It assumes the galaxy correlation function is a useful approximation to the mass correlation function on the larger scales of clustering that dominate the

14. William Irvine (1961) found equation (3.35) in his PhD dissertation, an investigation suggested by David Layzer. Later derivations are in Layzer (1963) and Dmitriev and Zel'dovich (1963).

integral, which means the estimate of the integral would take account of subluminal matter on the outskirts of galaxies, provided galaxies are useful tracers of mass on scales larger than their separations. But the galaxy correlation function $\xi(r)$ at large separation r, where it could be a useful mass tracer, was not well measured then.

Davis, Geller, and Huchra (1978) had a better measure of $\xi(r)$, which they used in a reconsideration of Fall's method along with the redshift space statistic discussed on page 87. Their conclusion (p. 1) is that

> Cosmic virial theorems which are cast in terms of the correlation functions and the peculiar velocity dispersion are used to estimate the contribution, Ω_G, of matter associated with galaxies to the critical density. The statistical virial theorems suggest $0.2 \lesssim \Omega_G \lesssim 0.7$.

Although the arguments for the critical Einstein–de Sitter density $\Omega_m = 1$ were less popular then, we see that these authors took care to distinguish Ω_G from mass that may be present and not associated with galaxies. But the integral in equation (3.35) depends on the mass correlation function on large scales, where the considerations in footnote 18 on page 94 suggest galaxies might be expected to be reasonably good mass tracers. And their bounds include the value $\Omega_m \sim 0.3$ that was eventually convincingly established.

Lynden-Bell et al. (1988) revisited Fall's approach. To get galaxy peculiar velocities, they used the $D_n - \sigma$ relation between the stellar velocity dispersion σ in an elliptical galaxy and its physical linear size D_n. (This method was introduced by Tonry and Davis 1981; Djorgovski and Davis 1987; and Dressler et al. 1987.) The observed value of σ gives a measure of the physical size D_n of a galaxy, and that with the observed angular size gives a measure of the distance r. The difference between the observed galaxy redshift and the Hubble streaming speed $H_0 r$ is a measure of the radial component of the galaxy peculiar velocity. Lynden-Bell et al. (1988) found that their systematic study of galaxy peculiar velocities sampled out to ~60 Mpc distance could only agree with the mass density in the Einstein–de Sitter model if the potential W were much less negative than the value obtained by assuming galaxies trace mass on the large scales relevant for the integral in equation (3.35). That could be because mass actually is more smoothly distributed than galaxies or because Ω_m is well below unity. The Lynden-Bell et al. (1988, 19) conclusion is that "The value of Ω_0 from these observations is still indeterminate." The Einstein–de Sitter density had become popular by this time, but the authors do not mention it. And note that, if galaxies trace mass, then their result indicates Ω_m is well below unity, in line with measures on smaller scales.

Seldner and Peebles (1977) measured the cross-correlation of Abell (1958) cluster positions with Lick galaxy positions. Following equation (2.10) for the galaxy-galaxy function, the cluster-galaxy spatial cross-correlation function $\xi_{cg}(r)$ is defined by the probability of finding a galaxy in the volume element δV at distance r from a cluster center:

$$\delta P = n(1 + \xi_{cg}(r))\delta V. \tag{3.36}$$

As before, n is the mean galaxy number density. If galaxies usefully trace mass on the scales of size and density of clusters, then ξ_{cg} is a useful measure of the cluster-mass cross-correlation function. The mean mass density at distance r from a cluster in a fair sample of clusters is then $\rho(r) = \rho_m(1 + \xi_{c\rho}(r))$, where ρ_m is the cosmic mean mass density. This can be compared to the mean run of mass density with radius derived from the velocity dispersion of the galaxies in the cluster. In the simple approximation that the velocity dispersion is isotropic and independent of radius, we can write

$$\rho(r) = \rho_m(1 + \xi_{cg}(r)) = \frac{v^2}{2\pi G r^2}, \tag{3.37}$$

where v is the mean line-of-sight galaxy velocity dispersion in the cluster. The result in row 7 of Table 3.2 is not far below $\Omega_m = 1$. At the time, I took this to be moderately encouraging evidence for the Einstein–de Sitter model. But the better data and more detailed application of this method in Peebles (1986), a decade later, yielded the significantly lower density in row 16.

Rows 9 through 12, along with rows 20 and 31, are derived from Virgocentric flow. This is the mean peculiar motions of the nearby galaxies toward the nearest large concentration of galaxies and presumably mass in and around the Virgo cluster at some 17 Mpc distance, or the analog for other clusters. (I defer consideration of the relation between mass, peculiar gravitational acceleration, and peculiar flow to the discussion leading up to equation (3.43).) These entries assume galaxies usefully trace mass at distances of 10 to 20 Mpc.

Silk (1974) discussed the departure from pure Hubble flow to be expected from the gravitational attraction of the mass in excess of homogeneity in a cluster, and the promise of an interesting constraint on cosmological parameters when the galaxy redshift and distance data improve. Peebles (1976a) presented a fit of the Sandage and Tammann (1975) compilation of distances and redshifts of galaxies and groups of galaxies to a model for the mass distribution and implied peculiar velocity field around the Virgo cluster. It yielded a reasonably secure detection of Virgocentric flow but not a serious constraint on Ω_m. Sandage (1975) also found evidence of Virgocentric flow, and an indication that the mass density parameter is less than unity, but his "uncertain data for field spirals" makes it difficult to judge the weight of this measurement. Yahil, Sandage, and Tammann (1980) had still better data, the complete redshift sample in the Revised Shapley-Ames Catalog of Bright Galaxies (Sandage and Tammann 1981). Their constraint on the mean mass density from Virgocentric flow is entered in row 9 in Table 3.2.

Davis et al. (1980) used an early version of the CfA redshift sample (shown in Figure 3.1) for their result in row 10. Tonry and Davis (1981) used the correlation between luminosity and stellar velocity dispersion in early-type galaxies to get measures of distances to compare to observed redshifts and so to estimate peculiar velocities (in a variant of the $D_n - \sigma$ method discussed earlier in this section). Their 160 peculiar velocities improved the measurement of Virgocentric flow and yielded the density parameter entered in row 11 in Table 3.2.

Aaronson et al. (1982) found peculiar velocities of 306 late-type galaxies from distances derived from the Tully and Fisher (1977) relation between absolute magnitude and 21-cm H I line width. They had infrared apparent magnitude measurements, an important advantage over observations at shorter wavelengths, because it reduces the effect of obscuration by dust. Their result from Virgocentric flow is shown in row 12.

Tully and Shaya (1984) added to the Virgocentric flow models the constraint from the allowed expansion time of the universe. Shaya, Peebles, and Tully (1995) continued this approach, now using redshifts and distances of 1,138 galaxies and concentrations of galaxies treated as mass tracers. The analysis used the numerical action method (Peebles 1989). The result in row 31 is labeled "Virgocentric flow," but it does not assume spherical symmetry around the Virgo cluster, and it samples the local variations of the peculiar velocity field out to twice the distance to the Virgo cluster.

Regős and Geller (1989) demonstrated the mean flow of galaxies toward the mass concentrations in other rich clusters of galaxies. The Virgocentric flow toward the relatively nearby Virgo cluster is observed from inside; Regős and Geller showed the analog of Virgocentric flow around other clusters observed from outside the flow. Their bound on the mass density parameter is given in row 20. It is a significant addition to the weight of the evidence, and it suggests that further progress in this direction is feasible and certainly would be welcome.

Virgocentric flow was well measured by the early 1980s, and it showed that Ω_m is well below unity if galaxies trace mass on scales of tens of megaparsecs. Similar results followed from the estimates in Table 3.2 labeled "relative velocity dispersion" and "mean flow convergence," which are derived from catalogs of galaxy angular positions and redshifts and can deal with less precisely determined distances by the statical measures in redshift space as follows.

Galaxy angular positions and redshifts define a three-dimensional space. A measure of the galaxy distribution in this space is the two-point correlation function $\xi_v(r_p, \pi)$, where the argument r_p is the projected separation of galaxies perpendicular to the line of sight, and π is the radial separation in redshift space: the difference of measured redshifts. This two-point function is defined as in equation (2.10). In a sample of galaxies in a redshift survey to a chosen limiting distance estimate or limiting apparent magnitude, let the mean number density of galaxies sampled at distance r be the selection function $\phi(r)$. The probability that galaxies in the sample are found in each of two volume elements in redshift space at separation r_p normal to the line of sight and $\pi = cz_1 - cz_2$ along the line of sight is[15]

$$dP = \phi(r_1)\phi(r_2)(1 + \xi_v(r_p, \pi))dV_1 dV_2. \tag{3.38}$$

This assumes the two galaxies are at roughly the same distance $r_1 \sim r_2$. The perpendicular separation r_p is sometimes written σ, following the traditional usage, *senkrecht*. The separations π and r_p may be expressed in units of velocity, km s^{-1}, direct from the measured redshifts cz, or megaparsecs, by dividing by a value for Hubble's constant, which cancels out of Ω_m.

At small r_p, the relative motions of galaxies close to each other stretch out the separations π in redshift, causing $\xi_v(r_p, \pi)$ to be elongated along the line of sight. At larger r_p, the dominant distortion of positions in redshift space relative to real space is the gravitational growth of clustering by the mean motion of gathering of the galaxies relative to the homogeneous Hubble flow. This flattens $\xi_v(r_p, \pi)$ along the line of sight. All this of course assumes we are seeing a statistically isotropic random process.

Geller and Peebles (1973) demonstrated the expected elongation along the line of sight at small separations in a preliminary version of the ξ_v statistic, using the modest redshift sample available then in the de Vaucouleurs and de Vaucouleurs (1964) *Reference Catalogue of Bright Galaxies*. The measure of the galaxy masses required to produce this elongation indicated that the galaxy mass-to-light ratio is $M/L \sim 300h\, M_\odot/L_\odot$ (in photographic magnitudes). This is much greater than the mass observed in the luminous parts of galaxies, but it is comparable to what Gott and Turner (1976) found from their group catalog (row 6 in Table 3.2). Peebles (1976b) applied another version of the ξ_v statistic to about the same data Geller and Peebles used; it again showed promise for better from a sample with clearer control on completeness. Davis, Geller, and Huchra (1978) reported application of yet another statistical measure of the galaxy relative velocity dispersion from their ongoing program of galaxy redshift measurements, combining the de Vaucouleurs and de Vaucouleurs (1964) data with their measurements of redshifts of galaxies in the Zwicky et al. (1961–1968) catalog, in a redshift sample that had grown to 955 redshifts of relatively nearby galaxies. They concluded that these data

15. Kaiser (1987) introduced the description of the two-point correlation in redshift space by the relation between the Fourier amplitudes $\delta(\vec{k})$ of positions in real space and $\delta_s(\vec{k})$ in redshift space,

$$\delta_s(\vec{k}) = \delta(\vec{k})(1 + f(\Omega_m)\cos^2\theta),$$

where θ is the angle between the line of sight and the wavenumber $\vec{k}$, and the function $f(\Omega_m)$ is defined in equation (3.41) below.

again seemed to require the large mass-to-light ratio, but with mass density likely in the range $0.2 \lesssim \Omega_m \lesssim 0.7$.

Peebles (1979) at last wrote out the full definition of the two-point correlation function $\xi_v(r_p, \pi)$ in redshift space and applied it to the Kirshner, Oemler, and Schechter (1978) redshift measurements in eight fields, each about 4° square. They had a total of just 166 redshifts, but their fields are deeper and well separated on the sky, which offered the chance of a reasonably close to fair sample. The result entered in row 8 in Table 3.2 is of historical interest as a demonstration of the value of a still-better sample.

The Harvard-Smithsonian Center for Astrophysics (CfA) redshift survey (Davis et al. 1982) was the wanted much-better sample. Marc Davis led this project. He was well aware of its scientific value that we discussed in Davis and Peebles (1977), though in personal conversations, Davis recalls that the faculty at Harvard and the Center for Astrophysics did not seem all that excited about the project at the time.

The CfA redshift survey used for their finding list the Zwicky et al. (1961–1968) catalog of galaxy apparent magnitudes and angular positions. The CfA catalog has 1,840 redshifts of galaxies in the northern galactic zone that are more luminous than absolute magnitude $M_B = -18.5 + 5\log h$. The survey is close to complete to distance $40h^{-1}$ Mpc. This was an important advance, the first good approximation to a fair sample of galaxy peculiar motions surveyed to a useful limiting luminosity and distance.

The density parameter[16] derived from the elongation of $\xi_v(r_p, \pi)$ along the line of sight is (row 13 in Table 3.2)

$$\Omega_m = 0.2e^{\pm 0.4} \text{ at projected separations } 0.2 \lesssim r_p \lesssim 2 \text{ Mpc.} \tag{3.39}$$

This value of the density parameter fits the indicated range of separations, which is considerable. Bean et al. (1983) obtained a similar result from their application of this statistic to their more-distant redshift sample: $\Omega_m = 0.14 \times 2^{\pm 1}$ (row 14 in Table 3.2). Both are one standard deviation below the best estimate in 2018, $\Omega_m = 0.31 \pm 0.01$. More to the present point, both are significantly less than the mass density in the Einstein–de Sitter model.

16. A technical point: The computation uses the galaxy spatial three-point function, which measures the mean concentration of galaxies near a pair of galaxies, and perhaps also the mean relative concentration of mass. This determines the relative gravitational acceleration of the pair (Peebles 1976b). Bartlett and Blanchard (1996) point out that the representation of the relative amount of dark mass around a pair of galaxies by the positions of galaxies close to the pair may well be questioned, particularly at small separations, where there may or may not be a third catalog galaxy nearby. But any error due to this might be expected to be smaller at larger separations, where the pairs have more neighbors. Thus our best evidence in support of this approach is the consistency of the model with the measured run of relative velocity dispersion at separations in the range $r_p \simeq 30$ kpc to 3 Mpc. This considerable range of sampling suggests the model likely is not far off. And the measurement, $\Omega_m = 0.2e^{\pm 0.4}$, is close to the value established at the turn of the century.

The coincidence argument discussed on page 58 (equations (3.28) to (3.30)) for $\Omega_m = 1$ seemed reasonable to many in the years around 1990, and it still does in principle. It was responsible for the commonly expressed opinion from this CfA result and the accumulation of other dynamical measurements that mass must be less strongly clustered than galaxies, consistent with the larger Einstein–de Sitter density (as discussed in Section 3.5.3). Prior to the CfA measurement in equation (3.39), the argument led me to expect that the mass density likely is the Einstein–de Sitter value. I was surprised by the CfA result, and disconcerted, but I did not trust the biasing argument for the reasons reviewed on page 68. In particular, why would a significant bias be so insensitive to length scales that range from $r_p \sim 0.2$ to $r_p \sim 2$ Mpc? That question persuaded me to accept as a working hypothesis that the mass density likely is less than the elegant Einstein–de Sitter prediction.

At larger galaxy separations, the growing clustering of matter, at a rate that depends on the mean mass density, produces a mean convergence of flow toward denser regions. This shrinks the width of the redshift space-correlation function $\xi_v(r_p, \pi)$ in the radial, π, direction. The effect on $\xi_v(r_p, \pi)$ is computed as follows.

The gravitational acceleration produced by the departure $\delta\rho(\vec{r}, t)$ of the mass distribution from its mean value, ρ_m, is

$$\vec{g}(\vec{r}, t) = G \int d^3 r' \frac{\vec{r}' - \vec{r}}{|\vec{r}' - \vec{r}|^3} \delta\rho(\vec{r}', t), \quad \delta\rho(\vec{r}, t) = \rho(\vec{r}, t) - \rho_m(t). \tag{3.40}$$

This is the peculiar acceleration, relative to the Hubble flow, in the Newtonian approximation. In linear perturbation theory, this peculiar acceleration produces the peculiar velocity (Peebles 1976a; 1980, §14):

$$\vec{v}(\vec{r}, t) = \frac{H_0 f(\Omega_m)}{4\pi G \rho_m} \vec{g}(\vec{r}, t), \quad f = \frac{a}{D} \frac{dD}{da} \simeq \Omega_m^{0.6}. \tag{3.41}$$

Here $D(t)$ is the growing mode of the departure from a homogeneous mass distribution in linear perturbation theory. At redshifts of interest, this relation is not very sensitive to the value of Λ. To estimate $\vec{g}$, one can express the observed galaxy distribution as the departure $\delta j(\vec{r}, t)$ of the starlight luminosity density from the mean value, $\bar{j}(t)$, as in equation (3.40) for the mass density. The translation from $\delta j(\vec{x}, t)$ to the departure from a homogeneous mass distribution, $\delta\rho(\vec{r}, t)$, traditionally takes account of the possibility that galaxies are significantly biased tracers of mass. The simple linear bias model is

$$\delta_m(\vec{r}) \equiv \frac{\delta\rho(\vec{r}, t)}{\rho_m(t)}, \quad \delta_g \equiv \frac{\delta j(\vec{r}, t)}{\bar{j}(t)} \simeq b\, \delta_m(\vec{r}). \tag{3.42}$$

The constant b is the bias parameter. Equations (3.40) to (3.42) give (Peebles 1980, §14)

$$\vec{v}(\vec{r}) = \frac{\beta H_0}{4\pi} \int d^3 r' \frac{\vec{r}' - \vec{r}}{|\vec{r}' - \vec{r}|^3} \delta_g(\vec{r}', t), \quad \beta \approx \Omega_m^{0.6} / b. \tag{3.43}$$

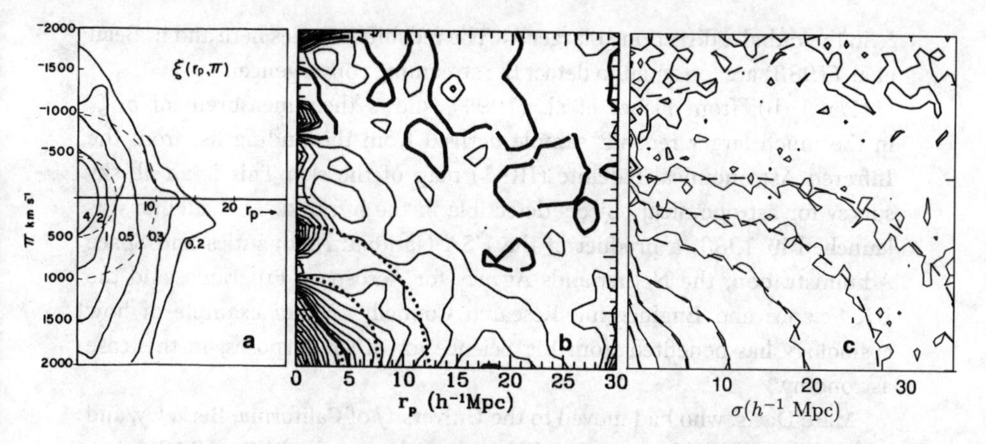

FIGURE 3.3. Progress in detecting the mean convergent flow in the galaxy two-point correlation function $\xi_v(r_p, \pi)$ in redshift space. The horizontal axis is the separation r_p normal to the line of sight. The vertical axis is the difference, π, of redshifts. The broken circles in Panels (a) (Davis and Peebles 1983a) and (b) (Fisher et al. 1994) are the shapes of contours of constant two-point function in redshift space when there is no peculiar motion. In Panel (c), from Hawkins et al. 2003, the dashed contours are the model fit to the streaming motion. Panel (a) © AAS. Reproduced with permission. Panels (b) and (c) reproduced by permission of Oxford University Press on behalf of the Royal Astronomical Society.

The lengths in this integral, from positions in redshift space, scale as H_0^{-1}. This makes the expression for $\vec{v}(\vec{r})$ independent of H_0. The two observables, $\delta_g(\vec{r})$ and $v(\vec{r})$, thus constrain β. Early use of the expression β to indicate how mass estimates scale with the difference between the distributions of galaxies and mass in the linear bias model is in Strauss et al. (1992); it became a standard way to express results. I set $b = 1$ in this review of the measurements for the reason presented in Section 3.6.3: It simplifies presentation of the systematic pattern in the measurements.

Figure 3.3 illustrates the progress in measuring the mean streaming flow of the galaxies relative to the Hubble expansion that accompanies the gravitational growth of the clustering of mass. The effect is seen as the flattening of the galaxy two-point function in redshift space in the line-of-sight direction π at larger projected separations r_p. (The label for the projected separation reverts to the older notation, σ, in Panel (c)). Panel (a) shows velocity units, km s^{-1}, and the other two panels show the result of conversion to distance units. Panel (a) shows positive and negative radial separations, π. Panels (b) and (c) fold positive and negative π.

The contours of constant $\xi_v(r_p, \pi)$ in Panel (a), from the Davis and Peebles (1983a) analysis of the CfA redshift catalog, are plotted only where the measurement is fairly secure, to $r_p \sim 1{,}500$ km s^{-1} $\sim 15h^{-1}$ Mpc at π close to zero. The elongation along π caused by relative motions of galaxies in tight concentrations is clearly displayed, but there is little sign of the shrinking of

ξ_v in the radial π direction at larger r_p. The redshift samples here and in Bean et al. (1983) are too small to detect the streaming convergence.

Panel (b), from Fisher et al. (1994), shows their measurement of ξ_v in the much larger redshift sample derived from the finding list from the Infrared Astronomical Satellite (IRAS) map of the sky. This is an all-sky survey for astronomical objects detectible in the infrared. The satellite was launched in 1983, a product of the USA National Aeronautics and Space Administration, the Netherlands Agency for Aerospace Programs, and the UK Science and Engineering Research Council, and an example of how cosmology has benefited from Big Science for other purposes, in this case astronomy.

Marc Davis, who had moved to the University of California, Berkeley, and colleagues took the lead in identifying the galaxies among the IRAS detections of infrared sources. They avoided the infrared glow of dust streams in the Milky Way by selecting sources with angular sizes less than a few arc minutes, the angular resolution in the IRAS broadband photometry at 60 μ wavelength. This largely selects stars and galaxies. From this they selected objects by the ratio of flux densities in the broadband photometry at 12 and 60 μ wavelengths, which efficiently distinguishes between stars and galaxies. The follow-up optical spectroscopy readily confirmed identifications of galaxies and yielded their redshifts, because galaxies that are luminous in the infrared tend to have prominent emission lines. Obscuration by interstellar dust at the infrared wavelengths of IRAS is much less serious than in the optical, which allowed collection of a valuable sample of late-type galaxies with uniform coverage over 87.6 percent of the sky. The sample with IRAS flux-density limit 1.936 Jy at 60 μ wavelength yielded 2,636 galaxy redshifts at a characteristic depth of $35h^{-1}$ Mpc (Strauss 1989). This is the 1.9-Jy IRAS catalog. The extension to a fainter flux-density limit is the 1.2-Jy IRAS catalog of some 5300 galaxy redshifts at median distance $45h^{-1}$ Mpc (Fisher et al. 1995).

Fisher et al. (1994) used the 1.2-Jy IRAS redshift catalog to compute the redshift space two-point correlation function shown in Panel (b) in Figure 3.3. This is a considerable advance over the data we had for Panel (a), which is a considerable advance over what we had before that. The elongation in the line-of-sight direction at small separation is clearly seen in the lower left-hand corner of the plot. The effect of convergent streaming flow on larger scales is reasonably well detected by the compression of contours in the π direction at projected separations out to roughly $10h^{-1}$ Mpc. The authors caution that their bounds on Ω_m are uncertain, not least because the contours are noisy, but their tentative result merits recording in row 29 in Table 3.2. Hamilton (1993) used the earlier 2-Jy IRAS sample for a somewhat broader bound, $\Omega_m \lesssim 1$.

Saunders et al. (2000) compiled the PSCz 0.6 Jy IRAS Point Source Redshift Catalog[17] of 15,411 redshifts, with references to the surveys by several groups leading up to what Saunders et al. (2000, 55) properly term the "unparalleled uniformity, sky coverage and depth for mapping the local galaxy density field" of PSCz. Hamilton, Tegmark, and Padmanabhan (2000) used the PSCz catalog for their report of a clear detection of the flattening of contours of ξ_v in redshift space. They obtained the tight bound on the mass density in shown in row 39 of Table 3.2. Tadros et al. (1999) analyzed the measure of β from the redshift space correlation function following the idea of Kaiser's Fourier amplitudes (equation (3.38)) but using instead the amplitudes of the spherical harmonic expansion of the 0.75 Jy subset of PSCz, for the result in row 38. (The earlier analysis along the same lines by Ballinger, Heavens, and Tayler (1995) gives a substantially larger mass density, but I omit this result, because it is a preliminary version of Tadros et al. 1999.)

The redshift space correlation function in Panel (c) in Figure 3.3 is from Hawkins et al. (2003). (I have taken the liberty of displaying only one quadrant of this figure, to match Panel (b)). It is derived from some 220,000 redshifts in the optically selected Two-Degree Field Galaxy Redshift Survey (2dFGRS) catalog. This survey at the Anglo-Australian Telescope is a beautiful example of how astronomy has made use of advances in technology. The instrument has optical fibers set up to measure up to 400 redshifts at a time in 2° fields. The earlier Durham/UKST Galaxy Redshift Survey has optical fibers for 50 to 100 redshift measurements at a time (Ratcliffe et al. 1998). These numbers are to be compared to the total of 527 redshifts measured one at a time on far less sensitive photographic plates that had been compiled by the early 1970s and Geller and Peebles (1973) used for their detection of the redshift space distortion due to the relative motions of galaxies on small scales.

Panel (c) in Figure 3.3 shows the stretching of $\xi_v(r_p, \pi)$ along the line of sight at small projected separations, and on larger scales the measurements clearly map out the effect of the convergent flow of matter from the gravitational growth of departures from a homogeneous mass distribution. The Hawkins et al. (2003) measure of the mass density from the model fit to $\xi_v(r_p, \pi)$—always assuming galaxies adequately trace mass—is in row 41 of Table 3.2. Peacock et al. (2001) found a similar result from an earlier version of the 2dFGRS sample.

Let us pause to consider that linear perturbation theory applied on scales ~ 30 Mpc to the redshift space correlation function in Panel (c) in Figure 3.3 indicates $\Omega_m = 0.31 \pm 0.09$, while the estimate in row 13 (Table 3.2) from the galaxy relative velocity dispersion on scales down to ~ 0.2 Mpc, where galaxies

17. PSCz is the Infra-Red Astronomical Satellite Point Source Catalog of galaxy redshifts.

are strongly clustered, is $\Omega_m = 0.2e^{\pm 0.4}$. This assumes galaxies are useful mass tracers on both scales. The assumption is encouraged by consistency within the uncertainties under two quite different conditions: strong and weak clustering of matter.[18] I count this as serious empirical evidence that Ω_m is well below unity.

Row 15 in Table 3.2 shows the result of an application of one of the classical cosmological tests, the count of galaxies as a function of redshift. If one can properly account for the evolution of galaxies, then the count $N(<z)$ of galaxies with redshifts less than z, as a function of z, is a measure of the volume of space in the curved spacetime of general relativity. The theory is discussed in Tolman (1934a, §§181 and 182). To reduce statistical fluctuations in the counts, the application requires measurements of redshifts of large numbers of galaxies. Loh and Spillar (1986a) took an important step in this direction by developing Baum's (1957) method: Fit apparent magnitudes, or flux densities, measured through broad-band filters across optical wavelengths to what would be expected from the redshifted spectrum of a galaxy. Loh and Spillar (1986a, 156) explain that "To identify an object as a star or a galaxy at redshift z, we match its fluxes against those of a set of fiducial objects for which fluxes in our filter bands have been computed from published spectra."

Koo (1981) had applied this approach to broadband spectra derived from photographic plates exposed through different wavelength passbands. Loh and Spillar developed the use of digital CCD (charge-coupled device) detectors that enabled more precise photometric redshifts of larger numbers of galaxies. The approach has grown to be widely applied.

Tinsley (1967 and 1972) had emphasized the serious challenge of correcting for the evolution of galaxy luminosities. Loh and Spillar (1986b) addressed

18. For a simple model of the evolution of bias in linear perturbation theory, consider the fractional departures from homogeneity in the distributions of mass and galaxies, with mass and number densities $\rho(\vec{x}, t)$ and $n(\vec{x}, t)$:

$$\delta_\rho(\vec{x}, t) = \frac{\rho(\vec{x}, t)}{\bar{\rho}} - 1, \quad \delta_n(\vec{x}, t) = \frac{n(\vec{x}, t)}{\bar{n}} - 1.$$

Imagine the galaxy positions are set at time t_i. Suppose that at $t > t_i$, the mass and galaxies have the same peculiar velocity field $\vec{v}(\vec{x}, t)$. This seems reasonable, because gravity pulls equally on mass and galaxies. Then in linear perturbation theory, the two density contrasts grow as

$$\frac{\partial \delta_\rho}{\partial t} = -\frac{\nabla \vec{v}}{a(t)} = \frac{\partial \delta_n}{\partial t}.$$

This indicates $\delta_n(\vec{x}, t) = \delta_\rho(\vec{x}, t) + \delta_n(\vec{x}, t_i) - \delta_\rho(\vec{x}, t_i)$. In this linear approximation, we see that as the mass density fluctuations $\delta_\rho(\vec{x}, t_i)$ grow, they may be expected to draw the galaxy distribution closer to the mass distribution: $\delta_n(\vec{x}, t) \to \delta_\rho(\vec{x}, t)$. The argument, which is elaborated in Tegmark and Peebles (1998), offers some reason to suspect that galaxies are useful tracers of mass on scales ~ 30 Mpc. And consistency with the measures of Ω_m on scales ~ 0.3 Mpc argues that galaxies are useful mass tracers on small scales, too.

this concern. Under the assumption that the luminosities of all galaxies evolve at the same fractional rate, they fitted the counts to a fixed functional form of the luminosity distribution, with normalization as a function of redshift to be adjusted along with the density parameter, to fit the counts as a function of redshift and apparent magnitude. The Loh and Spillar (1986b) measurement of Ω_m under the assumption $\Lambda = 0$ is shown in row 15 of Table 3.2. It is consistent with the Einstein–de Sitter model that many at the time felt is the likely case, but their lower error flag almost reaches the established value, $\Omega_m \simeq 0.3$. The Loh-Spillar approach is important, but is now applied to the study of how the galaxies evolve.

In a variant of the Loh-Spillar count-redshift test, Fukugita et al. (1990) concluded that the accumulation of galaxy counts down to very faint apparent magnitudes favors a cosmologically flat universe with low mass density, $\Omega_m \sim 0.1$, $\Omega_\Lambda \sim 0.9$. This result is in the right direction, and the evidence is interesting, but it is not entered in Table 3.2, because the result depends on a model for galaxy evolution that was carefully considered but difficult to check. Again, this approach is best to be counted as a probe of cosmic evolution.

Fukugita et al. (1992) analyzed the rate of observation of multiple images of quasars at high redshift caused by gravitational lensing by galaxies at lower redshift that happen to be close to the line of sight. The rate depends on the cosmological parameters, but it also depends on the distributions of subluminal matter around galaxies, and it is important now as a probe of the mass structures of galaxies.

The departures from a homogeneous mass distribution on large scales are expected to produce a gravitational acceleration field relative to the Hubble flow that would cause the Local Group of galaxies[19] to stream through the sea of microwave radiation, the CMB discussed in Chapter 4. (The CMB itself would be little disturbed by peculiar gravitational accelerations on scales small compared to the Hubble length.) The Doppler effect due to the streaming motion would cause the CMB temperature to be larger than the mean in the direction of streaming, and smaller in the opposite direction, in the dipole pattern $\delta T/T \propto \cos\theta$. This dipole anisotropy was detected by Smoot, Gorenstein, and Muller (1977). It would be produced by the Doppler effect of motion of the Local Group at about 600 km s^{-1} relative to the rest frame defined by the radiation. The measure of the mass density required to produce this motion is particularly interesting, because the gravitational acceleration of the Local Group is determined by the integral in equation (3.40), which is dominated by the small-amplitude departures from homogeneity on large scales that seem likely to be reasonably good tracers of mass. But the sensitivity to the galaxy distribution at large distances makes the numerical estimate of the integral difficult.

19. The Local Group is the gravitationally bound collection of two large spiral galaxies, the Milky Way and the Andromeda Nebula M31, and some dozens of smaller galaxies. It is usually reckoned to be about 2 Mpc across.

Yahil, Walker, and Rowan-Robinson (1986) used the angular distribution of their compilation of IRAS galaxies, pruned to eliminate sources in the Milky Way, to estimate the integral for the peculiar gravitational acceleration of the Local Group (equations (3.40) and (3.43)). This, along with the CMB dipole anisotropy, yields the factor β and their estimate of the density parameter entered in row 17 of Table 3.2. Strauss and Davis (1988) used their 1.9-Jy IRAS galaxy angular positions, and were able to add the constraint from the measured redshifts, for their estimate of β from this phenomenon. With Yahil et al., they found that the computed direction of gravitational acceleration is reasonably close to the direction of motion of the Local Group relative to the sea of thermal radiation, an indication that the IRAS sample may offer a useful approximation to the large-scale mass distribution. The Strauss and Davis mass density needed to account for the motion of the Local Group is entered in row 18 of Table 3.2. Strauss et al. (1992) used the larger 1.2-Jy IRAS catalog of 5288 galaxy redshifts, close to the data in Panel (b) of Figure 3.3, for the result in row 24. Lynden-Bell, Lahav, and Burstein (1989) used their catalog of redshifts of optical selected galaxies for row 21. The Schmoldt et al. (1999) result in row 37 is from the PSC*z* catalog that yielded an elegant measurement of the two-point correlation function in redshift space (Hamilton, Tegmark, and Padmanabhan 2000). But the Schmoldt et al. result is only two error flags from $\Omega_m = 1$, and one error flag above $\Omega_m = 0.3$. This is an important probe of the mean mass density and the large-scale distribution of the departures from the mean. The results add to the case that the mass density is less than Einstein–de Sitter, but at the turn of the century, this measure of the mass density had modest weight.

The challenge for the measurement of Ω_m from the peculiar motion of the Local Group is to preserve a stable distance calibration across the sky, because a variation appears as a systematic error in peculiar velocities as a function of angular position, which confuses the comparison to the gravity field. Well after the revolution, Springob et al. (2007, 599) resolved this challenge in their systematic survey across a large part of the sky. They report that their assembled "peculiar velocity catalog of 4861 field and cluster galaxies is large enough to permit the study not just of the global statistics of large-scale flows but also of the details of the local velocity field."

Davis and Nusser (2016, 310) used these data with improved redshift surveys to establish that "the gravity field is seen to predict the velocity field … to remarkable consistency. This is a beautiful demonstration of linear perturbation theory and is fully consistent with standard values of the cosmological variables." The standard values of cosmological variables that Davis and Nusser mention were well supported by this time. This advance stemming from control of the distance standard, brings to mind the importance of the first well-controlled galaxy redshift sample, the CfA catalog, by Davis and colleagues (Davis et al. 1982), which enabled the Davis and

Peebles (1983a) measurement of redshift space distortion and row 13 in Table 3.2.

Groth, Juszkiewicz, and Ostriker (1989) reported another approach, based on the measurement of the galaxy peculiar velocity correlation function.[20] They used optical samples of galaxy redshifts and distances drawn from Aaronson et al. (1982) and Lynden-Bell et al. (1988). Groth et al. concluded that the mean streaming flow speed is large compared to the dispersion around the mean. Ostriker and Suto (1990) put it that the flow is cool, with a large cosmic Mach number. An earlier term, applied to observations on smaller scales, is that the flow of galaxies is "quiet," that is, close to Hubble's law (e.g., Sandage and Tammann 1975). The flow is quiet even at distances ~ 10 Mpc, where the galaxy distribution is decidedly clumpy, as illustrated in Figure 2.2. This striking phenomenon is discussed further in Section 5.2.4, where it is considered as a constraint on ideas about how the present distribution of mass grew.

The Groth, Juszkiewicz, and Ostriker (1989, 564) report is that

> From our analysis of the observed large-scale flow, we conclude that at least one of the following statements must be seriously in error: (1) the velocity data are correct, (2) the standard (Davis et al. 1985) biased CDM model with $\Omega = 1$ and $b = 2.5$ is correct, or (3) the CBR frame defines the local standard of rest.

The reference is to the pioneering Davis et al. (1985) numerical simulations discussed in Section 8.3. The accumulation of evidence, of which this consideration of the "quiet" flow of the galaxies is a significant part (though not readily summarized in Table 3.2), is that statement (2) is in error: The mass density parameter proves to be well below unity with a modest degree of biasing, $b \sim 1.2$.

As with many of the density probes discussed in this section, one can ask whether this measure is biased by the difference between the distributions of galaxies and mass. Ostriker and Suto (1990, 381) put it that

> As long as the observed galaxies sample the velocity field homogeneously, the results reflect the dynamical information properly. They

20. In a statistically homogeneous and isotropic sample of galaxy positions and peculiar velocities, the mean value of products of velocity component differences v^α of the pairs of galaxies that are separated by distance components r^α defines the tensor

$$\langle v^\alpha v^\beta \rangle = \Sigma(r)\delta^{\alpha\beta} + [\Pi(r) - \Sigma(r)]\,\frac{r^\alpha r^\beta}{r^2}. \qquad (3.44)$$

The mean product of velocity components along the line connecting the galaxies is $\Pi(r)$, mean product of the velocity components in a direction normal to the line is $\Sigma(r)$. The form is familiar in other branches of physics; Davis and Peebles (1977) introduced it to extragalactic astronomy.

are not affected by how luminous objects (galaxies) trace the underlying dark mass distribution. Therefore, a possible biasing effect in galaxy formation should not affect the results.

The entry in row 19 of Table 3.2, with similar results presented lower down in the table, assumes cosmic structure grew by gravity out of primeval Gaussian adiabatic nearly scale-invariant departures from exact homogeneity. This process is discussed in Section 5.1 and incorporated in variants of the cold dark matter, or CDM, cosmology introduced in Peebles (1982b) and reviewed in Section 8.2. This is a lot of assumptions to be added to the starting one that the mass density can be estimated from the peculiar motions driven by the gravitational acceleration of the mass distribution. But these extra assumptions had already been found to be promising when this probe was applied in the late 1980s. In the following comments, let us agree that sCDM refers to the 1982 version, which assumes the Einstein–de Sitter parameters, while the more generic CDM includes such variants as a mass density less than the Einstein–de Sitter value.

Under all these assumptions, departures of the mass density from the mean are correlated at separations less than about $r_c \sim 50(\Omega_m h^2)^{-1}$ Mpc, and anticorrelated on larger scales, thus fixing the separation at the zero of the mass autcorrelation function (Peebles 1980, §92). This phenomenon would have grown by the gravitational evolution of the mass distribution through the epoch $z_{eq} = 3400$ of equality of mass densities in matter and in the sea of primeval radiation discussed in Section 4.1. The contributions of the cosmological constant and space curvature to the expansion rate at z_{eq} would have been small, so the values of Λ and space curvature would have had little effect on the spatial separation at the transition from correlation to anticorrelation.[21] And since this situation developed by the pull of gravity that would have moved the galaxies along with the mass, it seems quite reasonable to expect that the galaxy positions are correlated out to about the same scale, r_c, as the mass (as argued in footnote 18). Since measured distances to galaxies scale as h^{-1}, redshift measurements of the separation r_c at the transition constrain the product $\Omega_m h$ under all these assumptions.

Blumenthal, Dekel, and Primack (1988) were the first to apply this test. Their assessments of measures of large-scale clustering suggested that the galaxy correlation function $\xi(r)$ is positive to larger scales than would be expected in the Einstein–de Sitter model and instead would fit the density parameter entered in row 19 in Table 3.2, as usual with Hubble parameter $h = 0.7$. The title of their paper is "Very Large Scale Structure in an Open Cosmology of Cold Dark Matter and Baryons." Blumenthal, Dekel, and Primack (1988,

21. This is an aspect of the geometrical degeneracy spelled out in footnote 2 on page 334.

540) remark that this "conflicts with the dogma of cosmic inflation [but that] Another way out would be to invoke a nonzero cosmological constant."

Efstathiou, Sutherland, and Maddox (1990) measured the galaxy position correlation function $\xi(r)$ to separations r large enough for an independent and direct measurement of the scale of positive correlation of galaxy positions. They found the power law form on smaller scales seen in Figure 2.5, and they found clear evidence of the break downward from the power law toward anticorrelation at larger separations. Their conclusion was that the mass density parameter is $\Omega_m \sim 0.3$ (row 22, Table 3.2), and that this indicates the presence of Einstein's Λ. They did not mention the equally good interpretation of curved space sections without Λ.

It is worth noting the role of technology in the important result in row 22. Groth and Peebles (1977, Fig. 5) saw evidence of this transition toward anticorrelation in the Lick catalog of galaxies (Shane and Wirtanen 1967). It was more clearly seen in these data by Soneira and Peebles (1978, Fig. 4), though they did not see the significance of the break at a larger scale than expected in sCDM. The galaxies in the Lick catalog were identified by visual scans of photograph plates with a traveling microscope, largely by Donald Shane and Carl Wirtanen. Shane and Wirtanen (1967, 3) report that "the total number of images counted was 1,257,091. Due to the overlapping of the plates, many galaxies were counted more than once. The number of separate galaxies represented is 801,000."

Efstathiou, Sutherland, and Maddox (1990) used the University of Cambridge Automatic Plate Measuring machine (APM) for detection of galaxies on photographic plates. These data yielded a much better measurement of $\xi(r)$. The APM operation was a large effort, but far from the heroic work by Shane and Wirtanen. And the situation was changing yet again with the transition from photographic plates to far more efficient digital detectors.

In an independent check of the scale of correlation of galaxy positions, Saunders et al. (1991, 32) found from a catalog of redshifts of IRAS galaxies (a precursor to the PSCz catalog discussed on page 93) that "there is more structure on large scales than is predicted by the standard cold dark matter theory of galaxy formation." They did not discuss saving the CDM model by introducing a cosmological constant or space curvature.

Vogeley et al. (1992) added redshifts at greater distance to the CfA redshift catalog discussed earlier in this section. This new sample is less distant than the APM survey, but it has the great advantage of measured redshifts that suppress the noise in translating the observed angular function to the wanted spatial correlation function. They found a clear demonstration that the position correlation function has the familiar power law form at small separation but falls below the power law at larger separations. They concluded that the evidence favors "an open CDM model ($\Omega h = 0.2$)." This is entered in row 25 of Table 3.2. Einstein's cosmological constant Λ would do as well as space curvature, of course. Park, Gott, and da Costa (1992) added to the CfA sample measurements of redshifts in the southern sky. They also found $\Omega_{\rm m} h \simeq 0.2$. The Peacock and Dodds (1994) reconstruction of the primeval form of the mass fluctuation power spectrum from the measured form and the predicted nonlinear growth in the gravitational instability picture gave a similar result, $\Omega_{\rm m} h = 0.25$.

We see that the constraint on the mass density from the large scale of positive correlation of galaxy positions was well and thoroughly checked. The interpretation depends on—and tests by consistency—the assumption of initial conditions in the CDM-like models discussed in Sections 5.2.6 and 8.2. Footnote a in the compilation of mass density constraints in Table 3.2 points to this assumption.

The properties of clusters of galaxies offer important constraints on the cosmic mass density. Most analyses assume these mass concentrations grew by gravity from adiabatic nearly scale-invariant initial departures from homogeneity. With the added assumption that galaxies trace mass, the standard sCDM model with $\Omega_{\rm m} = 1$ overpredicts the cluster number density. Evrard (1989); Peebles, Daly, and Juszkiewicz (1989); Lilje (1992); and Bartlett and Silk (1993) reached this conclusion by semi-empirical arguments based on the Press-Schechter approximation (page 225) for the frequency of occurrence of the unusually large primeval mass fluctuations that would be expected to grow into rich clusters of galaxies. The problem is resolved if $\Omega_{\rm m}$ is well below unity or if the measure is biased low because the clustering of galaxies overestimates the normalization of the clustering of mass that grew into the great clusters.

Bahcall and Cen (1992) checked this by numerical N-body simulations that indicate that if $\Omega_{\rm m} = 1$, then no choice of the bias parameter b (in equation (3.42)) can account for the cluster number density and for the correlation of cluster positions. The latter is an aspect of the unacceptably short range of correlation of mass fluctuations if $\Omega_{\rm m} = 1$, as we have discussed. Bahcall and Cen showed that both problems are relieved by lowering the mass density in the CDM model to $\Omega_{\rm m} \simeq 0.2$ to 0.25 (row 23 in Table 3.2). This entry is marked by the footnote c to indicate that the argument is not sensitive to the assumption that galaxies are useful mass tracers. And note that consistency with other estimates of $\Omega_{\rm m}$ adds to the evidence that galaxies trace mass well enough for useful mass density estimates.

Yet another probe of the mass density follows from the evolution of cluster masses as a function of time or redshift. Bahcall, Fan, and Cen (1997) and Eke et al. (1998) reported numerical simulations that indicate the Einstein–de Sitter model predicts a rate of growth of cluster masses that is faster than observed. They found that this is remedied by adoption of the lower values of mass densities shown in rows 33 and 35 in Table 3.2. For a given value of the mass density, the rate of growth of clustering is slower if $\Lambda = 0$, but the effect of adding Λ is not large. The entries in the table assume the cosmologically flat case, $\Omega_m + \Omega_\Lambda = 1$. These entries also are marked by footnote c, because they are not seriously sensitive to how mass is distributed relative to galaxies.

Carlberg et al. (1997) found the mass density in row 34 of the table by applying the galaxy mass-to-light ratio M/L measured in clusters of galaxies to the mean luminosity density. The approach is similar to that of Gott and Turner (1976), who estimated the value of M/L entered in row 6 from the masses needed to gravitationally bind the groups of galaxies in their sample. Carlberg et al. considered larger systems: clusters of galaxies. This offers a cleaner case for stability and a larger source of data for a closer analysis. Carlberg et al. (1997, L10) conclude that

> The virial mass calculated from the full sample, empirically adjusted for its measurement biases, will correctly estimate the mass enclosed ... we have now tested each step in the chain of logic that supports our corrected, population-adjusted value of $\Omega_0 = 0.19 \pm 0.06$. There is no compelling evidence for any remaining systematic errors of the cluster M/L as an estimator of the field value over the redshift range of this sample.

This result does not depend on initial conditions, which is valuable. It does depend on their evidence that the mass-to-light ratios measured in the near infrared can be compared for galaxies in clusters and in the field. In support of this, Carlberg et al. observe little correlation of M/L in the r-band with color, contrary to earlier thinking that early-type galaxies are more massive than spirals.

Rows 26 and 27 in Table 3.2 are based on the assumption that clusters of galaxies are large enough to have grown by the gravitational collection of a close-to-fair sample of the ratio of baryonic to total mass, overcoming the stresses in the baryon matter that may tend to resist collection. The mass would of course include the nonbaryonic dark matter, discussed in Chapter 7. Under this assumption, the mean mass density is

$$\rho_m = \rho_{baryon} M_{tot}/M_{baryon}. \tag{3.45}$$

The baryon mass density ρ_{baryon} may be constrained by the condition that the nucleosynthesis of elements in the early stages of expansion of the hot big bang cosmology fits the observed light-element abundances, particularly the ratio of deuterium to hydrogen (as discussed in Sections 3.6.2 and 4.6). The mass M_{baryon} of baryons in a cluster is the sum of the plasma mass derived from the cluster X-ray spectrum and luminosity and the modest addition of baryons in stars in galaxy cluster members. The total cluster mass M_{tot} is derived from the gravitational potential needed to confine the motions of the galaxies and the pressure of the plasma. These quantities in equation (3.45) give the mean mass density. White (1991) and White and Frenk (1991) considered this argument. Briel, Henry, and Böhringer (1992) independently applied it;

they found $\Omega_m = (0.12 \pm 0.06)h^{-1/2}$, taking account only of the baryons in the intracluster plasma.[22] White et al. (1993) added the mass in stars and presented a check of the assumption that clusters contain a close-to-fair sample of ρ_{baryon}/ρ_m by numerical simulations of the gravitational dynamics of plasma and nonbaryonic dark matter in the growth of clusters in an expanding universe. The White et al. (1993, 429) conclusion is that "Either the density of the universe is less than that required for closure, or there is an error in the standard interpretation of element abundances."

White et al. (1993) termed their low value of Ω_m a "Challenge to Cosmological Orthodoxy," and indeed at the time, many felt that the Einstein–de Sitter model with $\Omega_m = 1$ is the reasonable and sensible case. But the evidence from the cluster baryon mass fraction for a lower mass density is in line with what was being revealed by the many other ways to probe the mass density chronicled in Table 3.2. And we may also note an earlier remark by White and Frenk (1991, 58) that an open universe is avoided and "The flat geometry predicted by the inflationary model could then be, rather inelegantly, preserved by introducing a cosmological constant."

This analysis of the cluster baryon mass fraction depends on the theory of nucleosynthesis in the early universe, which in turn is tested by the degree of consistency with other measures of ρ_{baryon}. It assumes clusters grew by capturing a fair sample of the ratio of baryons to subluminal matter, which White et al. (1993) checked by numerical simulations. It does not require that galaxies are good mass tracers, as indicated by the footnote c in Table 3.2. And it is worth pausing to note that the result is in line with many other measures. I mention the importance of such cross-checks of consistency throughout the book; an assembled picture is offered in Chapter 9.

Bertschinger and Dekel (1989) introduced an elegant way to derive the transverse components of the large-scale streaming velocity $\vec{v}(\vec{r})$ from the radial velocity component observed as the Doppler shift away from the Hubble relation. In the gravitational instability picture discussed in Chapter 5, the peculiar velocity field $\vec{v}(\vec{r})$ (that is, the velocity relative to the homogeneous Hubble expansion), appropriately smoothed over nonlinear clumping on smaller scales, is irrotational prior to orbit crossings. This follows from the assumption that the streaming velocity $\vec{v}(\vec{r})$ grew by gravity, which preserves irrotational flow, $\nabla \times \vec{v} = 0$. It means the velocity may be expressed as the gradient of a potential, $\vec{v} = \nabla\Phi$, hence the name later given to this approach: POTENT. In locally Cartesian coordinates, with the observed

22. It is an interesting exercise for the student to check that the dynamical estimate of the cluster mass scales with the cluster distance derived from its redshift as $M_{tot} \propto h^{-1}$, and that the mass in plasma derived from the observed X-ray flux density scales as $M_{baryon} \propto h^{-5/2}$. Since the baryon mass in stars is a small correction, the derived value of the density parameter scales about as $\Omega_m \propto h^{-1/2}$.

radial direction placed along the z-axis, the transverse components of the irrotational flow are

$$v_x = \int^z \frac{\partial v_z}{\partial x} dz, \quad v_y = \int^z \frac{\partial v_z}{\partial y} dz. \tag{3.46}$$

The divergence of this estimate of the peculiar velocity field $\vec{v}(\vec{r})$ is a measure of the departure of the mass density from the mean, in linear approximation,

$$\nabla \cdot \vec{v} = -\beta H_0 \delta_g(\vec{r}), \tag{3.47}$$

where δ_g is the fractional departure from homogeneity of the luminosity density of starlight, or of galaxy counts, as in equation (3.42), and $\beta = \Omega_m^{0.6}/b$ (equation (3.43)).

Bertschinger et al. (1990) found that application of the POTENT method yields a sensible map of the large-scale regions of high and low galaxy density out to distances $\sim 60h^{-1}$ Mpc, as indicated by the derived function $\delta_g(\vec{r})$. Dekel et al. (1993) used radial peculiar velocities derived from galaxy redshifts and distances based on the Tully-Fisher and $D_n - \sigma$ relations for late- and early-type galaxies (pages 85 and 87), which POTENT translates to the three-dimensional velocity field. The velocity field was translated to the mean mass density by an estimate of the degree of nonlinear departure from homogeneity, or else by using the 1.9-Jy IRAS redshift sample to model the mass distribution. The results indicate that Ω_m is in the range 0.5 to 3 at 95 percent confidence (row 28 in Table 3.2). The range of possible values is relatively large, and indeed, equation (3.46) requires numerical differencing of radial streaming velocity estimates that amplifies noise. Hudson et al. (1995) used the deeper 1.2-Jy IRAS redshift sample and a merged set of optical redshifts and distances to model the mass distribution. Their result in row 30 of the table agrees with $\Omega_m = 1$ at two error flags.

The reasonable degree of consistency of the two POTENT results with the Einstein–de Sitter mass density, which was the clear nonempirical favorite (Section 3.5.1), certainly attracted attention. But it was challenged by most other measurements in Table 3.2, which indicate that the density is considerably less than Einstein–de Sitter value. In particular, analyses based on fits of measured galaxy radial peculiar velocities to the peculiar gravitational acceleration of the galaxy distribution by Davis, Nusser, and Willick (1996), who used data similar to that of the POTENT analysis, and by Shaya, Tully, and Pierce (1992), who used a shallower but well-established peculiar velocity sample, argued against a mass density as large as Einstein–de Sitter.

The POTENT and Davis, Nusser, and Willick (1996) probes of the mass density are labeled "velocity and gravity fields" in the table, because they use a redshift catalog to map out the spatial distribution of the galaxies that is used to compute the gravitational acceleration as a function of position, up to the normalizing factor β. The value of β is found by fitting the gravitational acceleration to the peculiar velocities derived from the differences of recession velocities cz from cosmological redshifts H_0r for galaxies with measured distances as well as redshifts. The larger numbers of redshifts without distances enable finer resolution in the spatial distribution. Davis, Nusser, and Willick (1996) used the radial peculiar velocity sample in the Mark III catalog of 2,900 late-type galaxies with Tully-Fisher distances, and they used the 1.2-Jy IRAS redshift sample. Their "most likely value lies in the range" (Davis, Nusser, Willick 1996, 22) entered in row 32 of Table 3.2. Their exploration of why their result does not agree with POTENT, despite the similar data, yielded no conclusion other than the advice to treat with caution measurements of Ω_m based on smoothed velocity and gravity fields. But the detailed statistical analysis of velocity and gravity fields by Willick and Strauss (1998), who used the 1.2-Jy IRAS redshifts and an enlarged Mark III catalog of peculiar velocities, yielded the measurement in row 36 with the conclusion by Willick and Strauss (1998, 64) that "the data are consistent with a model in which the cosmological density parameter $\Omega \approx 0.3$ and IRAS galaxies are unbiased, $b_I = 1$." The subscript in the bias parameter b_I refers to the infrared IRAS sample, which is far less troubled by obscuration by dust in the galaxies and the Milky Way than are optical observations. The Willick and Strauss conclusion is in line with the trend to convergence of evidence that the mass density parameter is close to the value $\Omega_m \approx 0.3$.

The COBE satellite measurement of the large-scale CMB anisotropy (Smoot et al. 1992; with the 4-year data in Bennett et al. 1996) offers a normalization of the mass-fluctuation power spectrum, assuming adiabatic initial conditions and scale invariance, but without the assumption that galaxies trace mass. This was used to find measures of the mass density from cluster masses and their spatial distributions and from the comparison of galaxy density and velocity fields (as in Bahcall and Cen 1992; Bartlett and Silk 1993; Eke et al. 1996; Viana and Liddle 1996; and Willick et al. 1997). The results, under the assumption of scale-invariant initial conditions, are similar to what is found by normalization to the galaxy space distribution.

It is to be noted that the COBE microwave background anisotropy is measured on angular scales $\sim 10°$, which translates to the amplitude of primeval mass fluctuations on scales ~ 500 Mpc. It is a long extrapolation down to scales ~ 10 Mpc relevant for clusters and ~ 1 Mpc for the relative motions of galaxies. The consistency of COBE-normalized and galaxy-normalized mass density estimates at $\Omega_m \sim 0.3$ is to be taken as evidence in favor of near scale-invariance. But if one preferred $\Omega_m = 1$, it would invite exploration of an alternative, a departure from scale invariance. This is the tilted cold dark matter cosmological model discussed in Section 8.4.1.

The advances in the art of galaxy distance measurements—the Tully-Fisher indicator for spirals and the D_n-σ indicator for ellipticals—allowed the Feldman et al. (2003) measurement of the mean relative radial velocities of galaxies as a function of the pair separation. They found consistent results from four carefully compiled catalogs and the estimate of Ω_m entered in row 42 of Table 3.2.

The last of the entries in Table 3.2 to be discussed is the Percival et al. (2001) report of a reasonably clear detection of the effect of acoustic, or pressure, oscillations of the plasma-radiation fluid in the early universe. Percival et al. (2001) used 2dFGRS data similar to that for Panel (c) in Figure 3.3, here a sample of 160,000 redshifts, for the mass density in row 40 in the table. The reality of their detection was checked by Cole et al. (2005), who showed reasonable consistency with their clear detection of the effect of acoustic oscillations in the galaxy distribution derived from a version of the 2dFGRS sample with 221,414 redshifts.

The phenomenon is termed "BAO," for baryon acoustic oscillations. The oscillations produce a pattern in the galaxy distribution that appears as distinctive ripples in the power spectrum of the large-scale spatial distribution. The origin of the pattern is illustrated in Figure 5.2 in Section 5.1.3. The effect also appears as a pattern in the angular distribution of the thermal sea of radiation remnant from the early universe (to be discussed in Chapter 4). All this depends on the assumption that cosmic structure grew by gravity out of the adiabatic initial conditions discussed in Section 5.2.6, as applied in the CDM cosmological model and its variants introduced in Chapter 8. The assumption of all these elements is tested and well supported by the detection of the patterns. And the patterns point to the standard low mass density.

The BAO effect produces a train of waves in the power spectrum of the galaxy distribution that translates to a peak in its Fourier transform, the galaxy position two-point correlation function.[23] Eisenstein et al. (2005) found a clear and convincing demonstration of the BAO peak in the galaxy position correlation function. They used data from the Sloan Digital Sky Survey, an important part of the progress in cosmology after the revolution.

3.6.5 MASS DENSITY MEASUREMENTS: ASSESSMENTS

Let us consider now assessments of the results from the rich variety of lines of research reviewed in Section 3.6.4. Figure 3.4 shows the measurements of

23. Bashinsky and Bertschinger (2002) showed that one can think of the peak in terms of a Green's function. In linear perturbation theory, the primeval departures from homogeneity can be represented as an integral over single pointlike departures from homogeneity. The disturbance to the plasma-radiation fluid discussed in Section 5.1.3 from one of these pointlike departures propagates away from the point as a spherical wave that ends at dynamical decoupling of plasma and radiation at the redshift $z_{\rm dec}$ in equation (4.13). The distance traveled sets a characteristic length: the diameter of the spherical wave in the Green's function. This sets a peak in the position correlation function.

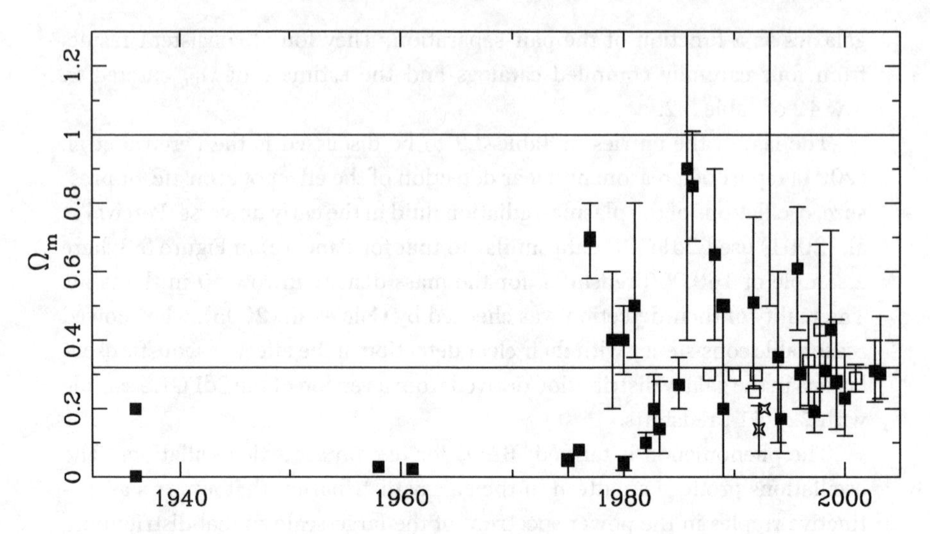

FIGURE 3.4. The history of measurements of the mass density parameter Ω_m, under the assumption that galaxies usefully trace mass.

Ω_m listed in Table 3.2 as a function of the year of publication. The measurements based on dynamics, within the assumptions of the relativistic model of the expanding universe, are plotted as filled squares. The two measures plotted as open crosses add the assumption that the baryon mass density is usefully constrained by the hot big bang theory of light-element formation that is reviewed in Chapter 4. The results plotted as open squares assume elements of the CDM cosmological picture introduced in Section 8.2. These open symbols are marked by the footnotes in Table 3.2. The upper horizontal line in the figure marks the Einstein–de Sitter mass density, $\Omega_m = 1$. This was the community favorite from the early 1980s through to the late 1990s. The lower horizontal line is the mass density in the cosmology established at the turn of the century.

It is notable that Hubble's (1936) larger estimate, based on the mean mass per galaxy in clusters of galaxies, is reasonably close to the value established much later. He had a measure of the mass in dark matter around galaxies from the condition that the mass must be large enough that gravity contains the motions of the galaxies in clusters, because much of the dark matter is draped around the galaxies. Hubble's lower estimate makes sense too, as a measure of the mass in stars.

Fall's (1975) pioneering statistical approach also is sensitive to dark matter, because this mass is driving the large-scale velocity field in his measure. But his estimate depends on the degree of clustering of mass on larger scales that might be traced by the clustering of galaxies, which was not yet well measured. Gott and Turner (1976) had a reasonable account of the mass in dark matter around galaxies from their consideration of the mass needed to hold groups of galaxies together, but their estimate of Ω_m is low, because they underestimated the mean luminosity density. Carlberg et al. (1997) had better data for their application of this approach, which yielded $\Omega_m = 0.19 \pm 0.06$, reasonably close to the standard fixed a few years later.

Figure 3.4 shows that the scatter among measurements of Ω_m was shrinking toward the turn of the century, approaching $\Omega_m \sim 0.3$. This is part of the convergence of evidence that drove the community to the consensus cosmology. One might wonder whether another reason for the decreasing scatter was that the value of Ω_m was no longer expected to be unity but instead was becoming "known" to be somewhere near 0.3. There may be something to this suspicion; there are many examples of the unconscious reluctance to stray outside accepted error flags. But there also was a clear objective factor leading in this direction: The samples of galaxy redshifts and distances were growing larger and better.

Some of the measurements in the mid-1980s to mid-1990s contain the Einstein–de Sitter mass density, and the error flags of others are approaching it. Papers and discussions of these and related estimates presented at the 1993 conference, "Cosmic Velocity Fields," led the conference summary speaker, Sandra Faber (1993, 491), to conclude that "a major highlight of this meeting — why it may even someday be remembered as a watershed — is that so many people with so many different methods said for the first time that Ω might actually be close to 1." Faber is a knowledgeable astronomer, and this was a serious remark. It was exciting for those who felt that $\Omega_m = 1$ surely is the likely result, for the reasons reviewed in Section 3.5. But the evidence largely came from the POTENT method, which is elegant but difficult to apply. And we see that the POTENT measurements are not seriously inconsistent with the considerable variety of other kinds of measurements that tend to fall well below $\Omega_m = 1$. Thus in the mid-1990s, the Bahcall, Lubin, and Dorman (1995, L84) assessment of the broader array of estimates led them to conclude that the evidence "suggests a low mass-density universe: $\Omega \simeq 0.15 - 0.2$, with most of the dark matter residing in large galaxy halos."

Figure 3.5 shows the results in Table 3.2 sorted by the categories of measurements listed in Table 3.3. The second column in the table offers rough estimates of the range of length scales sampled. The row numbers in the entries in Table 3.2 in each of the categories are entered at the end of each row of Table 3.5.

The measurements of Ω_m in category A are based on the reduction of redshift survey data to the galaxy two-point correlation function in redshift space (equation (3.38)). Those labeled A_1 sample relative velocities of galaxies at smaller separations, in the range of a few hundred kiloparsecs to a few megaparsecs, and they are weighted to galaxies outside rich clusters. These mass densities assume statistically stable clustering on what are relatively small scales. The main early samples that are dense enough for use in A_1 are the

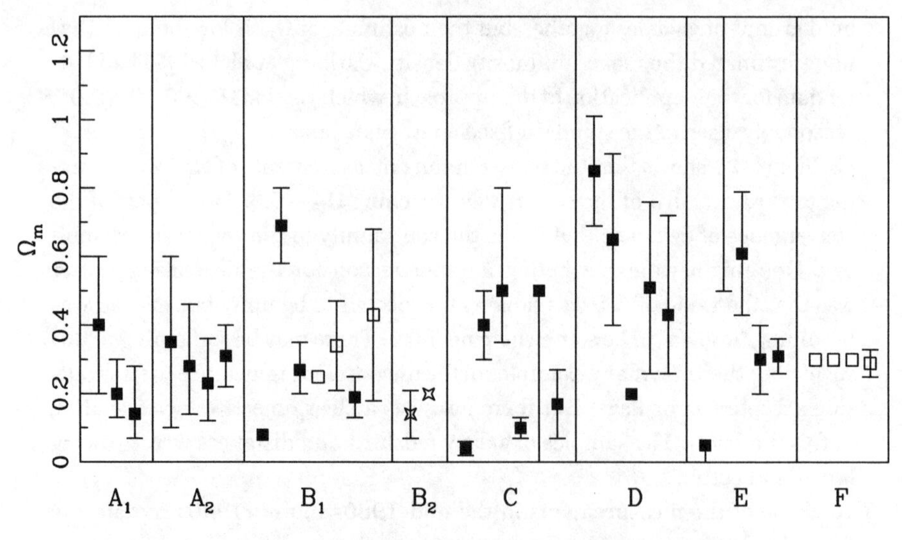

FIGURE 3.5. Measurements of Ω_m sorted by the categories listed in Table 3.3.

Table 3.3. Categories of Mass Density Probes

Category	Ω_m	Comment
A_1	0.2–2 Mpc	field galaxy relative velocity dispersion: 8, 13, 14[a]
A_2	3–30 Mpc	mean flow convergence: 29, 38, 39, 41
B_1	1 Mpc	cluster masses, distribution, evolution: 6, 7, 16, 23, 33, 34, 35
B_2	1 Mpc	cluster baryon mass fraction: 26, 27
C	10–30 Mpc	Virgocentric and clustercentric flow: 9, 10, 11, 12, 20, 31
D	30–50 Mpc	motion of the Local Group: 17, 18, 21, 24, 37
E	10–40 Mpc	galaxy velocity and gravity fields: 5, 28, 30, 32, 36
F	50–100 Mpc	large-scale correlation; BAO: 19, 22, 25, 40

[a] The final numbers in each row refer to the row numbers in Table 3.2.

CfA survey to $\sim 40h^{-1}$ Mpc distance (row 13 in Table 3.2) and the redshifts in five smaller well-separated fields to about three times that depth (row 14). The estimate in row 8 would not stand alone, but its consistency with the other two estimates, and the consistency of the measure of Ω_m over an order of magnitude spread of galaxy separations in row 13, argue for a reliable measurement.

The measurements at larger separations in A_2 sample the mean flow of galaxies toward the nearest significant mass concentration, a consequence of the gravitational growth of departures from a homogeneous mass distribution. Figure 3.3 illustrates the progress in detecting and measuring this phenomenon in the galaxy two-point correlation function in redshift space. In the early 1980s, the CfA sample was too small to show the effect; it was detected in the early 1990s, and the effect was well and clearly mapped out by the turn of the century. The results in both parts of category A consistently point to mass density in the neighborhood of $\Omega_m \sim 0.3$, sampled over the considerable range of scales from ~ 0.3 Mpc to ~ 30 Mpc.

As I have been arguing, this insensitivity of the apparent value of Ω_m to the length scale sampled is consistent with the conclusion that galaxies are useful mass tracers.

Category B in Table 3.3 is the measures based on clusters of galaxies. They are observed on length scales comparable to A_1 but in regions of much greater density. The measures in B_1 sample cluster mass values and evolution along with the spatial correlation of cluster positions. All but row 34 in Table 3.2 assume assume CDM-like initial conditions. The computations of the predicted evolution from these initial conditions to the formation of clusters of galaxies seem reliable, and the consistency internally and with other mass density probes is evidence that these initial conditions are a useful approximation to reality. The mass density estimate in B_2 (Table 3.3) from the cluster baryon mass fraction assumes the theory of light-element production in the hot big bang. Other than row 7 in Table 3.2, which is superseded by row 16, the entries in B scatter around $\Omega_m \sim 0.25$.

The probes in B_2 and those from cluster evolution and spatial correlation in B_1 are not sensitive to the assumption that galaxies trace mass. The similarity to the evidence from A, which does depend on this assumption, is a valuable check.

The estimates in category C, from Virgocentric flow and the Regős and Geller detection of the same effect around other clusters, scatter more broadly. This is largely a result of the difficulty of measuring the degree of clustering of galaxies in the Local Supercluster around the Virgo cluster. I expect measurements of the redshift space cluster-galaxy cross correlation function based on more recent and much better samples, together with the cluster-mass function from gravitational lensing, would considerably improve this measurement. There is now a well-supported value for Ω_m, but as always, a new check would be valued.

In the standard cosmology, the motion of the Local Group relative to the sea of microwave radiation was produced by the gravitational acceleration caused by the departure from a homogeneous mass distribution. The acceleration is dominated by mass density fluctuations of small amplitude on large scales, which are difficult to measure. In particular, the peculiar-velocity estimates are subject to systematic error from the difficulty of keeping the distance standard constant across the sky. This is the likely cause of the large scatter of estimates of Ω_m in Category d. Later developments in this approach are reviewed on page 96.

The analyses in category E in Table 3.3 use redshift samples to map the galaxy spatial distribution, which is used to compute the peculiar gravitational acceleration up to the normalizing factor, the mean mass density, which is adjusted to fit galaxy peculiar velocities as a function of position derived from measures of galaxy distances and redshifts. Again, the measurement is difficult, and the scatter of estimates of Ω_m is large.

Category F includes three measurements of the scale of transition from positive to negative correlation of galaxy positions. The distances are large, but the phenomenon is distinctive and was clearly detected by 1990. The interpretation relies on the assumption of nearly scale-invariant adiabatic primeval departures from homogeneity. This passes the test from the COBE-normalized probes discussed on page 104. It assumes galaxy positions are positively correlated to the same separation as the mass distribution. I argue in footnote 18 on page 94 that this assumption seems reasonable.

I take the revolution in cosmology to have ended in the year 2003, but I certainly do not mean to suggest that progress in the cosmological test stopped then. In particular, I put in category F the first detection of the BAO phenomenon, the remnant effect of acoustic oscillation of the coupled plasma and radiation in the early universe. This detection was an important but still modest start. The great progress in BAO measurements since then is illustrated in the bottom figure in Plate II.

All of this discussion assumes Newtonian mechanics, but again the reasonable degree of consistency of results from its applications to a considerable range of situations argues that Newtonian mechanics has proved to be a useful approximation.

The slow accumulation of these data, with the inevitable uncertainties and systematic errors—and the reasonable feeling that the mass density likely is Einstein–de Sitter—tended to obscure the key point: the development of the pattern seen in Figure 3.5. Few of the estimates are consistent with the Einstein–de Sitter prediction $\Omega_m = 1$; most instead indicate the mean density is about a third of Einstein–de Sitter. The dynamical estimates in categories A_1 and A_2 and C–E assume galaxies are useful mass tracers. In the mid-1980s through the mid-1990s, it was easy to imagine that galaxies could be biased. It was less easy to see how the bias could be so similar on scales ranging from about 200 kpc, where galaxy clustering is seriously nonlinear, to about 50 Mpc, where the fluctuations around the mean are small. Also difficult to explain is how probes that depend on different assumptions could be so consistently low if the mass density were Einstein–de Sitter.

The straightforward conclusion is that the evidence accumulated by the turn of the century from the considerable variety of mass estimates summarized in Figure 3.5 and Table 3.3 makes a good case that the cosmic mean mass density is about a third of the Einstein–de Sitter value. The philosophy behind this assessment might be expressed by an adaption of a comment by Einstein: nature is subtle but not usually malicious. It would be remarkable to find that this array of estimates by some eight different methods that sample a broad range of scales is so systematically biased low. But nature can be so subtle as to seem to border on malicious on occasion, so it is important that there are still more constraints on Ω_m from the other test to be considered in the rest of this book.

3.7 *Concluding Remarks*

The elegance of the 1948 steady-state cosmology is perhaps better appreciated in hindsight. Its very few assumptions—standard local physics in a spacetime that is globally stable to homogeneous, continual, and steady creation of matter—yielded predictions that offered fixed targets for observations. That was an important stimulus to research when there was a meager empirical basis for this subject. A prime example is Ryle's program of radio source counts, with the important unintended contribution to early empirical evidence of Einstein's cosmological principle (Section 2.3). A prime example of the capricious ways of research in natural science is the steady-state community's neglect of Gamow's (1954a) challenge: account for the steady-state prediction that the ages of nearby galaxies spread by a factor of 20 between the youngest 10 percent and the oldest 10 percent. As Gamow remarked, this does not seem to be at all in line with the observed near-uniformity of spectra of nearby galaxies. As it happened, interest in the 1948 steady-state model waned in the mid-1960s for other reasons, most notably the pressing challenge to be discussed in Chapter 4: account for the presence of the near-uniform, near-thermal sea of microwave radiation.

In the 1960s, most people active in the small community of empirical cosmologists chose to frame their research in terms of the Friedman-Lemaître solution to Einstein's field equation in general relativity. And some were taken by the thought that it might be possible to determine how the world will end, whether the mass of the universe is large enough to (theoretically) stop the expansion, whether the world might end in a big crunch or a big freeze. In hindsight, we may count it as curious that there were not many complaints about the exceedingly modest empirical basis for this application of general relativity on extragalactic scales. This assessment was implicitly pragmatic: In the 1960s, it was not evident that one could hope to do much better.

The promise of an improving empirical situation in the 1970s is illustrated by Figure 3.2 on page 76. Gott et al. (1974) could assemble five independent constraints from measurements of quite different phenomena to be fitted by the two free parameters in their model, the cosmic mean mass density and the cosmic expansion rate. If at this point constraints and parameters had proved to be inconsistent, we may be reasonably sure that the problem would have been advertised. The demonstration of consistency was not much celebrated, I suppose because the community by and large expected it, for nonempirical reasons. But this result, though modest, is to be counted as an early positive outcome of a test of the relativistic hot big bang theory.

The first thorough examination of the relativistic hot big bang cosmological model was the great program of measurements of the cosmic mean mass density. The list of probes of the mass density in Table 3.2 is not complete, but it is a large enough sample to show the effort that was devoted to compilations of galaxy types, environments, angular positions, redshifts, and distances. It also demonstrates the variety and ingenuity of analytic, numerical, and statistical methods devised to reduce these data to measures of the mass density. Though the test of cosmology was the nominal goal for this program, I suspect that, for many, the main attraction was that the search for meaningful measures of the mass density is scientifically interesting and just possibly feasible. It was essential to this enterprise that there are a lot of galaxies that could be characterized and their angular positions, redshifts, and distances measured with the increasing precision and numbers allowed by advances in technology. Consider, for example, that Alam et al. (2017, 2617) report that their "final sample includes 1.2 million massive galaxies over 9329 $\deg^2$ covering $0.2 < z < 0.75$." This is an attractive situation to those inclined to take up such serious challenges, and we see that such people do exist.

Might the community in the 1970s have turned instead to a program of probes of time scales derived from stellar evolution theory; radioactive decay ages; histories of element formation; explorations of the properties of stars, supernovae, AGNs, star clusters, and galaxies; for the purpose of establishing the distance scale and comparing astronomical and cosmological scales of time and distance? This is a rich topic for research; there are many stars and quite a few nearby galaxies to study in considerable detail. We would have learned a lot from more systematic analyses of these phenomena in the 1970s and 1980s. But the greater attention to the mass density seems to have been inevitable, because data in the literature in the 1970s already allowed reasonably interesting statistical analyses of galaxy dynamics, and telescopes of modest size with efficient detectors could be used to get much more. There were useful data on time scales in the 1970s, but I think it was less abundant, less readily improved, and so less likely to have attracted more intensive research.

Growing in parallel with the search for the value of the mean mass density was the search for properties of the sea of microwave radiation reviewed in Chapter 4. The two lines of research interacted, because the theoretical interpretations are related, as already considered for the COBE anisotropy. But the interaction became really serious at the turn of the century, when the constraints from these and other investigations were seen to be converging.

Research along the lines of the last of the entries in Table 3.2 remained active and productive after the revolution, along with other directions of research to be discussed. The difference was that the community had settled on a consensus normal cosmology. Before the revolution, it was interesting to cast about for viable ideas and varieties of parameters that might better fit the data. An example was the fashion of indicating how measurements of Ω_m

would be affected by a systematic difference between the distributions of galaxies and mass in the linear bias model in equations (3.42) and (3.43). If galaxies had not proved to be useful mass tracers, it seems doubtful that the situation would have been as simply described as in this biasing model, but the model was important in another way, a reminder of an open issue. Of course galaxies are biased tracers of mass, but the effect proves to be at a level too modest to be seen in Figure 3.5. Bias remains a consideration, now most important for tests of the theory of how galaxies formed.

Might community opinion have converged on the conclusion that the mass density is well below Einstein–de Sitter in the absence of the evidence for this result from the redshift-magnitude and CMB anisotropy measurements that reached fruition at the turn of the century? Consensus would have been difficult, because this earlier evidence was broadly scattered through the literature and salted with systematic errors. But I conclude that the case was there.

CHAPTER FOUR

Fossils: Microwave Radiation and Light Elements

IF OUR UNIVERSE expanded from a big bang, then one might expect to find fossils left from the very different conditions in the universe as it used to be. Lemaître (1931e) recognized this possibility, and suggested the cosmic rays might be such a fossil. That now seems unlikely, but we have other ideas.

Space is filled with a nearly uniform sea of radiation, the cosmic microwave background radiation, usually termed the CMB. It is detected at wavelengths from millimeters to tens of centimeters. The intensity spectrum, energy as a function of wavelength, is very close to that of radiation that has relaxed to thermal equilibrium. The universe now is transparent at CMB wavelengths, which means it cannot have forced radiation to relax to this distinctive thermal condition. The thermal spectrum thus makes an excellent case that this radiation is remnant from a time when our universe was quite different from now, dense and hot enough to have caused relaxation to nearly exact thermal equilibrium.

The thought that our universe is evolving is remarkable, and the scientific case for it is persuasive only if supported by considerable well-checked evidence. The detection of this radiation and an accompanying fossil—helium—were important to building this case.

The story of how the CMB was recognized to be a fossil from the early universe includes the recognition that the abundance of helium seemed to be much larger than expected from production in stars, and that it instead might be a remnant from well-understood thermonuclear reactions operating in the early hot and dense stages of expansion of the universe. Helium formation in the hot big bang would have been accompanied by deuterium, the stable heavy isotope of hydrogen. The amount of deuterium left over from the early thermonuclear reactions would depend on the density of ordinary baryonic matter: the greater the baryon density, the more complete the nuclear reactions that converted deuterons to helium nuclei and the lower the predicted residual deuterium abundance. Accounting for the abundances of helium and deuterium requires that the baryon mass density is less than the mass density indicated by the measures summarized in Figure 3.5. That calls for a postulate: most mass is not baryonic, or at least acts that way. The standard model includes the hypothetical nonbaryonic dark matter discussed in Chapter 7. This would be an awkward case of adjusting parameters to fit the measurements were it not for the fact that the baryon mass density required to fit the deuterium abundance also fits the constraint from the detected pattern in the angular distribution of the CMB discussed in Chapter 9. This is part of the long path to the establishment of a standard cosmology.

The behavior of a sea of microwave radiation in an expanding universe is reviewed in Section 4.1. Section 4.2 offers considerations of how George Gamow and his colleagues, Ralph Alpher and Robert Herman, hit on the main elements of the hot big bang cosmology, including the sea of microwave radiation and the large helium abundance, but failed to capture the interest of the community. Section 4.3 reviews how it came to be seen that the abundance of helium is much larger than expected from production in stars but is readily understood as the result of thermonuclear reactions in the hot big bang cosmology. This attracted little attention prior to the recognition of a second fossil: the sea of microwave radiation. As recalled in Section 4.4, this discovery was widely advertised, and its interpretation as a fossil from the early universe rightly questioned: Might it have been produced in objects in the universe as it is now? The key demonstration that the radiation intensity spectrum is very close to thermal, and so almost certainly not from local sources, is discussed in Section 4.5. This chapter concludes with the steps to a persuasive measurement of the primeval abundance of deuterium and the implied baryon mass density (Section 4.6), followed by an assessment of lessons from this story in Section 4.7.[1]

4.1 *Thermal Radiation in an Expanding Universe*

The distinctive property of the cosmic sea of microwave radiation, the CMB, is its close-to-thermal energy spectrum. The starting idea about how to account for this traces back to Tolman's (1934a) demonstration that free thermal radiation in a homogeneous and isotropic universe cools as the universe expands, but the thermal spectrum of the radiation is preserved without the need to

1. The developments in Gamow's group are reviewed in greater detail in Peebles (2014). More details of subsequent developments, with recollections by people who were there, are presented in the book, *Finding the Big Bang* (Peebles, Page, and Partridge 2009). Dicke's role in this story is discussed at length in the paper *Robert Dicke and the Naissance of Experimental Gravity Physics, 1957–1967* (Peebles 2017).

postulate the traditional thermalizing grains of dust. So let us consider the behavior of a uniform sea of radiation in an expanding universe.

In a conversation at Princeton University in 1964, Robert Dicke told Peter Roll, David Wilkinson, and me, junior members of his Gravity Research Group, to think of a box with perfectly reflecting walls inside and out. The box is expanding with the general expansion of the universe. There is a homogeneous freely propagating sea of thermal radiation inside and outside the box. The box is expanding with the universe. The walls reflect incoming photons, but that is compensated by photons in the box that are reflected back inside, so the presence of the box has negligible effect on the radiation. And Dicke said that everyone knows the expansion of the box cools the radiation it contains. This cooling must happen wherever the box is placed, or even if there is no box. That is, the radiation is cooling everywhere, as befits a homogeneous universe.

To see how the expansion affects the radiation energy spectrum, consider the photon occupation number $\mathcal{N}$ in a mode of oscillation of the electromagnetic field in the box.[2] At thermal equilibrium at temperature T, the occupation number—the mean number of photons—in a mode with wavelength λ is given by Planck's expression:

$$\mathcal{N} = \frac{1}{e^{hc/(k_B T\lambda)} - 1}. \tag{4.1}$$

Boltzmann's constant is written as k_B (to avoid confusion with the wavenumber k). The expansion of the universe is measured by the parameter $a(t)$ (as in eqs. (2.3) and (3.1)–(3.13)). The box is expanding with the universe, so its width, and the mode wavelength, are increasing in proportion to $a(t)$. If the expansion rate $\dot{a}/a$ is much less than the mode frequency (it is very much less at interesting frequencies, unless we trace back to the exceedingly early universe), the occupation number $\mathcal{N}$ is conserved. Since $\mathcal{N}$ is constant, and the mode wavelength is increasing as $\lambda \propto a(t)$, the mode temperature in equation (4.1) must vary as

$$T \propto \lambda^{-1} \propto a(t)^{-1}. \tag{4.2}$$

Since this mode temperature changes in the same way for all mode wavelengths, we see that freely propagating radiation remains thermal as the universe expands and the radiation cools.

The volume of the box is increasing as $V \propto a(t)^3$. If the universe contains particles with number density $n(t)$, and the particles are not created or destroyed, then the particle number density has to evolve as $n(t) \propto V^{-1}$

2. To be technical, the modes of oscillation of the electromagnetic field are discrete inside the box, and continuous or exceedingly close to it outside the box. But we can take the box to be large enough that the effect of discreteness is quite negligible at interesting wavelengths.

$\propto a(t)^{-3}$. Using equation (4.2), we see that the temperature and density are related by an equation familiar from thermodynamics with radiation pressure one third of the radiation energy density:

$$T \propto n^{1/3}. \tag{4.3}$$

The Planck thermal blackbody radiation intensity spectrum—the radiation energy per unit volume and unit increment of frequency ν at temperature T (measured from absolute zero, in Kelvins)—is

$$\frac{du}{d\nu} = \frac{8\pi h\nu^3}{c^3} \frac{1}{e^{h\nu/k_B T} - 1}, \tag{4.4}$$

where Planck's constant is $h = 2\pi\hbar$. The Rayleigh-Jeans expression valid at long wavelengths is

$$\frac{du}{d\nu} = \frac{8\pi k_B T \nu^2}{c^3} \quad \text{at } h\nu = \frac{hc}{\lambda} \ll k_B T. \tag{4.5}$$

The Planck expression at temperature T places half the energy at wavelengths greater than

$$\lambda_h = 0.29 \frac{hc}{k_B T} = 1.5 \text{ mm at } T_f = 2.725 \text{ K}, \tag{4.6}$$

at the measured CMB temperature T_f. This wavelength is in the range of microwaves, 1 mm $< \lambda <$ 100 cm. At wavelengths $\lambda \sim 3$ mm and longer, the atmosphere is transparent enough in bands of wavelength to allow ground-based detection of this sea of microwave radiation. Measurements at shorter wavelengths require lifting the detector above much or all of the atmosphere. Results obtained in the 1990s are displayed in Figure 4.7 on page 174.

There is excellent reason to expect that a thermal intensity spectrum agrees with Planck's prediction, the blackbody function in equation (4.4), but the last experimental tests were done in the 1920s (as reviewed by Crovini and Galgani 1984) and for completeness ought to be repeated with the much better technology developed since then. In practice, most CMB intensity spectrum measurements compare the CMB energy flux density to calibration sources that are very close to black and are at known temperatures. Thus it is correct to say that the sea of microwave radiation is observed to be thermal to the accuracy of tight measurements.

The expansion of the universe causes of free nonrelativistic matter and the temperature of radiation to evolve as[3]

$$T_{\rm m} \propto a(t)^{-2}, \quad T \propto a(t)^{-1}. \tag{4.7}$$

3. The considerations that led to equation (4.2) can be generalized by considering the de Broglie wavelength of a freely moving particle, whether relativistic or not. In the expanding box, the de Broglie wavelength is stretched in proportion to the expansion

Since matter tends to cool faster than radiation, this is one of the reasons the CMB that fills our universe cannot be at complete thermal equilibrium. It can be close to it, however, because the heat capacity of the radiation observed now as the CMB is much larger than the heat capacity of the matter. The latter modeled as an ideal monatomic gas is $C_{\rm m} = \frac{3}{2}nk_{\rm B}$, where n is the particle number density. The energy density in the radiation is $\rho_{\rm r}c^2 = u_{\rm r} = a_{\rm S}T^4$, where $a_{\rm S}$ is Stefan's constant (and the subscript avoids confusion with the expansion parameter $a(t)$), so the heat capacity of the radiation is $C_{\rm r} = 4a_{\rm S}T^3$. The ratio of heat capacities is then

$$\frac{C_{\rm m}}{C_{\rm r}} = \frac{3nk_{\rm B}}{8a_{\rm S}T^3} \sim 10^{-10}. \tag{4.8}$$

The numerical value assumes baryon number density comparable to the measured value, $n \sim 10^{-6}$ cm^{-3}. This ratio is nearly independent of time and is large enough that the interaction of matter and radiation need not have had much effect on the radiation. But the binding energy available from nuclear burning of hydrogen in stars is comparable to the energy in the CMB and could have seriously disturbed it.

The conversion of hydrogen to heavier elements by nuclear reactions in stars releases mass fraction ~ 0.7 percent as radiation and neutrinos. Some of this is lost to the kinetic energy of violent stellar winds and the neutrinos produced in nuclear reactions in stars, but we can ignore such losses for an order-of-magnitude estimate. Imagine a fraction f of the hydrogen with present number density $n \sim 10^{-6}$ cm^{-3} had been converted to heavy elements in nuclear reactions in an early generation of stars at redshift $z_{\rm s}$. Suppose these stars produced enough dust to have absorbed their starlight and reradiated the energy at wavelengths that were redshifted to microwave frequencies at the present epoch. This would have perturbed the energy density $u_{\rm r}$ observed at microwave wavelengths by the fractional amount

$$\frac{\delta u}{u_{\rm r}} \sim \frac{0.007 f \rho_{\rm m} c^2}{a_{\rm S}T^4} \sim \frac{10f}{1+z_{\rm s}}. \tag{4.9}$$

If all this had happened at a modest value of $z_{\rm s}$ and converted an appreciable fraction f of the hydrogen to heavy elements, the CMB intensity spectrum could have been seriously disturbed. But we know this did not happen, because

---

parameter $a(t)$, and since the canonical momentum is inversely proportional to the de Broglie wavelength, the momentum decreases as $p \propto a(t)^{-1}$ as the universe expands. In nonrelativistic matter, the occupation number in equation (4.1) is replaced by a constant times $\exp(-p^2/(2mk_{\rm B}T_{\rm m}))$, where m is the particle mass. So we see that for freely moving nonrelativistic matter, the kinetic temperature scales as $T_{\rm m} \propto a(t)^{-2}$. This takes account only of the kinetic energy of translation. Internal energy, such as rotation and vibration of molecules, would not evolve in the absence of coupling to the radiation or interactions among particles that change internal states.

the measurements discussed Section 4.5.4 show that the CMB spectrum is quite close to thermal.

There has been significant release of energy from stars and by relativistic collapse, as in the formation of the compact masses in the centers of galaxies (known as active galactic nuclei, or AGNs, which likely are centered on massive black holes). As it happens—and to our good fortune for the goal of testing cosmologies—this radiation is present at shorter wavelengths. It is known as the CIB, the cosmic infrared background; the science is reviewed by Hauser and Dwek (2001).

Note also that the universe is not exactly homogeneous and isotropic, so the radiation we observe must be a mixture of slightly different temperatures. It is an important indication of the nature of our universe that this temperature-mixing effect is not large enough to have been detected in the intensity spectrum (as discussed in Sec. 4.5.4). By this measure, the universe is quite close to exact homogeneity and isotropy. (A small variation of the thermal temperature across the sky is detected. This measure of the departure from homogeneity is the subject of Section 9.2.)

The CMB sets two interesting epochs in the evolution of the universe: the (1) redshift $z_{\rm eq}$ at which the mass densities in matter and the CMB are equal, and (2) the redshift $z_{\rm dec}$ at which the primeval plasma combined to neutral atoms, leaving traces of free electrons and molecular hydrogen. The redshift $z_{\rm eq}$ sets the time at which gravity can start to grow mass concentrations that can become galaxies. The effect is implicit in Lifshitz's (1946) analysis of the gravitational instability of the expanding universe, but not worked out. Gamow (1948a) recognized this effect. He knew Lifshitz's paper, which contains hints to it. Or maybe this is an example of Gamow's intuition. The other redshift, $z_{\rm dec}$, sets the time at which the baryonic matter is released from drag by the radiation, and gravity can start gathering baryons to begin structure formation, as Peebles (1965) pointed out.

The ratio of mean mass densities in the thermal radiation and matter is

$$\frac{\rho_{\rm r}}{\rho_{\rm m}} = \frac{a_{\rm S}T^4}{\rho_{\rm m}c^2} = \frac{2.5\times 10^{-5}}{\Omega_{\rm m}h^2}(1+z). \tag{4.10}$$

The third expression uses the CMB temperature in equation (4.6) and the dimensionless matter density and Hubble parameters $\Omega_{\rm m}$ and h in equations (3.13) and (3.15). In the 1960s, the product of the two was thought to be in the range $0.01 \lesssim \Omega_{\rm m}h^2 \lesssim 1$. We now have the remarkably well-established value $\Omega_{\rm m}h^2 = 0.143 \pm 0.002$ obtained from the research reviewed in Chapter 9. If the neutrinos thermally produced with the radiation in the early universe have rest masses well below 1 eV, then their presence near $z_{\rm eq}$ brings the mass density in radiation plus relativistic neutrinos to $\rho_{\rm rel}c^2 = 1.68a_{\rm S}T^4$, and the densities of relativistic and nonrelativistic matter were equal at redshift

$$z_{\rm eq} = 3400. \tag{4.11}$$

At $z > z_{\rm dec}$, the baryons in the big bang cosmology were thermally ionized by the radiative reactions

$$p + e \leftrightarrow H + \gamma, \tag{4.12}$$

for hydrogen, with similar reactions for helium and the traces of other elements. Scattering of the radiation by the free electrons, and scattering of the electrons by ions, caused the baryonic plasma and radiation to act as a fluid with high pressure (from the radiation) and viscosity (from diffusion of the radiation through the plasma). When the temperature had fallen to the point that there were too few ionizing photons to keep the baryons ionized, the primeval plasma combined to almost pure neutral atoms, and the baryonic matter decoupled from drag by the radiation.

The thermal equilibrium ionization set by the balance of rates of photodissociation and radiative formation of atoms is expressed by the Saha relation.[4] At the number densities of atoms in situations of interest here, the balance swings quite abruptly from near fully ionized plasma to near fully neutral hydrogen atoms as the temperature moves through $T_{\rm dec} \simeq 3000$ K (and a little higher temperature for helium).

Computation of the growth of the neutral fraction when the Saha relation allows it is complicated by the effect of the recombination photons. Capture of an electron to the ground level of a hydrogen atom releases a photon carrying the binding energy, which can ionize a hydrogen atom. Capture to an excited level releases a resonance photon that can place another atom in an excited level, where it is more readily photoionized. But some of these ionizing and recombination line photons are redshifted out of the resonance, and some are lost by radiative settling to the metastable 2s level of hydrogen, which decays by emitting two photons that have no effect on ionization. This was worked out independently by Peebles (1968) and Zel'dovich, Kurt, and Sunyaev (1968). The theory indicates that the fractional ionization is

$$x_{\rm e} = 0.5 \text{ at redshift } z_{\rm dec} \simeq 1200, \quad x_{\rm e} = 0.01 \text{ at redshift } z \simeq 900. \tag{4.13}$$

4. For the reactions in equation (4.12), the Saha relation for the ionization at thermal equilibrium is

$$\frac{n_{\rm e} n_{\rm p}}{n_{\rm h}} = \frac{(2\pi m_{\rm e} k_{\rm B} T)^{3/2}}{h^3} e^{-B/k_{\rm B}T},$$

where $n_{\rm e}$ and $n_{\rm p}$ are respectively the number densities of free electrons and protons, $n_{\rm h}$ is the number density of hydrogen atoms, and B is the atomic hydrogen binding energy. Under standard conditions at decoupling, $T \sim 3000$ K, the factor multiplying the exponential is

$$(2\pi m_{\rm e} k_{\rm B} T/h^2)^{3/2} \sim 10^{20}\ {\rm cm}^{-3}, \quad n_{\rm baryon} \sim 10^3\ {\rm cm}^{-3}.$$

The second quantity is the baryon number density extrapolated back to $z \sim 1{,}000$. The much larger value of the first density explains why $k_{\rm B}T$ is well below the binding energy B when the plasma combines and baryons and radiation decouple, and it explains why the equilibrium changes quite sharply from ionized to neutral as the temperature moves through ~ 3000 K.

The remnant abundance of free electrons and their ions figures in the formation of molecular hydrogen. This is important, because collisional excitation of molecular hydrogen followed by radiative decay would have been a primary means for baryons to dissipate energy during formation of the first generations of stars, before the stars could have produced the heavier elements that more efficiently dissipate energy (as analyzed by Abel, Bryan, and Norman 2002 and others). Radiative formation of molecular hydrogen is slow, because the two-atom system has no dipole moment. Saslaw and Zipoy (1967) pointed out that free electrons serve as catalyst in the reaction $e^- + \mathrm{H} \to \mathrm{H}^- + \gamma$ followed by $\mathrm{H} + \mathrm{H}^- \to \mathrm{H}_2 + e^-$, where H^- is a negative hydrogen ion: two electrons bound to a single proton.

Two interesting features to look for in the CMB energy spectrum are produced by free electrons that scatter the radiation. If the photon energies are well below the electron rest mass, then scattering exchanges energy between photons and electrons without changing the number of photons. At thermal (statistical) equilibrium with conserved photon number density, the radiation intensity spectrum is

$$u_\nu = \frac{8\pi}{c^3} \frac{h\nu^3}{e^{(h\nu+\mu)/kT} - 1}. \tag{4.14}$$

The constant μ is the chemical potential (but for convenience, here it is written with the opposite sign from usual). If μ in this expression were negative, it would signify an approach to Bose-Einstein condensation, pushing radiation energy to long wavelengths, where it would be dissipated by plasma waves, eliminating the excess photons. If μ were positive, it would reduce the energy density at frequency $\nu < \mu/h$. The CMB spectrum measurements so far are consistent with $\mu = 0$.

In a related situation, one that is observed, the spectrum of thermal radiation passing through hotter plasma is perturbed by electron scattering that tends to shift photons toward higher frequencies. In the limit that the plasma is much hotter than the radiation, the perturbation to the photon occupation number $\mathcal{N}$ in equation (4.1) is

$$\begin{aligned} \frac{\delta\mathcal{N}}{\mathcal{N}} &= -2y \text{ at } x \ll 1, \\ &= yx^2 \text{ at } x \gg 1, \end{aligned} \tag{4.15}$$

where

$$x = \frac{h\nu}{k_\mathrm{B} T}, \quad y = \sigma_\mathrm{T} n_\mathrm{e} c\, t \, \frac{k_\mathrm{B} T_\mathrm{e}}{m_\mathrm{e} c^2}. \tag{4.16}$$

The radiation temperature is T; the plasma temperature and free electron number density are T_e and n_e, σ_T is the Thomson scattering cross section; and the radiation moved through the plasma for time t. (For simplicity, this expression ignores the complication of variable plasma temperature and density along the path through the plasma.) Weymann (1965, 1966) introduced the application of this physics in cosmology. Zel'dovich and Sunyaev (1969) introduced the notation and the application to intracluster plasma; it is known as the thermal Sunyaev-Zel'dovich effect. A derivation is in Peebles (1971a, 202–209). The effect is observed in the CMB spectrum measured in directions where the radiation has passed through the hot plasma in a cluster of galaxies. This has become a powerful tool for the discovery of clusters and the exploration of their properties. The main role of the Sunyaev-Zel'dovich effect in cosmology prior to the turn of the century was in the considerations of apparent departures from a thermal CMB spectrum, discussed in Section 4.5.3.

4.2 Gamow's Scenario

George Gamow was a genius at creative physical intuition. This is exemplified by his thoughts, to be reviewed here, about how the chemical elements may have formed in the hot early stages of expansion of the universe. Perhaps great intuition must be accompanied by indifference to details. We must consider how this aspect of his character contributed to the delayed community recognition of these ideas.

Earlier Chandrasekhar and Henrich (1942) and others had considered the thought that the abundances of the chemical elements were set in the early stages of expansion of the universe by relaxation to thermal equilibrium ratios of abundances. These analyses used the Saha relation; an example is in footnote 4 on page 120. The relation assumes that matter relaxed to equilibrium in a sea of thermal radiation, as in equation (4.12). Interesting values of abundance ratios require radiation at temperatures on the order of 10^9 K. That is, the physical situation Chandrasekhar and Henrich and others had in mind requires conditions similar to the hot big bang picture, thermal radiation and all. Chandrasekhar and Henrich (1942), along with similar studies, count as the first discussions after Tolman (1934a) of physical processes involving a sea of thermal radiation that would have cooled to become the CMB. But I have not encountered any evidence that anyone at the time noticed that the thermal radiation implicitly assumed in their considerations would still be present and might even be observable.

Gamow (1946) pointed out that in general relativity, the early stages of expansion of the universe had to have been very rapid, because the young universe would have been very dense.[5] That means analyses of element

5. In the early universe, when the mass density ρ would have to have been the dominant source term in the Friedman-Lemaître expansion equations (3.3), the expansion rate is

$$\frac{1}{a}\frac{da}{dt} = \left(\frac{8}{3}\pi G\rho\right)^{1/2}.$$

The large mass density, which includes the mass equivalent of the radiation, implies rapid expansion.

abundance ratios at thermal equilibrium are questionable. Gamow (1946, 573) put it that "*the conditions necessary for rapid nuclear reactions were existing only for a very short time*, so that it may be quite dangerous to speak about an equilibrium-state which must have been established during this period." Gamow remarked that neutrons, being electrically neutral, can enter nuclear reactions in the short time allowed by the rapid expansion of the universe. His proposal (Gamow 1946, 573) was that free neutrons "were gradually coagulating into larger and larger neutral complexes which later turned into various atomic species by subsequent processes of β-emission." This is a first step to the theory of light element production in the hot big bang.

In the last edition of Gamow's book on nuclear physics, *Theory of Atomic Nucleus and Nuclear Energy-Sources*, coauthored with Charles Louis Critchfield (Gamow and Critchfield 1949), Gamow presented a modified version of his notion of the coagulation of neutrons. He proposed that the elements were built up in the early universe by successive radiative captures of neutrons, beta decays serving to keep the elements in the valley of stability. A footnote in this book (Gamow and Critchfield 1949, 213) states that "More detailed calculations on this non-equilibrium process are being carried out by R. A. Alpher." Alpher (in Alpher and Herman 2001, 70) recalls that he and Gamow had decided on a "dissertation topic: developing Gamow's rather cursory 1946 ideas on primordial nucleosynthesis."

Alpher's dissertation results were published, before his dissertation was accepted, in Alpher, Bethe, and Gamow (1948). This paper is wrong, for interesting reasons that are reviewed in Section 4.2.3. But let us consider first George Gamow's remarkable demonstration of intuition in Gamow (1948a) that corrects the problem with the Alpher, Bethe, and Gamow version, without clearly acknowledging that there is a problem. It sets out key parts of the established hot big bang theory for the CMB along with the origin of helium and considerations of lasting interest about how the galaxies formed.

4.2.1 GAMOW'S 1948 PAPERS

Gamow's (1948a) argument starts with the assumption that the universe in its early stages of expansion contained a uniform distribution of neutrons in a sea of thermal radiation. (And, although it was not stated, he implicitly took it that the baryon number density when element buildup started was small enough that the strong inequality in footnote 4 applies.) Gamow (1948a, 505) argued that

> Since the building-up process must have started with the formation of deuterons from the primordial neutrons and the protons into which some of these neutrons have decayed, we conclude that the temperature at that time must have been of the order $T_0 \simeq 10^9$ °K (which corresponds to the dissociation energy of deuterium nuclei).

I wish Gamow had been more explicit about what he meant by "dissociation energy." It seems reasonable to expect that he remembered that the deuteron binding energy is an order of magnitude larger than the dissociation energy $k_B T_0$ associated the temperature he mentions in this paper. The evidence is that he was referring instead to the characteristic temperature from the Saha relation in statistical mechanics for the balance between radiative recombination and its time reversal, radiative dissociation, in the reactions for neutrons n, protons p, and the bound state of a neutron and proton, the deuteron d, in the reactions

$$\mathrm{n} + \mathrm{p} \leftrightarrow \mathrm{d} + \gamma. \tag{4.17}$$

At the Saha dissociation temperature, the balance in this equation swings from suppression of deuterons to their accumulation. This dissociation temperature is close to $T_0 \simeq 10^9$ K, or energy $k_B T_0 \simeq 0.1$ MeV. The value of the deuteron binding energy, 2.3 MeV, is discussed in all three editions of Gamow's book on nuclear physics, including the last one coauthored with Critchfield. This edition also reviews applications of the Saha expression (though not so named) for nuclear abundance ratios in radiative nuclear reactions similar to equation (4.17). The evidence seems clear: by dissociation energy, Gamow had in mind the condition for thermal dissociation set by the Saha relation.

The Saha relation indicates that, at temperature T larger than Gamow's critical temperature $T_0 \simeq 10^9$ K, photodissociation strongly suppresses the abundance of deuterons. The balance in equation (4.17) abruptly changes as T falls below T_0 to negligible photodissociation and accumulation of deuterons, because there are too few photons energetic enough to dissociate the deuterons. As the deuteron number density increases, more massive atomic nuclei can be formed by particle-exchange reactions such as $\mathrm{d} + \mathrm{d} \rightarrow {}^3\mathrm{He} + \mathrm{n}$ and $\mathrm{t} + \mathrm{p}$, where t is the nucleus of tritium. And this buildup can go on up to ^{4}He. These reactions are more rapid than the radiative transitions in equation (4.17) that depend on the weaker electromagnetic interaction. The growth of the elements in this picture thus would be expected to have started at Gamow's temperature T_0, as he stated.

Gamow wrote down the order-of-magnitude condition that a significant but not excessive fraction of the nucleons ends up in deuterons and heavier elements. It takes account of the interval of time Δt the reactions would have happened; the product σv of the neutron-proton relative velocity and the radiative capture cross section; and the nucleon number density $n(t_0)$ at the time element buildup commenced. Gamow's condition, which assumes comparable numbers of neutrons and protons, is

$$\sigma v \, n(t_0) \Delta t \simeq 1 \text{ at temperature } T_0 \simeq 10^9 \text{ K}. \tag{4.18}$$

If this product were much smaller, there would have been uninterestingly little element buildup; if much larger, most of the protons would have been captured in heavier elements. Gamow knew the latter would not do, because he was aware of evidence that hydrogen is the most abundant element in the universe, or at least close to it.

Let us consider now Gamow's estimates of the quantities in equation (4.18). The cosmic expansion time t_O at $T \sim T_O$ is a good measure of the exposure time: $\Delta t \sim t_O$. In the early universe, the mass density is the only important term in the Friedman-Lemaître equation (3.3) for the expansion time, and at $T \sim T_O$ the mass density in thermal radiation, $a_S T^4/c^2$, would be much larger than interesting values of the mass density in matter. So the expansion time works out to

$$t = \left(\frac{3c^2}{32\pi G a_S T^4}\right)^{1/2} \sim 200\ \text{s} \simeq \Delta t \text{ at } T = T_O \simeq 10^9\ \text{K}. \tag{4.19}$$

Gamow had an estimate of the radiative capture rate σv from Bethe (1947),

$$\sigma v \sim 10^{-20}\ \text{cm}^3\ \text{s}^{-1}, \tag{4.20}$$

which is not very sensitive to the velocity v. This quantity in Gamow's equation (4.18) gives the product of the exposure time and the nucleon number density,

$$n(t_O)\,\Delta t \sim 10^{20}\text{s cm}^{-3}. \tag{4.21}$$

With Δt from equation (4.19), we see that the baryon number density when deuterium could start to accumulate would have been

$$n(t_O) \simeq 10^{18}\ \text{cm}^{-3} \text{ at } T_O = 10^9\ \text{K}. \tag{4.22}$$

Having the number density $n(t_O)$ at temperature T_O, we can use equation (4.3) to find the temperature at a later density. Gamow (1948a) used this in his considerations of galaxy formation to be discussed in Chapter 5. Alpher and Herman (1948a) used it with an estimate of the present baryon number density to arrive at the first estimate of the present CMB temperature. The somewhat complicated story of this development is reviewed in Section 4.2.2.

Panel (a) in Figure 4.1 shows Gamow's (1948b) computation of the first step in the formation of elements heavier than hydrogen: the radiative capture of neutrons by protons to form deuterons (equation (4.17)). The vertical axis on the left-hand side shows the temperature as a function of time, passing through 10^9 K about 3 minutes after expansion from a far denser state. The vertical axis on the right-hand side of Panel (a) shows fractional numbers X of neutrons and Y of protons, with the rest of the nucleons bound up in deuterons. This differs from the usual convention, which I use from now on, that X, Y, and Z are the cosmic mass fractions in hydrogen, helium, and heavier elements, respectively. (To avoid confusion, note also that the left-hand label for Panel (c) shows abundances by particle number density, not mass fraction.)

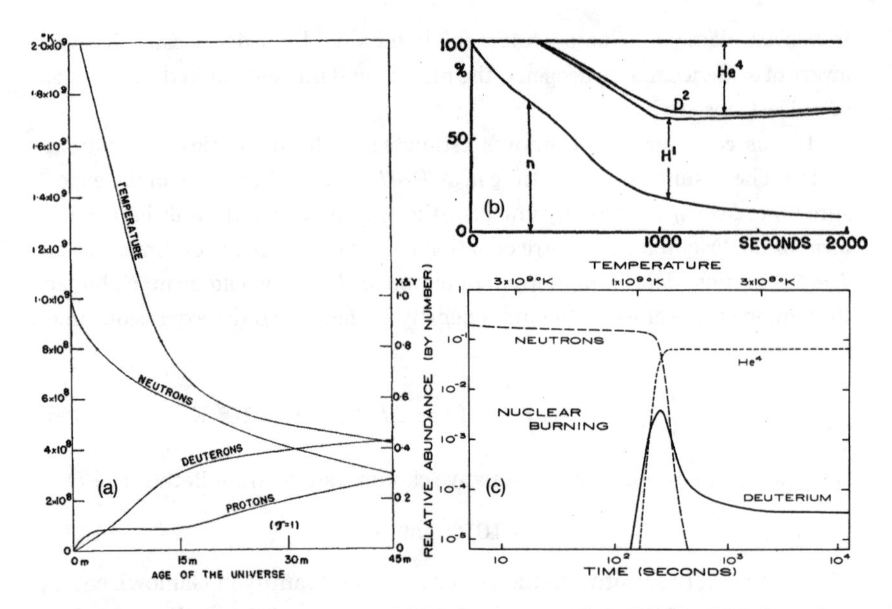

FIGURE 4.1. Gamow's scenario. Panel (a) is from Gamow (1948b). Panel (b) shows the computation by Fermi and Turkevich reported in Gamow (1949). Panel (c) is from Peebles (1966a). The time unit on the horizontal axis is minutes in Panel (a), seconds in Panels (b) and (c). Panel (a) reproduced by permission from Springer Nature. Panels (b) and (c) reproduced with the permission of AIP Publishing.

Deuterium, the stable heavy isotope of hydrogen, is not abundant. Thus we must suppose that Gamow took it, without discussion, that reactions such as $\mathrm{d} + \mathrm{d} \leftrightarrow {}^3\mathrm{He} + \mathrm{n}$ and ${}^3\mathrm{He} + \mathrm{d} \leftrightarrow {}^4\mathrm{He} + \mathrm{p}$ would be fast enough to convert the deuterons to isotopes of helium and maybe beyond that to the heavier elements. At the right-hand side of Panel (a), at 45 minutes, Gamow has it that about 30 percent of the nucleons are protons and 30 percent are neutrons, leaving 40 percent in deuterons. Allowing for a little more formation of deuterons after that, and assuming the deuterons mostly ended up as helium nuclei, we might take it that in Gamow's model, the mass fraction in hydrogen ends up at $X \approx 0.5$ with helium mass fraction $Y \sim 0.45$ and the rest in heavier elements.

Gamow (1948b, 681) stated that he chose the matter density in his model so his reaction rate equations "should yield $Y \simeq 0.5$ for $\tau \rightarrow \infty$ (since hydrogen is known to form about 50 per cent of all matter)." Here τ is a measure of the expansion time. This statement would be expected to be in his notation, meaning that in standard notation, his hydrogen mass fraction is $X \simeq 0.5$.

In his two 1948 papers, Gamow offered no explanation of why he thought the hydrogen abundance is about 50 percent. It may be significant that Gamow and Critchfield (1949) discuss Martin Schwarzschild's (1946) solar model, in which the abundances are

$$X = 0.47, \quad Y = 0.41, \quad Z = 0.12, \tag{4.23}$$

in hydrogen, helium, and heavier elements. If Gamow had this result in mind, it does not matter much whether he meant Y to be the mass fraction in hydrogen or helium: they would be comparable and well above what is in heavier elements.

Schwarzschild's (1946) solar model has a convective core below a radiative transfer envelope. The Schwarzschild, Howard, and Härm (1957) model has the now-established radiative transfer interior with a convective envelope. It indicates that, depending on the mass fraction Z in heavier elements, the helium abundance must be about $Y = 0.20$. This less than Schwarzschild's earlier result and what Gamow seems to have had in mind, but it still is a considerable mass fraction. It is in line with the abundance estimates in the 1960s that caused some to wonder where all this helium came from, and some to remember Gamow's idea, which is discussed in Section 4.3.

In his PhD dissertation, Alpher (1948a) pointed out that the absence of a reasonably long-lived isotope with atomic mass 5 is a barrier to Gamow's idea of element buildup by successive captures of neutrons. This led to a search for other nuclear reactions that might bridge the mass-5 gap. Alpher (in Alpher and Herman 1988, 30) recalls that

> As a result of attending a colloquium by Alpher in late 1948, Fermi, in collaboration with Anthony Turkevich, collected published values (or made new estimates) of cross sections for 28 thermonuclear reactions among nuclei up to atomic weight 7. Fermi and Turkevich used these cross sections together with the early-time approximations for the cosmological model to solve the rate equations among these 28 reactions, subject to the conditions that the reactions began at about 300 seconds into the expansion, when the temperature had fallen to about 0.07 MeV and photodissociation reactions would no longer be important; that the neutrons and protons were initially in the ratio 7:3, as would result from starting with neutrons only at 1 second.

Alpher and Herman (1950) explained that the starting time, 300 seconds after expansion began, is set by the critical temperature at which deuteron photodissociation becomes unimportant and deuterons can accumulate, as in equation (4.19). This is the first paper in Gamow's group to mention the Saha relation by name, here in connection with the older statistical equilibrium theory of relative element abundances. I cannot find its use in the Alpher and Herman (1950) discussion of the nonequilibrium big bang theory.

Fermi and Turkevich did not publish their computation. Gamow (1949) reported their result shown in Panel (b) in Figure 4.1, and Alpher and Herman (1950) reviewed the computation in some detail. The mass fractions in Panel (b) are plotted in a cumulative way. The computation begins at the left-hand side of the plot with only neutrons. The neutron mass fraction decreases as neutrons decay and protons accumulate, as indicated by the space labeled H^1. The mass fraction in deuterons remains small, because as they form, they are radiatively dissociated early on, and later they fuse to form helium. The computation ends at the right-hand side of the panel at mass fraction $Y \sim 0.3$ in helium with almost all the rest hydrogen, with only traces of elements heavier than helium. The latter is what Alpher (1948a) had warned might be the consequence of the gap at atomic mass 5. This computation also is an early explicit demonstration that the hot big bang picture naturally leads to a large primeval abundance of helium. Whether Gamow's Panel (a) is an earlier implicit demonstration is less clear.

In the Fermi and Turkevich computation, the residual abundance of deuterium relative to hydrogen is $\mathrm{D/H} \sim 0.01$ by number. This is not far from the computations a decade and a half later by Peebles (1966a,b) and Wagoner, Fowler, and Hoyle (1967).

The computations for Panels (a) and (b) assume that the nucleons began as neutrons that decayed to produce protons. Hayashi (1950) showed that the neutron-proton abundance ratio prior to element buildup is determined by the interaction of nucleons with the sea of thermal neutrinos and electron-positron pairs that at temperatures $T \gtrsim 10^{10}$ K are in close to thermal equilibrium with the radiation. The main reactions that determine the neutron-proton ratio are

$$\mathrm{n} + \mathrm{e}^+ \leftrightarrow \mathrm{p} + \bar{\nu}, \quad \mathrm{n} + \nu \leftrightarrow \mathrm{p} + \mathrm{e}^-, \quad \mathrm{n} \leftrightarrow \mathrm{p} + \mathrm{e}^- + \bar{\nu}. \tag{4.24}$$

Panel (c) in Figure 4.1 illustrates Gamow's scenario with Hayashi's process. Here the vertical axis is the relative abundance by number. I made this figure in 1965, shortly after recognition that there is a microwave sea that proves to be Gamow's thermal radiation. The present radiation temperature used in this figure, 3.5 K, is from Penzias and Wilson (1965). The present mass density, 7×10^{-31} g cm^{-3}, implicitly taken to be baryons of course, is from van den Bergh (1961).

At temperature $T = 3 \times 10^9$ K in Panel (c), the neutron abundance is close to the thermal equilibrium ratio of neutrons to protons, $n/p \simeq \exp -Q/kT \sim 0.13$ by number. The decay energy Q is the mass difference between a neutron and a proton plus electron, converted to energy. As the universe expands and the temperature drops below 3×10^9 K, the neutron abundance decreases, staying near the thermal equilibrium value. As the temperature approaches $T_0 \sim 10^9$ K, the photodissociation of deuterons slows, because ever fewer photons are energetic enough for photodissociation, which allows the accumulation of deuterons. The deuteron abundance is plotted as the solid curve in Panel (c), and the logarithmic scale allows us to see the rapid increase as the photodissociation rate slows followed by the decrease in the deuteron abundance as charge-exchange reactions convert deuterons to the two isotopes of helium. This leaves some tritium, the unstable isotope of hydrogen, which decays to ^{3}He. The computation does not follow the abundances of elements beyond helium.

The present baryon density fixes the density at $T_0 = 10^9$ K, given a value for the present radiation temperature. The greater the baryon density is at T_0, the more complete the conversion of deuterons to helium and the smaller the residual amount of deuterium will be. This means that the abundance of deuterium relative to hydrogen prior to element creation and destruction in stars is a measure of the mean mass density in baryons, always assuming the hot big bang theory. Early applications of this consideration were discussed in Section 3.6 and are shown as the constraints on the mass density at the curves labeled D/H in Figure 3.2. It became an important measure of the baryon density to compare to what is indicated by the pattern of the CMB radiation anisotropy discussed in Sections 4.6 and 9.2.

At interesting values of the baryon density, most of the neutrons present at 10^9 K are predicted to end up in helium. Since the neutron abundance at this temperature would have been close to the thermal value and not sensitive to the baryon density, we see that the helium mass fraction Y coming out of the hot early universe is not very sensitive to the baryon mass density. But the computation of Y depends on the assumption of the relativistic hot big bang theory, so the early indications of consistency of observed and computed helium abundances meant that the hot big bang cosmology passed a serious cosmological test. It was not much celebrated at the time, but perhaps this result contributed to the implicit community feeling that the hot big bang cosmology actually may be empirically interesting.

The steady-state cosmology—not having a big bang—helped drive research into how elements may have been formed in stars and helped motivate the foundational papers on the subject by Burbidge et al. (1957); and Cameron (1957). One sees in Gamow's writing the evolution of his thinking about the possible role of stars in the origin of the elements. The record of Gamow's (1953a, 19) lectures at the June 29 to July 24, 1953, Symposium on Astrophysics at the University of Michigan reports that

> Gamow considered the possibility that population II stars have original abundances of elements, and that population I stars have a mixture of elements, which includes the original abundances and the abundances of elements formed in stars. This theory is excluded, however, by the observation that not enough stars have contributed much to interstellar matter during the age of the universe. The interstellar matter is of original pre-stellar composition.

In a later article on modern cosmology in the magazine *Scientific American*, Gamow (1954c, 62) discussed Alpher's gap at atomic mass 5 that suppresses element buildup past helium and wrote that

> If no way is found to bridge the gap, we may have to conclude that the main bulk of the heavier elements was formed not in the early stages of

> the Universe's expansion but some time later, perhaps in the interiors of fantastically hot stars.

In a still later issue of *Scientific American*, among ten articles by leading authorities on cosmology and extragalactic astronomy, Gamow (1956b, 154) conceded that

> since the absence of any stable nucleus of atomic weight 5 makes it improbable that the heavier elements could have been produced in the first half hour in the abundances now observed, I would agree that the lion's share of the heavy elements may well have been formed later in the hot interiors of stars.

This is the direction evidence has taken for the origin of elements heavier than helium, along with traces of lithium formed in the early universe and cosmic ray spallation reactions in interstellar matter, that have added to the abundances of lithium, beryllium, and boron.

4.2.2 PREDICTING THE PRESENT CMB TEMPERATURE

Gamow (1948a) certainly understood that the thermal radiation present at light element formation would remain in the universe as it expanded and cooled. He recognized—perhaps by intuition, perhaps from the analysis by Lifshitz (1946) to be discussed in Section 5.1.2—that gravitational formation of the mass concentrations in galaxies might be expected to have started when the rate of expansion of the universe had become dominated by the mass density in low-pressure matter. Prior to that, the rate of expansion of the universe driven by the mass density in radiation would have been rapid enough to have suppressed the gravitational growth of departures from a near-homogeneous mass distribution. Lifshitz did not consider the particular situation Gamow had in mind, but hints were there if Gamow had chosen to consider them.

Gamow (1948a) estimated that the mass densities in matter and the thermal radiation would have passed through equality, and structure formation would have started, when the temperature and the matter density had fallen from the conditions at light-element formation to

$$T_{\rm eq} \simeq 10^3 \text{ K}, \quad \rho_{\rm eq} \simeq 10^{-24} \text{ g cm}^{-3}. \tag{4.25}$$

Knowing present conditions, we see that this is at redshift $z_{\rm eq} \sim 300$, depending on how you compute, and about a tenth of what was established much later (equation (4.11)). This is close enough, considering the rough nature of Gamow's calculations. Alpher and Herman (1988) recall that in the late 1940s, Gamow doubted that the remnant thermal radiation is of much experimental interest, because it seemed likely to him that the primeval radiation would be obscured by all the radiation produced later by stars. The evolution of his thinking on this point is considered in Section 4.4. As it happens, the accumulated energy in starlight, including that absorbed by dust and reradiated at longer wavelengths, is comparable to the energy in the CMB but is present at shorter wavelengths.

Alpher and Herman (1948a) took the bold step: estimate the present radiation temperature from the baryon density at $T_0 \simeq 10^9$ K required for reasonable production of elements. They needed the present mass density. The value they used is not stated, but in other papers, Gamow and colleagues consistently mention the present density

$$\rho_p = 10^{-30} \mathrm{g\ cm}^{-3}. \tag{4.26}$$

In his thesis, Alpher (1948a) attributes this to Hubble (1936 and 1937). It is the smaller of Hubble's two estimates. And they of course took this mass to be baryonic.

Alpher and Herman (1948a) listed corrections to Gamow's (1948a) computation of how element buildup would have started in a hot big bang. Corrections are not unexpected; Gamow tended not to be careful about details. Their paper refers to Gamow (1948b), which was published a few days after receipt by the journal of the Alpher and Herman (1948a) paper. Their remarks apply equally well to Gamow (1948a). Their conclusion (Alpher and Herman 1948a, 775) is that "the temperature in the universe at the present time is found to be about 5° K." We have then the present values

$$T_f = 5 \mathrm{\ K\ predicted}, \quad T_f = 2.725 \mathrm{\ K\ measured}. \tag{4.27}$$

The agreement is to be celebrated as a wonderfully prescient application of Gamow's scenario, and as a check of the standard physics of the rates of nuclear reactions and the expansion of the universe applied when the radiation temperature was some nine orders of magnitude greater than now.

But there are complications. Following a list of corrections to Gamow's computation, Alpher and Herman (1948a, 774) wrote that "In checking the results presented by Gamow [and] correcting for these errors, we find" an estimate of the mass density at Gamow's critical temperature T_0. Their result, converted to Gamow's notation in equation (4.22), is

$$n(t_0) = 8.2 \times 10^{16} \mathrm{\ cm}^{-3} \mathrm{\ at\ } T_0 = 10^9 \mathrm{\ K}. \tag{4.28}$$

This is an order of magnitude smaller than Gamow's estimate (equation (4.22)). It would put the CMB temperature at the group's standard present mass density at

$$T_f = 19 \mathrm{\ K\ at\ } \rho_f = 10^{-30} \mathrm{\ g\ cm}^{-3}, \tag{4.29}$$

well above what Alpher and Herman stated in their paper. The explanation is to be seen as follows

In their paper published the following year, Alpher and Herman (1949) present their computation of element buildup in an expanding universe with mass density dominated by radiation. They again assume reactions other than neutron capture carry the production of elements past helium at about the rate given by the fit of the reaction rate coefficients $(\sigma v)_i$ to a smooth function of atomic weight i. Their prediction of the present CMB radiation temperature requires the baryon density when element buildup starts. Alpher and Herman (1949, 1092) write that

> We believe that a determination of the matter density [at $T_0 = 10^9$ K] on the basis of only the first few light elements is likely to be in error. Our experience with integrations required to determine the relative abundances of all elements[6,7] indicates that these computed abundances are critically dependent upon the choice of matter density.

The references 6 and 7 in this extract are to Alpher (1948b) and Alpher and Herman (1948b). At about the same time as the Alpher and Herman (1949) paper, Gamow (1949) had reported that Fermi and Turkevich concluded that there is no way for nuclear reactions to bridge the mass-5 gap at a significant rate. We see with the aid of hindsight that this meant the elements heavier than helium had to have been formed later, mostly in stars, and that, contrary to the Alpher and Herman (1949) assertion, one could only use the abundances of the first few light elements to determine the matter density at T_0.

Alpher and Herman (1949) reported the conditions they considered reasonable for element formation: the temperature would be 0.6×10^9 K at mass density 10^{-6} g cm^{-3}. This extrapolates to present temperature 6 K at present density 10^{-30} g cm^{-3}. (This is their equations (12b) and (12d) with the temperature directly below equation (12d).) Alpher and Herman (1949, 1093) reported that their matter density "corresponds to a temperature now of the order of 5 K." This is a reasonable rounded value from their numbers. And it is to be noted that their first announcement of the present temperature, in Alpher and Herman (1948a, 775), reads that "the temperature in the universe at the present time is found to be about 5°K." The wording is similar and the temperature is the same.

I conclude that the CMB temperature Alpher and Herman (1948a) first reported was based on the right general picture—a radiation-dominated early universe—but that instead of fitting to the condition that there be a reasonable abundance of the elements heavier than hydrogen, they fit to the general run of abundances of the elements heavier than helium. It is questionable behavior for Alpher and Herman (1948a) to have listed corrections to Gamow's (1948a) computation of production of the elements heavier than hydrogen and then, without telling the reader, presented their result from a different calculation, that for the formation of the elements heavier than helium.

Peebles, Page, and Partridge (2009, 27–30) point out the discrepancy, and Peebles (2014) explains it. Both papers remark on another curious situation in the first of the papers by Alpher and Gamow. This is to be discussed next.

4.2.3 THE ALPHER, BETHE, AND GAMOW PAPER

The paper Alpher, Bethe, and Gamow (1948) is to be celebrated for its introduction of important ideas about the physics of the early stages of expansion of the universe, but the paper is wrong. In his doctoral dissertation, Alpher (1948a) pointed out the inconsistency in the computation, and Gamow (1949) explained the error and how to correct it. But as far as I have been able to discover, the error seems to have gone unremarked in print since then until Peebles, Page, and Partridge (2009).

Starting thoughts about the Alpher, Bethe, and Gamow paper appear in the book on nuclear physics by Gamow and Critchfield (1949). On page 315 in the book, the expression

$$\frac{dN_i}{dt} = N_0 v(\sigma_{i-1}N_{i-1} - \sigma_i N_i), \quad (i = 1, 2, 3, \ldots), \tag{4.30}$$

expresses Gamow's picture for the buildup of the elements by successive neutron captures. Here N_i is the abundance of atomic nuclei with atomic weight i. The value of N_i is increased by captures of neutrons by nuclei with atomic weight $i-1$, and diminished by captures that promote the atomic weight from i to $i+1$. Alpher's (1948a) doctoral dissertation project was to find numerical solutions to this set of equations. Although the book Gamow and Critchfield (1949) was published after Alpher's thesis, we know from the remark in the book noted on page 123 that equation (4.30) had been written down when Alpher started to work on this problem.

Alpher's project required the reaction rate coefficients $(\sigma v)_i$, the products of relative velocities and radiative neutron capture cross sections, as in equations (4.18) and (4.30). Helge Kragh (1996) reports that Alpher attended a talk by Donald J. Hughes at the meeting of the American Physical Society on June 20–22, 1946. Hughes reported measurements of neutron capture cross sections. This led Alpher to a declassified document on these measurements that used "1-Mev pile neutrons." The measurements were published by Hughes, Spatz, and Goldstein (1949).

In his thesis and the published version, Alpher (1948a,b) pointed out that the absence of a reasonably long-lived atomic nucleus at mass $i = 5$, and then at $i = 8$, prevents element buildup by successive neutron capture beyond helium. Alpher made the sensible working assumption that other nuclear reactions might carry element buildup past the mass gaps. He could then use his smoothed trend of the Hughes et al. measurements of the $(\sigma v)_i$ with atomic weight i in his numerical solutions of the buildup equations (4.30), with the mass density and exposure time to be adjusted to fit the observed trend of element abundances with atomic weight. Finding numerical solutions with the numerical computers of the time was an impressive accomplishment.

Alpher's solutions showed that atomic nuclei with larger neutron capture cross sections end up with lower abundances, depleted by promotion to larger atomic weight. Hughes, Spatz, and Goldstein (1949, 1784) mention this anticorrelation in their publication of the cross section measurements Alpher had used:

> Recently, Alpher, Bethe, and Gamow[2] have shown an interesting relationship between the fast cross sections and the abundance of the elements and have interpreted the relationship in terms of a neutron-capture theory of the origin of the elements.

The reference 2 is to Alpher, Bethe, and Gamow (1948) and Alpher (1948b). The neutron-capture theory became known as the r-process for formation of the more massive elements by radiative capture of fast neutrons, now thought to happen in exploding stars or merging neutron stars rather than in the early universe.

The second problem Alpher (1948a) encountered in his thesis is a vast inconsistency between the wanted rates of the relevant nuclear reactions and the rate of expansion of the universe. Alpher assumed that the mass density in the early universe was dominated by low-pressure matter, which meant he needed to postulate a starting time t_0 for element build up, because in this model, the integral of the buildup equation diverges at $t \to 0$. With his smooth trend of $(\sigma v)_i$ with atomic weight, which ignores the mass gaps, Alpher (1948a, eq. (78); 1948b) found a reasonable match of the computed element buildup to what was observed if at starting time t_0, and lasting for time Δt, the baryon number density satisfies

$$\int_{t_0}^{t_0+\Delta t} n(t)\,dt \simeq n(t_0)\,\Delta t \sim 0.8 \times 10^{18}\ \mathrm{s\ cm^{-3}}. \tag{4.31}$$

Gamow's estimate from equations (4.19) and (4.22) is $n(t_0)\,\Delta t \sim 10^{20}$ s cm^{-3}, two orders of magnitude larger. The difference does not seem unexpected, given that Gamow made a rough estimate for the production of deuterium, and Alpher made a numerical fit to the curve of abundances of the heavier elements. Thus we can take it that equation (4.31) is a well-based order-of-magnitude condition on the product of number density and exposure time for a reasonable degree of element buildup, given the measured neutron capture rates and the interpolation past the mass gaps.

Alpher, Bethe, and Gamow (1948, 803) took the sensible step of taking the exposure time Δt to be comparable to the cosmic expansion time t_0 when element buildup began. They state that

> In order to fit the calculated curve with the observed abundances[3] it is necessary to assume the integral of $\rho_n dt$ during the building-up period is equal to 5×10^4 g sec./cm^3.
>
> On the other hand, according to the relativistic theory of the expanding universe[4] the density dependence on time is given by $\rho \simeq 10^6/t^2$. Since the integral of this expression diverges at $t = 0$, it is necessary to assume that the buildingup process began at a certain time t_0 satisfying the relation:
>
> $$\int_{t_0}^{\infty} (10^6/t^2)\, dt \simeq 5 \times 10^4, \qquad (2)$$
>
> which gives us $t_0 \simeq 20$ sec.

References 3 and 4 in the above extract are to the measured element abundances and to Tolman (1934a).

The problem is that "the integral of $\rho_n dt$ during the building-up ... 5×10^4 g sec./cm^3" divided by the mass m_n of a nucleon is 3×10^{28} s cm^{-3}. This is some ten orders of magnitude larger than the value in equation (4.31) that Alpher had found gave a reasonable fit to measured abundances based on measured nuclear reaction rate coefficients $(\sigma v)_i$.

This enormous discrepancy calls for a check. Alpher, and likely Gamow before him, took it that the expansion time is that of a universe with mass dominated by nonrelativistic baryons with mass density $\rho_{\rm mat}(t)$. In this case, we have

$$\begin{aligned} \rho_{\rm mat} &= (6\pi G t^2)^{-1} = 0.79 \times 10^6 t^{-2} \text{ g cm}^{-3}, \\ \int_{t_0}^{\infty} \rho_{\rm mat}(t)\, dt &= 0.79 \times 10^6 t_0^{-1} \text{ s g cm}^{-3}. \end{aligned} \qquad (4.32)$$

The expansion time t has units of seconds. The first line is close enough to their "density dependence on time." The integral of this expression in the second line, divided by equation (2) in the extract, is $t_0 = 16$ s. This is close enough to their $t_0 = 20$ s to count as a second check. And we see that the integral in equation (4.32) with $t_0 = 16$ s and divided by a nucleon mass to get the time integral of the nucleon number density is

$$\int_{t_0}^{\infty} \frac{\rho_{\rm mat}}{m_{\rm n}}\, dt \simeq 3 \times 10^{28} \text{ s cm}^{-3}. \qquad (4.33)$$

As before, this is ten orders of magnitude larger than what is called for from the measured cross sections.

This large inconsistency arose from the assumption that the mass density is dominated by nonrelativistic baryons. Gamow (1948a) showed that the inconsistency is removed by the assumption that there was a sea of thermal radiation, which would have set the start of nucleosynthesis at $T \simeq 10^9$ K (Section 4.2.1). In Gamow's new model, the subdominant baryon mass density is a parameter to be adjusted to account for a reasonable build up of elements heavier than hydrogen at the known nuclear reaction rates.

What were the authors of the Alpher, Bethe, and Gamow paper thinking? Hans Bethe likely gave the matter no thought, because his name was added only to complete an approximation to α, β, and γ in the first letters of these authors' names. In his citation to this paper, Gamow (1949) added the name Delter to the list of authors. Kragh (1996, 113) presents more of the details of this example of Gamow's sense of humor.

It is not clear whether Alpher and Gamow had any notion of the inconsistency when they submitted this first paper for publication, but there are many hints to what happened later because Alpher, Gamow, and Herman published frequently; the details are explored in Peebles (2014). Along with Gamow's impressive physical intuition, we see an example of his distinct lack of interest in details. In his paper resolving the problem with the Alpher, Bethe, and Gamow paper, Gamow's (1948a, 506) only comment about it is that the time integral of ρ_{mat} (equation (4.31)) "was given incorrectly in the previous paper[2] because of a numerical error in the calculations."

The reference (2) is to Alpher, Bethe, and Gamow (1948). The phrase, "numerical error in the calculations," seems odd for the change of physical situation from a cold to hot big bang. In a review article the next year, Gamow (1949) did explain the problem and show how the presence of thermal radiation resolves it.

In his PhD thesis and the published version, Alpher (1948a,b) remarked that neutrons moving at the kinetic energy ~ 1 MeV of the measured cross sections he was using might be expected to have been accompanied by thermal radiation. He wrote in his thesis (Alpher 1948a, 55) that

> It seems reasonable to suppose that the temperatures during the process were well above the resonance regions of the elements. On the other hand, temperatures of an order of magnitude higher than 1 Mev correspond to neutron energies larger on average than the binding energy per particle in nuclei. A temperature of 10 Mev, therefore, must be well above the temperatures during the period of formation. A temperature above 10^3 ev, and less than 10 Mev, perhaps of the order of 10^5 eV (about 10^9 K) appears to be approximately the correct one.

Alpher's lower bound is from the condition that the neutrons must be moving fast enough to avoid the large neutron-capture resonances that would upset the trend of decreasing element abundances with increasing neutron-capture cross section. Alpher's $T_0 \simeq 10^9$ K is the geometric mean of the two bounds mentioned in this quote; perhaps his choice of temperature is simply a convenient round number. Or perhaps Alpher chose the temperature to agree with Gamow's (1948a) thermal dissociation temperature. In the published version, Alpher (1948b, 1587) wrote that

> if radiation were present, then the radiation density exceeded the density of matter by many orders of magnitude. It would therefore appear that radiation was dominant in determining the behavior of the universe in the early stages of its expansion, and the cosmological model which has been introduced is probably incorrect.... Preliminary calculations of a cosmological model involving black body radiation only[24] (the effect of matter on the behavior of the model being negligible in the early stages because of the great difference between radiation and matter density) indicate that at some 200 to 300 seconds after the expansion began the temperature would have dropped to about 10^9 °K, at which time the neutron-capture could have begun.

The reference [24] is to Tolman (1931). At this point, did Alpher understand Gamow's (1948a) use of the Saha relation to find the start of element buildup? The comment that "neutron-capture could have begun" at 10^9 K certainly resembles Gamow's (1948a) balance of photo-dissociation and radiative capture. But in his thesis, Alpher (1948a, 41) wrote that in the early universe, "the mean thermal energies per particle still exceeded the binding energy per particle in nuclei, so that no nuclei could be formed." That is an approximation to the situation, but Gamow's dissociation energy is an order of magnitude below the deuteron binding energy. Beyond such hints, I have found no evidence that Alpher understood Gamow's argument from Saha equilibrium in statistical mechanics prior to the discussion of this physics in Alpher and Herman (1950).

In papers published later in the 1950s to the mid-1960s, I have not found recognition of how Gamow used the thermal radiation to determine the onset of element buildup. Note in particular the paper by Alpher, Follin, and Herman (1953) on a careful and well-done computation of the evolution of the neutron-proton ratio, taking account of Hayashi's processes (see equation (4.24)). They did not analyze element buildup to deuterons and beyond, but their review (Alpher, Follin, and Herman 1953, 1347) of earlier work explains that

> by the time the universal temperature had decreased to a value where nuclei would be thermally stable, an appreciable number of protons had been generated. Then the capture of neutrons by protons provided the first step in the formation of the successively heavier elements. More

> specifically, the temperature for the beginning of building-up reactions was taken to be 0.1 Mev. . . . This choice was dictated by the magnitude of the binding energy of the deuteron on the one hand and by the lack of evidence in the abundance data for any resonance neutron capture on the other hand.

The remark about thermal stability is on the right track, but the discussion following that in this quote, which parallels Alpher's considerations quoted on page 136, is not. This is curious, because Alpher and Herman (1950, §III (b) 2) had already reviewed the Saha relation.

A decade later, Hoyle and Tayler (1964) discussed the increasing evidence that the abundance of helium is much larger than might be expected from production in stars. At Hoyle's request, John Faulkner[6] computed the evolution of the neutron-proton ratio, along the lines of the Alpher, Follin, and Herman (1953) computation. Hoyle and Tayler (1964, 1109) reported that the initial condition for this computation was set by the consideration that "The concentration of deuterium used in establishing this conclusion was just that which exists for statistical equilibrium in $n + p \leftrightarrow D + \gamma$." This is Gamow's (1948a) point. Later computations in Peebles (1966a,b) and Wagoner, Fowler, and Hoyle (1967) took full account of the concepts that Gamow (1948a) and Hayashi (1950) introduced, along with the chains of nuclear reactions that produced isotopes of the light elements.

Gamow's disinterest in details may account for his later use of a strange ansatz for the numerical values of the parameters that fix the rates of evolution of the thermal radiation temperature and the mass density in matter in the relativistic hot big bang cosmology (Gamow 1953a,b, 1956a). Gamow's new ansatz makes no mention of the physical considerations in Gamow (1948a) that led to the relation between the radiation temperature and matter density in equation (4.22), and Alpher and Herman's (1948a) prediction of the present CMB temperature. The starting assumption (which is implicit, not stated) is instead that when the mass densities in matter and radiation were equal, there was an equal contribution to the expansion rate equation (3.28) by the space curvature term. That is, he assumed there was a time when all three components—matter, radiation, and space curvature—made equal contributions to the expansion rate. With his usual choices of Hubble's constant, $H_0 \sim 1.8 \times 10^{-17}$ s$^{-1} \sim 500$ km s^{-1} Mpc^{-1}, and the present mass density, 10^{-30} g cm^{-3}, his ansatz indicates that the present radiation temperature is reasonably close to Alpher and Herman's value (equation (4.27)): Gamow (1956a) mentioned the present radiation temperature as being 7 K, and Gamow (1953b) mentioned 6 K. (I get 4 K from these numbers, which is close enough.) The ansatz implies that the mass density parameter is $\Omega_m \sim 0.002$.

6. Faulkner (2009) recalls this in the book, *Finding the Big Bang*, by Peebles, Page, and Partridge (2009), hereafter "PPP" in this chapter.

This agrees with his consistently stated impression, beginning with Gamow and Teller (1939), that the present mass density is well below Einstein–de Sitter. Apart from this, I cannot think of any physical argument or intuition that might suggest Gamow's ansatz.

4.3 *Helium and Deuterium from the Hot Big Bang*

The story of how helium came to be seen to be a likely remnant from the hot big bang is best told in two parts. First is the slow accumulation of evidence that the helium abundance is much larger than might have been expected from helium production in stars. Next is the developments in 1964 and 1965 that made a case many found hard to resist.

4.3.1 RECOGNITION OF FOSSIL HELIUM

The idea that primeval helium, a relic from the hot early universe, may amount to a considerable cosmic mass fraction was displayed without comment in the Fermi and Turkevich computation. It showed that there would be little element production past helium, confirming Alpher's warning about the absence of a reasonably long-lived isotope at mass 5. And it showed in Panel (b) in Figure 4.1 the prediction that about a third of the baryons would end up as helium in the hot big bang cosmology.

Gamow (1948b) did not explain why he felt the abundance of elements heavier than hydrogen is roughly 50 percent, but we saw on page 126 that the book on nuclear physics by Gamow and Critchfield (1949) has a discussion of Schwarzschild's (1946) argument for a solar mass fraction of about this amount. Gamow and Critchfield have a reference to Donald Menzel's estimates of abundances of heavier elements in the solar atmosphere. Since Gamow knew about Menzel's research on this subject, he may have known about the prewar series of papers by Menzel, Lawrence Aller, and colleagues on measurements of spectra and estimates of element abundances in interstellar gaseous nebulae.[7] In a summary paper, Aller and Menzel (1945) reported that in planetary nebulae, the mean ratio of helium to hydrogen by number is $N_{\rm He}/N_{\rm H} \sim 0.1$. If the abundance of heavier elements is small, this gives a helium mass fraction $Y \sim 0.3$. It is less than the 50 percent Gamow (1948b) had in mind, but close enough. Later evidence is that the primeval helium mass fraction is $Y \simeq 0.24$, after a modest correction for production of helium by stars—again, close enough to agree with Gamow's impression.

7. Because of the high excitation potential of helium, its spectral lines can only be observed in very hot stars, and interpreting them is difficult. Interstellar matter, or the matter shed by evolving stars in planetary nebulae, may be illuminated and ionized by radiation from hot stars. The recombination lines in the plasma offer reasonably direct measures of the relative abundance of ionized hydrogen and singly ionized helium.

The discussions of element production in a hot big bang cosmology could have added to astronomers' interest in measurements of the abundances of helium and deuterium in systems young and old, and the results could have added to their interest in the hot big bang cosmology. But, as with with the notion of microwave radiation remnant from a hot big bang, there is no evidence that those equipped to check ideas about elements remnant from the early universe took serious notice prior to the 1960s.

Fred Hoyle (1949, 196) presented an early statement of the proposition that the production of helium by stars is quite modest:

> in spite of the steady conversion of hydrogen to helium and higher elements, taking place within the stars, which amounts to the transformation in 10^9 years of about 0.1 percent of the hydrogen present in the nebulae, hydrogen still constitutes about 99 percent of all material.

This of course assumes our Milky Way galaxy formed out of nearly pure hydrogen, which would agree with the consideration by Hoyle, and with Bondi and Gold, that continual creation of neutrons or protons with electrons would be natural in their new steady-state cosmology.

Geoffrey Burbidge (1958) reviewed the astronomical evidence that the helium mass fraction is at least $Y \sim 0.1$, and he argued along lines similar to Hoyle (1949) that it is difficult to see how this much helium could have been produced in stars. He pointed out that the binding energy that would have been released as starlight in converting hydrogen to the helium mass fraction $Y \sim 0.1$ is significantly greater than the energy released as starlight in a Hubble time at the present luminosity of the galaxy. He noted that the young galaxy might have been much more luminous, and so able to have produced more helium, but argued against the idea (perhaps, he suggested, because magnetic fields with flux density increased by collapse of the protogalaxy could have tended to slow the contraction of matter to form early generations of stars). Burbidge pointed out that in the steady-state cosmology, the Milky Way could be much older than the Hubble time, $H_0^{-1} \sim 10^{10}$ y, allowing enough time for stellar helium production by stars in our galaxy shining at about its present luminosity. He did not mention Gamow's (1954a) objection: the Milky Way does not look much older than other nearby $L \sim L^*$ galaxies with ages predicted to be on average a third of the Hubble time.

For this history, it is particularly notable that Burbidge did not mention the idea of helium production in a hot big bang. He attended the 1953 Symposium on Astrophysics at the University of Michigan, where Gamow spoke about this idea. (The photograph of Gamow, Burbridge, and the other participants in this meeting is in Plate IV.) The recorded version of Gamow's lectures (Gamow 1953a) includes his curious ansatz for the values of the cosmological parameters, but it also outlines his picture of how helium might form in the early universe, and his typically enigmatic statement (noted on page 129) that the conversion of hydrogen to heavier elements in stars is far less than the 50 percent or so he mentioned in 1948. But we see the point, if that is what Gamow had in mind.[8] Burbidge might not have attended Gamow's lectures, or he might not have taken them seriously or even remembered them.

Schmidt (1959), and Truran, Hansen, and Cameron (1965) presented models for how the stars in our Milky Way galaxy could have produced a large present helium abundance, provided the galaxy is old enough and was sufficiently more luminous in the past. Schmidt's model assumes the star formation rate evolves as about the square of the surface density of gas, which he assumed would be decreasing as star formation consumed the gas. This would produce a luminous early phase. He showed that his model parameters could be adjusted to produce present interstellar helium mass fraction $Y \sim 0.3$ out of initially pure hydrogen. Truran, Hansen, and Cameron took more detailed account of models for stellar evolution, including the limited amount of helium thought to be ejected from massive stars, but still were able to find a viable picture for production of helium mass fraction $Y \sim 0.3$ by stars in a galaxy about 10^{10} years old. The notion of primeval helium was not mentioned in either paper, but Truran, Hansen, and Cameron (1965) did publish just in time to enter a reference to the paper by Hoyle and Tayler (1964) on evidence that the Milky Way had a large helium abundance at formation.[9]

In the second edition of his book, *Cosmology*, Bondi (1960, 58) adds the cautionary remark that

> the abundance of helium may conceivably be greater than would be accounted for by ordinary stellar transmutation and so might have to be explained on a cosmological basis, but the evidence as yet is far too slight to merit serious consideration now.

Bondi does not explain why he entered this revision of the comments in the first edition on whether the chemical elements were produced in stars, or maybe in the hot early universe, if there were one. He may have been aware of the arguments that helium might have been produced in stars, provided the young galaxy was considerably more luminous. But his remarks are a reasonable indication of the thinking, or lack of it, about helium in 1960.

8. Gamow (1953a) also discussed cosmic ages from radioactive decay, geology, and the Hubble expansion time; and he gave a perceptive discussion of the relevance of the Jeans stability condition and the epoch at equality of the mass densities in matter and radiation for the formation of galaxies. These topics have become part of the established ΛCDM cosmology.

9. The Truran, Hansen, and Cameron paper is marked as received March 31; the Hoyle and Tayler paper was published September 12. But the tradition has been to circulate preprints, then by mail and now by posting on the web, sometimes well in advance of publication.

If stars in an early, more luminous phase of the Milky Way had produced much of the helium, and this were typical of galaxies, it would have been a problem for the 1948 steady-state model: Why don't we see galaxies in the luminous early phase of helium production? The postulate of continual creation of helium as well as hydrogen is a way out, of course, but perhaps not one that would inspire confidence. Considerable production of helium in stars in young galaxies would have been a problem for Gamow's hot big bang model, too: Isn't there supposed to be considerable production of helium in the early universe? But parameters could be adjusted, as by introducing a sea of degenerate neutrinos, to save the big bang theory without primeval helium.

Osterbrock and Rogerson (1961) offered an empirical assessment of an important issue: if stars produced the helium in the Milky Way, then older stars might be expected to have smaller helium mass fractions. Osterbrock and Rogerson discussed measurements of solar heavy-element abundances that tightened constraints from solar models on the mass fraction Z in heavy elements in the Sun, which in turn tightened the constraint on the solar helium abundance, to $Y \simeq 0.28$. They remarked that planetary nebulae have similar helium abundances (as discussed in footnote 7). They put the ages of the Sun and planetary nebulae at roughly $\sim 5 \times 10^9$ y. And they pointed out that the helium abundance in the Orion Nebula, which might be a fair measure of helium in present interstellar matter, is about the same (Mathis 1957). Osterbrock and Rogerson (1961, 133) concluded that

> It is of course quite conceivable that the helium abundance of interstellar matter has not changed appreciably in the past 5×10^9 years, if the stars in which helium was produced did not return much of it to space, and if the original helium abundance was high. The helium abundance $Y = 0.32$ existing since such an early epoch could be at least in part the original abundance of helium from the time the universe formed, for the build-up of elements to helium can be understood without difficulty on the explosive formation picture.[21]

The reference [21] is to Gamow (1949). Donald Osterbrock (2009, in PPP) recalled attending Gamow's (1953a) lectures at the Michigan Symposium on Astrophysics. He was impressed by the range of interesting ideas Gamow discussed, and he remembered Gamow's remarks about helium.

Osterbrock and Rogerson (1961) presented serious evidence that the helium abundance is about the same in old stars and in the present interstellar medium. It suggested that the galaxy formed out of matter already enriched in helium. And we see that they pointed out that helium present when the galaxies formed might be explained by Gamow's hot big bang scenario, discussed in Section 4.2. To my knowledge, Osterbrock and Rogerson were the first to suggest the possible detection of a fossil from the early hot universe. But their argument attracted little attention until recognition of another fossil, the sea of microwave radiation.

Fred Hoyle also came to recognize the case for helium formation in the early universe. The evidence I know is that Hoyle's recognition was independent of Osterbrock and Rogerson, and that he was thinking about the issue of helium well before their paper was published. This is seen in the discussion following one of Hoyle's (1958, 283) lectures at the 1957 Vatican Conference on Stellar Populations. The recorded exchange of comments on whether supernovae may have dispersed the elements produced in stars includes this excerpt:

> Hoyle: The difficulty about helium still remains, however.
>
> Martin Schwarzschild: The evidence for the increase in heavy elements with the age of the galaxy supports [element formation in stars]. However, it does not necessarily mean that He production occurs mainly in stars. Gamow's mechanism may work up to mass 4.
>
> Hoyle: That is why a knowledge of the He concentration in extreme population II is so important.

The old Population II stars mentioned in this exchange have low heavy element abundances. As Osterbrock and Rogerson (1961) later argued, if the helium present in the Milky Way were produced by stars along with the heavier elements, then these old stars would be expected to have low abundances of helium as well as of the heavier elements. Consistent with this thinking, Hoyle's (1959) computations of stellar structure and evolution took the helium mass fraction to be $Y = 0.009$ in Population II stars, and $Y = 0.25$ in the relatively young Population I stars. Hoyle (1959) did not discuss how the stars in the Milky Way could have produced its present abundance of helium, but the paper was on stellar structure. We are left to wonder whether at the 1957 Vatican conference Hoyle and Schwarzschild were thinking about where the large helium abundance in the Sun came from if not from Gamow's hot big bang.

4.3.2 HELIUM IN A COLD UNIVERSE

Prior to recognition of the sea of microwave radiation, it was natural to consider the idea that the early universe might have been cold, with zero entropy, an arguably elegant initial state. At high redshift in this universe, when the electron number density was greater than about 10^{30} cm^{-3}, the electron degeneracy energy could have been large enough to convert protons to neutrons by $e + p \to \nu + n$. As the universe expanded, the decreasing electron degeneracy energy would allow these neutrons to decay to protons, which would then be radiatively captured by other neutrons, leading to an unacceptable accumulation of elements heavier than hydrogen. In the Soviet Union, Yakov Zel'dovich had become interested in cosmology and in Gamow's ideas, but Zel'dovich thought that the helium abundance in old stars is too small to be consistent with Gamow's hot big bang. Zel'dovich (1962a, 1963a,b) offered a way around this problem: suppose the cold early universe contained equal number densities of protons, electrons, and neutrinos, all initially uniformly distributed. Since electrons are allowed two spins states and neutrinos just one, the neutrino degeneracy energy in each family would be much larger than for the same number density of electrons in the early dense universe. Thus almost all the nucleons would be protons.[10] There would be little formation of helium prior to the first generation of stars, which would agree with Zel'dovich's impression that old stars have little helium. His thought that the universe might have begun in a particularly simple state, with zero entropy and exact homogeneity, is an early example of the idea of spontaneous homogeneity breaking discussed in Section 5.2.5.

In 1964, Bob Dicke charged me to think about theoretical implications of the Princeton search for a sea of thermal radiation that might have been left from a hot early universe. He didn't say it, but that of course includes the possibility of a negative outcome. A seriously low upper bound on the present temperature could mean that the matter density at the onset of nucleosynthesis was large enough to have converted an unacceptably large fraction of the hydrogen to helium. I did not know about Zel'dovich's thinking, but naturally also thought about lepton degeneracy energies in a cold early universe. The remark about lepton number in Panel (a) in Figure 4.2 in Section 4.3.3 was meant to indicate that this figure assumes a hot big bang model in which the lepton number is small enough that degeneracy can be ignored. I had also analyzed the evolution of a cold early universe with high degeneracy energies. I estimated that at high redshift, where the electron mass and the neutron-proton mass difference may be neglected, energy is minimized at neutron-to-proton number densities that would result in mass fraction heavier than hydrogen:

$$Y = 2 - \frac{4}{3}\frac{L}{N}. \tag{4.34}$$

Here L is the lepton number, imagined to be electrons plus neutrinos in a single family, and N is the number of protons plus neutrons. These considerations were published only in a comment in Dicke and Peebles (1965, 449):

> If it were assumed that the lepton number abundance as defined above is greater than 1.3 times the nucleon abundance, the cold universe would be a possibility, because the neutrino degeneracy in this case

10. At number density n, neutrinos and relativistic electrons have degeneracy energies $\varepsilon \propto (n/g)^{1/3}$, with $g = 2$ spin states for electrons and $g = 1$ for a neutrino family. The larger value of g for electrons means that at large n and equal number densities, neutrinos would have the greater degeneracy energy. Thus the reaction $p + e^- \leftrightarrow n + \nu$ minimizes energy by converting most of the neutrons to protons.

would force a very low neutron abundance. As mentioned above we find this possibility philosophically unappealing because, in contrast to the nucleon abundance, an excess neutrino abundance does not appear to be essential to the formation of the universe.

I cannot explain the anthropic-like character of this comment. Our paper was submitted in March 1965. That was late enough that we could add a note in proof about detection of the sea of microwave radiation. But the paper was written before we knew about it.

Not long before Dicke's invitation to think about a hot big bang, I had published a study of the structures of the planets Jupiter and Saturn (Peebles 1964), also at Dicke's invitation. He and his Gravity Research Group were casting about for probes of gravity physics. I concluded that the constraints on the properties of these planets were best fit by a model in which the bulk of the heavy elements had settled to a dense core, and above that the helium abundance was about that of the Sun, which I took to be about 20 percent by mass. Among my references for the helium abundance were Osterbrock and Rogerson (1961). As far as I can recall or reconstruct, I had quite forgotten about helium in gas giant planets when I wrote the comment about a cold universe in Dicke and Peebles (1965). Now I cannot recall why I mentioned $L/N = 1.3$, which would indicate $Y \sim 0.3$ in equation (4.34). Maybe there was some conscious or subconscious connection to Osterbrock and Rogerson.

As we have seen, prior to the recognition of the microwave background, Zel'dovich felt that the primeval light-element abundances likely are significantly lower than might be expected in Gamow's hot big bang picture. Thus Zel'dovich (1962a, 1102) wrote that "In the prestellar stage they [Gamow, Alpher and Herman] obtain a large amount of helium (about 10–20%) and deuterium (about 0.5%). ... These deductions are incompatible with the observations."

It is not clear what observations of deuterium Zel'dovich had in mind. Interpreting the terrestrial abundance is complicated by chemical isotope separation that tends to concentrate the more massive isotope, deuterium. Zel'dovich referred to evidence of stars with helium mass fractions as low as $Y \sim 0.025$. This would challenge the standard hot big bang model and agree with his initially cold degenerate universe. He cited Minnaert's (1957) survey of element abundances. But the low helium abundance entered in this study was based on stellar optical spectra that are much more difficult to interpret than are recombination lines in diffuse plasma. Anne Underhill (1958, 128), in a commentary on the uncertainties of helium abundance estimates from stellar spectra, wrote that "the situation is deplorable."

Yuri Smirnov recalls in PPP that Zel'dovich invited him to check light element production in Gamow's scenario. His computation of the chain of nuclear reactions leading to the final abundances of the isotopes of hydrogen and helium, along lines close to the modern theory, led Smirnov (1965, 867) to conclude that

> The theory of a “hot" state for pre-stellar matter fails, then, to yield a correct composition for the medium from which first-generation stars formed: for $\rho_1 \leq 10^{-6}$g cm^{-3} several percent of deuterium is obtained, in conflict with observation [9], while for $\rho_1 \geq 10^{-6}$g cm^{-3} too high a He^4 content is found.

Here ρ_1 is the nucleon mass density when the mass density in radiation was 1 g cm^{-3}. The reference [9] is to Zel'dovich (1963b). Smirnov's paper is marked as submitted in February 1964. The papers announcing evidence of detection of radiation from a hot big bang were published in July 1965.

4.3.3 DEVELOPMENTS IN 1964 AND 1965

In the winter of 1964, Hoyle had concluded that the helium abundance in old stars likely is large. Malcolm Longair and John Faulkner recall this in PPP (pp. 238–243 and 243–257). Both attended a course of lectures by Hoyle on stellar evolution and related issues at the University of Cambridge. Hoyle's thinking at that time may have been influenced by a recent paper by O'Dell (1963), who reported measurements of helium abundances $N_{\rm He}/N_{\rm H} \simeq 0.1$ to 0.2 ($Y \sim 0.3$ to 0.4) in nine planetary nebulae. This added to the evidence for high helium that traces back to what Aller and Menzel (1945) had found (page 139).

The recollections by Longair and Faulkner suggest that Hoyle was even more seriously influenced by another paper, O'Dell, Peimbert, and Kinman (1964). It was published later, in July, but was available at Cambridge as a preprint the previous winter. It reported helium mass fraction $Y = 0.42 \pm 0.08$ in a planetary nebula in the globular star cluster M 15. This was particularly interesting, because O'Dell, Peimbert, and Kinman concluded from stellar spectra that the stars in this old cluster have heavy element abundances that are only a few percent of solar. It suggests that these stars formed out of nearly primeval matter that had not been seriously polluted by heavy elements produced in earlier generations of stars. But one of the stars in this cluster produced a planetary nebula with a high helium abundance. In PPP, Faulkner points out that the paper by Hoyle and Tayler (1964) on the helium abundance discussed the observations of M 15, including the importance of the low heavy element abundances, but attributed it to O'Dell (1963), not O'Dell, Peimbert, and Kinman (1964). It seems clear that Hoyle and Tayler meant the latter.

O'Dell, Peimbert, and Kinman (1964) reviewed what had grown to be a considerable variety of other measures of the helium abundance, including that of Osterbrock and Rogerson (1961, 128). They conclude that

> These facts, taken together with the observations that the helium abundance in the nebulae surrounding highly evolved stars in the field

> planetary nebulae and the planetary in M 15 is high, argue strongly that the original helium abundance in the Galaxy, or at least the abundance since the time of formation of the stars that we now observe, was non-zero and probably about N(He)/N(H) = 0.14.

This is $Y \sim 0.36$. O'Dell, Peimbert, and Kinman (1964) did not comment on where this helium might have come from. But Longair and Faulkner recall that Hoyle's lectures included discussion of the work by Gamow and colleagues on helium production in a hot early universe.

Hoyle recruited Faulkner to compute the early evolution of the ratio n/(n + p), the neutron fraction of the nucleons. His results were published in the paper by Hoyle and Tayler (1964) with the title "The Mystery of the Cosmic Helium Abundance." Faulkner's numerical computation took account of Hayashi's (1950) processes of conversion between neutrons and protons by thermal neutrino and electron-positron pairs (equation (4.24)), but it did not keep track of the formation of deuterons and their conversion to isotopes of helium. Hoyle and Tayler took it that when Saha equilibrium had switched to allow accumulation of deuterons, all the neutrons present in Faulkner's computation were incorporated in helium. This is a reasonable first approximation, but it does not take account of the neutrons lost by decay before being captured by protons.

Hoyle and Tayler give a clear statement of the important points: the helium abundance is observed to be large and close to the same in objects that are young and old, containing large and small abundances of the heavier elements that plausibly were produced in earlier generations of stars. And this would fit helium production in a hot big bang. Their references to the earlier work on this idea by George Gamow and colleagues were Alpher, Bethe, and Gamow (1948) and Alpher, Follin, and Herman (1953). The first is wrong, for the reasons discussed in Section 4.2.3, but that was little noticed. The latter present a carefully detailed computation of Hayashi's effect on the evolution of the ratio n/(n + p). With Faulkner, they did not examine the production of deuterium and heavier elements.

The 1948 steady-state model was challenged by the evidence of a large helium abundance in old objects, and I can find no suggestion that one might postulate continual creation of helium as well as protons or neutrons. Hoyle and Tayler (1964) pointed out that helium might have been produced by an early generation of explosions of massive objects that expanded from suitably large maximum densities and temperatures. This could happen without adding much to the abundance of elements heavier than helium, just as in the hot big bang. It would fit the observations but leave a serious problem for the steady-state model: If these hypothetical "little bangs" produced helium in the Milky Way when it was young, then in the 1948 steady-state picture there ought to be quite visible "little bangs" in nearby galaxies that are not as far along in their evolution, which is not observed. But the steady-state model

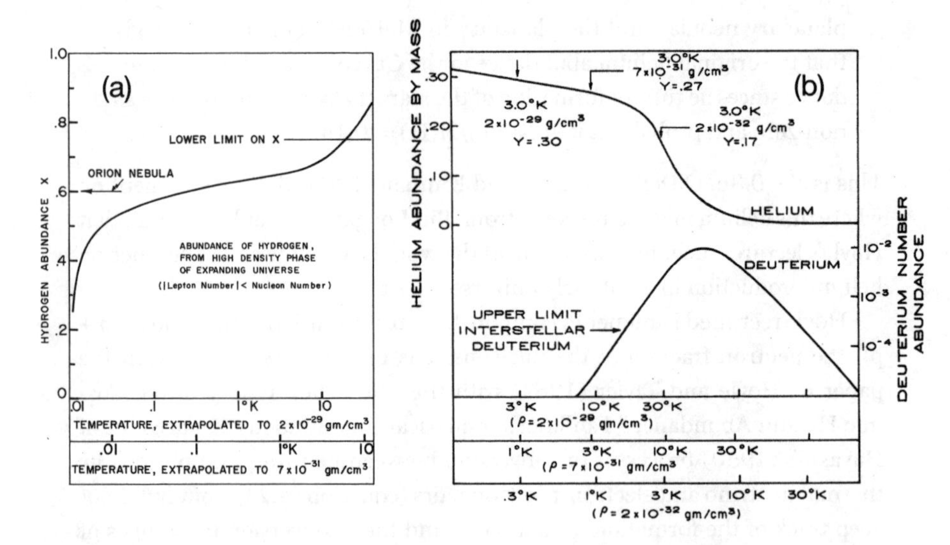

FIGURE 4.2. Computations of helium and deuterium abundances as functions of the present temperature, given the present baryon mass density. I made Panel (a) before recognition of the CMB and Panel (b) shortly after (Peebles 1966a). Panel (b) reproduced with the permission of AIP Publishing.

could be adjusted to account for this by replacing the postulate of continual creation by creation in bursts well enough spaced in time that the galaxies seen around us all had formed some time in the past, leaving them with similar present ages.

Hoyle and Tayler (1964, 1109) argued that measurements of the helium abundance in old objects might challenge the relativistic big bang model: "if the Universe originated in a singular way, the He/H ratio cannot be less than about 0.14." This is helium mass fraction $Y = 0.36$. But they had assumed that all the neutrons present when element buildup commenced at $T_0 \sim 10^9$ K ended up in helium. Attention to the time elapsed during formation of deuterons while the free neutrons were decaying changes the picture, as illustrated in Figure 4.2.

I made Panel (a) prior to recognition of the 3 K cosmic microwave background radiation (CMB). I showed it at a colloquium at Wesleyan University, Connecticut, on December 2, 1964; it was not published (until much later, in PPP). The calculation took account of conversions between neutrons and protons in an approximation I later learned was better worked out by Hayashi (1950). The vertical axis in Panel (a) is the mass fraction in hydrogen, $X = 1 - Y$, and the horizontal axis is the present radiation temperature for two given values of the present mass density (all baryons, of course). In the shallow central part of the curve, most of the neutrons that had not decayed before the radiation temperature approached the critical value $T_0 \sim 10^9$ K (equation (4.22)) were captured by protons, and most of the resulting deuterons ended up in helium. This is roughly in line with Hoyle and Tayler's picture, except that the curve has nonzero slope: Given the present baryon density, a larger present temperature would mean a lower baryon density when the temperature passed through T_0, a lower baryon density would mean a slower rate of nuclear reactions, and a slower nuclear reaction rate would mean more neutrons decay, leaving a lower residual mass fraction in helium (and a larger fraction X in hydrogen). That is, there is freedom to adjust the baryon density in the relativistic hot big bang model to fit a considerable range of values of the primeval helium abundance.

I published Panel (b) in Peebles (1966a) after recognition of the microwave sea, when there were two radiation temperature measurements: Penzias and Wilson (1965) reported $T_f = 3.5 \pm 1$ K at 7.4 cm wavelength, and Roll and Wilkinson (1966) found $T_f = 3.0 \pm 0.5$ K at 3.2 cm. The upper curve shows the computed helium abundance $Y = 1 - X$. It is labeled at three choices of the present baryon density. We see again how the baryon density could be adjusted to get a considerable range of values of the primeval helium abundance. But there is a constraint from the residual deuterium abundance shown in the lower curve in Panel (b). This computation shows that the deuterium abundance would have been maximum, and large, if the present baryon density were $\rho_{\rm baryon} \sim 10^{-33}$ g cm^{-3} at $T_f = 3$ K. Under these conditions, the density at T_0 would have been large enough for significant production of deuterons but not large enough for efficient depletion of the deuteron abundance by conversion to helium. At still lower present density, fewer deuterons would have been created; at larger density, more deuterons would have been created but more efficiently converted to helium. The label indicates that if $\rho_{\rm baryon} = 7 \times 10^{-31}$ g cm^{-3}, van den Bergh's (1961) estimate of the mass density, and $T_f = 3$ K, then the primeval helium abundance would be predicted to be $Y = 0.27$, with deuterium abundance of about 10^{-5} by number.

A letter dated September 1965 that Zel'dovich wrote to Dicke (from the papers of Robert Henry Dicke, Princeton University Archives) is a prompt response to news of the detection, considering the delay in allowing such papers into the USSR. Zel'dovich wrote that

> I am not more so cock-sure in my colduniverse hypothesis: It was based on the assumption that the initial helium content is much smaller than 35% by weight. Now I understand better the difficulty of helium determination ... I sincerely congratulate you and your team on a success.

Zel'dovich was quick to accept that the CMB may be remnant from a hot big bang, but he continued to be cautious about the primeval helium abundance. Thus in a review of 'The "Hot" Model of the Universe' (in the English translation), Zel'dovich (1967, 602) wrote that "According to private communications, old stars with low helium content have been observed very recently. The reliability of these communications is uncertain."

Outside the USSR, we had broader access to experts on light-element abundances but did not always make use of it. Dicke's invitation to think about theoretical implications of a hot big bang naturally led me to think about the primeval abundances of deuterium and helium. The present helium abundance had figured in my study of the structures of Jupiter and Saturn, and my references for that included Osterbrock and Rogerson (1961), who argued for a high pre-stellar helium abundance. I am sure the comment about the "explosive formation" picture mentioned on page 142 meant nothing to me when I was writing about the gas giant planets, and when I started thinking about light-element production in a hot big bang, the helium abundance in the solar system had faded from my mind. I do not recall paying much attention to what was known about the abundance of deuterium, apart from the fact that the terrestrial abundance is hard to sort out. I did ask the local experts, Bengt Strömgren and Martin Schwarzschild, about what was known of the pre-stellar helium and deuterium abundances. They were friendly and instructive but not specific. I now suspect they were too polite to indicate in more than an implicit way that I must evaluate the literature myself.

My January 1965 draft of a paper on light-element production in a hot big bang, not published, shows that I had learned that Gamow, Alpher, and Herman had already covered much the ground in my earlier computations; my reference to their work in the draft is to the review paper Alpher and Herman (1953). I did not yet know Hayashi's (1950) paper. My first computations took account of the reactions with thermal leptons that tend to drive the neutron-proton ratio toward the equilibrium value, $n/p = e^{-Q/kT}$, where $Q = 1.28$ MeV is the energy difference, but my method was clumsy. I was not aware of Zel'dovich's (1962a) considerations of cold initial conditions, and a considerable part of my draft was devoted to my own thoughts about a cold universe with degenerate leptons.

In the January draft and a circulated preprint, I announced that if degeneracy were unimportant, then "the present cosmic blackbody radiation temperature should exceed 10°K." This followed, in an approximate way, from the condition that the CMB temperature should be large enough that when element synthesis commenced at T_0, the matter density was low enough to avoid overproduction of helium. That depended on a reliable upper bound on the primeval helium mass fraction Y. In the computations shown in Panel (a) in Figure 4.2, the line marked "lower limit on X" would require the present temperature to be a little larger than 10 K if the present baryon density were 2×10^{-29} g cm^{-3}, and a little lower than 10 K at 7×10^{-31} g cm^{-3}. My hopeful lower bound on the present temperature may have had something to do with an unconscious hope that my colleagues' search would detect the CMB.

The results of my later computation shown in Panel (b) in Figure 4.2 (Peebles 1966a) more directly take account of Hayashi's effect. The analysis is close to the computation of the neutron fraction n/(n + p) by Alpher, Follin, and Herman (1953), but it also includes computation of the nuclear reactions that convert neutrons and protons to the final abundances of the stable isotopes of hydrogen and helium. The figure indicates that at CMB temperature 3 K and baryon matter density 2×10^{-32} g cm^{-3}, the helium abundance would be $Y \sim 0.17$. This seriously violates the Hoyle-Tayler argument mentioned on page 148, but it is unacceptable for another reason: the unacceptably large remnant deuteron abundance. That is, once the present temperature had been fixed at about 3 K (and of course, pending evidence that the radiation is reasonably close to thermal, so we can be reasonably sure it is from the hot big bang), the detection of objects with helium abundance $Y \lesssim 0.2$ would challenge the hot big bang cosmology because of the large predicted deuterium abundance.

The Hoyle and Tayler (1964) paper on helium could have been the stimulus that drew community attention and research to the theory and observations of the primeval abundances of helium and deuterium. But attention was diverted by the realization that there is another promising phenomenon: We are in a sea of microwave radiation.

4.4 Sources of Microwave Radiation

In the 1950s, Gamow was still thinking about a sea of microwave radiation left from a hot big bang, though perhaps only occasionally. Virginia Trimble (in PPP,[11] 62) recalls that in 1949, Joseph Weber told Gamow about his practical experience with microwave technology. Weber recalled that Gamow offered no suggestion about how his expertise might serve Gamow's research interests. But in his article, "Half an Hour of Creation," Gamow (1950, 18) offered an explanation in words of the evolution of the mass densities in matter and radiation in his cosmology, and he stated that "At the present epoch, in which the density of matter in the universe is about 10^{-30} g/cm^3 ... the temperature is only about 3° K."

As is typical of Gamow, the words in this paper are clear, but there is no indication of how he arrived at that radiation temperature, and as is typical, he cited no references. We have seen on page 138 other examples of Gamow's thinking about the present radiation temperature, along with his odd ansatz that also argues for a temperature of about this value. In the popular book, *The Creation of the Universe*, Gamow (1952a) may have confused those who were paying attention by mentioning the present radiation temperature as being ~ 50 K. But that was only an illustrative example, the temperature at about

11. I remind the reader that "PPP" is a reference to the recollections in the book, "Finding the Big Bang," Peebles, Page, and Partridge (2009).

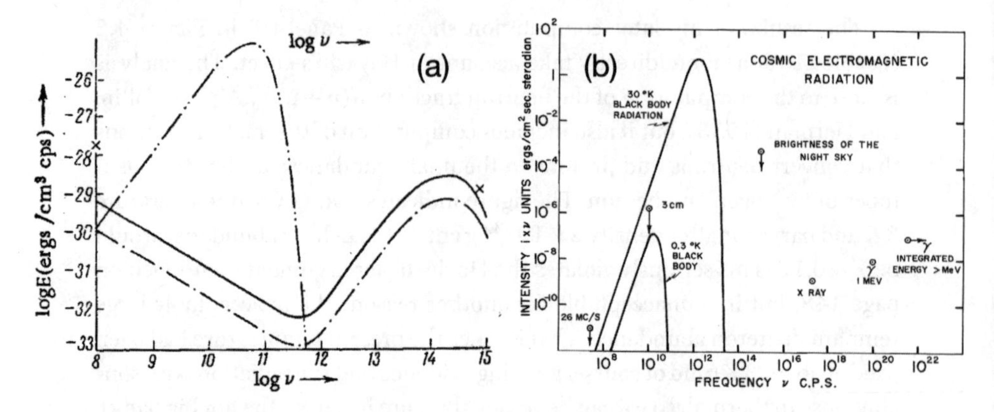

FIGURE 4.3. Two 1964 estimates of the electromagnetic radiation background. Panel (a), from Doroshkevich and Novikov (1964), shows the Planck thermal spectrum at temperature 1 K (the double-dot-dashed curve). The other two curves are estimates of the accumulated background radiation from stars and radio sources. I made Panel (b), unpublished, in 1964. It shows two thermal spectra and estimates of the measured cosmic energy densities at other bands of frequency. Panel (a) reproduced with the permission of AIP Publishing.

the present expansion time, 10^{17} s, in a Friedman-Lemaître model universe with mass density dominated by radiation up to the present.

Gamow's scenario discussed in Section 4.2.1 could have inspired an attempt to detect this CMB—the technology was available—but Alpher and Herman (1988) recall that

> In 1948 and 1949 he [Gamow] argued with us personally and in correspondence that even if the concept of a remnant cosmic background radiation was real, it was not useful because of the presence of starlight at the Earth of about the same energy density.

As Gamow argued, the energy density in the cosmic infrared background from stars is comparable to the expected energy density in the CMB. But as it happens, the two components are at reasonably well-separated wavelengths. This is illustrated in Figure 4.3.

The figure shows two estimates of the cosmic radiation energy density as a function of frequency. Both were made in 1964, independently, and prior to recognition of the microwave sea. Both show examples of Planck's thermal blackbody spectrum at microwave wavelengths. Panel (a) is from Zel'dovich's research group in the USSR and Panel (b) from Dicke's Gravity Research Group in the USA. The vertical axis in Panel (a) is the radiation energy density per Hertz, u_ν, and in Panel (b) it is νu_ν, which I liked, because the energy density is the integral of νu_ν over $\log \nu$, which is the horizontal axis. This difference in the vertical axes accounts for the difference in appearance of the contribution by radio galaxies.

To be discussed here is what was known or conjectured about the radiation background: starlight, light absorbed by dust and reradiated at longer wavelengths, and the contribution from galaxies that are luminous at radio frequencies. This is to be compared to what could be inferred from the results of communications research at the Bell Telephone Laboratories and astronomers' observations of the spin temperature of the interstellar cyanogen molecule CN, along with examples of the Planck spectrum to be expected of thermal radiation remnant from a hot big bang. I have attempted to clarify this account by a separation of topics in the following sections, presented in an approximation to the order of what happened, though this requires the inevitable switching back and forth in time.

4.4.1 INTERSTELLAR CYANOGEN

The first hint of the presence of the microwave background radiation, the CMB, came from observations of optical absorption lines produced by interstellar cyanogen, the diatomic molecule CN. The lines correspond to absorption from the ground level and from the first rotationally excited level of the molecule. Walter Sydney Adams (1941) reported this observation in the spectrum of the star ζ Ophiuchi. Andrew McKellar (1941) converted Adams's ratio of populations in the excited and ground levels, from the ratio of absorption line strengths, to an effective radiation temperature of 2.3 K.[12] That is, if the molecules were in a sea of thermal radiation at this temperature, the equilibrium ratio of populations would be the observed value.

There was the possibility that the CN molecules are excited by collisions rather than by microwave radiation. In his comments about this measurement, Gerhard Herzberg (1950, 496) suggested that there may be CN in the excited level, because the expected rate of decay to the ground level is particularly slow,

> on account of the smaller dipole moment as well as the smaller frequency.... That is why lines from the second lowest level ($K=1$) have been observed for CN. From the intensity ratio of the lines with $K=0$

12. At equilibrium with thermal radiation at temperature T, the ratio of populations in the first rotationally excited level and the ground level of CN is

$$n_1/n_0 = 3e^{-\varepsilon/k_B T},$$

where ε is the difference of energies of the levels, and the factor of three takes account of the three states with angular momentum quantum number 1 in the excited level and one state in the ground level. The energy difference defines the frequency $\nu = \varepsilon/h$ of microwave radiation whose absorption and emission would establish the ratio of populations. In the absence of equilibrium with thermal radiation, and given the ratio of populations, this equation defines an effective temperature, with modifiers such as spin, rotation, or excitation.

and $K = 1$, a rotational temperature of 2.3° K follows, which has of course only a very restricted meaning.

Earlier in the paragraph, Herzberg wrote of excitation "by collisions or radiation," so we can imagine he was open to the possibility of excitation by radiation as well as particle collisions. If the former, the radiation need not be thermal, of course; all that is required is radiation of the right intensity at the wavelength of the transition, 2.64 mm. But in any case, the effective spin temperature of interstellar CN offers a useful upper bound on the possible intensity of a microwave radiation background at wavelength 2.64 mm.[13]

Fred Hoyle was aware of this constraint. In his discussion of the possible effect of radiation pressure on the growth of concentrations of matter in an expanding universe, Hoyle (1949, 197) writes that "The work of Adams[5], as interpreted by McKellar[6], requires the present background temperature to be not greater than about 1°K." The references are to Adams (1941) and McKellar (1941); the lower bound on the temperature might be a convenient round number.

Hoyle (1950) referred to the CN excitation temperature in his review of Gamow and Critchfield's (1949) book on nuclear physics. Their appendix VI outlines the hot big bang considerations in Gamow (1948a,b). In his review, Hoyle (1950, 195) argues that Gamow's model "would lead to a temperature of the radiation at present maintained throughout the whole of space much greater than McKellar's determination for some regions within the Galaxy." If Gamow read Hoyle's review, we may wonder what Gamow made of what could have been to him a perfectly cryptic statement.

Hoyle (1981, 522) tells of a meeting with Gamow in 1956:[14]

> I recall George driving me around in the white Cadillac, explaining his conviction that the Universe must have a microwave background, and I recall my telling George that it was impossible for the Universe to have a microwave background with a temperature as high as he was claiming, because observations of the CH and CN radicals by Andrew McKellar had set an upper limit of 3 K for any such background. Whether it was the too-great comfort of the Cadillac, or because George wanted a temperature greater than 3 K, whereas I wanted a temperature of 0 K, we

13. Palmer et al. (1969) found excitation temperatures in interstellar formaldehyde, CH_2O, that are lower than the ambient radiation temperature. CN is a far simpler molecule, but it might be best to say that the CN spin temperature is not fully guaranteed to be no less than the radiation temperature.

14. The company General Dynamics would have provided the white Cadillac car. Industry and the military were deeply impressed by physicists' contributions to fighting the Second World War, and in the years after the war sought the occasional presence of accomplished physicists, perhaps in the hope of their dropping hints to what curiosity-driven research might come up with next.

> missed the chance of spotting the discovery made nine years later by Arno Penzias and Bob Wilson. For my sins, I missed it again in exactly the same way in a discussion with Bob Dicke at the 20th Varenna summer school on relativity in 1961.

In the published version of Hoyle's lecture, "Fifty Years in Cosmology," in Hyderabad on February 27, 1987, Hoyle's (1988, 5) recollection of that encounter with Dicke was that

> It must have been in 1964 that I was sitting beside lake Camo in Italy, with Bob Dicke from Princeton university. Dicke told me that his group at Princeton was setting up an experiment to look for a possible microwave background, and that they were expecting a temperature of about 20K. I said this was much too high, because a background, if there was one, could not have a temperature above 3K, the excitation temperature of molecular lines of CH and CN found by Mckellar in 1940. Shortly after that the background was found at the Bell telephone laboratories by Penzias and Wilson, and it had a temperature almost exactly on Mckellar's value. The big mistake Bob and I had made was not to realize we had it there beside Lake Camo, in our coffee cups. However carefully one guards against it, opportunities like this come and then slip away through one's fingers.

Hoyle and Dicke did attend a conference in Varenna on the shores of Lake Como in 1961. But a conversation then would have been before Bob Dicke had set us to work on the CMB, in 1964, and before I had started talking hopefully about a temperature of about 10 K, which is close enough to Hoyle's 20 K. Perhaps Hoyle was thinking about a conversation elsewhere in 1964; memories can be complicated. I do not know whether Bob Dicke understood Hoyle's point, which need not have been exactly what Hoyle remembered saying. If Bob had understood that the optical observation of absorption lines by interstellar CN presents an interesting hint to a microwave background, and if he had remembered it, it would have been quite unlike him not to have told us about it. But Hoyle's larger point certainly is right. Missed connections are common in this story, and I suppose common in every human endeavor.

Astronomers have a remarkable (to me) ability to remember such apparently obscure facts as McKellar's conversion of the ratio of populations in the lowest two energy levels of interstellar CN to the spin temperature 2.3 K, real or effective. Hoyle remembered it. When news of the 1965 Penzias and Wilson announcement of excess noise in the Bell communications receivers reached Moscow, Iosif Shklovsky (1966) remembered it and promptly pointed out that the measured spin temperature of interstellar CN could be the temperature of the microwave sea left from the hot early universe. At Princeton University, when Roll and Wilkinson had just started building their radiometer to search for detection of the microwave background, Dicke asked Neville Woolf what he knew about background radiation. Woolfe's recollection (in PPP, pp. 74–75) is:

> I said "Well, there were your own measurements in 1946." He grunted. And I said, "and then there are the interstellar molecules."
>
> He didn't say a word. "Oh", I thought, "I must have said something stupid" and I shut up.

Earlier at Princeton University, George Field also knew about the CN excitation. He recalled in PPP (p. 76) that while a member of the Princeton faculty, he had estimated the rate of decay of CN from its first excited level. He concluded that the decay is much faster than the likely rate of excitation by particle collisions. That is, it seemed likely that CN is excited by radiation present at the wavelength of the transition between levels, 2.64 mm. But he did not publish. Field was at Princeton University until 1965, and I profited from conversations with him, but I do not remember our discussing the search for the microwave background. Field had moved to the University of California at Berkeley when he learned of the Penzias and Wilson announcement of evidence of a sea of microwave radiation, and he recognized the corroborative evidence from the excitation temperature of interstellar CN. As it happened, John Hitchcock, in the office next to Field, was analyzing observations of the interstellar CN lines in plates that George Herbig had taken of the spectrum of ζ Ophiuchi, the star with the CN absorption lines that Adams (1941) had observed. Field, Herbig, and Hitchcock (1966, 161) announced at the December 1965 meeting of the American Astronomical Society the result of their analyses and their interpretation: "the only reasonable mechanism for the observed excitation is absorption of pure rotational quanta at λ2.6 mm." If this radiation has a thermal spectrum, then interstellar CN likely is excited by the microwave radiation detected in the communication experiments at Bell Telephone Laboratories to be discussed next.

4.4.2 DETECTION AT BELL LABORATORIES

Early Bell studies of the possibility of communication at microwave frequencies included measurements of radiation from the atmosphere by DeGrasse et al. (1959a,b). Their budget for the sources of the detected microwave noise includes the assessment (DeGrasse et al. 1959a, p. 2013) that "From the data, it is estimated that radiation from the environment into the side and back lobes of the antenna contributes $2 \pm 1°$K."

Another Bell System experiment detected 12.55 cm wavelength radiation sent from a station on Earth and reflected back by the Echo I satellite, a balloon with a metallized surface that reflected microwaves. Edward Ohm's (1961) assessment of the noise budget in this experiment led him to conclude that the measured sky temperature appears to be consistent with the noise expected from the sum of emission from the atmosphere, radiation originating in the system, and ground radiation entering through antenna sidelobes. But Ohm's sum was an effective temperature of 18.90 ± 3.00 K, and from noise source calibrators, Ohm (1961, 1080) concluded that the "most likely minimum total system temperature was therefore 21 ± 1°K [and that] the "+" temperature possibilities [in the noise budget] must predominate."

The sense of this statement seems to be that to account for the measured noise, Ohm had to suppose that he had systematically underestimated the noise sources. But Ohm's assignment of ground noise entering the horn-reflector antenna is again 2 ± 1 K, which seems to be a significant overestimate. The horn-reflector antennae in these experiments were designed to reject stray radiation better than that.[15]

We can conclude that some at Bell Laboratories knew of an anomaly: more electromagnetic noise detected than could be accounted for. This is true of the three systems that David Hogg reviews in PPP (pp. 70–73). We might say that the anomaly remained a "dirty little secret" until Arno Penzias and Robert Wilson, both new to the Bell Radio Research Lab at Crawford Hill, New Jersey, began a systematic search for local sources of the excess microwave noise.

In their paper announcing the microwave radiation noise excess, Penzias and Wilson (1965, 420) report that

> The backlobe response to ground radiation is taken to be less than 0.1° K for two reasons: (1) Measurements of the response of the antenna to a small transmitter located on the ground in its vicinity indicate that the average back-lobe level is more than 30 db below isotropic response. The horn-reflector antenna was pointed to the zenith for these measurements, and complete rotations in azimuth were made with the transmitter in each of ten locations using horizontal and vertical transmitted polarization from each position. (2) Measurements on smaller horn-reflector antennas at these laboratories, using pulsed measuring sets on flat antenna ranges, have consistently shown a back-lobe level of 30 db below isotropic response. Our larger antenna would be expected to have an even lower back-lobe level.

Thus it was acknowledged that the reports of earlier studies did overestimate the radiation entering the antenna from the ground, obscuring the anomaly.

Penzias and Wilson recall in PPP (pp. 144–176) that they had begun with the intention of using the Bell 20-foot horn-reflector antenna for precision

15. One can still see these horn-reflector antennas back-to-back on towers for microwave communication. One antenna receives radiation that is amplified and transmitted on to the next station by the other antenna. If radiation from the transmitting antenna leaked back into the receiving antenna through sidelobes, it would cause the electromagnetic version of the acoustic feedback howling by poorly arranged microphone and speaker. These antennae were designed to avoid that.

measurements of the energy flux densities from galactic and extragalactic radio sources. That led them to seek the source of the excess antenna temperature. They made painstakingly careful and thorough searches for possible sources of the radiation that might originate within the instrument or somehow enter it from the local environment. That included a variant of the Dicke switching technique to be discussed on page 161: compare the receiver noise when switched between the antenna and a reference load at the known temperature of liquid helium. But because their system was so stable and the system temperature so low, they could do this by hand, quite a different operation from the rapid switching required in the Roll and Wilkinson radiometer experiment discussed on page 162.

In their paper announcing detection of this sea of microwave radiation Penzias and Wilson (1965, 420) referred to the earlier communications experiments:

> DeGrasse et al. (1959) and Ohm (1961) give total system temperatures at 5650 Mc/s and 2390 Mc/s, respectively. From these it is possible to infer upper limits to the background temperatures at these frequencies. These limits are, in both cases, of the same general magnitude as our value.

This is a reference to DeGrasse et al. (1959b). The statement is fair enough but might be elaborated. Thus in PPP (p. 147), Penzias recalls that "we needed to resolve the uncertainty surrounding the seeming extraneous sources of system noise encountered by several of our Bell Labs colleagues."

Indeed, DeGrasse et al. (1959a,b) and Ohm (1961) had indications of an excess antenna noise temperature of a few degrees Kelvin, but they did not advertise it. Penzias and Wilson did the right thing: they did not give up the search for the noise source in the instrument or local surroundings, and they complained about their inability to account for it until someone heard. The idea of a sea of radiation left from the early universe went from Bob Dicke to me to Ken Turner to Bernie Burke to Arno Penzias to Bob Dicke. This part of the story continues in Section 4.4.4.

4.4.3 ZEL'DOVICH'S GROUP

Yakov Borisovich Zel'dovich was a leading theorist at Arzamas-16, the Russian nuclear weapons research center. While there, Zel'dovich became interested in cosmology. Rashid Sunyaev recalls in PPP (p. 113) that "at his request librarians at Arzamas-16 had searched everywhere for all of Gamow's old papers."

Doroshkevich and Novikov joined Zel'dovich's research group in Moscow, where they analyzed the expected accumulation of radiation from galaxies in an expanding big bang universe. Panel (a) in Figure 4.3 shows two of their estimates of the contributions by redshifted starlight, which produces the peak in the solid and dot-dashed curves at the right-hand side of the figure, and by extragalactic radio sources, such as those mapped in Figure 2.4, which contribute the sloping part of the spectrum on the left. The publication shows other estimates of these two contributions; they are removed for clarity.

Doroshkevich and Novikov recall this work in PPP (pp. 99–108). They knew the analyses of a hot big bang cosmological model by Gamow, Alpher, and Herman; it led them to show the blackbody spectrum in Panel (a). They knew Ohm's (1961) assessment of the noise budget in a Bell System exploration of communication by microwave radiation, at 12.55 cm wavelength. That led to their choice of the radiation temperature, 1 K, which they took to be what the uncertainties in Ohm's error budget would allow. The concluding paragraph in the Doroshkevich and Novikov (1964, 113) paper is

> Measurements in the region of frequencies 10^9–$5 \cdot 10^{10}$ cps are extremely important for experimental checking of the Gamow theory [12]. The astronomical corollaries of this theory are analyzed in detail in a paper by Zel'dovich [13]. According to the Gamow theory, at the present time it should be possible to observe equilibrium Planck radiation with a temperature of 1–10°K. A curve for $T = 1°$K has been plotted in Fig. 1, B. Measurements reported in [14] at a frequency $\nu = 2.4 \cdot 10^9$ cps give a temperature $2.3 \pm 0.2°$K, which coincides with theoretically computed atmospheric noise (2.4°K). Additional measurements in this region (preferably on an artificial earth satellite) will assist in final solution of the problem of the correctness of the Gamow theory.

Their reference [12] is to Gamow (1949), [13] is to Zel'dovich (1963a), and [14] is to Ohm (1961).

At the time, Zel'dovich felt that the indications of low helium abundances in some stars argued for a cold big bang (as reviewed on page 145 in Section 4.3.2). One certainly can imagine he would have welcomed further evidence, and the Doroshkevich and Novikov paper offered the possibility of something very interesting. As they anticipated, measurements from satellites were important in establishing the thermal spectrum of this radiation and later for the detection of the signature of early stages of structure formation in the pattern of its angular distribution. Doroshkevich and Novikov could not be expected to have realized that the Bell experiments, including the one discussed in their reference [14], had an anomalous excess noise amounting to a few degrees Kelvin, the CMB. Doroshkevich's conclusion in PPP (p. 108) is that "unfortunately, owing to very limited contacts between Soviet and Western astronomers, our publication ... remained unknown for many years."

Because of the interest in Gamow's science in the Soviet Union (though not in political circles; he had defected to the West) we can imagine that, absent developments elsewhere, the research by Zel'dovich and colleagues would have inspired experiments that detected the thermal sea of microwave radiation, maybe using satellites. But developments elsewhere were approaching detection.

4.4.4 DICKE'S GROUP

The interpretation of the anomalous noise in the Bell System communications experiments grew out of work in the Gravity Research Group at Princeton University. Robert Henry Dicke assembled the first members of this group in 1956 or 1957, for the purpose of improving the experimental basis for gravity physics. This research, reviewed at length in Peebles (2017), included laboratory experiments and investigations into the roles of gravity in geology; astronomy; and eventually, cosmology. The last led to a question Dicke liked to ask us: What might the universe have been doing before it was expanding? In 1964, he called together three young members of the Gravity Group, Peter Roll, David Wilkinson, and me, to explain his thought: maybe our universe "bounced" from a collapse following a previous epoch of expansion. Dicke pointed out that light radiated by stars in the last epoch of expansion would be blueshifted during contraction, maybe reaching frequencies high enough to photodissociate the heavy elements whose binding energies were released to produce starlight prior to the bounce. The formation of a helium nucleus out of four protons releases energy that is radiated as some million starlight photons. A few of these starlight photons, if sufficiently blueshifted during the last collapse, would be capable of photodissociating the helium nucleus, returning it to free nucleons. That would leave hydrogen to make the next generation of stars, and it would leave some million photons per nucleon to be thermalized in the dense conditions during the bounce. The entropy this produced would be largely in thermal radiation at a temperature in the neighborhood of $k_B T \sim 1$ MeV. In short, the bounce would have been a heavily irreversible process. The radiation would cool as the universe expanded, and just possibly present us with a detectible sea of microwave radiation: the CMB.

In an oscillating universe, the entropy would increase at each bounce, unless the new physics required for a bounce suppressed the entropy density. In Zel'dovich's 1965 letter to Dicke (mentioned on page 149) he cautioned us about the "unlimited growth of entropy" in a cyclic universe. Tolman (1934a) may have been the first to make this point. Actors in the two early CMB research groups, Zel'dovich and Novikov (1966) and Dicke and Peebles (1979), discussed it.

The romantic appeal of a cyclic universe along the lines of Dicke's thinking is durable. Lemaître's (1933a) feeling is recorded on page 71. Other examples and analogues are reviewed in the books *Particle Physics and Inflationary Cosmology,* Linde (1990); *Endless Universe: Beyond the Big Bang,* Steinhardt and Turok (2007); *Finding the Big Bang,* Peebles, Page, and Partridge (2009, 40–42); and in a review by Helge Kragh (2013). Such thoughts about a collapsing universe, with Dicke's beautiful elaboration of entropy production in a bounce, led the Princeton Gravity Group to study the theory and observation of a sea of thermal microwave radiation. But the group soon turned from thoughts of a bouncing universe to measurements of properties of this radiation and analyses of what the measurements might teach us.

Dicke invited Roll and Wilkinson to build a microwave radiometer that might detect a remnant sea of thermal radiation, and he suggested that I look into the theoretical implications of the experiment. Peter Roll went on to a career in education, bringing computers into classrooms and instruction laboratories. David Wilkinson and I followed Dicke's admonition for the rest of our careers. Our research on this subject commenced in 1964, the year of publication of the Doroshkevich and Novikov paper. To my knowledge, the move to cosmology by the two groups led by Zel'dovich and Dicke was independent, except perhaps from some sort of community feeling that the time was right for this line of research.

Dicke invented the radiometer he had in mind for Roll and Wilkinson as part of the war research at the Radiation Laboratory based at the Massachusetts Institute of Technology in Massachusetts. The radiometer synchronously detects the difference of receiver response on switching between the antenna and a stable reference source, or load, of radiation at known temperature. The mean of the difference of receiver response to antenna and load averages out the noise originating in the receiver, and the rapid switching removes slower drifts in the receiver gain and noise. The Bell receivers used in the measurements discussed in Section 4.4.2, with the result that Doroshkevich and Novikov (1964) mentioned, employed traveling-wave maser amplifiers that inserted so little noise and were so stable that Dicke-switching was not needed. But as part of their thorough investigation of the excess noise in the receivers, Penzias and Wilson (1965) used Dicke's switching technique in the manner described on page 158.

Dicke took his microwave radiometer to Leesberg, 40 miles from Orlando, Florida, to measure atmospheric microwave absorption at wavelengths 1–1.5 cm. This was a part of the considerations of development of radar at these shorter wavelengths. Absorption by water vapor was a particular concern, and the state of Florida was chosen for its high humidity. The radiometer measures atmospheric emission, which time reversal relates to atmospheric absorption. Dicke found the atmospheric emission by measuring the energy flux density per unit solid angle, $f(\theta)$, incident on the horn antenna as a function of its angular distance θ from the zenith. Dicke fitted $f(\theta)$ to a term proportional to $(\cos\theta)^{-1}$, which represents the path length through the atmosphere in the approximation of a flat Earth. And he added a constant that would represent isotropic incident radiation. Ohm (1961) at Bell Laboratories followed the same procedure. Dicke's result, reported in Dicke et al. (1946, 340), yielded an upper bound on the isotropic term: "there is very little (< 20° K) radiation from cosmic matter at the radiometer wave-lengths."

In 1964, when Dicke offered us our experimental and theoretical assignments, he had forgotten about his upper bound on an isotopic background; we had to remind him.

4.4.5 RECOGNITION OF THE CMB

Fred Hoyle knew the Gamow, Alpher, and Herman exploration of ideas about thermal radiation and light-element abundances, notably deuterium and helium. He discussed the presentation of these ideas in Gamow and Critchfield (1949), and he recalled discussing the idea of a sea of microwave radiation with Gamow (page 154). Hoyle and Tayler (1964) reviewed the evidence of a large and apparently universal helium abundance and its possible explanation by the Gamow et al. theory. Hoyle knew that thermal radiation would accompany the helium in this theory; Hoyle and Tayler noted its role in fixing the ratio of neutrons to protons leading up to formation of the light isotopes (page 138). But it is not mentioned in the published paper. Tayler (1990, 372) recalls that

> When Steven Weinberg (1977) wrote his book *The First Three Minutes* he asked why Hoyle and I did not mention it [the CMB] in 1964. In fact we did in the first draft of our paper, which I still possess, and I gave two talks at Cambridge and Manchester in 1964 in which I commented that it should be there but with such a low temperature that it was not surprising that it had not been discovered. For reasons which I do not now understand, it was not mentioned in our final published paper. Unknown to us and to Penzias & Wilson, Dicke had decided that it could probably be seen and had started an experiment to look for it, and he and colleagues (Dicke et al. 1965) were able to provide an immediate interpretation of the observations of Penzias & Wilson.

While Hoyle and Tayler were writing their paper, I was reinventing the wheel: Alpher, Gamow, and Herman already had a theory of light-element production that suggests a CMB radiation temperature large enough to be detectible by the radiometer Roll and Wilkinson were building. My scattered records suggest that I realized the connection to the earlier work by Gamow et al. in December of 1964 or early January 1965, the latter fixed by the date on an unpublished draft paper. Meanwhile, Bell laboratories already had reproducible though not recognized evidence of the radiation Roll and

Wilkinson were aiming to detect. Penzias and Wilson had made a clear case for detection of unexpected radiation, but they were not equipped to interpret the detection.

I made the summary of what was known about cosmic radiation backgrounds in Panel (b) in Figure 4.3 on page 152, in the year 1964. It grew out of my assignment to think about what the search for a microwave background might teach us. The arrow pointing to the right on the right-hand side of the figure is a bound from the condition that very high energy cosmic ray protons not be unduly slowed by interaction with the radiation. There were detections of nearly isotropic and hence likely cosmic X-ray and gamma-ray backgrounds, and useful upper bounds on the cosmic background from starlight at optical frequencies and from radio sources at longer wavelengths. I cannot discover my source of the upper bound on isotropic radiation at 3 cm wavelength shown in this figure. In a paper also written before recognition of the CMB (but with a note added in proof about this development) Dicke and Peebles (1965) referred to Hogg and Semplak (1961). This is another paper reporting the Bell System microwave communications experiments. We judged that it limited the background temperature to 10 K. But this was from measurements at 5 cm wavelength, not the 3 cm indicated in Panel (b) in Figure 4.3. Maybe I mistook the wavelength. Clearly, in 1964, we at Princeton would have done well to discuss the situation with Ohm, Hogg, and others at Bell Laboratories in Holmdel, New Jersey, not far from Princeton.

In 1964, while Roll and Wilkinson were building their radiometer and I was thinking about the possible consequences, I was invited to present colloquia at Wesleyan University in Connecticut and the Applied Physics Laboratory at Johns Hopkins University in Maryland. Wesleyan was thinking of offering me a job. I did not know then that Alpher and Herman were at the Applied Physics Laboratory in the late 1940s; I never asked whether that is why I was invited to speak there. I asked David Wilkinson whether he would be comfortable with my speaking about the instrument he and Peter Roll were building that could detect the CMB, if warm enough, and my thinking about the theoretical implications. His reply, as it is fixed in my mind, was that it would be no problem: "no one could catch up with us now." We did not consider the possibility that the radiation had already been detected but not interpreted.

In my lecture at Wesleyan University on December 2, 1964, I showed Panel (a) in Figure 4.2, which shows how the primeval helium abundance depends on the CMB temperature and matter density, and Panel (b) in Figure 4.3, which shows what was known and what we had conjectured about the cosmic radiation intensity spectrum. As far as I can tell, this lecture attracted no significant notice.

At the Applied Physics Laboratory on February 19, 1965, I showed an earlier version of Panel (b) in Figure 4.2 and a different version of the cosmic radiation intensity spectrum. A friend from graduate student days in

Dicke's Gravity Research Group, Keneth Turner, attended my lecture. Ken told Bernard Burke about it. Both were then at the Carnegie Institution of Washington, Department of Terrestrial Magnetism. Arno Penzias had already told Bernie about his and Bob Wilson's attempts to track down the source of excess noise in the Bell receiving system they were using. Ken's news led Bernie to advise Arno to contact Bob Dicke. Arno's telephone call led Dicke, Roll, and Wilkinson to visit Penzias and Wilson at Bell Laboratories. The Princeton people soon saw that Penzias and Wilson had a convincing case of detection of radiation that could not be traced to sources in or around the receiver. That led to publication of a paper by Penzias and Wilson (1965) on the detection and the companion paper by Dicke et al. (1965) on an interpretation: the Bell communications systems may have detected thermal radiation remnant from the hot early stages of expansion of the universe.

The Dicke et al. paper referred to Alpher, Bethe, and Gamow (1948) and Alpher, Follin, and Herman (1953). The latter was appropriate: they had presented a careful and close-to-modern numerical integration of the evolution of the nucleon abundance ratio n/(n + p). I was responsible for the first reference; I had not yet realized that this paper is wrong. I should have referred to Gamow (1948a,b).

Photographs of actors in the part of the story around the mid-1960s are in Plates V and VI.

4.5 *Measuring the CMB Intensity Spectrum*

In 1990 it was established that the spectrum of the CMB is very close to that of thermal radiation. The universe now is close to transparent at CMB wavelengths and so is not capable of thermalizing radiation. The thermal spectrum thus offers an excellent case that our universe evolved from a hotter denser state, one in which conditions forced relaxation to thermal equilibrium. The implication is deep: the physical state of the universe has changed. The evidence is simple: a thermal spectrum. But the measurements reviewed here were difficult.

4.5.1 THE SITUATION IN THE 1970s

Figure 4.4 shows two illustrations of the early progress in measurements of the intensity spectrum of the CMB. Panel (a) is from Howell and Shakeshaft (1967). (The units are W m^{-2} Hz^{-1} steradian^{-1}.) Dautcourt and Wallis (1968) show this figure with additions to the measurements. Panel (b) is from *Physical Cosmology* (Peebles 1971a; the references for the measurements are listed on p. 134 in that book). The solid line in Panel (a) is the long-wavelength Rayleigh-Jeans power law spectrum at temperature $T = 3$ K (equation (4.5)). The solid curve in (b) is the Planck blackbody spectrum at $T = 2.69$ K.

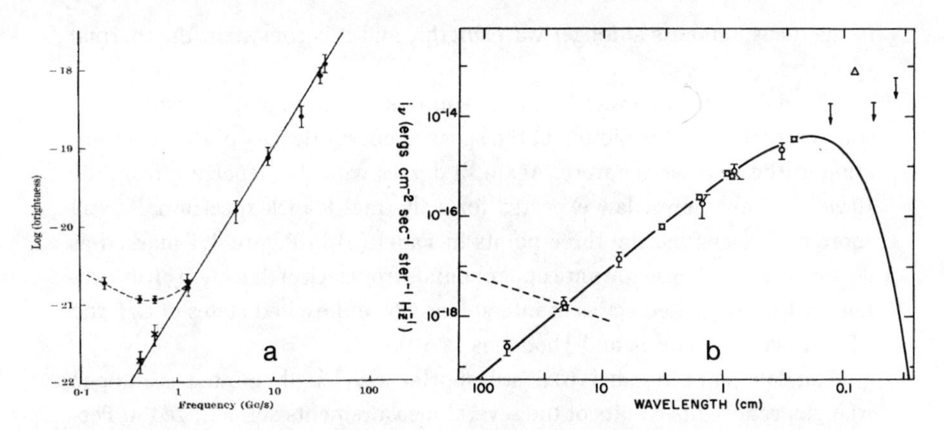

FIGURE 4.4. CMB intensity spectrum measurements in the late 1960s presented by Howell and Shakeshaft (1967) in Panel (a) and Peebles (1971a) in Panel (b). Panel (a) reproduced by permission from Springer Nature.

The dashed lines in both panels are estimates of the contribution to the background radiation by known extragalactic radio sources (added to the thermal spectrum in Panel (a), plotted separately in Panel (b)). The earlier Doroshkevich and Novikov (1964) estimates of this contribution to the radiation background are plotted as the low-frequency parts of the two curves in Panel (a) in Figure 4.3. Radio sources are resolved by large telescopes or interferometers. This means one can separate a source from the near-isotropic CMB by differing the receiver response on and off the source. Howell and Shakeshaft (1967) used the mean radio source spectrum for a notable achievement: detection of the presence of the CMB at long wavelengths, 50 cm and 75 cm. They used two antennae with sizes scaled to the same antenna pattern. Separating the two surface-brightness measurements (defined in footnote 4 in Section 2.4) into one component with the measured spectrum of radio sources and a second component with an assumed thermal spectrum in the Rayleigh-Jeans limit results in the presence of isotropic radiation at effective temperature $T_f = 3.7 \pm 1.2$ K. This is a valuable extension of the CMB spectrum measurements to long wavelengths. I plotted the result as one point in Panel (b), because it assumes the CMB has a thermal spectrum.

The measurements in Panel (a) show that by 1967, 2 years after recognition of the CMB, there already was a good case that this radiation has a thermal spectrum at wavelengths $\lambda \lesssim 1$ cm, in the Rayleigh-Jeans limit. Most of the later measurements in Panel (b) add to the case for the Rayleigh-Jeans power law, and a few offer evidence for the break away from the power law expected in the full Planck spectrum. The measurement by Boynton, Stokes, and Wilkinson (1968) at 3.3 mm wavelength, from a high-altitude site where atmospheric emission is tolerable, found that the spectrum is significantly below an extrapolation of the Rayleigh-Jeans power law fitted to the measurements at longer wavelengths, and it is consistent the thermal spectrum.

The data point in Panel (b) at 2.64 mm is the measured interstellar CN spin temperature. It is plotted at the spectral energy density of thermal radiation at the spin temperature. Again, it agrees with the departure from the Rayleigh-Jeans power law expected for a thermal Planck spectrum. At still shorter wavelengths, the three points in Panel (b) in Figure 4.4 marked as downward-pointing arrows are upper bounds from lack of detection of absorption by higher excited states of interstellar CN and excited states of CH and CH^+ (Bortolot, Clauser, and Thaddeus 1969).

The data point in Panel (b) at still shorter wavelength, plotted as an open triangle, represents results of the several measurements summarized in Peebles (1971a, 133–138). Notable are measurements from a balloon (Muehlner and Weiss 1970) and a rocket (Pipher et al. 1971) suggesting that the CMB spectrum is close to a power law that happens to agree with the Rayleigh-Jeans power-law form but without the break away from the power law in the Planck spectrum. The triangle is above the upper bounds from interstellar molecules, but we nevertheless had to consider the possibility of a radiation background confined to a frequency band that happens to miss the molecule resonances. The situation was troublesome and remained so until 1990, when the anomalies were seen to be systematic errors. But meanwhile there were other interpretations of the CMB to consider.

4.5.2 ALTERNATIVE INTERPRETATIONS

It was important that, in the 1970s, some were exploring alternatives to the community favorite, the relativistic hot big bang cosmology; the community could have been wrong. There were interesting ideas about alternatives, but as will be discussed, all were challenged by what was known in the 1970s, and none has proved to be of lasting interest. I take the lesson to be that the community acceptance of the hot big bang in the 1970s was motivated in part by the promising evidence from measurements of the CMB and the helium abundance, and in part by the simple lack of a promising-looking alternative.

Extragalactic sources were known to make an appreciable contribution to the cosmic radiation energy density at radio wavelengths, as illustrated by the dashed lines in Figure 4.4. Might there be another class of sources with spectra that happen to approximate the Rayleigh-Jeans power law? Could they account for the CMB in a steady-state cosmology? Early considerations of the thought include Sciama (1966), Pariiskii (1968), and Wolfe and Burbidge (1969). One test is observation of the CMB at higher angular resolution, which might resolve it into discrete sources. An early probe for this effect by Penzias, Schraml, and Wilson (1969) used a 36-foot antenna to establish that the rms fluctuation of the CMB surface brightness is $\delta f < 0.024$ K in beam solid angle

$\Omega = 1.4 \times 10^{-3}$ square degrees. If the sources are randomly distributed, this requires a source number density comparable to the galaxy number density.[16] This familiar number density might have seemed marginally promising, but observed radio sources are much less common than ordinary galaxies, and if there were another more abundant class of sources with nearly Rayleigh-Jeans spectra, one would wonder why so few examples had been detected.

The 1948 steady-state cosmology was seriously challenged in the 1970s. An example is the issue of helium discussed in Section 4.3; others are reviewed in Section 3.4. But if these problems could be addressed, one might consider accounting for the CMB by postulating continual creation of radiation as well as matter. If microwave absorption and emission by intergalactic matter could be ignored, the background spectrum that this would create would be an integral over the redshifted spectrum at creation. In the 1970s, the CMB spectrum was known to be close to a power law. That would be produced by the same power law spectrum at creation. The measured break away from the CMB power law at wavelengths approaching the peak of the Plank spectrum might be fitted by a modest adjustment of the spectrum at creation; it would have been a viable (if ad hoc) idea in the 1970s. But the idea that the spectrum of continually created radiation convolved over redshift happens to be close to the measurements would become exceedingly awkward in 1990, when the measured spectrum was shown to be close to thermal well over the peak.

Soon after identification of the microwave background, Layzer (1968) considered the idea that absorption and emission of radiation by intergalactic dust may have caused radiation from galaxies or other sources to relax to a thermal spectrum. The idea could be applied to steady-state as well as evolving cosmologies. Hoyle and Wickramasinghe (1988, 255) explored this line of thought in some detail. They considered needle-like particles that may have been produced in supernovae and driven into intergalactic space. In their model,

> the opacity for a unit column of cosmologically significant length, $\sim 10^{28}$ cm, would thus be about 10, more than ample to produce thermalisation of radiation in the far infrared. The situation remains quite translucent in the visual range, however, so that an astrophysical situation which produces much infrared radiation would explain the

16. Suppose identical sources are distributed across the sky as a homogeneous and isotropic random Poisson process at about the Hubble distance, $d = cH_0^{-1} \sim 4000$ Mpc. If there were closer sources in this class, we can assume they would have been detected and removed as actual sources. Then the rms fractional fluctuation in the radiation background flux density observed in solid angle Ω is $\delta f/f \sim N^{-1/2}$ with $N \sim \Omega \bar{n} d^3$ for source number density $\bar{n}$. The Penzias et al. numbers indicate the mean source number density is $\bar{n} \sim 1$ Mpc^{-3}. This is an order of magnitude or so larger than the local abundance of large galaxies similar to the Milky Way.

> background. We remark again that it is not hard to invent such a situation. Supporters of big-bang cosmology will no doubt feel they have other props to support their beliefs, but at least one erstwhile prop is clearly suspect, as indeed others may be.

There need be no objection to their picture for the properties and space distribution of the dust; the authors point out that nature is quite capable of surprising us about such things. And the large microwave opacity might in fact convert radiation from galaxies, or maybe radiation continually created with the matter in the 1948 steady-state model, to an adequate approximation to the measured CMB spectrum in 1988.

The problem with this picture is that the large optical depth for radio and microwaves, $\tau \sim 10$ at the Hubble length, would prevent detection of radio galaxies at redshifts $z \gtrsim 1$. The Spinrad et al. (1985) program of optical identifications of radio sources in the *Revised Third Cambridge Catalog* found 20 radio sources at redshifts $1.2 \leq z \leq 2.0$. There is the possibility that the Hoyle and Wickramasinghe (1988) picture may allow occasional rifts in the dust clouds that would have passed radio radiation from the occasional high-redshift radio source. But patchy dust would produce a mixture of thermal spectra, some of it starlight coming through the rifts. The hotter components would produce a submillimeter excess in the CMB spectrum. There was some evidence of this effect at the time, as in Figure 4.4 (and Figure 4.5 in Section 4.5.3), but to my knowledge, this was not discussed in 1988. Two years later, the excess was shown to be a systematic error and the spectrum close to thermal, quite inconsistent with the notion of rifts in a cloudy universe.

The transparency of space to microwaves would not be a problem for a model in which the CMB was produced by annihilation of matter and then was thermalized at redshifts larger than that reached by radio source detections. Rees (1978) presented a carefully argued case for this process operating at redshift $z \sim 100$. He pointed out that if the radiation were produced by nuclear burning in massive or supermassive stars at this early redshift, the stellar remnants could serve as the subluminal matter (to be discussed in Chapter 6). Might this picture account for the large helium abundances in old stars in the Milky Way? Rees (1978, 37) argued that

> The arguments pertaining to helium and deuterium are thus our only real clues to the entropy and dynamics at much earlier eras; and the options of non-zero lepton numbers or non-Friedmannian dynamics complicate this issue (as do the uncertainties about nucleosynthesis in the pregalactic objects themselves).

But recall that O'Dell, Peimbert, and Kinman (1964) had demonstrated high helium and low abundances of heavier elements in the globular star cluster M 15. It influenced Hoyle's thinking (see Section 4.3.3, page 146). If the helium in old stars had been produced along with the radiation that was thermalized to become the CMB, it would have to be assumed that whatever was annihilating matter to produce radiation converted hydrogen to helium but produced little of the heavier elements, while at the same time producing enough heavy elements to make the dust to thermalize the radiation. Rees's picture had problems, but it was worth exploring prior to the turn of the century.

I present these considerations to show that, in the years around 1990, there were alternatives to the hot big bang cosmology, as Rees (1978) and Hoyle and Wickramasinghe (1988) showed. There was little discussion of the problems with these alternatives, however, and even less discussion of how the problems might be resolved. The implicit pragmatic community assessment was that we may as well frame our theories and observations in terms of the relativistic hot big bang model, because it looks promising and we do not know an alternative model that looks similarly promising. This philosophy is not a guarantee of success, of course, which we may take to be the larger point made by Hoyle, Narlikar, Rees, and Wickramasinghe.

Another option is a quasi-steady-state model along the lines considered in Section 3.4. Perhaps matter and radiation were created in bursts, between which matter and radiation expanded and cooled in a similar way to the relativistic expanding universe. In his memoir, Fred Hoyle (1988) remarked that such a quasi-steady-state picture of this kind may be compared to the cosmological inflation scenario. Indeed, if the bursts of creation in a quasi-steady-state model were confined to intervals during which the temperature reached well above 10^{10} K, and if the creation were homogeneous enough, it could account for the CMB thermal spectrum, the abundances of the light elements, and all the rest of the hot big bang model, including the signatures of acoustic oscillations in the early universe (to be discussed in Chapter 9). This would be phenomenologically equivalent to the standard ΛCDM theory. In this sense, we can conclude that Hoyle and Narlikar (1966) had anticipated elements of the philosophy of eternal inflation.

4.5.3 THE SUBMILLIMETER ANOMALIES

Since the CMB has been interacting with the rest of the universe in a way that is far from statistical equilibrium, the CMB energy intensity spectrum cannot be exactly thermal. But since the radiation has a large heat capacity (equation (4.8)), a considerable energy release from annihilation of matter or gravitational collapse would be needed to produce a measurable departure from a thermal CMB spectrum. There are violent events such as supernovae and whatever is happening in active galactic nuclei, or quasars, and they could have done interesting things to the CMB intensity spectrum. Checking for the possible effect on the CMB spectrum was difficult, because atmospheric emission at wavelengths shorter than 3.3 mm requires CMB measurements at the still-higher altitudes of balloons, rockets, and satellites—difficult arts. It led to the troublesome data point plotted as the open triangle in Figure 4.4. Such

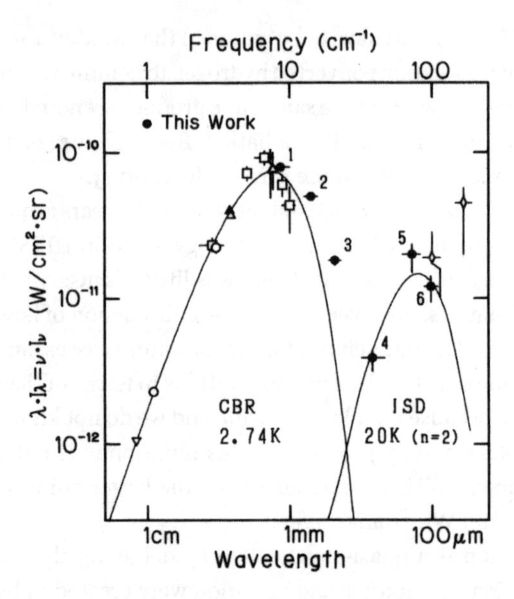

FIGURE 4.5. State of CMB intensity measurements near the thermal peak in 1988 (Matsumoto et al. 1988). © AAS. Reproduced with permission.

confusing indications of departures from a thermal spectrum at millimeter wavelengths persisted for the next two decades.

Figure 4.5, from Matsumoto et al. (1988), shows the state of measurements of the background intensity spectrum toward the end of the 1980s. The radiation detected near 100μm wavelength is from interstellar dust in our galaxy. Its intensity varies across the sky, with the maximum in the plane of the Milky Way. The curve on the right side of the figure is a model for this source of radiation. The Matsumoto et al. (1988) data point at 1.2 mm plotted as the filled circle labeled "1" is close to the thermal spectrum plotted as the left curve with temperature fitted to the measurements at longer wavelengths. But the Matsumoto et al. measurements at shorter wavelengths labeled "2" and "3" appear to be true anomalies, not to be attributed to radiation from dust.

The apparent indications of a departure from a thermal spectrum, beginning with the triangle in Figure 4.4, had to be taken seriously. Scattering of CMB photons by electrons in hot plasma might have shifted photons to shorter wavelengths (that is, higher energy $h\nu$) by the Sunyaev-Zel'dovich effect discussed in Section 4.1 (equation (4.15)). Radiation by dust created in abundance and heated by young stars, maybe along the lines reviewed in Section 4.5.2, could have disturbed the spectrum. Carr (1988) presents a summary of these thoughts; Danese et al. (1990) present a fuller analysis.

The anomalies were interesting; they could have been pointing to something new. To some they were discouraging, because events energetic enough to have produced a perceptible disturbance to the intensity spectrum surely would have disturbed the angular distribution of the CMB. That could have obscured signatures of what was happening in the still-earlier universe. (It caused me to turn my attention from the CMB to statistical analyses of the space distributions and motions of extragalactic objects, for which I was better suited anyway.) But the anomalies proved to be systematic errors in exceedingly difficult measurements from the balloons and rockets needed to bring detectors above interference by the atmosphere. For example, identification of systematic errors in an experiment subsequent to Matsumoto et al. (1988) caused Bernstein et al. (1990) to suspect there was a similar problem with the Matsumoto et al. experiment. Bernstein et al. (1990, 111) conclude that "there is no longer any compelling evidence for an excess flux near the 2.8 K blackbody peak."

The use of scanning Michelson Fourier transform interferometers helped resolve these problems. It was pioneered by groups at Queen Mary College in the UK (Beckman et al. 1972); the University of California, Berkeley, in the USA (Mather 1974); and the University of British Columbia in Canada (Gush 1974). Two programs that grew out of this work at last convincingly showed that the CMB intensity spectrum is very close to thermal.

4.5.4 ESTABLISHING THE CMB THERMAL SPECTRUM

The two teams who independently established that the microwave sea has a thermal spectrum are pictured in Plate VIII.

The origin of one of the programs that established the thermal spectrum is marked by a 1974 proposal to NASA for a Cosmological Background Radiation Satellite. The proposal names the authors: John Mather and Patrick Thaddeus, NASA Goddard Institute for Space Studies; Rainer Weiss and Dirk Muehlner, Massachusetts Institute of Technology; David Wilkinson, Princeton University; and Michael Hauser and Robert Silverberg, NASA Goddard Space Flight Center. It proposed three experiments. Differential microwave radiometers (DMR) would search for departures from an intrinsically isotropic CMB at wavelengths longward of the peak of a thermal spectrum. A far infrared absolute Fourier transform spectrophotometer (FIRAS) would measure the CMB intensity spectrum covering wavelengths near the intensity peak in Figure 4.4, with particular attention to the apparent departures from a thermal spectrum indicated in Figure 4.5. And the diffuse infrared background experiment (DIRBE) would measure the radiation energy density at still shorter wavelengths, in bands from 1μm to 300μm, with the goal of detection and characterization of the cosmic infrared background (CIB), the accumulated radiation energy density from starlight and active galactic nuclei. Michael Hauser recalls in PPP (pp. 418–420) negotiations among other groups who naturally had similar ideas for exploration of the CMB and CIB. Combining the three experiments was bold, but all were successful.

The satellite was named COBE, for Cosmic Background Explorer. Of direct interest in this section is FIRAS, for which John Mather was principal investigator. His recollections are in *The Very First Light* (Mather and Boslough 1996).

The other program that successfully measured the CMB intensity spectrum over the peak, at essentially the same time, was led by Herbert Gush at the University of British Columbia. He had experience with other imaginative uses of Fourier transform interferometers. He used this technology carried in balloons and rockets to measure the spectrum of airglow as a function of time of day, showing the effect of sunlight excitation of emission lines. In the year that Mather and colleagues submitted their proposal to NASA, Gush (1974) reported an attempted Fourier transform spectrometer measurement of the CMB intensity spectrum over the thermal peak. It failed because of inadequate shielding of radiation from Earth. Gush's (1974, 561) conclusion is that "a larger rocket must be used because introducing more elaborate entrance optics [to screen ground radiation] will approximately double the physical size of the instrument. Design studies are underway to this end."

Gush (1981, 746) reported a measurement with well-shielded optics in a rocket launched from Churchill, Manitoba, in 1978. It illustrates the hazards of measurement from a rocket:

> When the observation door was first opened expected signals were obtained, but shortly thereafter a perturbation at 1-hz frequency suddenly appeared, synchronized with the spin of the payload. It subsequently slowly decayed. Recent information supplied by the rocket manufacturer reveals that it arose from the rocket motor, which slowly overtook the payload because of a residual after-burnout thrust.

The rocket motor in the field of view of the instrument beclouded interpretation of this measurement. More details of what happened up to the successful COBRA rocket flight are to be found in the recollections by Mark Halpern and Edward Wishnow in PPP (pp. 416–418).

The COBE satellite carrying FIRAS and the two other experiments was launched in November 1989 from Vandenberg Air Force Base, California. The rocket carrying the COBRA experiment was launched from the White Sands Missile Range, New Mexico, on January 20, 1990. The Mather et al. (1990) publication of first results in May includes the spectrum shown in Panel (a) in Figure 4.6. The Gush, Halpern, and Wishnow (1990) publication of their measurement in July includes the spectrum shown in Panel (b). Both programs were under active development in 1974, some 15 years earlier. Within the space of a few months, both successfully demonstrated the thermal nature the CMB spectrum over the Planck peak.

The FIRAS satellite measurements in Figure 4.6 were taken in 9 minutes. The COBRA rocket flight allowed measurements for a total of 5 minutes. Both

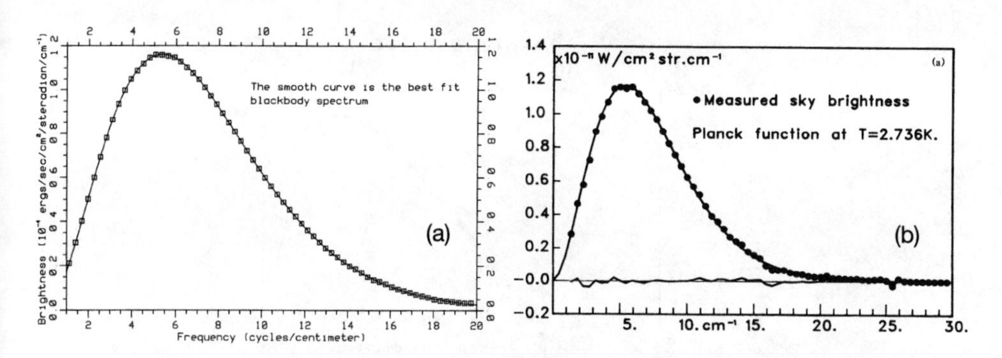

FIGURE 4.6. The 1990 CMB intensity spectrum measurements by FIRAS (Mather et al. 1990) in Panel (a) and by COBRA (Gush, Halpern, and Wishnow 1990) in Panel (b). Panel (a) © AAS, reproduced with permission. Panel (b) reproduced with the permission of AIP Publishing.

made the case that we are in a sea of microwave radiation that is very close to thermal.

Figure 4.7, made by Alan Kogut, Goddard Space Flight Center, summarizes measurements of the CMB intensity spectrum. (Plate I in the color pages shows this figure printed in color, which better shows the provenances of the data.) The thin dotted curve is a thermal spectrum. The solid curve running over the peak shows the FIRAS and COBRA measurements. The former is more accurate, because the satellite allowed measurements for many years that enabled reduction of noise and exploration of possible systematic errors. The measurements at discrete frequencies shown in Figure 4.7 extend the range of data to three orders of magnitude in frequency. The earlier indications of spectrum anomalies were systematic errors; at the time of writing, there is no significant evidence of a departure from a thermal spectrum measured over an impressively broad range of frequencies.

Apart from the CN absorption line observations and the radio measurements at longer wavelengths, the data in Figure 4.7 are measurements of departures from calibrating thermal sources at well-controlled temperatures. This means the measurements establish that the CMB spectrum is wonderfully close to that of radiation at thermal equilibrium (with the caution about the Planck spectrum noted in Section 4.1).

Figure 4.7 deserves a special place in the history of science for the deep significance of a simple figure (showing measurements that were far from simple). It shows a clear case that our universe is not permanent, but rather has evolved from a state dense and hot enough to have relaxed the radiation to thermal equilibrium. This is a profound extension of the empirical reach of natural science. Important also is the evidence displayed here that violent events, such as exploding stars and gravitational collapses, have not been so frequent as to have appreciably disturbed the spectrum. There are energetic

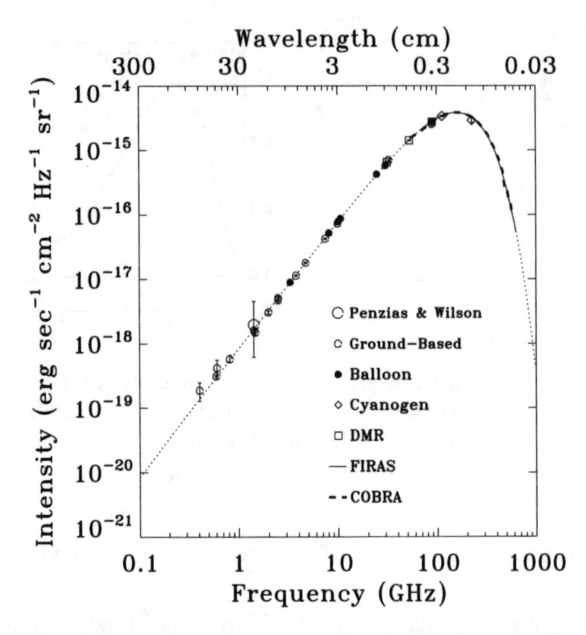

FIGURE 4.7. Measurements of the CMB intensity spectrum (by Alan Kogut, Goddard Space Flight Center).

events, but by and large, our universe has been a tranquil place. The figure also demonstrates that our universe cannot be sufficiently far from exact homogeneity and isotropy to have presented us with a mix of radiation temperatures broad enough to have been detected.

The reach of evidence from this one figure is limited, of course. It makes an excellent case that as the universe expanded and cooled, it preserved the thermal form of the spectrum in Figure 4.7. This preservation does not require general relativity; it follows from standard local physics and the assumption of the metric geometry of a nearly homogeneous and isotropic spacetime, along with the condition that our universe has not been a particularly violent place. There are other lines of evidence to review that do probe general relativity.

4.6 *Nucleosynthesis and the Baryon Mass Density*

The relativistic hot big bang model parameters are constrained, and the theory tested, by the comparison of observed and predicted abundances of the stable isotopes of the light elements. The theory of element formation in the early stages of expansion of the universe, sometimes termed big bang nucleosynthesis (BBNS) figures in the discussion in Section 3.6.2; the origin of the theory is reviewed in Section 4.2. In this theory, the present baryon mass density ρ_{baryon} is a parameter, to be adjusted to fit the abundance measurements, but it is constrained to be no less than the baryons observed in stars and no more than the dynamical bounds on the total mass density (which has come to include the nonbaryonic dark matter, discussed in Chapter 7). The first check of BBNS is whether ρ_{baryon} can be adjusted within these limits to account for the observations of the light elements that seem unlikely to have been produced in significant amounts in the hot big bang. The chains of nuclear reactions from the neutrons and protons that would have been present with the radiation at $T \gtrsim 10^9$ K were first assembled and the reaction rates computed by Fermi and Turkevich, and reported by Gamow (1949) and Alpher and Herman (1950). Better data on nuclear reactions and more capable computers allowed Peebles (1966a,b) to redo the computation in more detail to atomic weight 4, and Wagoner, Fowler, and Hoyle (1967) to carry the computation to the trace abundances at larger atomic weights.

Wagoner, Fowler, and Hoyle (1967) gave the first systematic assessment of the observational constraints on light element abundances. They concluded that the interesting isotopes that would be produced in the early universe are ^{2}H, ^{4}He, ^{3}He, and ^{7}Li, but that comparisons to observed abundances of ^{3}He, the light isotope of helium, and the lithium isotope ^{7}Li, are complicated by their creation and destruction in stars. That leaves the primeval abundances of helium and deuterium for interesting tests of the BBNS theory. Stars create helium and destroy deuterium, but as far as is known, production of helium in stars is modest, and production of deuterium can be neglected. This can be checked by comparing helium and deuterium abundances in galaxies that have high and low heavy-element abundances, on the assumption that a lower heavy-element abundance means there has been less star formation and less chance for stars to have affected the light-element abundances.

Figure 4.8 shows Wagoner's (1973) BBNS computation of the light-element abundances as a function of the present baryon mass density, the parameter ρ_{baryon}. In the several years since the computations published in 1966 and 1967, Wagoner had collected better information on nuclear reaction cross sections.

The top curve in Figure 4.8 shows the abundance of helium. It is not very sensitive to ρ_{baryon}, because, as we have noted, the neutron-to-proton ratio when element buildup can start at $T \sim 10^9$ K is close to the thermal value, which does not depend on ρ_{baryon}, and in most of the range of densities in the figure, most of these neutrons end up in helium. The helium abundance is distinctly lower at the left-hand side of the figure, where the mass density is so low that the nuclear reaction rates are slow enough to allow most of the neutrons to decay. Associated with this, we see in the figure that the slower rates of reactions at lower present densities leave a larger residual deuterium abundance.

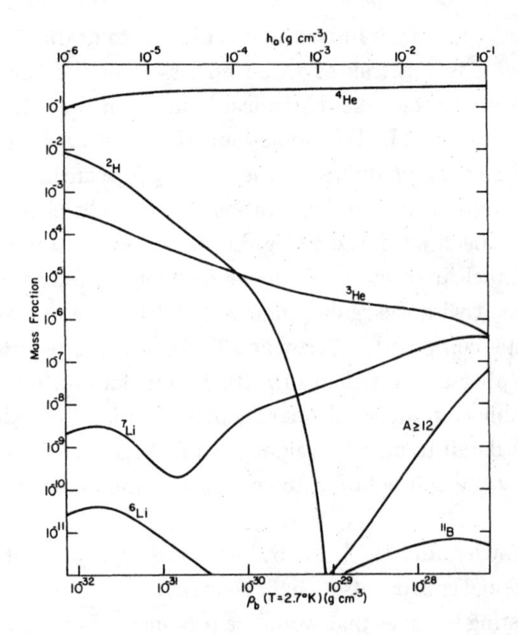

FIGURE 4.8. Wagoner's (1973) computations of primeval isotope abundances as functions of the present baryon mass density. © AAS. Reproduced with permission.

Wagoner's (1973, 356) estimate is that

> The present evidence appears to argue strongly for a primordial source of at least ^{2}H and ^{4}He. We have thus taken the viewpoint that their pregalactic mass fractions were in the ranges $3\times10^{-5} \leq X(^2\mathrm{H}) \leq 5\times10^{-4}$ and $0.22 \leq X(^4\mathrm{He}) \leq 0.32$.

Section 4.2.1 recalls Gamow's (1948b) remark that "hydrogen is known to form about 50 per cent of all matter." He was not explicit about it, but most of the rest of the baryons would have been expected to be in helium, because Alpher's (1948a) mass gap seriously slows element buildup past that. Thus we see in Panel (b) in Figure 4.1 that the Fermi and Turkevich computation indicated primeval helium mass fraction $Y \sim 0.3$, with a much smaller mass fraction in deuterium and even less in elements heavier than helium.

Table 4.1 samples the history of applications of three ways to estimate the primeval helium abundance. The first line in the table is an attempt to interpret Gamow's (1948a,b) thinking about the observed abundance of hydrogen and his BBNS computation of the evolution of element abundances illustrated in Panel (a) in Figure 4.1. Panel (b) in this figure shows the more complete Fermi and Turkevich computation, with $Y_\mathrm{p} \sim 0.3$.

The table lists astronomical evidence considered by Osterbrock and Rogerson (1961); Boesgaard and Steigman (1985); Pagel (2000); and Aver, Olive, Skillman (2015). The results are in the neighborhood of primeval helium mass

Table 4.1. Estimates of the Primeval Helium Mass Fraction

Source	Year	$Y_{\rm p}$	Method
Gamow	1948b	0.4 to 0.5	BBNS, astronomy
Fermi and Turkevich	~1949	~0.3	BBNS
Osterbrock and Rogerson	1961	~0.32	Astronomy
Wagoner	1973	0.22 to 0.32	BBNS
Boesgaard and Steigman	1985	0.239 ± 0.015	Astronomy
Sarkar	1996	0.25 ± 0.01	BBNS
Pagel	2000	0.23 ± 0.01	Astronomy
Aver, Olive, Skillman	2015	0.246 ± 0.010	Astronomy
Aver, Olive, Skillman	2015	0.2485 ± 0.0002	BBNS
Komatsu et al.	2011	0.33 ± 0.08	CMB anisotropy
Planck Collaboration	2018	0.241 ± 0.025	CMB anisotropy

fraction $Y_{\rm p} \sim 0.25$, after a modest correction for helium produced in stars (derived from the observed correlation of the abundances of helium and heavier elements). These later measurements are close to the Fermi and Terkovich computation and are not far from what Gamow seems to have had in mind. I count this as an example of his impressive physical intuition.

The last two of the entries that are labeled "BBNS" in Table 4.1 take the baryon density from Table 4.2 below (Sarkar 1996; Aver, Olive, and Skillman 2015). The helium abundance and the baryon mass density are related in the BBNS theory; Figure 4.8 shows the modest sensitivity of the predicted helium abundance coming out of the big bang to the baryon mass density. The results from BBNS are entered in both tables to be compared to two quite different observational situations, in Table 4.1 the astronomical art of helium abundance measurements. We see reasonable consistency.

The bottom two entries in Table 4.1 are derived from yet another phenomenon observed in yet a different way: the adjustment of parameters in the ΛCDM cosmological model to fit the pattern in the angular distribution of the CMB remnant from acoustic oscillations of the coupled baryons and radiation in the early universe (see Section 5.1.3). The helium abundance, along with the baryon mass density, determines the velocity of sound in the coupled plasma and radiation fluid that sets the anisotropy pattern. The table shows the result from the first detection of this bound on helium from the Komatsu et al. (2011) analysis of the Wilkinson Microwave Anisotropy Probe, WMAP, seven-year measurements with other CMB anisotropy data.[17] The even tighter

17. The MAP satellite project was renamed the Wilkinson Microwave Anisotropy Probe in recognition of David Wilkinson's signal contributions to this science up to his untimely death. The WMAP collaboration included Brown University, the University of Chicago, NASA/Goddard Space Flight Center (GSFC), Princeton University, UCLA, and

Table 4.2. The Mean Baryon Mass Density

Source	10^5D/H	$\Omega_{\text{baryon}}h^2$
Wagoner, Fowler, and Hoyle (1967)	10	0.01
Wagoner (1973)	2 to 35	0.004 to 0.02
Reeves et al. (1973)	5 ± 2	0.015 ± 0.005
Gott et al. (1974)	0.7 to 3	0.03 ± 0.01
Yang et al. (1979)	$\gtrsim 2$	$\lesssim 0.025$
Boesgaard and Steigman (1985)	1.6 to 4	0.03 ± 0.01
Burles and Tytler (1998)	3.40 ± 0.25	0.019 ± 0.001
Kirkman et al. (2003)	$2.78^{+0.44}_{-0.38}$	0.021 ± 0.002
Cooke et al. (2016)	2.55 ± 0.03	0.0216 ± 0.0002
Spergel et al. (2003), WMAP first year	—	0.0224 ± 0.0009
Planck Collaboration (2018), full mission	—	0.0224 ± 0.0001

measurement in the last line is from the Planck Collaboration (2018) analysis of the Planck spacecraft full mission measurements.

The importance of establishing consistency of parameter constraints from independent phenomena has been mentioned and is worth repeating. Here we have reasonably consistent constraints on the primeval helium mass fraction based on the ΛCDM theory of BBNS at redshift $z \sim 10^9$, the ΛCDM theory of the CMB anisotropy produced at $z \gtrsim 10^3$, and the astronomical observations at $z \sim 0$.

Figure 4.8 shows the pronounced sensitivity of the predicted abundance of deuterium, ^{2}H, to the present value of the baryon mass density ρ_{baryon}. This sensitivity, along with the expected negligible creation of deuterium by stars, makes the deuterium abundance an interesting probe of the baryon mass density. Table 4.2 illustrates the history of its application. Some of these studies took account of other isotopes, but the primeval deuterium abundance is the important constraint. The deuterium abundance measurements are listed in the second column as the ratio D/H of the number densities of deuterium to hydrogen atoms. The BBNS theory predicts the present value of the baryon mass density given D/H. This mass density is entered as the combination of cosmological parameters $\Omega_{\text{baryon}}h^2$ (as in equations (3.13) and (3.15)).

The terrestrial deuterium abundance was known in the 1960s, but its interpretation is complicated, because it may be expected to have been affected by chemical fractionation that preferentially captures the more massive deuterons, depending on temperature. Wagoner, Fowler, and Hoyle (1967) proposed that they may as well ignore this complication and take the

the University of British Columbia, in a partnership between NASA/GSFC and Princeton University.

abundance found on Earth and in meteorites to be roughly primeval. The result is entered in the first row of Table 4.2.

Wagoner's (1973) estimate of reasonable values of the deuterium abundance yields the range of baryon densities in the second row. He compared it to Shapiro's (1971) estimate of the mean mass density in the luminous parts of galaxies:

$$\begin{aligned} \rho_{\text{baryon}} &= 0.8 \text{ to } 5\times 10^{-31} \text{ g cm}^{-3} \quad \text{Wagoner (1973)} \\ \rho_{\text{m}} &= 2\times 10^{-31} h^2 \text{ g cm}^{-3} \quad \text{Shapiro (1971)} \end{aligned} \tag{4.35}$$

Shapiro had improved the measurement of the luminosity density, but there remained an uncertain conversion from luminosity to mass, an unknown correction for the baryons outside galaxies, and the later complication of nonbaryonic dark matter. But these are relatively modest details at this point in the history. We should pause once again to admire the rough consistency of an application of the theory of nuclear reactions in the early stages of expansion of the universe and the results of galaxy counts and the galaxy masses needed for gravitational confinement of their stars. It bears constant repetition that such checks, though in this case quite approximate, are pieces of the evidence that the theory is on about the right track.

The third entry in Table 4.2, from Reeves et al. (1973, 918, 927), used a deuterium abundance "based on molecular equilibrium effects and on solar-wind $^3\text{He}/^4\text{He}$ ratio," a delicate art. They used Wagoner's (1973) BBNS computation and concluded that "The suggested universal density is then $\rho \simeq 3 \pm 1\times 10^{-31}$ g cm^{-3} (with $L_e = 0$), well within the range of the observational limits though less than the critical density necessary to close the universe."

The condition $L_e = 0$ means the lepton number density is assumed to be small enough that there is negligible effect of the neutrino degeneracy discussed in Section 4.3.2. We see again the interest in whether the mass density is large enough to close the universe. At the lowest value of the Hubble parameter then under discussion, $h \sim 0.5$, the baryon density would not be much larger than a tenth of the critical value $\Omega_{\text{m}} = 1$.

The entry in Table 4.2 from Gott et al. (1974) was discussed in Section 3.6. Yang et al. (1979) cited evidence for considerable variability of the deuterium abundance relative to hydrogen in interstellar matter along different lines of sight. That might be measurement error, but Yang et al. considered that it might be a signature of variable destruction of deuterium in stars. They accordingly assigned only a lower bound on D/H, and the corresponding upper bound on the baryon density.

Boesgaard and Steigman (1985) reviewed the many ideas about possible departures from the simple standard BBNS theory considered by Peebles (1966a,b) and Wagoner, Fowler, and Hoyle (1967). Some may prove to be relevant for a better cosmology than ΛCDM, but there has been no evidence of that so far. Before reviewing these ideas, let us consider later progress in the measurements.

The 1993 start of astronomical observations with the Keck 10-meter telescope on Mauna Kea in Hawaii enabled measurements of absorption lines of hydrogen and deuterium in galaxies at redshift $z \sim 3$ that happen to be along the lines of sight to background quasars at still higher redshift (Songaila et al. 1994; Burles and Tytler 1998). The Lyman series resonance lines of deuterium are shifted to wavelengths shorter than the hydrogen lines because of the greater deuteron mass. The Rogerson and York (1973) detection of interstellar deuterium absorption lines in the spectrum of a relatively nearby star was an important advance (see page 76 in Section 3.6.2), but there was the uncertain effect of cycling through stars on the abundance of interstellar deuterium in our own Milky Way galaxy. Tytler, Fan, and Burles (1996) reported clear detection of the Lyman-series absorption lines of hydrogen and deuterium in an object at redshift $z = 3.57$. The high redshift of this object, and its low abundances of heavier elements, argue for a young system in which stars are just starting to build elements and have had little time to reduce the deuterium abundance. The authors (Tytler, Fan, and Burles, 1996, 207) conclude from these observations that "We calculate a baryon density that is 5% of the critical density required to close the Universe."

The similar Burles and Tytler (1998) result is entered in Table 4.2. The Kirkman et al. (2003) measurement shows the situation at the end of the revolution in cosmology presented in Chapter 9. The entry from well after the revolution is the Cooke et al. (2016) report of measurements of the hydrogen and deuterium Lyman series absorption lines produced by a gas cloud at redshift $z = 2.85$ with particularly low heavy element abundances that suggest the deuterium abundance is particularly close to primeval.

The last two entries in Table 4.2 are derived from the fit of parameters in the ΛCDM cosmological model to the pattern in the CMB anisotropy. It was noted in connection with the last two entries in Table 4.1 that this depends on the velocity of acoustic (pressure) oscillations of the coupled plasma and radiation, which depends on the helium abundance, as indicated in Table 4.1, and on the baryon mass density entered in the last two lines in Table 4.2. The first of these is from the first-year WMAP CMB anisotropy measurements (Spergel et al. 2003), and the second is from the full-mission anisotropy measurements by the Planck spacecraft (Planck Collaboration 2018).

Within the measurement error flags, which have grown impressively tight, we see consistent results from astronomical observations of deuterium and helium and measurements of the CMB analyzed in quite different ways. There is yet another test of BBNS from the baryon mass fraction in clusters of galaxies. Estimates of the cosmic mean mass density based on the assumption that clusters contain a close to fair sample of the ratio of baryonic to nonbaryonic matter, together with the baryonic mass density derived from BBNS fitted to the deuterium abundance, are entered in rows 26 and 27 of Table 3.2. They are not out of line with the dynamical estimates of the total matter density. At the time, many felt that the dynamical estimates are likely biased low, because the mass density seems likely to be the Einstein–de Sitter value. In retrospect, we see instead an addition to the evidence for the hot big bang model from the consistency of the cluster baryon mass fraction and the BBNS prediction based on observations of deuterium.

We should note possible complications in this relatively simple BBNS story. In the late 1970s, two lepton families were known, and evidence was emerging for the third tau family. To be considered then was the effect on BBNS of still more kinds of neutrinos. Might the thermal sea of one of the neutrino families be degenerate or close to it? This was a serious issue in the 1960s, when the idea of thermal radiation from the hot big bang was still being debated (Section 4.3.3), and it continues to be discussed (e.g., Simha and Steigman 2008). Could the parameters of physics have been different when the universe had been expanding for just a few minutes? Dicke's fascination with the idea is discussed on page 49. Dicke (1968) showed that if the strength of gravity were evolving at about the rate he expected, then the early universe would have expanded too rapidly to have allowed any significant production of helium or deuterium. Yang et al. (1979) reconsidered this issue. Could the baryons in the early universe have been created in a clumpy fashion, producing a spatially inhomogeneous ratio of baryon to photon number? Or could a first-order phase transition in the early universe have deposited the entropy from the irreversible transition in an inhomogeneous way? Could the light element abundances produced in the early universe under such conditions fit the observed abundances and allow a mean baryon density large enough to close the universe? If not, could the primeval baryon inhomogeneity at least offer observable signatures of the difference from homogeneous nucleosynthesis? Early considerations include Gisler, Harrison, and Rees (1974); Applegate and Hogan (1985); and Alcock, Fuller, and Mathews (1987). Searches through the space of parameters to describe a clumpy baryon distribution suggested to Jedamzik et al. (1994) a simple postulate: The standard initially homogeneous sea of baryons might be accompanied by tight packets of baryons, in which there may be produced a different variety of abundances of the isotopes of light elements, maybe along with interesting amounts of heavier ones. The isotopes from dense spots might be observable and interesting, or where in apparent conflict with the observations may be supposed to have been sequestered in the compact central masses in galaxies.

These are serious considerations. None so far has proved to be useful in improving the fit of theory to measurements, though that certainly might happen as the tests grow tighter. We now have a reasonable degree of consistency of measures of $\rho_{\rm baryon}$ based on three quite different phenomena: light element abundances, the anisotropy pattern in the CMB, and the cluster

Table 4.3. Citations of Gamow's 1948 Publications, to 1966

Year	Author	Title
1949	Mayer and Teller	On the Origin of the Elements
1950	Ter Haar	Cosmogonical Problems and Stellar Energy
1961	Clayton et al.	Neutron Capture Chains in Heavy Element Synthesis
1965	Peebles	The Black-Body Radiation Content of the Universe and the Formation of Galaxies
1965	Smirnov	Hydrogen and He^4 Formation in the Prestellar Gamow Universe
1966	Davidson and Narlikar	Cosmological Models and Their Observational Validation
1966a, b	Peebles	Primordial Helium Abundance and the Primordial Fireball

baryon mass fraction. Despite the complexity of what we have imagined might have happened, nature seems to have taken a simple way to produce the light elements.

4.7 *Why Was the Hot Big Bang Cosmology Reinvented?*

In the late 1940s, Gamow presented three ideas of lasting interest. First was the buildup of elements by neutron capture. Gamow and Critchfield (1949) introduced the picture, Gamow's graduate student Ralph Alpher implemented the computation, and astronomers added it to the standard model for element formation in stars: the s- and r-processes for slow and rapid neutron capture rates, respectively, relative to nuclear beta decay rates. The second idea, Gamow's picture of the physics of galaxy formation, is reviewed in Chapter 5. The third idea is Gamow's (1948a) intuitive but physically sensible theory of the role of thermal radiation in setting the start of light element formation in the early stages of expansion of the universe, as reviewed in Section 4.2.1.

These pioneering ideas were not very influential. This is illustrated by the citations of the two papers by Gamow (1948a,b) that set down these basic considerations. Table 4.3 shows all citations of these two papers listed in NASA's Astrophysics Data System in papers published in 1966 or earlier, and excluding citations by Gamow's group. I count nine more references to Gamow (1948a,b) published in the following year, 1967. Five of them are from the Princeton Group, but the other four show that others were taking notice, too.

Through 1966, and again excluding self-citations, there are 18 citations to Alpher, Bethe, and Gamow (1948). Even though this paper is wrong, we could add its citations to those in Table 4.3 for a measure of Gamow's influence. Also, the data system cautions us that their citations need not be complete; in fact,

Dicke et al. (1965) referred to the Alpher, Bethe, and Gamow paper, but the data system missed it. But still the sum of citations is modest.

Apart from Alpher (1948a,b) and Gamow (1949), I have not been able to find in the literature until much later notice taken of the problem with the computation in the more heavily cited Alpher, Bethe, and Gamow paper (Section 4.2.3). The failure to recognize this includes the paper by Dicke et al. (1965) on the idea that the Bell Laboratories' excess antenna temperature is thermal radiation remnant from the hot early universe. I only examined the Alpher, Bethe, and Gamow (1948) paper with the care it deserves much later. The 1965 paper by Smirnov in Table 4.3 is from Zel'dovich's Moscow group. Zel'dovich had become interested in the big bang cosmology, quite independent of the Princeton group, and Zel'dovich had given thought to Gamow's pioneering contributions. But I have seen no indication that the Moscow group checked the Alpher, Bethe, and Gamow (1948) paper. My 1965 paper in Table 4.3 was on galaxy formation. It was easier to see that Gamow was on the right track about this.

The research in the late 1940s by Gamow, Alpher, and Herman was remembered by some but seldom cited and not critically examined until the ideas had been reinvented. I take this to be a consequence of the modest level of research in cosmology prior to the late 1960s, and this in turn to be a consequence of the data-starved nature of the subject. Other lines of research offered far richer grounds for empirically driven research that tested and stimulated theoretically driven advances in physics. Cosmology in the 1950s certainly was a real physical science; there were theories, observations, and lines of research aimed at improving both. But the exceedingly slow pace of research allowed ample time to lose sight of what had come before.

CHAPTER FIVE

How Cosmic Structure Grew

HOW DID THE observed large-scale homogeneity of the universe come to be broken on smaller scales by the formation of cosmic structure: galaxies, their spatial clustering, and their internal constitutions? Ideas were informed in part by the gravity physics to be reviewed in Section 5.1, in part by the observations, and of course in part by nonempirical assessments of what seemed right and reasonable by intuition and experience. Gravity holds galaxies together, so it seems reasonable enough to expect that gravity also played an important role in assembling them by rearranging the primeval mass distribution. Nongravitational stresses, such as those produced by radiation and magnetic fields, must have been important, too, because they are observed to be rearranging matter within galaxies. But these thoughts leave considerable freedom to adjust initial conditions and introduce dynamical actors in addition to gravity when they seem to be needed to advance a story. This chapter reviews the resulting abundance and confusion of ideas leading up to the situation in the early 1980s. That is when the astronomers' subluminal matter discussed in Chapter 6 and the particle physicists' nonbaryonic matter considered in Chapter 7 helped bring order out of the many ideas reviewed here and reveal a path to a well-supported relativistic cosmological model.

A theory of how the galaxies formed in the big bang cosmology has to provide a physically consistent picture of how cosmic structure evolved from the very different conditions in the early stages of expansion. That consideration is absent in the 1948 steady-state cosmology, so thinking about structure formation had to be different. Gold and Hoyle (1959) considered the idea that the continually created matter might be hot enough to be thermally unstable. That is, a region in which the density of newly created matter is slightly greater than average might radiate and cool more rapidly than average, producing a pressure gradient that pushes more matter into the regions of higher density. Gold and Hoyle proposed that continually created neutrons would, by their decay energy, fill space with plasma hot enough for an interesting degree of thermal instability. But Gould and Burbidge (1963) pointed out that thermal bremsstrahlung radiation by this hot plasma would exceed the measured X-ray background. Thermal instability is important in astrophysics (Field 1965), but it does not appear likely to be important in the assembly of galaxy-size mass concentrations. Sciama (1955) proposed that in the steady-state cosmology motions of galaxies relative to the Hubble flow would produce wakes, caused by the gravitational attraction of the galaxy masses. Newly created matter falling into wakes could collapse to form young galaxies. Harwit (1961) argued that the effect is too weak to produce new galaxies, but he pointed out that one is free to postulate that the continual creation is sufficiently patchy to have produced clumps of matter dense enough for gravity to have pulled the matter together to make galaxies.

We see that the very simplicity of the prescription for the 1948 steady-state cosmology, which made it such an inviting target for some cosmological tests, did not equip it for usefully constrained analyses of how cosmic structure formed. The relativistic Friedman-Lemaître cosmology shares with the 1948 steady-state cosmology this need to make additions to the theory to account for the presence of galaxies. In the former, it may be inhomogeneous continual creation; in the latter, it may be primeval spacetime curvature fluctuations. This means both cosmologies are incomplete, which does not falsify them, of course. A topic for this chapter is how we learned to live with the need to stipulate initial conditions in what became the standard cosmology.

In the relativistic cosmology, gravity slows the expansion of the universe. The rate of expansion of matter in a region in which the local mass density happens to be slightly greater than average would be slowed by gravity at a slightly greater-than-average rate. This may eventually stop the expansion in overdense regions, causing the matter to settle into gravitationally bound objects that we might imagine become galaxies. In short, the relativistic expanding universe is gravitationally unstable to the growth of departures from a homogeneous mass distribution—provided the mean mass density is not too small, and radiation pressure and Einstein's cosmological constant are not too large.

This gravitational instability has a different character from that of fluid flow, which led to early confusion. To some, the gravitational instability seemed to be too weak to be effective. To others, images of spiral galaxies suggested fossil eddies of primeval turbulence rather than an effect of the simple pull of gravity. Observations of winds blown from galaxies by the energy released by massive stars, supernovae, and gravitational collapses suggested the thought that an early generation of seeds, maybe stars, might have triggered a hierarchy of explosions that drove structure formation on ever larger scales. And yet another thought given to us by condensed matter and particle physics is that the spontaneous formation of cosmic strings or other field defects might have broken the symmetry of an initially perfectly homogeneous universe and triggered structure formation. The physical and observational bases for the merits and problems with these ideas are the subject of this chapter.

5.1 *The Gravitational Instability Picture*

In the gravitational instability picture, the concentrations of mass in galaxies grew by the gravitational attraction of matter in the expanding, initially very close to homogeneous universe. This is a universe without edges. Before Einstein there were conceptual issues with the idea. Edmund Halley (1720, 22) wrote that if the system of the world were

> finite; though never so extended, [it] would still occupy no part of the *infinitum* of Space, which necessarily and evidently exists; whence the whole would be surrounded on all sides with an infinite *inane*, and the superficial Stars would gravitate towards those near the center.... But if the whole be Infinite, all the parts of it would be nearly *in equilibrio*, and consequently each fixt Star, being drawn by contrary Powers, would keep its place; or move, till such a time, as from such an *equilibrium*, it found its resting place; on which account, some, perhaps, may think the Infinty of the Sphere of Fixt Stars no very precarious Postulate.

Issac Newton seems to have had a different opinion about whether this equilibrium is stable. In his letter to the theologian Richard Bentley dated January 17, 1692/3, Newton wrote (in the normalized text from the Newton Project, Oxford): "And much harder it is to suppose that all the particles in an infinite space should be so accurately poised one among another as to stand still in a perfect equilibrium."

James Jeans (1902) considered this issue in analyses of the stability of distributions of gravitating matter in a broad variety of situations (in Newtonian mechanics, of course). Jeans, with Halley, remarked that if the distribution is bounded, filling only part of space, and pressure does not prevent it, then gravity would cause an initially static distribution to collapse toward its center of mass. But if a nearly homogeneous distribution of matter fills all infinite space, there is no way to define a center of mass, and no way to define the direction of the gravitational acceleration produced by this infinite distribution. And Jeans followed up on Newton's point: the demonstration that small local departures from exact homogeneity are unstable.

Let us follow Jeans's argument with a change to more recent notation. Let the mean mass density be the constant $\rho_{\rm m}$, and let the mass density as a function of position $\vec{x}$ and time t be $\rho(\vec{x}, t)$. The fractional departure from the mean—the density contrast—is defined as

$$\delta(\vec{x}, t) = \frac{\rho(\vec{x}, t)}{\rho_{\mathrm{m}}} - 1. \tag{5.1}$$

It is an easy exercise to check Jeans's conclusion that in this nearly static unbounded homogeneous mass distribution with density contrast δ small enough to be treated in linear perturbation theory, and assuming that matter pressure may be ignored, the density contrast grows as

$$\delta \propto e^{\pm t/\tau}, \quad \tau = (4\pi G \rho_{\mathrm{m}})^{-1/2}. \tag{5.2}$$

The solution with the positive sign in the exponential shows the situation is unstable: a small departure from exact static homogeneity grows much larger in a few multiples of the characteristic exponential growth time τ.

Jeans's calculation is sometimes termed a swindle, because it is difficult to see how to fit into Newtonian physics the picture of a nearly uniform distribution of gravitating mass that fills all infinite space. It is an example of good intuition, however. Einstein's (1917) static universe is exponentially unstable on the time scale given in equation (5.2).

Einstein showed how the picture of a nearly uniform unbounded distribution of matter readily fits into the relativistic cosmology. And Jeans's Newtonian calculation is readily adapted to describe perturbative departures from homogeneity in a relativistic expanding universe of low-pressure matter. This is because Newtonian mechanics is the limiting case of general relativity when relative velocities are small compared to the speed of light and the departure of spacetime from the flat Newtonian geometry is small across the region of interest. In particular, our universe is supposed to be close to homogeneous, so we can consider the behavior of a part of it that is small compared to the Hubble length H_0^{-1}. Then if the pressure is small, $p \ll \rho c^2$, Newtonian mechanics is a good approximation. This is why McCrea and Milne (1934) could use Newtonian mechanics to derive the Friedman-Lemaître equations that describe the general expansion of the universe (see page 45 in Section 3.2), and Bonnor (1957) could find the theory of the evolution of departures from exact homogeneity in an expanding universe without resorting to the full machinery of general relativity. The approach is limited by the assumptions, of course. We need to look further into general relativity to understand the predicted behavior of the early universe when it was dominated by the mass in radiation, the CMB.

The instability of Einstein's static model of the universe was first apparent from Lemaître's (1927) solution for an expanding universe that asymptotically traces back to Einstein's (1917) static solution in the remote past. That is, Lemaître's solution represents the evolution of a homogeneous perturbation to

Einstein's solution. A slight homogeneous disturbance in the other direction, to greater density, would cause Einstein's universe to collapse.

Eddington (1930, 670) looked at this in a slightly different way:

> Evidently Einstein's world is unstable. The initial small disturbance can happen without supernatural interference. If we start with a uniformly diffused nebula which (by ordinary gravitational instability) gradually condenses into galaxies, the actual mass may not alter but the equivalent mass to be used in applying the equations for a strictly uniform distribution must be slightly altered. It seems quite possible that this evolutionary process started off the expansion of the universe.

It is not clear what Eddington meant by "ordinary gravitational instability;" perhaps he had in mind Jeans's analysis, or perhaps just the intuitive notion that gravity tends to draw matter together.

Lemaître (1933c, eq. (3)) found that in an expanding universe, the mass density contrast defined in equation (5.1) grows as

$$\delta(x,t) \propto t^{2/3}, \tag{5.3}$$

in linear perturbation theory and assuming pressure, space curvature, and the cosmological constant all may be ignored. This is from his analytic solution to the Friedman-Lemaître equation for spherical symmetry with zero pressure (Lemaître 1933b). Lifshitz (1946) found this result without assuming spherical symmetry, in his linear perturbation analysis in general relativity. Bonnor (1957) found it in the Newtonian limit.

In the static situation, the mass density provides the characteristic time scale $(4\pi G\rho_{\rm m})^{-1/2}$ in equation (5.2) for the exponential growth of a departure from homogeneity. There also is a characteristic time scale for the exponential growth of departures from laminar flow in fluid dynamics with negligible viscosity. But the evolution of the mass density contrast δ in equation (5.3) could not have been exponential, because in this situation there is no time scale. If we are given that δ is a function of time in this situation, then we know the function has to be a power law (or a sum of power laws).

Three aspects of the gravitational instability picture for the formation of our clumpy universe caused early misgivings. The first is Lifshitz's demonstration that, in general relativity, the mass density perturbation represented by equation (5.3) is accompanied by a nearly constant perturbation to spacetime curvature. It is worth pausing to consider how this comes about. In a region of physical size r in which the matter density departs from the mean $\rho_{\rm m}$ by the fractional amount δ (equation (5.1)), the mass in excess of homogeneity is $\delta M \sim \rho_{\rm m}\, \delta\, r^3$. This excess mass has Newtonian gravitational potential per unit mass

$$\phi \sim G\delta M/r \sim G\rho_{\rm m} r^2 \delta \sim \text{constant}. \tag{5.4}$$

In the expanding model, the radius of the region is increasing as $r \propto a(t)$, and the mean mass density is decreasing as $\rho_{\rm m} \propto a(t)^{-3}$. In the Einstein–de Sitter model, the expansion parameter varies as $a(t) \propto t^{2/3}$, and we see from equation (5.3) that the density contrast is increasing as $\delta \propto a(t)$. The combination of these relations shows that the potential ϕ is constant as δ grows. This constant gravitational potential translates to a constant perturbation to spacetime curvature in general relativity. Lemaître's analytic solution to Einstein's equation that illustrates this situation is discussed in Section 5.1.1.

Apart from neutron stars and black holes, gravitational potentials ϕ in our universe are observed to be small. For example, a large galaxy with internal velocity dispersion $v \sim 300$ km s^{-1} has dimensionless gravitational potential $\phi \sim (v/c)^2 \sim 10^{-6}$. This is well into the Newtonian limit, and we can term the value of ϕ occasioned by the concentration of mass in this galaxy to be a dent in spacetime. But even though the perturbations δ to the mass distribution that would grow into galaxies are arbitrarily small at arbitrarily early time t, as far back in time as the standard model applies, spacetime has nearly permanent primeval departures from exact homogeneity: dents. In this theory, which was well tested by the turn of the century, galaxies are in effect built into the initial state of the expanding universe as dents in spacetime. An early discussion of this situation is in Peebles (1967a).

Even minor dents may seem unreasonable in the nominally perturbative gravitational instability picture for structure formation. The picture gained credibility from numerical simulations that show how small departures from homogeneity can grow into clumps of matter that appear to be suggestive of an approach to the observed natures of galaxies and their clumpy space distribution. And in the early 1980s, cosmological inflation at last offered an elegant idea of a pedigree for the dents: Quantum zero-point modes would be squeezed to classical curvature fluctuations during the nearly exponential expansion of the early universe.

The second troubling aspect of the gravitational instability picture is the considerable difference in character from the instability of laminar flow of a fluid with low viscosity. The development of turbulence out of laminar flow can be said to cause the fluid to "forget" its initial conditions: fluid elements initially close in phase space grow exponentially well separated. The power-law evolution of the density contrast in equation (5.3) does not share this feature: recall the dents. It means that the established ΛCDM cosmology has to include parameters that specify details of initial conditions. The freedom to adjust initial conditions to fit what is observed might only amount to shaping the theory to fit whatever is wanted. Thus it is important that there are enough independent constraints to rule out this interpretation, as will be argued in Chapter 9. The positive consequence of the memory of primeval conditions is that the established initial conditions offer us a fossil left from what happened in the still-earlier universe, before the ΛCDM theory could have been a useful approximation.

Lifshitz (1946) pointed out the third of the apparently troubling aspects of the gravitational instability picture for structure formation. He estimated that in an Einstein–de Sitter model universe dominated by nonrelativistic matter, the growth of the mass density contrast, $\delta \propto t^{2/3}$, from the time when the cosmic mean mass density was the density within atomic nuclei to the present time, when the mean density is $\sim 10^{-30}$ g cm^{-3}, is a growth factor of about 10^{15}. But there are $N \sim 10^{68}$ protons plus neutrons in a large galaxy such as the Milky Way. If the nucleons were initially distributed uniformly at random, the initial contrast would be $\delta_{\rm i} \sim N^{-1/2} \sim 10^{-34}$. That number multiplied by the growth factor $\sim 10^{15}$ is still a tiny departure from homogeneity, not the prominent mass concentrations seen in galaxies.

Lifshitz (1946, 116) concluded that perturbations to the mass distribution in an expanding universe "increase so slowly that they cannot serve as centers of formation of separate nebulae or stars." Bonnor (1957) reached the same conclusion by the same argument. Hawking (1966, 550) echoed Bonnor's conclusion, and added that "To account for galaxies in an evolutionary universe, we must assume there were finite, non-statistical, initial inhomogeneities."

Zel'dovich (1962b and 1965) made the same point, and in the second paper added that it is not rational to start with $\delta_{\rm i} \sim N^{-1/2}$ fluctuations anyway. That would require an arbitrarily chosen starting time. For that matter, why apply this initial condition at all? Zel'dovich nevertheless stressed the difficulty of understanding the origin of the primeval spacetime curvature fluctuations required by Lifshitz's analysis.

Georges Lemaître introduced one way around this problem. In Lemaître (1931d, 1934), he proposed that the mass density, space curvature, and cosmological constant may be just such that the universe expanded from high density and then passed through an epoch in which the universe hovered at close to the conditions of Einstein's static solution (as discussed on page 70 in Section 3.6). The universe would then return to expansion that approached exponential, driven by the cosmological constant. The hovering solution approximates the original 1927 solution in which the expansion traces back to Einstein's static universe, but the hovering solution instead traces back to expansion from Lemaître's primeval atom or big bang. It requires a delicate balance of parameters, for as we have seen, Einstein's static solution is exponentially unstable. But the near-static epoch can make the time of expansion from high density larger than the Hubble time H_0^{-1}, which seemed desirable in the 1930s, because H_0^{-1} had been seriously underestimated. And it also was arguably attractive that in the hovering or quasi-static epoch, small departures from homogeneity are close to exponentially unstable (equation (5.2)). Regarding this epoch Lemaître (1934, 13) wrote that "The hypothesis we wish to discuss is that collapsing regions must be identified with the extra-galactic nebulae and the equilibrium-regions with the clusters of nebulae."

But Lemaître's hovering universe has not proved to be of lasting interest. And it may be significant that the extended version of Lifshitz's (1946) analysis presented in Lifshitz and Khalatnikov (1963) does not restate the conclusion that galaxies cannot have formed by gravity.

A pragmatic assessment of the situation is that, in general relativity, small departures from a homogeneous mass distribution, $\delta_i \ll 1$, present at sufficiently early stages of expansion of the universe, can grow into mass concentrations that are much denser than the cosmic mean, provided their escape velocities are well below the speed of light, meaning that $|\phi| \ll 1$ in equation (5.4), so spacetime is only modestly dented. As will be discussed, this condition also means that the sizes of gravitationally grown mass concentrations are much smaller than the Hubble length, which certainly applies to galaxies. Early arguments for this way to think of the gravitational formation of galaxies are in Raychaudhuri (1952), who referred to the solution for a pressureless spherically symmetric mass distribution in Tolman (1934b); Novikov (1964a), who referred to Lifshitz's (1946) analysis; and Peebles (1967a), who referred to Lemaître's (1931d and 1933b) solution.

Visual illustrations can be influential. The numerical N-body methods used to study the behavior of stars in clusters (as in von Hoerner 1960), and the behavior of stars in the disks of galaxies (as discussed in Section 6.4), were used in Peebles (1972) to present a numerical illustration of the growth of departures from homogeneity in the relativistic model of an expanding universe. Davis et al. (1985) used a more competent computation to produce a clearer example. It was for a different purpose, the exploration of conditions for the formation of galaxies and their spatial distribution, because by the time of their paper, the community by and large had implicitly come to disregard the features earlier considered troubling and accept the gravitational instability picture as a promising idea.

Later, still better, numerical computations show in beautiful detail evolving distributions of baryonic and dark matter that start with small departures from homogeneity and grow into mass concentrations that look much like real galaxies. These numerical simulations require adjustment of parameters in skillfully crafted prescriptions meant to approximate how stars form and the effect of their formation on the matter around them. That is, the gravitational instability picture proves to be a useful working hypothesis for detailed studies of how the galaxies formed. This adds to the case for the relativistic ΛCDM cosmology, but the weight is difficult to judge. Could there be an alternative gravity theory that adequately accounts for the mean evolution of the universe and, with a lot of clever work on devising prescriptions and adjusting their parameters, can also produce realistic-looking simulations of galaxies? But it is difficult to imagine anyone being motivated enough for a thorough check of this question absent a serious challenge to general relativity.

Studies of how galaxies would have formed in the relativistic hot big bang model must take account of the addition to cosmology of the 3 K CMB radiation discussed in Chapter 4. It makes the picture of how departures from homogeneity grew much more interesting than the simple power law behavior of equation (5.3). Let us note five points.

1. The CMB sets the redshift z_{eq}, at which the mass density in radiation fell below the density in nonrelativistic matter. Gamow (1948a) recognized that at z_{eq}, the mass distribution (we now say the distribution of the nonbaryonic dark matter) became unstable to the gravitational growth of departures from homogeneity on scales much less than the Hubble length. He did not explain. We know that in the late 1940s, Gamow was aware of Lifshitz's analysis of the evolution of departures from homogeneity in general relativity (as explained on page 198). Perhaps Gamow extended Lifshitz's analyses to this consideration of the role of z_{eq}. Or perhaps it was another example of his intuition, here the feeling that when the rate of expansion of the universe is driven by radiation, the self-gravitation of the lower mass density in matter could not be important. The redshift at equality is $z_{eq} = 3400$ (equation (4.11)). Gamow's (1948a) estimate, $z_{eq} \simeq 300$, is not very far off, considering all the uncertainties. He was arguing for a basic element of the established picture.
2. The CMB gives the baryons a temperature history through the thermal coupling of matter and radiation. This sets the characteristic Jeans mass, the minimum mass for structures to have formed by gravity against the resisting pressure-gradient force. The Jeans mass is discussed in Section 5.1.4. It certainly is interesting, but at present there is no generally accepted empirical signature of this quantity.
3. The dynamical coupling between plasma and radiation in the early universe would have caused baryons and radiation to behave as a fluid with sound speed comparable to the speed of light, because the mass densities in plasma and radiation are comparable. The fluid viscosity is set by the diffusion of radiation through the plasma. Departures from a homogeneous distribution of this fluid were able to grow by gravity on scales larger than the Hubble length (which is in effect the Jeans length for this high-pressure fluid). An informal interpretation is that when different parts of the plasma-radiation fluid were separated by more than the Hubble length $ct \sim cH_0^{-1}$, they could not have "known" about or responded to a pressure difference. They instead would have behaved as homogeneous universes with slightly different parameters. As time advanced and cH_0^{-1} grew

larger than the physical scale of departures from homogeneity, the fluid would have responded to the pressure difference, oscillating as an acoustic or pressure wave. The amplitude of the oscillation would have been close to constant at longer wavelengths as the universe expanded, but it would decrease at shorter wavelengths, where the effective viscosity of the plasma-radiation fluid was significant. This is reviewed in Section 5.1.3.

4. The coupling of the baryons and CMB would have ended rather abruptly at redshift $z_{\rm dec} \simeq 1200$ when the sea of radiation cooled enough to allow the plasma to combine to a gas of largely neutral atoms (equation (4.13)). This releases the baryonic matter from radiation drag. The now neutral gas could flow through the radiation and gravitationally collect in the growing cosmic structures. In the model established at the turn of the century, there is also nonbaryonic dark matter that interacts with the radiation only by gravity. Its departures from homogeneity would have started to grow by gravity at the Lifshitz-Gamow redshift $z_{\rm eq}$, and the baryons would have moved to join the dark matter distribution after decoupling.
5. The fluid pressure (acoustic) waves present when the baryons and radiation were decoupling left patterns in the distributions of matter and radiation. These patterns are observed in the spatial distribution of the galaxies and the angular distribution of the CMB. The former phenomenon has come to be termed "baryon acoustic oscillations," or BAO. But of course the same physical processes determine the CMB anisotropy. This is reviewed in Section 5.1.3. The BAO signatures are an important part of the cosmological tests.

The ideas about the gravitational growth of cosmic structure outlined here have come to be accepted community knowledge. Let us consider now some of the details of the considerations that helped establish this consensus, and the considerations that did not always make the physics seem so obvious and promising.

5.1.1 LEMAÎTRE'S SOLUTION

Lemaître (1931d and 1933a,b) found an analytic solution to Einstein's general relativity field equation for a model universe of matter with zero pressure that is inhomogeneous but spherically symmetric. It is sometimes termed the "Lemaître-Tolman-Bondi solution," and sometimes one or a combination of two of the names. Tolman (1934b) referred to the prior discussion by Lemaître (1933a). Bondi (1947) referred to Tolman (1934b) and Lemaître (1931d). (Dingle 1933a,b wrote out Einstein's equation for the general diagonal line element, which includes a spherically symmetric model, and examined properties of solutions, but as best I can tell, Dingle did not find the zero-pressure spherical solution.)

In this solution, the mass density and the motions of mass shells may be functions of distance from the center of symmetry as well as of time, but the mass shells are assumed not to cross. For our purpose, the point of this solution is that it gives us a quantitative example of the proposition that, in general relativity, the prominent but nonrelativistic mass concentrations in galaxies could have grown by gravity out of initially small departures from a homogeneous mass distribution, although accompanied by the primeval dents in spacetime discussed in Section 5.1.

The coordinate labeling in this solution, in notation (adapted from Peebles 1967a) that invites comparison to the Robertson-Walker line element in equation (3.1), is

$$ds^2 = dt^2 - \frac{\left[\partial(ax)/\partial x\right]^2 dx^2}{1 - x^2 R(x)^{-2}} - a^2 x^2 \left(d\theta^2 + \sin^2\theta \, d\phi^2\right). \quad (5.5)$$

Each shell of the pressureless matter has a fixed coordinate label x. The function $R(x)^{-2}$, which is independent of time, can be negative or zero. A positive value is assumed in this discussion, because it describes the growth of a mass concentration. The expansion factor is a function of time and coordinate radius: $a = a(x,t)$. The world time, t, at any point x is the physical time kept by an observer moving with the matter at x.

We read from the line element (5.5) that the physical length $\delta l_\perp$ of the segment of a circle at fixed x that subtends the segment of angle $\delta\theta$ at the origin, and the physical length $\delta l_\parallel$ in the radial direction between the mass shells at x and $x + \delta x$ are

$$\delta l_\perp = x a(x,t)\delta\theta, \quad \delta l_\parallel = \frac{(\partial ax/\partial x)\delta x}{\sqrt{1 - x^2 R(x)^{-2}}}. \quad (5.6)$$

The product $l_\perp = x a(x,t)$ is said to be the angular size distance from the origin to coordinate position x at time t.

The expansion parameter $a(x,t)$ and the mass density $\rho(x,t)$ measured by an observer at fixed x and keeping physical time t at that position satisfy the equations

$$\frac{\ddot{a}}{a} = -\frac{4\pi}{3}\frac{GD}{a^3}, \quad \left(\frac{\dot{a}}{a}\right)^2 + \frac{1}{a^2 R(x)^2} = \frac{8}{3}\pi G \frac{D}{a^3}, \quad \frac{\rho}{a}\frac{\partial(ax)}{\partial x} = \frac{D}{a^3}. \quad (5.7)$$

Here D is a constant, independent of x and t. The overdots are first and second partial derivatives of $a(x,t)$ with respect to the physical time t kept by an observer at the fixed value of x attached to a mass shell. These equations can be compared to the Friedman-Lemaître equations (3.3) for pressureless homogeneous matter with mass density $\rho(t) = D a(t)^{-3}$.

The form of the third of equations (5.7) can be understood by writing the first equation as

$$\ddot{l}_\perp = -\frac{GM_{\rm eff}}{l_\perp^2}, \text{ where } l_\perp = a(x,t)x \text{ and } M_{\rm eff}(x) \equiv \frac{4\pi}{3}Dx^3. \tag{5.8}$$

The mass $M_{\rm eff}$ is a function of x alone. Its differential, with the third of equations (5.7), is

$$\delta M_{\rm eff} = 4\pi\rho(x,t)l_\perp^2 \frac{\partial(ax)}{\partial x}\delta x = 4\pi\rho(x,t)l_\perp^2\, \delta l_\parallel (1 - x^2R(x)^{-2})^{1/2}. \tag{5.9}$$

The last expression is the product of the physical mass density and the physical volume between the neighboring mass shells, with a correction factor. If $x^2R(x)^{-2}$ is small, this correction reduces $M_{\rm eff}$ by the fractional amount

$$\frac{\delta M_{\rm eff}}{M_{\rm eff}} = -\frac{1}{2}x^2R(x)^{-2} \equiv \phi(x). \tag{5.10}$$

We see that the radius of curvature of space sections plays the role of the mass deficit from the gravitational binding energy $\phi(x)$. (Here ϕ is not the polar angle in equation (5.5)!) Another way to see this is to write the second of equations (5.7) as

$$\frac{\dot{l}_\perp^2}{2} - \frac{GM_{\rm eff}}{l_\perp} = \phi(x). \tag{5.11}$$

Here $\phi(x)$ serves as the effective energy of the mass shell, kinetic plus potential.

The solution to equations (5.7) is the parametric form

$$\begin{aligned} a(x,t) &= A(x)\,(1-\cos\eta), \\ t &= A(x)R(x)\,(\eta - \sin\eta), \\ A(x) &= \frac{4}{3}\pi GR(x)^2D. \end{aligned} \tag{5.12}$$

A constant of integration can be added to the second line to represent a variable starting time. We see that each mass shell labeled by a fixed coordinate value x evolves in the same way as a homogeneous Friedman-Lemaître model universe. This recalls Birkhoff's theorem: the empty exterior of a spherically symmetric system has the unique geometry of the Schwarzschild solution characterized by one parameter, the mass. Here the motion of a mass shell is determined by $R(x)$ and the parameter D, along with the freedom to specify a starting time $t_{\rm i}$. But the motion of the shell is independent of what is happening inside or outside the shell. This assumes the shells do not cross, along with the starting assumptions of negligible pressure and spherical symmetry.

In this model, the central region has just stopped expanding and is starting to collapse when $\eta = \pi$ at $x = 0$ in equations (5.12). At this point, the time of expansion of the center, $t_{\rm c}$, the central mass density, $\rho_{\rm c}$, and the physical radius of curvature $a(0,t_c)R(0)$ of space sections of constant time at the center work out to

$$t_c = \left(\frac{3\pi}{32G\rho_c}\right)^{1/2}, \quad \frac{\rho_c}{\rho_m(t_c)} = \frac{9\pi^2}{16}, \quad a(0,t_c)R(0) = \frac{2t_c}{\pi}. \tag{5.13}$$

Here ρ_m is the mass density in the zero-pressure Einstein–de Sitter model,

$$\rho_m(t) = (6\pi G t^2)^{-1}. \tag{5.14}$$

The central mass density at the point of maximum expansion, at time t_c, is on the order of the homogeneous Einstein–de Sitter density at time t_c. The physical radius of curvature of space at the center of the mass concentration is comparable to the expansion time t_c. This is not unexpected: Given the units, how else might the density and radius of curvature be approximated?

The evolution of the mass density when it is only slightly perturbed from Einsten–de Sitter is found by writing out the expressions for the expansion parameter and time in equation (5.12) to the lowest two powers of η. The result for the central region, $x \to 0$, is

$$\delta(0,t) = \frac{3}{20}\left(\frac{6\pi t}{t_c}\right)^{2/3} + 2\,\frac{t_i(0)}{t}. \tag{5.15}$$

The density contrast is $\delta(\vec{x},t) = \rho(\vec{x},t)/\rho_m - 1$ (equation (5.1)); this is the central value. The decaying term follows by adding the constant of integration $t_i(x)$ to the second of equations (5.12). It can be interpreted as a disturbance to the starting time that becomes unimportant as time goes on. In the growing term on the right-hand side of equation (5.15), t_c is the time at which the central region stops expanding and starts to collapse, as in equation (5.13). This term evaluated at $t = t_c$ is $\delta(0,t_c) = 3(6\pi)^{2/3}/20 = 1.06$. This linear perturbation approximation can be compared to the analytic solution in equation (5.13): $\delta_c = \rho_c/\rho_m(t_c) - 1 = 4.6$. It is greater than the extrapolation of the linear perturbation solution (5.15), but not by a large factor.

Let us return now to the observation that the velocity dispersions of stars and plasma in galaxies rarely exceed $v \sim 300$ km s^{-1}, or 10^{-3} times the speed of light, apart from the extreme concentrations of mass in the centers of galaxies and in stellar remnants. The relatively small velocities mean that galaxies and clusters of galaxies are nonrelativistic, and so well described by Newtonian mechanics. In the gravitational instability picture, these seriously nonlinear but nonrelativistic systems grew out of the small primeval dents in spacetime discussed on page 189. To see how this follows in the spherical solution, note that the line element (5.5) is only slightly perturbed from a homogeneous spacetime if $\phi(x) \equiv -x^2 R(x)^{-2}/2$ is small. This is the effective energy in equation (5.11). It is small in Newtonian mechanics, and it is small in galaxies (apart from the occasional tight mass concentrations in the centers of galaxies and stellar remnants): $\phi \sim (v/c)^2 \sim 10^{-6}$.

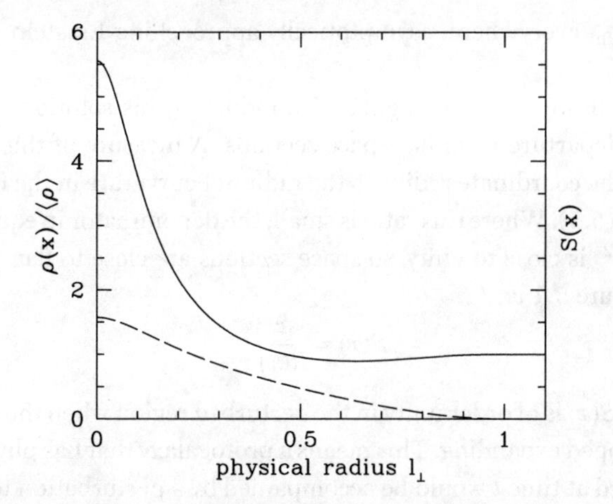

FIGURE 5.1. An example of Lemaître's solution at maximum expansion of the center. The solid curve is the density as a function of the physical angular size radius $l_\perp$ at fixed t. The dashed curve is the measure S of space curvature defined by equation (5.17).

We see from equation (5.13) that at the time t_c of maximum expansion of the central region of a mass concentration with physical size $r = ax$, the gravitational potential is

$$\phi = -\frac{1}{2}x^2 R(x)^{-2} = \frac{\pi^2}{8}\left(\frac{a(0,t_c)x}{t_c}\right)^2. \tag{5.16}$$

If the size at formation is small compared to the Hubble length ct_c, then this is a nonrelativistic object: $|\phi| \ll 1$.

Let us consider finally the example of Lemaître's solution in figure 5.1.[1] The geometry and mass density at radius $l_\perp > 1$ are the Einstein–de Sitter solution. The solid curve is the ratio $\rho(x)/\rho_m$ of the physical mass density to the Einstein–de Sitter value as a function of the physical angular size radius $l_\perp$. The fixed time t is chosen so the central region has just stopped expanding and the central mass density is given by equation (5.13). In this example, the mass density is less than Einstein–de Sitter at $0.5 \lesssim r < 1$, which compensates for the excess central mass. But of course one can choose initial conditions so

1. Convenient combinations of equations (5.12) for this computation are

$$b_x^3 = \frac{2}{9}\frac{(1-\cos\eta)^3}{(\eta-\sin\eta)^3}, \quad \frac{\rho_x}{\rho_m} = \frac{1}{b_x^2\partial(b_x x)\partial x}, \quad b_x = \frac{a_x}{\bar{a}}, \quad \bar{a}^3 = \frac{D}{\rho_m}.$$

Here ρ_m is the mass density in the Einstein–de Sitter solution. The choice of the parameter η as a function of x in the last expression is free, apart from avoidance of shell crossings. This example uses $\eta = \frac{\pi}{2}(1+\cos(\pi x))$, which produces a smooth density peak and a smooth transition to the Einstein–de Sitter solution at coordinate radius $x > 1$. The measure of space curvature in equation (5.17) is

$$S(x) = \frac{x\,t}{R(x)l_\perp} = \frac{\eta - \sin\eta}{1-\cos\eta}.$$

that $\rho(x) \geq \rho_m$ everywhere, asymptotically approaching Einstein–de Sitter at large radius.

The length unit for the angular size radius in this solution is set by the size of the departure from flat space sections. A measure of this is the ratio $xR(x)^{-1}$ of the coordinate radius to the radius of curvature in the line element in equation (5.5). Where this ratio is small, the denominator in equation (5.5), $1 - x^2/R(x)^2$, is close to unity, so space sections are close to flat. The dashed curve in Figure 5.1 is

$$S(x) = \frac{x}{R(x)} \frac{t}{r}. \tag{5.17}$$

We see that $S(x)$ is of order unity in the perturbed region when the central part has just stopped expanding. This means a protogalaxy that has physical radius r as it formed at time t would be accompanied by a perturbation to spacetime of order $x^2 R(x)^{-2} \sim (r/t)^2$. Since galaxies would have formed at $r \ll t$, the ratio of the size of a young galaxy to the radius of curvature of space sections is $x^2 R(x)^{-2} \ll 1$, a dent in spacetime, as we have seen from the considerations leading to equation (5.16).

This spherically symmetric solution does allow $x^2 R(x)^{-2}$ to approach unity, meaning that the mass concentration is approaching a spacetime singularity, a relativistic black hole in an unusual coordinate labeling. Black holes with masses comparable to that of a galaxy are not observed; this case is not of empirical interest. The massive compact objects observed at the centers of many galaxies are thought to be black holes that formed by dissipative contraction during and after gravitational assembly of the galaxies. If these central mass concentrations existed prior to galaxy formation, then we would have to conclude that primeval spacetime was only slightly dented for the most part, but that here and there, where galaxies were going to form, there were more serious departures from the Friedman-Lemaître solution.

Lemaître's solution gives us a useful measure of the primeval conditions needed for the gravitational assembly of galaxies in general relativity, but it is limited by the assumption of spherical symmetry and the neglect of pressure. Evgeny Mikhailovich Lifshitz introduced the full perturbative analysis of departures from homogeneity and isotropy, to be discussed next.

5.1.2 LIFSHITZ'S PERTURBATION ANALYSES

Gamow's interest in how the mass concentrations in galaxies grew traces back at least as far as Gamow and Teller (1939). Alpher recalls (in Alpher and Herman 2001, 70) that his first PhD dissertation project with Gamow was meant to follow up the Gamow and Teller discussion by analyzing in general relativity the gravitational growth of departures from homogeneity in an expanding universe. Alpher turned to the analysis of element formation in the early universe when Gamow showed him Lifshitz's (1946) paper on the subject. Alpher and Herman (2001) report that Alpher had the vivid recollection of Gamow coming into his office waving a copy of the journal and saying "Ralph, you have been scooped."

Lifshitz (1946) introduced the general analysis of the evolution of departures from the homogeneity of the Friedman-Lemaître solution to Einstein's equation in linear perturbation theory. Among the cases he considered, Lifshitz showed that if the matter pressure can be neglected, then the evolution of the mass density contrast $\delta(\vec{x}, t)$ defined in equation (5.1), as a perturbative departure from the Einstein–de Sitter model, is of the form

$$\delta(\vec{x}, t) = C_1(\vec{x})\, t^{2/3} + C_2(\vec{x})\, t^{-1}. \tag{5.18}$$

Lemaître (1933c) seems to have been the first to find the growing term in this equation, under the assumption of spherical symmetry. In Lifshitz's analysis, the coefficients C_1 and C_2 are freely chosen functions of position.

An observer at fixed coordinate position $\vec{x}$ sees that the local mass density $\rho(\vec{x}, t)$ as a function of time t kept by a physical clock departs from the Einstein–de Sitter model by the fractional amount given by equation (5.18). The second term is said to be coordinate dependent: it is changed by changing the starting time on the clock. This consideration gets more complicated when matter pressure cannot be ignored; the standard practice is to use Bardeen's (1980) gauge-invariant formalism. But for purposes of tracing the historical development of this subject, it is safe to use equation (5.18) and its generalizations.

We have learned how to simplify Lifshitz's calculation for cases of interest in cosmology. Newtonian physics is a good approximation for analysis of the growth of a nonrelativistic mass concentration, such as a galaxy (as discussed on page 187). In this approximation, and in linear perturbation theory, the equations of conservation of mass and momentum are

$$\frac{\partial \delta}{\partial t} = -\frac{1}{a(t)} \nabla \cdot \vec{v}, \qquad \frac{\partial}{\partial t}\vec{v} + \frac{\dot{a}}{a}\vec{v} = -\frac{1}{a}\nabla \phi, \qquad \vec{v} = a\frac{d\vec{x}}{dt}. \tag{5.19}$$

The coordinates are the physical world time t and the coordinate position $\vec{x}$ that is comoving with the matter in the unperturbed, spatially homogeneous cosmological model. An element of coordinate displacement $d\vec{x}$ translates to the element of physical displacement $d\vec{r} = a(t)d\vec{x}$. Thus matter moving relative to the Hubble flow with coordinate velocity $d\vec{x}/dt$ has the proper peculiar velocity $\vec{v} = a(t)d\vec{x}/dt$ in equations (5.19) The source for the gravitational potential ϕ is the departure of the mass distribution from homogeneity, and the Poisson equation for ϕ is

$$\frac{1}{a^2}\nabla^2 \phi = 4\pi G \rho_{\rm m} \delta(\vec{x}, t). \tag{5.20}$$

The proper gravitational acceleration, meaning the acceleration relative to the unperturbed background flow, is $\vec{g} = -\nabla\phi/a$. If $\nabla\phi = 0$, the second of equations (5.19) indicates that the peculiar velocity field is decreasing as $\vec{v}(\vec{r}, t) \propto 1/a$. This is because the flowing matter is overtaking comoving observers that are moving away with the Hubble flow. A more careful explanation of this setup is presented on page 45 in Section 3.2 and in even more detail in Peebles (1980, §6).

We can combine the time derivative of the first of equations (5.19) with the divergence of the second and use equation (5.20) to get[2]

$$\frac{\partial^2 \delta}{\partial t^2} + 2\frac{\dot{a}}{a}\frac{\partial \delta}{\partial t} = 4\pi G\rho_{\rm m}\delta. \tag{5.21}$$

In the Einstein–de Sitter model, where $a(t) \propto t^{2/3}$ and the mass density is $\rho_{\rm m} = (6\pi G t^2)^{-1}$, the growing solution to equation (5.21) is $\delta \propto t^{2/3}$, as in Lemaître's equation (5.3) and Lifshitz's equation (5.18).

Lifshitz found the condition on curvature fluctuations shown in equation (5.16). An alternative derivation in Peebles (1980, §86) shows yet again that the nonlinear but nonrelativistic mass concentrations in galaxies only modestly perturb spacetime, and that on scales larger than the Hubble length, the departure of the density from the mean $\rho_{\rm m}(t)$ must have been small to have avoided large spacetime curvature fluctuations that would have formed relativistic mass concentrations.

In the hot big bang model discussed in Chapter 4, the mass density in the early universe is dominated by radiation. Lifshitz analyzed this situation under the assumption that radiation, or relativistic matter, can be treated as an ideal fluid with pressure $p = \rho/3$. He showed that when the characteristic length scale of a perturbation from homogeneity is much larger than the Hubble length $\sim t$, the evolution of a small fractional departure of the mass density from the cosmologically flat model with $\Lambda = 0$ allows three modes:

$$\delta_r(\vec{x}, t) = D_1(\vec{x})\, t + D_2(\vec{x})\, t^{1/2} + D_3(\vec{x})\, t^{-1}. \tag{5.22}$$

The effect of the pressure gradient force does not appear in this solution. We can say that regions with different densities are too well separated to "know" about the pressure gradient.

In the opposite limit, where the length scale is small compared to t, Lifshitz showed that the perturbation oscillates as an acoustic or pressure wave with velocity of sound $1/\sqrt{3}$ times the velocity of light. The following

2. It is worth pausing to note that in this equation, which assumes linear theory and negligible pressure, the evolution of $\delta(\vec{x}, t) = \delta\rho/\rho$ at coordinate position $\vec{x}$ depends only on the mass density contrast at $\vec{x}$. The perturbation to the mass at $\vec{x}$ is a source for the velocity field $\vec{v}$ outside $\vec{x}$, but in linear theory, gravity gives this component of the perturbed flow zero divergence, so it has no effect on the mass density.

sections show how this and related effects of the CMB make the cosmic evolution of departures from homogeneity much more interesting than for gravity alone.

Lifshitz showed that the gravitational growth of the clustering of matter is suppressed when the rate of expansion of the universe is dominated by a negative space curvature term in the Friedman-Lemaître equation (3.3). Gamow and Teller (1939) offered an intuitive explanation: The universe in effect is expanding with greater than escape velocity, meaning a perturbation from homogeneity also would have escape velocity, so perturbed regions would not be expected ever to stop expanding.

Similar considerations follow when the expansion rate is dominated by Einstein's cosmological constant, Λ, or by the mass density in the CMB. Let us leave it as an exercise for the student to check that the two solutions for the evolution of the matter density contrast in a universe of pressureless matter that interacts with a homogeneous sea of radiation only by gravity is, at $a(t) \ll a_{\rm eq}$,

$$\delta_1 \propto e^{1.5a(t)/a_{\rm eq}}, \quad \delta_2 \propto \log \tau/t, \tag{5.23}$$

with τ a constant. The growing solution approaches a constant at $a(t) \to 0$, meaning that the matter in a density perturbation can expand with the general expansion without significant gravitational evolution of the density contrast until the expansion parameter approaches $a_{\rm eq}$, at redshift $z_{\rm eq}$. Gamow (1948a, 506) recognized this effect:

> The epoch when the radiation density fell below the density of matter has an important cosmogonical significance since it is only at that time that the Jeans principle of "gravitational instability" could begin to work. In fact, we would expect that as soon as the matter took over the principal role, the previously homogeneous gaseous substance began to show the tendency of breaking up into separate clouds.

Jeans (1902) did not demonstrate this. Lifshitz showed that the gravitational growth of departures from a homogeneous mass distribution is suppressed on scales smaller than the Hubble length when the mass density is dominated by coupled radiation and matter that act as a fluid with velocity of sound comparable to the velocity of light. But Lifshitz did not consider Gamow's implicit assumption that the matter can slip through the radiation. Gamow did not present an analytic demonstration of his effect; that was not his style. Guyot and Zel'dovich (1970) and Mészáros (1974) rediscovered it, with derivations. But we can conclude that Gamow had anticipated that in the standard ΛCDM cosmology, the departure from a homogeneous distribution of nonbaryonic dark matter on scales small compared to the Hubble length began to grow when the mass density of matter approached and then exceeded the density in radiation. This would have happened at redshift $z_{\rm eq} \simeq 3400$.

5.1.3 NONGRAVITATIONAL INTERACTION OF BARYONS AND THE CMB

In the early universe, the thermal and dynamical interactions of the baryons with the radiation observed now as the CMB were dominated by momentum transfer by Thomson scattering of the radiation by free electrons.

Consider a free electron moving with nonrelativistic speed v through a homogeneous and isotropic sea of thermal radiation at temperature $T \ll m_e c^2/k$, where m_e is the electron mass. This is the interesting nonrelativistic case valid at redshifts $z \lesssim 10^{10}$. In the electron rest frame, the radiation approaching from the direction in which the electron is moving relative to the sea of radiation is hotter than the radiation approaching from the opposite direction. This is the effect of the Doppler shift. Scattering of this anisotropic radiation in the electron rest frame produces a drag force F on the electron,[3]

$$\vec{F} = -\frac{4}{3}\frac{\sigma_T a_S T^4 \vec{v}}{c}, \tag{5.24}$$

where σ_T is the nonrelativistic Thomson cross section for scattering radiation by a free electron.

The radiation drag force causes the electron to lose kinetic energy at the rate Fv. The mean kinetic energy of a freely moving electron at thermal equilibrium at temperature T_m is $m_e\langle v^2\rangle/2 = 3k_B T_m/2$. For simplicity, let us allow one free ion per free electron. Since Coulomb scattering keeps ions and electrons close to thermal equilibrium, the mean kinetic energy per free electron is then $3k_B T_m$. With this and the rate Fv at which the electron is doing work on the radiation, we get the rate of change of the plasma temperature:

$$\frac{dT_m}{dt} = \frac{8\sigma_T a_S T^4}{3m_e c}\frac{x}{1+x}(T - T_m). \tag{5.25}$$

This allows for partial ionization: the fraction x of the electrons are free and the fraction $1 - x$ are bound in neutral atoms without significant interaction with the radiation. The second term in the parentheses is the effect of radiation drag, which slows the motion of the electron. The first term represents the

3. In the electron rest frame, the Doppler shift causes the radiation energy flux density incident on the electron to vary with angle θ from the direction of motion in proportion to $1 + 4(v/c)\cos\theta$. One may say that the factor of four is the sum of the Doppler shift in frequency of each photon, which contributes anisotropy v/c to this expression, while the Doppler shift in the rate of arrival of photons contributes v/c, and the aberration of solid angles contributes $2v/c$. The rate of transfer of momentum to the electron by the scattered radiation is then, in polar coordinates,

$$F = \sigma_T a_S T^4 \int \frac{\sin\theta d\theta d\phi}{4\pi}\cos(\theta)\left(1 + 4\frac{v}{c}\cos(\theta)\right) = \frac{4}{3}\frac{\sigma_T a_S T^4 v}{c}.$$

effect of the thermal fluctuating force of the radiation on the electrons. The two terms drive the matter temperature to the radiation temperature.[4]

Now let us compare the characteristic thermal relaxation time $\tau_{\rm th}$ for the matter to the cosmic expansion time t:

$$\tau_{\rm th}^{-1} = \frac{8\sigma_{\rm T} a_{\rm S} T_f^4 (1+z)^4}{3 m_{\rm e} c} \frac{x}{1+x}, \quad t^{-1} \simeq \frac{3}{2} H_0 \Omega_{\rm m}^{1/2} (1+z)^{3/2}. \tag{5.26}$$

The second expression is a good approximation at redshifts large enough that space curvature and Λ may be neglected but are still well below $z_{\rm eq}$. The ratio is

$$\frac{t}{\tau_{\rm th}} \approx \frac{0.0056}{\Omega_{\rm m}^{1/2} h} \frac{x}{1+x} (1+z)^{5/2} \approx 0.015 (1+z)^{5/2} \frac{x}{1+x}. \tag{5.27}$$

The last expression uses the parameter $\Omega_{\rm m} h^2 = 0.14$ established by much later tests.

At redshift $z_{\rm dec} \simeq 1200$, when the balance between photoionization and recombination of hydrogen, $\rm p + e \leftrightarrow H + \gamma$, switched to allow combination of the primeval plasma to neutral atoms (equation (4.13)), the thermal relaxation time is much shorter than the expansion time:

$$\tau_{\rm th} \sim 10^{-6} t \text{ at decoupling}. \tag{5.28}$$

This means plasma and radiation can be expected to have been quite close to thermal equilibrium up to decoupling. At redshift $z = 100$, when the radiation temperature was ~ 300 K, the residual ionization would have been about enough so that $\tau_{\rm th} \sim t$ (Peebles 1993, Sec. 6). After that, baryons would be expected to have started cooling at about the adiabatic rate for low-pressure matter, $T_{\rm m} \propto a(t)^{-2}$ (equation (4.7)), until stars formed and heated the matter.

The interval from $z \sim 1000$ to about $z = 30$ has been termed the "cosmic dark ages," because all that was happening was the growth of still-small departures from homogeneity as the matter cooled somewhat more rapidly than the radiation. In the standard cosmology, the dark ages are a consequence of the small effect of the primeval dents in spacetime. The dark ages would have ended as structure formation grew advanced enough to have produced ionizing radiation, perhaps from early generations of massive stars, changing the nearly complete neutrality of matter during the dark ages to the present nearly complete ionization of intergalactic matter.[5]

4. In a little more detail, the rate of dissipation of the mean electron kinetic energy occasioned by radiation drag has to be balanced by thermal fluctuations that tend to increase the mean kinetic energy. Since dissipation and fluctuation serve to keep the mean kinetic energy at the equipartition value, we know the radiation temperature T has to be placed in the parentheses in the right-hand side of equation (5.25), so that $dT_{\rm m}/dt$ vanishes at thermal equilibrium, $T_{\rm m} = T$.

5. The Ly-α resonance absorption line of atomic hydrogen for objects at redshift $z \gtrsim 2$ is shifted into the visible part of the spectrum. Gunn and Peterson (1965) pointed out

One of the factors determining the growth of the mass concentrations that were responsible for reionizing the baryonic matter is the radiation drag on the electrons that tends to make the primeval plasma move with the streaming flow of the radiation. The characteristic dynamical time, $\tau_{\rm dyn}$, for the relaxation of the streaming of the plasma relative to the streaming flow of the radiation follows by replacing the electron mass in equation (5.26) with the mean ion mass, which we can take to be the mass of a proton. This gives the ratio of expansion time to dynamical relaxation time:

$$\frac{t}{\tau_{\rm dyn}} = \frac{8\sigma_{\rm T} a_{\rm S} T_f^4 (1+z)^{5/2}}{9 H_0 \Omega_{\rm m}^{1/2} m_{\rm p} c} \approx 4 \times 10^{-6} (1+z)^{5/2}. \quad (5.29)$$

This expression follows after simplifying by the assumptions that the baryons are fully ionized and are not in clumps compact enough for self-shielding of the matter from the radiation. At decoupling, $z_{\rm dec} \simeq 1200$, the dissipation time is $\tau_{\rm dyn} \sim 0.01$ times the expansion time t. Thus at the approach to decoupling, $\tau_{\rm dyn}$ would have been short enough that plasma and radiation behaved as a fluid at wavelengths comparable to the expansion time. But the viscosity of the fluid would have been appreciable, because $\tau_{\rm dyn}$ was not that much smaller than t.

In the standard ΛCDM cosmology, the departures from homogeneity that grew by gravity into galaxies and concentrations of galaxies were initially adiabatic (where, as discussed in Section 5.2.6, the entropy per baryon number is homogeneous). Prior to decoupling, when baryons and radiation acted as a fluid, the adiabatic perturbations behaved as pressure or acoustic waves. At $z \sim z_{\rm dec}$, the primeval plasma combined to form neutral atoms (Section 4.1), causing $\tau_{\rm dyn}$ to grow larger than t. This set the baryon distribution free to grow under the effect of gravity, joining the inhomogeneous distribution of the nonbaryonic dark matter that had been growing since $z_{\rm eq} \sim 3400$.

This decoupling process is illustrated in a little more detail by ignoring the mass in the nonbaryonic matter (to be discussed in the next chapters) and assuming that the mean space curvature can be neglected. Then in linear perturbation theory, we can expand the mass density contrast in plane waves:

$$\delta(\vec{x}, t) = \int d^3k\, \delta_{\vec{k}}(t) e^{i\vec{k}\cdot\vec{x}}. \quad (5.30)$$

---

that the absence of pronounced absorption of the radiation from quasars at $z \gtrsim 2$ that has passed through the resonance indicates that in intergalactic space, the number density of neutral hydrogen atoms and molecules is well below the cosmic mean density. The detection of the Gunn-Peterson resonant scattering effect at redshift $z \sim 6$ in quasars discovered by the Sloan Digital Sky Survey (Becker et al. 2001) shows the presence of at least a modest neutral fraction at this redshift. The ionized fraction of intergalactic matter had to have been small at $z \gtrsim 20$ to account for the slight effect of free electron scattering on the CMB anisotropy discussed in Section 9.2.

If the physical wavelength $\lambda = 2\pi a(t)/k$ (where $a(t)$ is the expansion parameter in equation (3.1)) is small compared to the Hubble length, then up to decoupling the evolution of the Fourier amplitude $\delta_k(t)$ of the mass density contrast is reasonably well approximated as

$$\delta_k(t) \propto \frac{\exp \int^t (i\omega - \gamma)dt}{(1+R)^{1/4}}, \quad R = \frac{3}{4}\frac{\rho_{\rm m}}{\rho_{\rm r}}. \tag{5.31}$$

The wave is oscillating at frequency ω and decaying at the rate γ, where

$$\omega = \frac{kc}{a}\frac{1}{[3(1+R)]^{1/2}}, \quad \gamma = \frac{k^2 c}{6\sigma_{\rm T} n_{\rm e} a(t)^2}\frac{R^2 + 4(R+1)/5}{(R+1)^2}. \tag{5.32}$$

The factor 3/4 in the ratio R of mean mass densities in baryons and radiation simplifies the expressions a little. The mean number density of free electrons is $n_{\rm e}$. The frequency ω is close to that of a fluid with the equation of state $p = \rho/[3(1+R)]$. The Thomson scattering cross section $\sigma_{\rm T}$ enters equation (5.24), which describes the radiation drag on the plasma. It enters equation (5.32), which describes the effect of radiation diffusion through the plasma in determining the rate γ of dissipation of a pressure wave. Equations (5.31) and (5.32) are obtained in the short mean path limit of the Boltzmann equation for the radiation. The notation is from Peebles (1971a, eqs. [92.33–35]).[6]

Panel (a) in Figure 5.2 is an example of the evolution of the Fourier amplitude $\delta_k(t)$ of a plane wave in the baryon distribution in a numerical solution of the Boltzmann equation for the radiation interacting with the motion of the matter (all baryonic). The epoch of decoupling of baryons and radiation as the plasma combines is indicated by the vertical dashed line. This is an early computation, unpublished. (I don't remember making it, but it's in our file, marked October 1968, so who else?). The wavelength $\lambda = 2\pi a(t)/k$ was chosen so the amplitude $\delta_k(t)$ is growing away from zero at decoupling, which produces a relatively large amplitudes well after decoupling. At other wavelengths, the amplitude at decoupling is growing away from zero with the opposite sign. Between these peaks with opposite signs, there is a wavelength whose amplitude approaches zero after decoupling. The result is that $\delta_k(t)$ as a function of k oscillates through zero. This is shown in Panel (b), from Peebles and Yu (1970). These are numerical solutions in linear perturbation theory, where the radiation is described by the Boltzmann equation interacting with the baryons by free electron scattering.

6. When the CMB was recognized in 1965, it soon inspired interest in the dynamical behavior of departures from an exactly homogeneous sea of radiation. The first discussions treated plasma and radiation as a viscous fluid (Peebles 1965 and 1967b; Silk 1967 and 1968; Bardeen 1968; Michie 1969; Sunyaev and Zel'dovich 1970; Weinberg 1971). The fluid model was replaced by analytic and numerical approximations to the Boltzmann equation (Peebles 1967b; Peebles and Yu 1970; Field 1971; Chibisov 1972). Peebles and Yu (1970) introduced the starting elements of the now standard method of numerical solution of the Boltzmann equation for the radiation interacting with the baryonic matter.

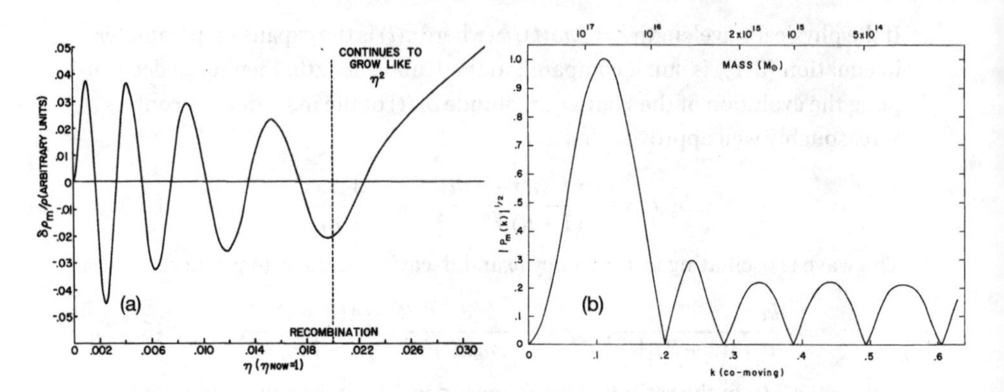

FIGURE 5.2. Panel (a): evolution of the amplitude of a plane wave with wavelength chosen so the amplitude is growing at decoupling. Panel (b): the resulting power spectrum defined in equation (5.33) (Peebles and Yu 1970). Panel (b) © AAS. Reproduced with permission.

The power spectrum $P(k)$ of the mass distribution well after decoupling is the Fourier transform of the mass two-point correlation function $\xi(r)$ defined in equation (2.10):

$$P(k) = \int d^3r\, \xi(r) e^{i\vec{k}\cdot\vec{r}}, \quad \langle \delta_{\vec{k}} \delta_{\vec{k}'} \rangle = \frac{\delta^3(\vec{k}+\vec{k}')}{(2\pi)^3} P(k). \tag{5.33}$$

In the second expression, $P(k)$ is the square of the Fourier amplitudes $\delta_{\vec{k}}$ in equation (5.30). This form follows from the assumption that the mass distribution is a spatially stationary and isotropic random process, an application of the cosmological principle discussed in Section 2.5.

Wilson and Silk (1981) presented results similar to those in Figure 5.2. Computations since then have analyzed the evolution of the distributions of matter and radiation from high redshift through decoupling and the dark ages in increasing detail, as it has become increasingly interesting to compare theory and measurements. The early computations assumed all matter is baryonic. This produces the zeros in the power spectrum $P(k)$ of the present matter distribution, which would have been easy to detect. The addition of the nonbaryonic dark matter discussed in the following chapters removes the zeros and smooths $P(k)$ (Section 9.2), but that leaves the well-observed ripples in the power spectrum shown in the bottom figure in Plate II. The effect was first detected by Percival et al. (2001), with the result discussed on page 105 in Section 3.6.4 and in Chapter 9.

The Fourier transform of a line of equally spaced bumps in the power spectrum $P(k)$ is a single bump in the two-point function $\xi(r)$. The roughly equally spaced bumps in Panel (b) in Figure 5.2 accordingly produce a bump in the matter position correlation function. The effect is illustrated in Figure 5.3. Another way to think of it is outlined in footnote 23 on page 105. The bump

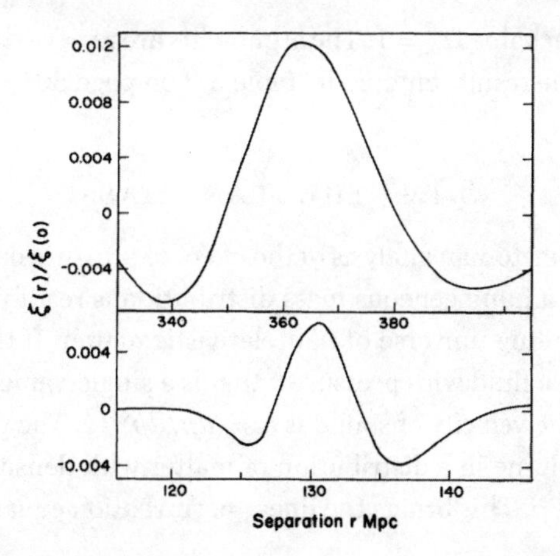

FIGURE 5.3. Early examples of the BAO feature in the matter correlation function (Peebles 1981a). © AAS. Reproduced with permission.

in the correlation function was first detected in the Sloan Digital Sky Survey galaxy redshift catalog by Eisenstein et al. (2005).

The residual distribution of the radiation after decoupling is set by a combination of the compression of matter and radiation at wavelengths that reached maximum amplitude at decoupling, the Doppler shift from motions of the baryons in waves passing through zero amplitude, and the smoothing effect of scattering of the radiation by free electrons, both those present during decoupling (Sunyaev and Zel'dovich 1970) and those present after reionization of the intergalactic medium (Spergel et al. 2003).

The decaying part, $\exp(-\int \gamma dt)$, of the mass density contrast δ in equation (5.31) sets a characteristic wavelength λ_S of plasma-radiation waves that survive through decoupling with only modest dissipation by diffusion of the free electrons. The effect of the decay on the CMB anisotropy is termed Silk damping, after Silk's (1967 and 1968) pioneering discussion. It too is observed in the CMB anisotropy spectrum.

Under the adiabatic and nearly scale-invariant initial conditions assumed in Figures 5.2 and 5.3 and discussed in Section 5.2.6, the matter power spectrum approaches zero at $k \to 0$, which is to say at long wavelengths. Since the power spectrum is the Fourier transform of the two-point position correlation function (equation (5.33)), the limit $k \to 0$ gives $\int d^3r\, \xi(r) = 0$. The correlation function is positive on relatively small scales, so it must pass through zero to negative values at a larger radii. The zero is approximately

$$\lambda_x \simeq 50(\Omega_m h^2)^{-1} \text{ Mpc}. \tag{5.34}$$

This was used in the early arguments that the mean mass density represented by the density parameter Ω_m (equation (3.13)) is less than the elegant

Einstein–de Sitter value $\Omega_m = 1$. The arguments are reviewed in Sections 3.6.4 and 3.6.5, and the results entered in Table 3.2 on page 80.

5.1.4 THE JEANS MASS

Jeans's (1902) Newtonian analysis of the effect of pressure on the evolution of departures from a homogeneous mass distribution is readily generalized to a relativistic expanding universe of nonrelativistic matter. If the matter can be approximated as a fluid with pressure p that is a single-valued function of the density ρ, then the velocity of sound is $c_s = (dp/d\rho)^{1/2}$. The pressure gradient force per unit volume in a distribution of matter with density contrast $\delta(\vec{x}, t)$ is $-\nabla p = -c_s^2 \rho_m \nabla\delta$. This brings the linear perturbation equation (5.21) to

$$\frac{\partial^2 \delta}{\partial t^2} + 2\frac{\dot{a}}{a}\frac{\partial \delta}{\partial t} = 4\pi G\rho_m \delta + \frac{c_s^2}{a^2}\nabla^2\delta. \tag{5.35}$$

For a plane wave, $\delta \propto \cos \vec{k}\cdot\vec{x}$, with physical wavelength $\lambda = 2\pi a/k$, the right-hand side vanishes at the physical Jeans wavelength

$$\lambda_J = \left(\frac{\pi c_s^2}{G\rho_m}\right)^{1/2} \simeq \left(\frac{\pi k_B T}{G\rho_m m}\right)^{1/2}. \tag{5.36}$$

The last expression uses the simple isothermal derivative of the ideal gas law, $p = \rho_m k_B T/m$, where the mean gas particle mass is m. At wavelength shorter than the Jeans length, the wave oscillates; at longer wavelengths gravity dominates, and the wave amplitude can grow. This behavior is illustrated in Figure 1 in Peebles (1969b).

The CMB radiation sets the thermal history of the baryons in the early universe through the thermal coupling of plasma and radiation (equation (5.27)). Since the CMB heat capacity is much larger than that of the matter (equation (4.8)), matter and radiation would be expected to have been cooling at very close to the rate for radiation alone, $T \propto a(t)^{-1}$ (equation (4.2)), continuing well past decoupling thanks to the expected modest residual abundance of free electrons. The temperature sets the Jeans length in equation (5.36), which sets the Jeans mass, in order of magnitude:

$$M_{\rm Jeans} \sim \rho_m \lambda_J^3 \sim (k_B T/Gm)^{3/2} \rho_m^{-1/2}. \tag{5.37}$$

Since the matter density is decreasing as $\rho_m \propto a(t)^{-3}$, we see that as long as the baryons cool as $T \propto a(t)^{-1}$, the Jeans mass does not change as the universe expands and cools. This is interesting, because it imprints a characteristic physical mass on the distribution coming out of the early universe. The estimate in Peebles and Dicke (1968) and Peebles (1969b) is

$$M_{\rm Jeans} \simeq 10^6 M_\odot, \tag{5.38}$$

depending somewhat on the form of the spectrum of primeval density fluctuations.[7]

This picture is complicated by the nonbaryonic dark matter that in the standard cosmology is pressureless initially and interacts with matter and radiation only by gravity. Primeval disturbances to its distribution on scales smaller than the Jeans mass for the baryons would continue to grow and contribute to mass density fluctuations on scales smaller than the baryon Jeans mass. But the baryon mass density is a significant part of the total, so we might expect to find that there is an observable signature of the characteristic mass imposed on the baryons, and through them, on the total mass distribution.

Gamow and Teller (1939) put the Jeans length in an expanding universe at the radius at which the speed $H_0 r$ of cosmological expansion is comparable to the thermal velocity dispersion of the particles in the gas. Apart from numerical factors this works out to the Jeans length in equation (5.36). They proposed that the Jeans mass accounts for the typical mass of a galaxy. But their considerations could not go very far, because Gamow and Teller had no way to account for the temperature needed for an interesting value of the Jeans mass.

Gamow (1948a) had a way to estimate the matter temperature, from his thought (discussed in Section 4.2) that the universe is filled with a sea of thermal radiation at a temperature that would allow conversion of an interesting fraction of hydrogen into heavier elements. Gamow recognized that if the matter temperature remained the same as the radiation as the universe expanded and cooled, then the Jeans mass would be independent of time. He did not discuss why the temperatures would be the same, rather than the matter cooling as the kinetic energy of a free gas, $T_{\rm m} \propto a(t)^{-2}$, but it is not atypical of him to have ignored this detail in the publication. Whether he thought about it at all is a mystery.

Gamow's estimate of the baryon density at the critical temperature $T_0 = 10^9$ K (equation (4.22)) for a reasonable buildup of elements heavier than hydrogen set the relation between the baryon temperature and density. His estimate of the resulting characteristic mass for the gravitational formation of gravitationally bound systems is $M_{\rm Jeans} \sim 10^7 M_\odot$. This is well below the baryon mass $\sim 10^{11} M_\odot$ typical of the luminous parts of a large galaxy such as the Milky Way, but it was an interesting start.

Identification of the CMB in 1965 as a promising candidate for a thermal remnant from the hot big bang (Section 4.4.4) renewed interest in the Jeans

7. The first concentrations of baryonic matter to break away from the general expansion might be expected to have the Jeans mass. To see this, suppose the mass fluctuation power spectrum at decoupling can be approximated as $P(k) \propto k^n$, with $n > -3$. Then equation (5.50) below shows that the mean-square mass fluctuations smoothed through a Gaussian window of radius X varies as $\bar{\delta}_X^2 \propto X^{-(3+n)}$. Under this condition, the mass fluctuations are largest on the smallest scale, which for baryons would be the Jeans length.

mass (Peebles 1965; Doroshkevich, Zel'dovich, and Novikov 1967; Peebles and Dicke 1968; and Rees and Sciama 1969). The similarity to typical masses of globular star clusters led Peebles and Dicke (1968) to speculate that these clusters were the among the first generation to have formed after z_{eq}, at about the Jeans mass. The thought continues to be discussed, but without a lot of community enthusiasm. It remains to be seen whether nature thought to leave observable traces of this physically well-defined characteristic mass.

5.2 *Scenarios*

The gravity physics discussed in the last section is informative, always assuming we have the right fundamental physics. But given the physics, there is considerable room for ideas about initial conditions and the interplay of gravity with nongravitational stresses, which can include the pressures of matter and radiation, shocks and explosions, magnetic fields, and turbulence and dissipation. To this, we can add the possibility of interesting properties of the nonbaryonic dark matter beyond the minimal assumptions of the standard cosmology. Nongravitational stresses are observed to be rearranging matter in and around galaxies, and certainly can be expected to have figured in galaxy formation. To be considered in this section are motivations and challenges for leading ideas about this possibility up to the mid-1990s. That was when the observations started to yield more serious and informative constraints on ideas. I begin with a deep issue: whether the universe we see around us grew out of primeval chaos or order.

5.2.1 CHAOS AND ORDER

Consider a bounded, initially close-to-homogeneous distribution of pressureless matter, dust, in flat spacetime and Newtonian mechanics. The matter is initially at rest and allowed to collapse freely. Since the rate of collapse is faster in regions of greater density (because gravity is pulling harder and speeding the collapse), the mass distribution grows clumpy as it collapses. Since the physics is invariant under time reversal, there is another solution to the equations of motion in which the direction of time is reversed. In this time-reversed case, we would see that the initially clumpy mass distribution grows closer to homogeneous as it expands. But this time-reversed expanding situation is quite unrealistic, because it requires an impossibly tight relation between the initial clumpy mass distribution and the initial velocity field. A tiny error in these initial conditions would produce a growing departure from homogeneity that would become a clumpy distribution that is not at all like the near homogeneity of the original initial conditions for the collapsing case.

The situation is exemplified in an expanding universe of pressureless matter by the linear perturbation expression for the evolution of the density contrast $\delta(\vec{x},t) = \rho(\vec{x},t)/\rho_m - 1$ in equation (5.15). Gravity physics allows an expanding universe that has the pure decaying mode, $\delta \propto t^{-1}$. But the slightest error in initial conditions would add the term $\delta \propto t^{2/3}$ that would grow to dominate the decaying term in an expanding universe.[8]

Press (1976, 311) remarked that in the standard cosmological model, "our particular universe comes from a singularity with very special initial conditions, chosen from a set of measure zero (in some sense) which is not typical of the generic case." This is well put, but how can we find a useful approximation to an observationally viable generic case? One approach uses the analytic solutions to Einstein's field equations for a universe that is homogeneous but anisotropic. These solutions allow flows with shear and vorticity in a spacetime that is seriously different from the standard cosmological model. Lifshitz and Khalatnikov (1963) examined this approach as a possible step toward modeling a fully chaotic—inhomogeneous as well as anisotropic—early universe. Misner (1969) termed the situation a mixmaster universe, after the churning of shear in time-varying directions of expansion and contraction in these solutions. The idea has been influential, in part because of the thought that these solutions might mimic a fully chaotic early universe, and I expect also in part because they allow analytic examinations of chaotic relativistic gravity physics.

Although the mixmaster universe presents us with a valuable example of complex evolution, it does not address the central question: Why is the universe observed to be close to homogeneous? It has been suggested that the mixmaster universe may describe churning in local patches at rates and directions of churning that differ from place to place in an initially inhomogeneous universe, and that the random mixing might drive the universe to nearly spatial isotropy and homogeneity. Misner (1967) considered mixing by efficient energy transfer by neutrinos, or maybe gravitational waves. But also to be considered is the production of entropy by shear. Barrow and Matzner (1977) showed that primeval shear and vorticity in the mixmaster universe would involve considerable dissipation that produced entropy, perhaps even more than is observed. But the central point is that if dissipation of local shear and vorticity were different in different patches of the universe, then we would have to expect that such dissipation deposited a distinctly inhomogeneous entropy density. This is quite contrary to the small anisotropy of the CMB that was already known in the 1970s.

It might be objected that the lessons in Section 5.1 drawn from linear perturbation theory and the spherically symmetric Lemaître solution need not apply to a seriously inhomogeneous very early universe. Could the universe

8. There are examples of strictly decaying departures from homogeneity in an expanding universe. In the absence of a peculiar gravitational acceleration the peculiar velocity of a freely moving test particle decays as $v \propto a(t)^{-1}$ (equation (5.19)). Lifshitz (1946) showed that vortical flow without divergence decays in the same way. But in these examples, the mass distribution remains homogeneous.

have begun in chaos that decayed to near homogeneity and isotropy, after which the galaxies started to grow? The argument against this is that the degree of homogeneity of matter and radiation at redshift $z_{\rm dec}$, indicated by the observed close-to-thermal CMB spectrum and its small anisotropy, suggests the conditions at $z_{\rm dec}$ would be well described by linear perturbation theory. And perturbation theory argues for evolution to growing inhomogeneity from a more nearly homogeneous and isotropic past.

These lines of thought suggest a simple but fundamentally important physical principle:

$$\text{our universe has been evolving from order to chaos.} \tag{5.39}$$

This is not a theorem, or a prediction of the general theory of relativity, but rather a subjective assessment of plausibility (as offered in Peebles 1967a, 1972). It can be debated, as is right and proper for a far-reaching principle.

Jones and Peebles (1972) point to early examples of the thought that, on general philosophical grounds, it is more natural to imagine that the world grew out chaos. Barrow (2017) presents an example of similar thinking in general relativity. The example to be reviewed in Section 5.2.2 is the thought that galaxies are remnants of primeval turbulence. Also to be considered is Rees's (1972, 1670) thought that "it is interesting to explore the possibility that the fluctuations on scales up to (say) protoclusters may already be of order unity when they come within the particle horizon."

The argument against this idea, based on Figure 5.1 on page 197, is that mass density fluctuations with contrast $\delta \sim 1$ as they pass through the scale of the particle horizon, or the Hubble length, would be associated with spacetime curvature fluctuations on the order of unity, which would be expected to produce relativistic mass concentrations. Maybe the mass concentrations in the centers of galaxies formed that way, but I think we can be sure that the galaxies and clusters of galaxies did not.

Note also that we now have observational evidence of evolution from order to chaos, from the detection of the acoustic patterns in the distributions of the CMB and the galaxies. They fit the theory of small-amplitude acoustic density fluctuations at decoupling, at redshift $z \sim 1000$ (as approximated by equation (5.32)), which grew into our present distinctly clumpy universe.

A consequence of evolution from order to chaos is that we need new physics to account for the very nearly homogeneous and isotropic state of the early universe. It seems perfectly reasonable to expect that the new physics did the right thing by not placing the initial departures from homogeneity almost entirely in the decaying mode. But that must remain a judgment, pending convincing establishment of a suitably deeper theory of the very early universe, maybe some variant of cosmological inflation.

Progress in analyzing how cosmic structure grew does not require that we specify the wanted new physics. We can instead postulate the nature of early departures from exact order and compare what would have grown out of that to what is observed. This was the approach in the 1970s, at the quite modest level of research in cosmology at the time. But in the early 1980s, the introduction of the inflation concept for the behavior of the very early universe (Section 3.5.2) soon became influential and gave the principle of chaos out of order an authoritative nonempirical pedigree. Chapter 8 presents an account of how the empirical basis for this principle grew in the late 1990s.

5.2.2 PRIMEVAL TURBULENCE

Thoughts about turbulence in the early universe were suggested in part by the resemblance of images of spiral galaxies to images of eddies in incompressible turbulent fluid flow. Perhaps, Carl Friedrich von Weizsäcker (1951a, 160) argued, "it is a consistent theory to think of the galaxies (or perhaps the clusters of galaxies) as the largest eddies of a cosmic turbulence that existed a couple of billion years ago."

Sebastian von Hoerner (1953, 58), at the time with von Weizsäcker at Göttingen, concluded from his study of how spiral galaxies might form out of primeval turbulence that "Da wir auf drei völlig verschiedenen Wegen qualitativ übereinstimmende Aussagen über den Verlauf der Flächendichte in Spiralnebeln erhalten haben, wollen wir dies Ergebnis als Argument betrachten für die Anwendbarkeit der vorausgesetzten Turbulenztheorie." Or, in my translation aided by Google, "Since we have obtained qualitatively consistent statements on the run of surface density with radius in spiral nebulae in three completely different ways, we will consider this result as an argument for the applicability of the assumed turbulence theory."

George Gamow (1952b, 251) argued that galaxies

> would never form in the originally homogeneous material, unless there are some additional physical factors favoring large-scale fluctuations of density. ... Thus, it seems at present that the only way of understanding the formation of gaseous protogalaxies lies in the assumption of extensive turbulent motion in the primordial expanding material.[3] ... Although Reynold's number for the universe is always sufficiently large to expect the presence of turbulent motion, it is, however, difficult to see how such a motion could originate in a uniformly expanding homogeneous material. Thus, it may be well to introduce the primordial turbulence on a postulatory basis along with the original density of matter and the rate of expansion.

As usual, Gamow did not explain. The reference (3) is to von Weizsäcker (1951b). Gamow and Hynek (1945) had reviewed von Weizsäcker's earlier thought that the planets in the solar system might have grown out of whirlpools driven by vortical motions in a rotationally supported disk initially consisting of diffuse matter, and Gamow's (1954b) article on primeval turbulence acknowledged a "private conversation with C. von Weizsäcker." We can then take it that Gamow was well aware of thoughts in this direction that may have influenced him. But another serious motivation may have followed from his impression, consistently stated, that the present mean mass density is well below the Einstein–de Sitter value.[9] Gamow (1954b, 481) wrote that

> It is clear that under such conditions ... any rudimentary condensation will disperse again, no matter how large in size it is. The only escape from this difficulty is to assume the existence of very large original density fluctuations in the primordial gas. Besides, in order to permit large variations of density, this turbulence must have been supersonic.

The physical considerations are correct, as discussed in Section 5.1.2. Gamow and Teller (1939) had addressed the problem with the apparently low value of the mass density by proposing that the galaxies formed at high redshift, when the mass density would have been closer to Einstein–de Sitter and gravity would have been more effective. Gamow (1954b) argued instead for primeval turbulence. Gamow's (1952b, 251) earlier paper added the thought that "The existence of such turbulent motion in the primordial material of the universe is, in fact, strongly suggested by the recent studies by H. Shapley and C. D. Shane of the space distribution of galaxies in the observed part of the universe."

Gamow's graduate student, Vera Rubin, devised and applied a statistical measure of the distribution of galaxies. Her conclusion (Rubin 1954, 548) is that the result "is physically reasonable if the galaxies have condensed from a turbulent gaseous medium." Later analyses of results from Donald Shane's program, along the lines of the statistics Rubin pioneered, are reviewed in Section 2.5. But Rubin's (1954) conclusion was suitably cautious, because it would be difficult to argue that her result is specifically suggestive of primeval turbulence.

Jan Oort attended the Paris conference at which von Weizsäcker (1951a) spoke on primeval turbulence; the proceedings record Oort's remarks following von Weizsäcker's presentation. That may have contributed to Oort's (1958, 1) later thought that

> The total angular momentum must have been present in the primeval clump of material from which the galaxy has contracted. The angular momentum together with the strength of the concentration of mass towards the center of a galaxy contain information on the proportion between *regular* large-scale rotation and the *irregular* currents in the part of the universe which contracted into that galaxy.

9. For example, Gamow (1946) took Hubble's constant to be 1.8×10^{-17} s^{-1}, which is close to Hubble's (1929) value, and his standard value of the present mass density is $\rho \simeq 10^{-30}$ g cm^{-3}. That translates to mass density parameter $\Omega_m = 0.002$ (equation (3.13)).

Oort (1970, 381) later wrote that "The universe must thus have had a high degree of turbulence on a galactic scale, as was first proposed by von Weizäcker."

At the time, Oort doubted that tidal torques on neighboring clumps of matter as they start to break away from the general expansion of the universe would transfer enough angular momentum to account for the rotation of spiral galaxies. The developments that made this idea of gravitational transfer of angular momentum look more promising are reviewed in Section 5.2.3.

Other discussions of the idea of primeval turbulence include Ozernoi and Chernin (1967); Tomita et al. (1970); and Dallaporta and Lucchin (1972). Ozernoi and Chernin offered two reasons to turn again to the picture of primeval turbulence. First, they argued that it is advisable to consider alternatives to more popular ideas. This is a good point: The observational and theoretical constraints on cosmic structure formation were not very informative then, a situation that certainly called for broad explorations of ideas. Their second reason was that the density fluctuations of statistical mechanics at strict equilibrium are far too small to have grown by gravity into the structures we see around us starting from any time that would seem reasonable. This is Lifshitz's (1946) point, and it is correct. But Section 5.1 reviews the turn of popular thinking to a pragmatic assessment of the situation: The relativistic cosmology nevertheless allows evolution of nonrelativistic mass concentrations, such as galaxies, out of small primeval departures from homogeneity. In the 1960s, that was largely hopeful thinking, of course.

Thus we can conclude that in the 1950s through the 1960s, people found persuasive reasons to consider primeval turbulence scenarios. But in general relativity, the notion of primeval turbulence is seriously problematic. To see this (expressed in the notation in Peebles 1980, §85, eq. (85.5)), suppose that the mass of the universe is dominated by an ideal fluid with pressure $p=\nu\rho$ (with units chosen so the velocity of light is $c=1$, as usual, so ν is a dimensionless constant). Imagine this fluid is streaming in rotational flow with zero divergence so that, in linear perturbation theory, there is no departure from a homogeneous mass distribution. In a generalization of Lifshitz's analysis, one finds that the proper streaming velocity varies as

$$v \propto a(t)^{3\nu-1}, \tag{5.40}$$

where $a(t)$ is the expansion parameter (as in equation (2.3)). We have $v \propto 1/a(t)$ in the nonrelativistic limit $\nu=0$ (as expected from the considerations in footnote 3 on page 117). In the early radiation-dominated universe, with $\nu=1/3$, the streaming speed v is constant.

Suppose now that a fluid element moving with speed $v(t)$ in rotational flow typically has to move through coordinate distance y before it encounters rotational flow moving in some other direction, which would cause a nonlinear disturbance. Whether that encounter happens in a cosmic expansion time t depends on the ratio (Peebles 1971b)

$$R(t) = \frac{v(t)t}{a(t)y} \propto a(t)^{(9\nu-1)/2} \begin{cases} \propto a(t)^{-1/2} \text{ if } p = 0, \\ \propto a(t) \text{ if } p = \rho/3. \end{cases} \tag{5.41}$$

In the early radiation-dominated universe, with $\nu = 1/3$, $R(t)$ grows as the universe expands. If $R(t)$ reaches unity while the pressure p is still close to $\rho/3$, then the flow is required to change direction. This means small primeval rotational streaming flows could grow into nonlinear turbulence that cascades to eddies that might grow into something like galaxies. But this turbulence would have to have developed before redshift $z_{\rm eq} = 3400$ (equation (4.11)). The matter density then would seem to be far too large for production of real galaxies. At redshifts less than $z_{\rm eq}$, the ratio $R(t)$ decreases as the universe expands. Thus if turbulence has not developed by redshift $z_{\rm eq}$, it will never happen. Arguments along these lines were presented in Peebles (1971b and 1972), Jones and Peebles (1972), and Chan and Jones (1975).

In the 1950s and 1960s, the idea of primeval turbulence was empirically motivated, which made it worth exploring. But it was also clear—though not always noticed—that the idea is not consistent with usual thinking about the properties of galaxies and the physics of the expanding universe.

5.2.3 GRAVITATIONAL ORIGIN OF GALAXY ROTATION

The standard model for the rotation of spiral galaxies replaces primeval turbulent eddies with the gravitational exchange of angular momentum between orbital and internal motions of protogalaxies. The idea that gravity caused galaxies to rotate traces back at least as far as Gustaf Strömberg (1934), and later, independently, Fred Hoyle (1951). The first paragraph in the abstract in Strömberg's (1934, 460) paper is

> *The origin of the rotation of the galaxy and of the spiral nebulae in general* is traced back to a time when the nebulae were recently formed from a common system of primordial gas. At this stage the diameters of the nebulae were of the same order as their mutual separations, the system as a whole had gravitational instability, and the mutual attractions and relative motions of the extended nebulae produced large angular momenta, but only very small angular motions. During the process of contraction in any nebula, the linear and angular velocities increase, leaving the total angular momentum constant.

In his report in the proceedings of the *Symposium on Cosmical Gas Dynamics* in Paris in 1949, Hoyle (1951, 195) pointed out that

> A condensation, so long as it is not of strictly spherically symmetric form, will in general acquire angular momentum during the early stages of the condensation process. For the principal moments of inertia at its mass centre are then not all equal, and any external gravitational field will in general produce a couple acting on the condensation.

The couple, or torque, would transfer angular momentum between relative motions of the protogalaxies and their rotational angular momenta. In the simplest approximation, the quadrupole moment of the protogalaxy couples to the tidal field of the neighboring mass distribution.

The considerations in these two papers are now well tested and accepted. But an indication that they were not always considered to be so intuitively clear is to be seen from Werner Heisenberg's recorded comment after Hoyle's (1951, 198) talk: "I feel that one really should take the concept of turbulence much more seriously than seems to be done."

The NASA Astrophysics Data System archive does not list any citations of Strömberg (1934) prior to 1995, or of Hoyle (1951) prior to 1970. Extragalactic astronomy and cosmology were small sciences until fairly recently. There were discussions of other aspects of the rotation of galaxies; a notable example is Leon Mestel's (1963) idea of comparing the distributions of specific angular momentum and mass in a spiral galaxy to what might have been present in a uniformly rotating protogalaxy. Another example, reviewed in Section 6.4, is the analysis of the gravitational instability of rotationally supported galaxies for clues to how spiral arms formed. But serious debate on the gravitational origin of the rotation of galaxies did not begin until the late 1960s.

Peebles (1971c) introduced a convenient dimensionless measure of the typical angular momentum of a protogalaxy once gravitationally assembled:

$$\lambda = \left(\frac{L^2 E}{G^2 M^5} \right)^{1/2} \sim 0.1. \tag{5.42}$$

The angular momentum of the gravitationally bound system is L, its gravitational binding energy is E (the negative of kinetic plus gravitational potential energy), and its mass is M, with Newton's gravitational constant G. In the scale-free Einstein–de Sitter cosmological model, this dimensionless number is of order unity, as we might expect from dimensional analysis.

Here are the main steps to the conclusion that galaxies rotate because gravity transferred to galaxies the internal angular momentum characterized by equation (5.42).

1. Strömberg (1934), and independently Hoyle (1951), introduced the picture of angular momentum transfer by tidal torques.
2. Peebles (1969a) reinvented the idea yet again and presented an analytic estimate of the magnitude of the effect, amounting to $\lambda \simeq 0.08$.

(This is the product of equations (35) and (37) in the paper; the measure λ was introduced in step 4.) Doroshkevich (1970) and White (1984) discuss other analytic ways to approach this analysis.

3. Oort (1970, 381) concluded from his estimate of the gravitational transfer of angular momentum that the effect is too small to account for the rotation of galaxies, and that "they must have been endowed with their angular momentum from the beginning." Michie (1966, 172) reached a similar conclusion ... "tidally induced torques are of no importance for most fragments in the protogalaxy cloud."
4. Oort's critique led Peebles (1971c) to change from analysis of the complex process of angular momentum transfer to simple numerical simulations. They indicated $\lambda = 0.10^{+0.1}_{-0.03}$. The simulations had only $N \sim 90$ to 150 particles, ludicrously small by later standards, but that was a different age. The result seemed reasonable to me, because the dimensionless number λ might be expected to be of order unity: What other value might it have in the dimensionless gravitational evolution of pressureless matter?
5. Jan Oort sought me out in about 1972 to explain that he withdrew his objection to the gravitational transfer of angular momentum. It was an edifying act.
6. Efstathiou and Jones (1979) found $\lambda = 0.07 \pm 0.03$ in numerical simulations of the expanding universe modeled by the motions of $N = 1000$ particles. It seemed quite clear at this point that the typical value of the angular momenta of gravitationally assembled protogalaxies is not likely to be very different from $\lambda \approx 0.1$.
7. Gunn et al. (1978) and White and Rees (1978) proposed that the luminous parts of a galaxy formed by the dissipative settling of diffuse baryonic gas or plasma into a massive halo of subluminal matter that may consist of remnants of early generations of stars, or perhaps nonbaryonic dark matter. The contraction would account for the observation that most of the subluminal matter is on the outskirts of the luminous parts of a spiral galaxy, and the contraction would increase the rotational angular velocity.
8. Fall (1979), Fall and Efstathiou (1980), and Efstathiou and Jones (1980) pointed out that dissipative settling of baryons in a preexisting massive subluminal halo by a factor of about 10 would naturally spin up the matter from $\lambda \simeq 0.1$ to rotational support in a spiral galaxy with $\lambda \simeq 1.0$.

The idea that dissipative settling sped up the rotations of the galaxies goes back at least to Strömberg (1934). The idea that dissipative settling of baryons took place in massive subluminal halos grew out of the growing realization that there is subluminal matter on the outskirts of galaxies; this phenomenon is reviewed in Chapter 6. The idea of a massive halo is also essential to the explanation of how spiral galaxies came to be rotationally supported, as we now discuss.

In a gravitationally bound cloud of matter that is supported by nearly random motions with typical speed v, the mean transverse speed of rotation is $v_\perp \sim \lambda v$, where λ is defined in equation (5.42). Suppose first that the entire mass of the cloud dissipatively contracts to rotational support while conserving angular momentum. Then as the radius r of the system decreases, the gravitational binding energy E increases, as $E \propto r^{-1}$. This, along with conservation of angular momentum, causes the angular momentum parameter to increase as $\lambda \propto r^{-1/2}$. To bring the initial value $\lambda \sim 0.1$ from gravitational transfer of angular momentum to $\lambda \sim 1$ for rotational support, the radius r of the cloud would have to contract by a factor of about 100. Efstathiou and Jones (1980) pointed out that this collapse by two orders of magnitude in radius would have the outer parts of the Milky Way, now at 10 kpc radius, contract from about 1 Mpc. But this would make the collapse time comparable to the present age of the expanding universe, which is absurd.

The way out was suggested by the evidence Gunn et al. (1978) and White and Rees (1978) were considering: The mass outside the central parts of a spiral galaxy is dominated by subluminal matter with velocity dispersion that changes little with radius. This approximates an isothermal gas sphere with mass density run $\rho \propto r^{-2}$, as discussed in Section 6.3. The diffuse baryons that settled within a massive subluminal halo would have to contract by a factor of about 10 to be spun up to rotational support, because the rotation speed needed for rotational support is not sensitive to radius while the transverse speed increases as $v_\perp \propto r^{-1}$. This more modest contraction seems much more reasonable, while still large enough that most of the stellar mass would end up near the center of the subluminal halo, as observed.

The contraction by about a factor of 10 to support a spiral galaxy by rotation is in line with the Eggen, Lynden-Bell, and Sandage (1962, 749) argument from analyses of astronomical clues to how the Milky Way galaxy formed. They concluded that "Our data suggest that the oldest stars with the lowest abundance of the heavy elements must have been formed in the collapsing protogalaxy when its size was at least ten times its present diameter."

This argument follows from the sorting of the stars in our galaxy into two types: Populations I and II. The former have greater abundances of heavy elements and tend to be supported by streaming in nearly circular orbits in the disk of the Milky Way. The latter are older, have lower heavy element abundances, and tend to be moving in more nearly random motions in the close-to-spherical stellar halo of high-velocity stars. Eggen et al. suggested that the Milky Way formed out of a cloud of gas or plasma that had settled or collapsed. Stars that formed during the collapse would be among the oldest. They would have low heavy element abundances, because not many stars would have already formed, evolved, and scattered heavy elements for incorporation in younger stars. And these old stars would be moving in orbits with large apogalacticons, from the large radii at which they formed. All this is characteristic of Population II. The dissipative contraction of the remaining diffuse matter would have been stopped by rotation, and the stars that formed after that would be younger, with greater element abundances, and they would be streaming in the disk, all characteristic of Population I.

The Eggen, Lynden-Bell and Sandage (1962) contraction by about an order of magnitude is what is needed to spin up a spiral galaxy in a dark halo from the initial value produced by gravity, $\lambda \approx 0.1$. Gunn et al. (1978) and White and Rees (1978) had proposed that the initial diffuse cloud of baryons contracted by about an order of magnitude in radius to form a normal galaxy. But Gunn, Rees, and White inform me that, to the best of their recollections, they were not thinking about the rotation of spiral galaxies when they presented their 1978 arguments for dissipative contraction in a dark halo. Although their arguments were based on other grounds, they were important in helping complete the steps to the accepted explanation of why spiral galaxies rotate.[10]

Later evidence is that the characteristic radius of the luminous part of a spiral galaxy is roughly λ times the radius of its dark matter halo (e.g., Somerville et al. 2018; Fall and Romanowsky 2018). This contraction factor $\sim\lambda^{-1}$ is consistent with the Eggen, Lynden-Bell and Sandage (1962) argument from high-velocity stars; the 1978 ideas about contraction within a massive halo; and the realization by 1980 that contraction accounts for rotational support of the baryonic matter of a spiral galaxy settling into its dark matter halo.

Fall and Romanowsky (2018) point out that large ellipticals behave as if their prestellar matter had contracted by about the same factor, $\sim\lambda^{-1}$, but that this matter was endowed with unusually small λ or else somehow lost much of its primeval angular momentum. The shapes of a normal large spiral and a normal large elliptical are quite different. There are transitional cases, but they are not common. We see a distinct bimodal phenomenon that is not an obvious result of a single-peak distribution of λ. It presents and an opportunity for more research.

10. A technical point to consider is that in the gravitational instability picture, gravity draws together matter without producing vorticity. That is, in the absence of shocks or other nongravitational disturbances, the streaming velocity driven by gravity satisfies $\nabla \times \vec{v} = 0$. The streaming flow of stars and diffuse baryons in the disk of a spiral galaxy has vorticity. There was some thought that this contradicts Kelvin's circulation theorem (Tomita 1973; Binney 1974). But orbit crossings of streams of noninteracting dark matter with zero vorticity generally produce mass-weighted mean streaming motion with vorticity. And the dissipative/viscous settling of diffuse baryons violates the conditions of the circulation theorem, again allowing formation of streaming flow with vorticity (Chernin 1972, Sunyaev and Zel'dovich 1972, Peebles 1973c).

5.2.4 EXPLOSIONS

Matter is observed to be rearranged on the scales of galaxies by explosions powered by nuclear burning in massive stars, supernovae, and massive gravitational collapse in the central parts of galaxies. One might then imagine that explosions in the early universe pushed matter into piles that grew into galaxies, or even that structure formation was initiated by the time-reversal of relativistic collapse, as in white holes.

Hoyle (1980, 35) put this last thought as

> The hot big-bang is an example of a white hole, a white hole that is postulated to be so enormous that its products encompass everything we observe. But there is nothing to prevent us from thinking of a larger universe containing very many individual white holes, just as relativists have become accustomed to thinking of the universe as containing many separate black holes.

Novikov (1964b) and Ne'eman (1965) proposed that delayed expansion of parts of the expanding universe may be observed as powerful radio sources in some galaxies and in the then-recently discovered quasistellar objects, or quasars. Lynden-Bell's (1969) argument that galactic nuclei might contain black holes that are remnants of inactive quasars might have added credibility to the thought of active quasars as white holes. It violates the argument in Section 5.2.1 that one might reasonably expect chaos to have grown out of order, but our notions of what is reasonable may be irrelevant in this situation. The idea of white holes has not attracted much attention since then, but that could change.

The idea that the more familiar explosions powered by nuclear reactions played a serious role in driving cosmic structure formation by pushing material around did attract serious attention. To see the energy available, consider that galaxies are observed to move relative to the general expansion of the universe with peculiar velocities typically ~ 300 km s^{-1}. In the gravitational instability picture, this kinetic energy was produced by the gravitational attraction of the growing cosmic structure, accompanied by the growing magnitude of the negative gravitational binding energy. But the production of a present cosmic mass fraction $Z \sim 0.01$ in heavy elements in stars by the conversion of hydrogen and primeval helium to heavier elements, releasing nuclear binding energy on the order of 1 percent of the rest mass, amounts to annihilation of 1 part in 10^4 of the rest mass of the baryons. This is equal to the kinetic energy $\rho v^2/2$ of these baryons moving at speeds ~ 3000 km s^{-1}. We see that, if efficiently employed, there is ample energy to account for the motions of the galaxies by explosions rather than by gravity.

Early explorations of the possible role of explosions in driving galaxy formation were presented by Doroshkevich, Zel'dovich, and Novikov (1967);

Ikeuchi (1981); and Ostriker and Cowie (1981); and Ostriker (1982). Ostriker and Cowie (1981, L127) articulate the program as follows:

> The explosive energy released at the death of massive stars in forming stellar systems will propagate into the intergalactic medium. There, under certain circumstances, a dense cooled shell will form with mass many times greater than the original "seed" system, whereupon gravitational instability and fragmentation of the shell can lead to the formation of new stellar systems.

The program is well motivated. Energy released by nuclear burning in stars, aided by gravity, is observed to rearrange baryons: Winds of plasma flow out of nearby star-forming galaxies and distant young galaxies at speeds of $\sim$ 100 to 1000 km s^{-1}. Ostriker, Thompson, and Witten (1986) added another thought: the cosmic strings discussed in Section 5.2.5 might be superconducting and magnetized. As the loops thrashed about, the fluctuating magnetic field could release energy sufficient to push plasma into piles that would fragment into young galaxies.

These explosion scenarios were challenged by the evidence that the peculiar velocities of galaxies relative to the Hubble flow have a broad coherence length, that is, the galaxies move in streaming flows with relatively small dispersion around the mean. This phenomenon is quite clearly observed; it has been termed the "quiet Hubble flow" and the "large cosmic Mach number." It is discussed on page 97 in Section 3.6.4, in connection with the nature of the mass density and distribution. Let us add here the conclusions by de Vaucouleurs and Peters (1968, 874), that "there is nothing in the bright galaxy data to contradict the assumption that the large scale cosmological red-shift is linear and isotropic," and by Sandage and Tammann (1975, 313), that "the local velocity field, mapped with the present material, is as regular, linear, and isotropic as we can measure it."

Lilje, Yahil, and Jones (1986), and Peebles (1988), used data from advances in the methods of galaxy distance measurements for improved measures of departures from Hubble's law among the relatively nearby galaxies. In these samples, at recession velocities $cz \lesssim 600$ km s^{-1}, the relative peculiar velocities are $\lesssim 100$ km s^{-1}. Lynden-Bell et al. (1988) obtained similar results at greater distances using an angular size measure of distances to early-type galaxies. These velocities relative to the mean streaming flows of the galaxies can be compared to the measured CMB dipole ($\delta T/T \propto \cos\theta$) anisotropy (Smoot, Gorenstein, and Muller 1977; Fixsen, Cheng, and Wilkinson 1983). After removing the dipole, the remaining CMB anisotropy was known in the 1980s to be at least an order of magnitude smaller. This suggests the dipole is not intrinsic but is instead the result of the peculiar motion of the Local Group at 600 km s^{-1} through the sea of radiation. Since the scatter of relative velocities is much smaller than this, the evidence indicates that the galaxies in our neighborhood, out to distances ~ 10 Mpc, are streaming through the CMB at nearly uniform speed.

The point of these observations is that it was and is difficult to see how explosions could have produced the distinctly clumpy distribution of the galaxies while leaving a small relative velocity dispersion and producing a much larger streaming motion of the entire region of the Local Group (Peebles 1988).

The explosion picture also faced a challenge of a sociological nature: Community attention had turned to numerical modeling of the growth of cosmic structure in a universe with mass dominated by nonbaryonic dark matter that moves only under the effect of gravity. Simulating the evolution of this situation is far simpler than simulating the effects of explosions. I expect this helped discourage attempts to craft a version of the explosion picture that might have been observationally viable. But later advances in the evidence have convincingly placed explosions in the category of a real but subdominant effect on the gathering of mass concentrations in protogalaxies, while remaining an essential factor in forming the internal structures of galaxies.

Lynden-Bell et al. (1988) pointed out that the local streaming motion of the galaxies relative to the CMB could be caused by the gravitational attraction of a suitably distant and large mass concentration they termed the "great attractor." But subsequent studies indicate that the CMB dipole is the integrated effect of gravitational attraction by the many large-scale, low-amplitude mass fluctuations scattered through space (Nusser, Davis, and Branchini 2014).

5.2.5 SPONTANEOUSLY BROKEN HOMOGENEITY

When a ferromagnetic material is cooled through the phase transition at the Curie temperature, it acquires magnetized domains. If there is no external magnetic field or internal structure to guide them, the domains spontaneously settle on directions, breaking the isotropy of spin states at higher temperature. The distinguished pedigree of this kind of phenomenon in condensed matter and particle physics helped attract interest in the thought that the exact spatial homogeneity and isotropy of the early universe similarly was spontaneously broken. The history of such thoughts is reviewed by Copeland and Kibble (2009).

Zel'dovich (1962b) presented an early example of how exact primeval cosmic homogeneity might have been spontaneously broken. Suppose the primeval universe had zero entropy. The idea is elegant; what could be simpler? The baryons would have been in an initially condensed compressed homogeneous quantum state at zero entropy. But this condensed system had to have cracked as the universe expanded. This is a spontaneously broken homogeneity that would have produced fragments of matter; maybe they would grow into galaxies. But it later became clear that the early universe almost certainly was hot.

Press and Schechter (1974) offered another thought: Perhaps the gravitational interaction of small nonlinear seed departures from homogeneity caused the growth of larger mass concentrations. These in turn gravitationally disturbed neighboring matter on larger scales and on up. This resembles Layzer's (1954, 171) vision:

> Consider a cosmic distribution of matter in which there are slight local irregularities. As the universe expands the irregularities become more and more pronounced until finally self-gravitating systems separate out. The newly formed systems play the role of particles in a new cosmic distribution, which will also have slight local irregularities, in general, and the stage is set for a repetition of the clustering process.

The situation Layzer seems to have had in mind, and Press and Schechter proposed, is that cosmic structure may form by a gravitational autocatalytic process. Maybe this process would have to have been seeded, or maybe it just traces back to arbitrarily small nonlinear clumps at arbitrarily early times. The idea is elegant, of historical interest, and worth considering, but wrong: It is not seen in numerical simulations of the growth of mass clustering in an expanding universe.

The Press and Schechter (1974) argument is worth reviewing. Consider an Einstein–de Sitter universe of pressureless matter that interacts only by gravity. Suppose the matter is concentrated in nonlinear gravitationally bound clumps on the length scale $r_{\rm nl}$, and imagine the clumps are quite smoothly distributed on larger scales. The variance of the mass contained in a randomly placed sphere of radius $R \gg r_{\rm nl}$ might be expected to contain the term

$$\delta M^2 \equiv \langle (M - \langle M \rangle)^2 \rangle \sim (R/r_{\rm nl})^2, \quad \delta M/M \propto R^{-2}. \tag{5.43}$$

This is because the surface of the sphere has area $\sim R^2$, so the surface cuts through $N \sim (R/r_{\rm nl})^2$ nonlinear clumps, and we are assuming clumps close to the surface might at random fall inside or outside the sphere. If this density contrast grew at the usual rate, $\delta M/M \propto t^{2/3} R^{-2}$, then $\delta M/M$ would reach unity at time $t \propto R^3$, bringing the mass in a typical nonlinear clump to M. This suggests that the mass that has become gravitationally concentrated in nonlinear clumps grows in proportion to the time. The mass within the Hubble length cH_0^{-1}, the greatest distance we can readily observe, also grows in proportion to the time, assuming an Einstein–de Sitter universe. The inviting prospect was that cosmic structure grows as the mass of the observable universe grows.

Zel'dovich (1965, 359) anticipated this:

> The problem is that if we act in a primitive way and take the volume V with a sharp boundary, then fluctuations of the number N [in V] will

be determined exactly by the fortuities of dispositioning galaxies to the right or left of the boundary.

The noise in the mass distribution on the surface of the sphere has little to do with the gravity of the mass that is pulling matter into or out of the sphere. The analytic argument in Peebles (1974a; 1980, §28) predicts that the rate of autocatalaytic growth of nonlinear mass clustering is $M_{\rm nl} \propto t^{4/7}$, too slow to do anything of interest. I am not aware of a check by numerical simulation, which would be interesting. It is to noted here that the Press and Schechter (1974) paper also introduced the widely used Press-Schechter form for the frequency distribution of masses of gravitationally produced concentrations.

Weinberg (1974) considered another more widely discussed idea for the spontaneous breaking of primeval homogeneity. Imagine a scalar field that came to behave as a classical field as the universe expanded and cooled. The field would have assumed different classical values in causally unrelated regions of space. As the Hubble length grew with the expansion of the universe, the different field values in different regions might have moved to produce a stable field structure—a domain wall, whose energy would have broken the homogeneous mass distribution. Weinberg attributed the thought to a discussion with Robert Schrieffer, who certainly was familiar with domain walls in condensed matter. Kibble (1976, 1387) put it this way:

> In the hot big-bang model, the universe must at one time have exceeded the critical temperature so that initially the symmetry was unbroken. It is then natural to enquire whether as it expands and cools it might acquire a domain structure, as in a ferromagnet cooled through its Curie point.

Vilenkin (1981a, 1169) argued that

> there seem to be only two natural choices of the initial state: (i) Chaotic universe with $\delta\rho/\rho \sim 1$ and (ii) exactly homogeneous universe with $\delta\rho/\rho = 0$. In the present paper we shall discuss only the second possibility. Then one has to assume that at the beginning the universe was exactly Friedmanian and that the density fluctuations were generated later by some physical process (e.g., a phase transition).

The thought is that the existence of galaxies can be traced back to a spontaneous breaking of homogeneity. Early discussions include Kibble (1980), Zel'dovich (1980), Vilenkin (1981a), Turok (1983), and Silk and Vilenkin (1984). It was an influential addition to the ferment of debate on how cosmic structure grew as the universe evolved.

For this review of the history of ideas, let us consider simple examples of the breaking of homogeneity by a multiplet of scalar fields, $\phi(\vec{x}, t)$, with potential energy density

$$V(\phi) = \lambda(\phi^2 - \eta^2)^2, \tag{5.44}$$

along with the usual field gradient energy. The factor λ is a dimensionless constant. The constant η and field $\phi(\vec{x}, t)$ have units of energy (where we lapse into units with $\hbar = 1 = c$, so an energy is the reciprocal of a length, and the potential $V(\phi)$ has units of energy to the fourth power). There were discussions of interactions of ϕ with ordinary matter and radiation, but that did not receive much attention and need not be considered here.

The expansion of the universe tends to dissipate gradients of $\phi(\vec{x}, t)$ in space and time, allowing the field to relax to minimize its energy, but that can be frustrated by the formation of stable configurations. Consider first a field with a single component. Its potential energy in equation (5.44) has two minima, $\phi = \pm\,\mu$. As ϕ assumed the character of a classical field in the hot and cooling early universe, it would have acquired a positive value in some regions and a negative value in others. As the now-classical field relaxed to minimize its energy, the field value would have approached $\phi = +\eta$ in some regions and $\phi = -\eta$ in others. Since the field value must pass through zero on going from one region to the next, it must relax to a domain wall in which $\phi = +\eta$ on one side, $\phi = -\eta$ on the other, and in between there has to be a sheet of positions where $\phi = 0$ and the potential energy density is $V = \lambda\eta^4$. This potential energy with the field gradient energy adds up to energy $\Sigma \sim \lambda^{1/2}\eta^3$ per unit area of the wall.

The wall behaves as a two-dimensional sheet of energy with surface tension equal to the energy per unit area. We know this because the work done by stretching the wall to increase its area by the amount δA is $\delta E = \Sigma\delta A$, which is the energy added to the wall with its increased area. The surface tension causes waves to propagate along the wall at the speed of light, and it causes bends in the wall to move about at similar speed toward flattening the wall. This would cause walls to find and annihilate one another, leaving only a few of them running across the Hubble length.

Zel'dovich, Kobzarev, and Okun' (1975) pointed out that the motions of walls driven by their surface tension could seriously disturb the CMB. Vilenkin (1981b) added the consideration of the gravitational effect of a wall. He showed that a flat and stationary wall produces gravitational acceleration $g = 2\pi G\Sigma$ away from the wall on each side.[11] This gravitational acceleration is independent of distance from an unbounded flat wall, meaning that the

11. Recall that pressure p has positive active gravitational mass density p/c^2, as in equation (3.3). Tension is a negative pressure that has negative active gravitational mass density. With tension in two directions on the wall, the negative active mass of the tension is twice the positive active mass, Σ. Hence the gravitational acceleration is away from the wall with magnitude the same as an ordinary wall with the same mass per unit area. A straight cosmic string with tension equal to the energy per unit length has negative active gravitational mass due to the tension in one direction that just cancels the active gravitational mass due to the energy per unit length. The string thus produces no gravitational acceleration. It does warp spacetime.

gravitational potential relative to the wall increases linearly with increasing distance until the approach to the next wall, perhaps at the Hubble distance. The large distances involved in cosmology, with the large velocities of bent walls, led Zel'dovich, Kobzarev, and Okun' (1975), and Kibble (1980), to consider this case unpromising.[12]

Kibble (1976) introduced consideration of the cosmic strings that would appear in a two-component field with the potential in equation (5.44) and $\phi^2 = \phi_1^2 + \phi_2^2$. Here the energy (potential plus field gradient) is minimized at $\phi_1 = \eta \cos\theta$ and $\phi_2 = \eta \sin\theta$ for any value of θ. If along a closed path, a loop, the value of θ increases by 2π, then the loop contains a thin tube, or defect line, with nonzero energy. For if the loop is shrunk without passing over the defect line, then θ still increases by 2π going around the loop. And since ϕ_1 and ϕ_2 must have definite values as the loop shrinks to zero radius, the field value must vanish there, leaving $V = \lambda\eta^4$ at the line. With the addition of the field gradient energy, the energy per unit length of the string is $\mu \sim \eta^2$. The string tension is the same as the energy per unit length, because the work to increase the length of the string goes into the energy of the increased length of string. Waves along the string move at the speed of light, and bends in the string flex at like speed.

In the expanding and cooling universe a two-component field would have relaxed to a tangle of cosmic strings that expanded with the general expansion of the universe on scales larger than the Hubble length. On smaller scales, the strings would have intersected, maybe producing an accumulation of string loops that tended to be dissipated by the dynamical drag of matter and the radiation of gravitational waves. That would leave some loops with a few long strings running across the Hubble length (Zel'dovich 1980; Vilenkin 1981b).

A straight string running across the Hubble length $\ell \sim cH_0^{-1}$ has mass $M_{\rm string} \sim \mu\ell$. In the Einstein–de Sitter model, the mass density is $\rho \sim H_0^2/G \sim c^2/G\ell^2$, so the total mass within this distance is $M_{\rm total} \sim c^2\ell/G$. The ratio is

12. But consider that if η and λ were taken to be free parameters rather than models taken from particle physics, then domain walls might have seemed more interesting. A flat wall running across the Hubble length, having surface mass density Σ, and with us near the wall, would produce a roughly quadrupole CMB anisotropy on the order of the dimensionless potential energy difference from here out to the Hubble length:

$$\frac{\delta T}{T} \sim \frac{G\Sigma}{H_0 c} \sim 10^{-4}, \quad \text{for} \quad \Sigma \sim 10^{11} M_\odot \, \text{Mpc}^{-2}.$$

(I have reinserted the velocity of light). The mass in the wall in the nearest megaparsec would amount to a few percent of the mass of the Local Group, which does not seem intolerable. At the time, there were indications of detection of a quadrupole CMB radiation anisotropy at about this value of $\delta T/T$ (Fabbri et al. 1980; Boughn, Cheng, and Wilkinson 1981). The latter group soon withdrew the anisotropy detection, but there still is room for a somewhat smaller effect of a domain wall.

$$\frac{M_{\text{string}}}{M_{\text{total}}} \sim \frac{G\mu}{c^2} \sim 10^{-6}. \tag{5.45}$$

This is the numerical value usually considered to produce an interesting disturbance from homogeneity.

Vilenkin (1981b) showed that a long straight string produces no gravitational attraction (though long strings appearing at the Hubble length cH_0^{-1} would be irregularly shaped, which would cause them to flail about and would give them active gravitational mass). The string instead gives spacetime a conical structure: In the string rest frame, a circle containing the string, and with physical radius r, has physical circumference

$$C = 2\pi r(1 - 4G\mu c^{-2}), \tag{5.46}$$

for $G\mu c^{-2} \ll 1$. Vilenkin pointed out that this angle deficit would produce two images with the same magnification of a distant object on a line of sight that passes close to the string. With the mass per unit length in equation (5.45) the angular separation of images produced by a string perpendicular to the line of sight, and at a distance halfway between the object and observer, would be a few seconds of arc. Vilenkin remarked that this is close to the angular separation of two quasar images with quite similar spectra that Walsh, Carswell, and Weymann (1979) had found. This certainly was interesting, but identification of a giant elliptical galaxy that likely is capable of producing the two images by gravitational lensing (with a third image lost in the light of the galaxy) argued against the string interpretation (Soifer et al. 1980). But Vilenkin's idea naturally inspired searches for double images with the equal magnification produced by cosmic strings (e.g., Cowie and Hu 1987).

A nearly straight string passing at close to the speed of light would produce a velocity impulse from the deficit angle in equation (5.46) amounting to about $G\mu/c^2 \sim 10^{-6}$, or perhaps ~ 1 km s^{-1}. The sudden equal and opposite velocity impulses produced by a string passing through objects on Earth could have interesting effects, but the chance of this happening would be exceedingly small. Kaiser and Stebbins (1984) pointed out a related effect. The angle deficit of a nearly straight string moving across the line of sight at near relativistic speed would produce different redshifts of radiation reaching us from directions on either side of the string. That would cause a steplike discontinuity of the CMB temperature, in the amount $\delta T/T \sim G\mu/c^2$. The idea continues to fascinate (e.g., Planck Collaboration 2014a).

The field structures in domain walls and strings would behave in a highly nonlinear way, which makes it difficult to analyze how they might trigger cosmic structure formation in enough detail to be compared to observations. Early examples of this challenging numerical art include Vachaspati and Vilenkin (1984), Albrecht and Turok (1985), and Bennett and Bouchet (1988).

Crittenden and Turok (1995) and Durrer, Gangui, and Sakellariadou (1996) showed how the spontaneous formation of global textures in a cold dark-matter universe with symmetry broken by a four-component field can produce a CMB anisotropy spectrum with features similar to (but in detail different from) what grows out of adiabatic initial conditions. Notable also is that spontaneous symmetry breaking would make the anisotropy pattern non-Gaussian, while squeezed states from inflation naturally are close to Gaussian and are consistent with what was observed later.

The signatures of what to look for in the CMB anisotropy in theories of homogeneity spontaneously broken by cosmic fields, and the comparison to what to look for in the theory of growing primeval adiabatic departures from homogeneity, had been well examined before the measurements that tested the theories (e.g., Albrecht et al. 1996, Hu and White 1996, Durrer and Sakellariadou 1997; Pen, Seljak, and Turok 1997). The bottom figure in Plate I (discussed in Chapter 9) illustrates the progress of measurements of the power spectrum in the pattern of variation of the CMB intensity across the sky. Durrer, Kunz, and Melchiorri (2002) confirmed the situation after clear detection of the CMB anisotropy pattern. These measurements made a very good case that structure grew out of primeval adiabatic departures from homogeneity, and they made an equally compelling case against the elegant idea that the homogeneity of the early universe had been spontaneously broken by cosmic fields.

There is good motivation from fundamental physics for the continued interest in the possibility that signatures of the spontaneous formation of cosmic field patterns may be observable in the CMB, in images of distant galaxies, in gravitational waves produced by tumbling strings, or maybe in the nature of the mass distribution. But the conclusion seems clear: such processes have had no more than modest effects on cosmic structure formation.

5.2.6 INITIAL CONDITIONS

While some were analyzing ideas about the spontaneous breaking of homogeneity, others were following the earlier tradition: Postulate initial conditions that describe primeval departures from homogeneity in the very early universe, consider simpler cases first, and explore the possible observational consequences.

Zel'dovich (1967) pointed out that we have a choice of adiabatic or isocurvature initial conditions. The latter assumes that a clumpy distribution of matter or cosmic fields in the early universe is accompanied by small disturbances in the dominant mass in radiation, thus serving to keep the net mass density homogeneous. There would be negligibly small primeval spacetime curvature perturbations. Zel'dovich (1967) termed this "isothermal" or "entropic" initial conditions; it is also known as the primeval isocurvature condition.

The more commonly discussed adiabatic initial conditions assume a homogeneous primeval entropy per baryon (and a homogeneous entropy per dark matter particle, if the dark matter can be described that way). It is the spatial distribution that would be produced if an exactly homogeneous universe were reversibly (that is, in a process that conserves entropy) slightly compressed or decompressed in different regions. Cosmic structure would be supposed to have grown by gravity out of these small departures from an exactly homogeneous early universe. It would have been accompanied by the primeval dents in spacetime curvature discussed in Section 5.1. Let us consider a measure of these dents and the postulate for initial conditions suggested by this measure.

The Newtonian gravitational potential belonging to the departure from a homogeneous mass distribution offers a useful way to measure the spacetime dents (equations (5.4) and (5.16)). With the mass distribution at fixed time represented by the Fourier integral in equation (5.30), the solution to Poisson's equation (5.20) for the potential is

$$\phi(\vec{x}) = -4\pi G\rho_{\rm m}\, a(t)^2 \int \frac{d^3k}{k^2}\delta_{\vec{k}}(t)e^{i\vec{k}\cdot x}, \tag{5.47}$$

as can be seen by differentiating out this expression. We get the potential $\bar{\phi}_X$ averaged over a chosen length scale, X, by convolving this expression with a Gaussian window function of width X:

$$\bar{\phi}_X(\vec{x}) = \int d^3x' \frac{e^{-(\vec{x}'-\vec{x})^2/(2X^2)}}{(2\pi)^{3/2}X^3}\phi(\vec{x}') = -4\pi G\rho_{\rm m}\, a(t)^2 \int \frac{d^3k}{k^2}\delta_{\vec{k}}(t)e^{i\vec{k}\cdot\vec{x}-k^2X^2/2}. \tag{5.48}$$

This smoothing suppresses wavenumbers $k \gtrsim X^{-1}$ that represent fluctuations on scales smaller than X, giving a measure $\bar{\phi}_X(\vec{x})$ of the typical size of potential fluctuations on the scale X, provided that the fluctuations are not dominated by scales much larger than X.

Let us model the mass fluctuation power spectrum defined in equation (5.33) as a power law:

$$P(k) = Ak^n. \tag{5.49}$$

Then the mean-square values of the fluctuations in the mass distribution and gravitational potential smoothed over scale X are

$$\langle\bar{\delta}^2\rangle_X = A\int \frac{d^3k}{(2\pi)^3}k^n e^{-k^2X^2}, \tag{5.50}$$

$$\langle\bar{\phi}^2\rangle_X = (4\pi G\rho_{\rm m})^2 a(t)^4 A\int \frac{d^3k}{(2\pi)^3}k^{n-4}e^{-k^2X^2}. \tag{5.51}$$

Suppose first that $n > 1$. Then the exponential $e^{-k^2X^2}$ in equation (5.51) cuts off the integral at wavenumber $k \sim X^{-1}$. Thus the smoothed potential fluctuations vary as $\langle\bar{\phi}^2\rangle_X \propto X^{1-n}$. The smaller the smoothing length X, the larger the potential energy fluctuations will be. If this continued down to the scale where $\langle\bar{\phi}^2\rangle_X \sim 1$, it would signify nonlinear spacetime curvature fluctuations in a chaotic spacetime "foam." Maybe spacetime structure on small enough scales is a quantum foam, but we must avoid anything like it on observationally accessible scales; it is not seen. That is, we must postulate that the power law model $P(k) \propto k^n$ with $n > 1$ is truncated on scales less than some characteristic length. If $n < 1$, the integral in equation (5.51) diverges on large scales, or $k \to 0$, as $\langle\bar{\phi}^2\rangle_X \propto X^{1-n}$. If the power spectrum power law model extended to small enough k, or large enough X, then curvature fluctuations would have tended to drive the part of the universe we can observe well away from Einstein–de Sitter. We must again postulate that the power law model is truncated, in this case by an upper cutoff at some length scale.

At the present level of understanding, there is no fundamental basis for rejecting a truncated power law power spectrum, apart from the consideration that it would require the introduction of a characteristic length and call for an explanation of where that length came from. We avoid that inconvenience by postulating $n = 1$, so the integral for $\langle\bar{\phi}^2\rangle_X$ diverges on large and small scales only as the logarithm of the length. This allows us to take it that $P(k) \propto k$ is truncated at uninterestingly large and small scales, and we have

$$P(k) \propto k, \quad \langle\bar{\delta}^2\rangle_X^{1/2} \propto X^{-2}, \quad \langle\bar{\phi}^2\rangle_X^{1/2} \sim |\log X|. \tag{5.52}$$

This has come to be termed "scale-invariant" initial conditions.

The argument from simplicity for scale-invariant initial conditions was introduced independently by Harrison (1970) and Peebles and Yu (1970). Zel'dovich (1972) took a particular interest in this case. The example of a power spectrum from adiabatic initial conditions in Panel (b) in Figure 5.2 uses scale invariance, with $n = 1$. It means that the fractional mass density fluctuations appearing on the scale of the Hubble length in the Einstein–de Sitter model are independent of time.[13] The plasma-radiation fluid amplitude remembers this; it causes the train of peaks in the power spectrum to have roughly constant amplitude.

The nonempirical case for scale-invariant initial conditions was reenforced by early interpretations of the picture of cosmological inflation (Section 3.5.2). The tendency was to suppose that the expansion of the universe during inflation was close to exponential, $a(t) \sim e^{\alpha t}$, with α close to constant. In this case, there is no characteristic length to set a physical scale for initial conditions.

13. To see this, consider a matter-dominated Einstein–de Sitter universe. Since the expansion parameter varies as $a(t) \propto t^{2/3}$, the physical wavelength that is equal to the Hubble length at time t_k is given by $a(t_k)/k \sim t_k \propto t_k^{2/3}/k$. This gives $t_k \propto k^{-3}$. The Fourier amplitudes grow as $\delta_k \propto t^{2/3}$ (equation (5.3)), so the Fourier amplitude that is initially proportional to $k^{1/2}$, assuming $n = 1$, grows to $\delta_k \propto k^{1/2} t_k^{2/3} \propto k^{-3/2}$ at Hubble length crossing. The mean-square mass fluctuation on the scale of the Hubble length is proportional to $k^3 \delta_k^2$, which we see is independent of k. It is a good exercise for the student to check that the same line of argument applies in a radiation-dominated universe.

Absent a length scale, it follows that the initial conditions generated during this exponential expansion, by the squeezing of quantum fluctuations, could only be scale invariant. This can be changed by assuming that the expansion significantly differs from a pure exponential, which would produce a tilt away from scale invariance. The nonempirical appeal against this adjustment is simplicity, and nature more or less agrees: The cosmology established at the turn of the century was seen to require a real but slight departure from scale invariance.

In the early 1980s, improving bounds on the anisotropy of the CMB, after removing the dipole attributed to the effect of our motion through the sea of radiation, made it clear that this sea of radiation is far smoother than the clumpy distribution of matter in galaxies. This was a challenge for adiabatic initial conditions: How could gravity have pulled matter together into the great clumps that are galaxies and clusters of galaxies without more seriously disturbing the CMB? A way around the problem is the cold dark matter, or CDM, introduced by Peebles (1982b) and discussed in Section 8.2. In this model, most of the mass of the universe is the nonbaryonic dark matter discussed in Chapters 6 and 7. This hypothetical matter would be transparent to baryonic matter and radiation, allowing gravity to pull together mass with acceptably little disturbance to the radiation. The original CDM model assumed scale-invariant initial conditions, because they are simple and served to show how the problem of reconciling the smooth CMB with the clumpy matter might be resolved.

The CDM model proves to be on the right track, but in the 1980s it was easy to think of other reasonably simple ways to account for the smooth CMB. The minimal isocurvature model developed in Peebles (1987a,b) assumes the only dynamical actors are baryons, massless neutrinos, and the CMB. The mass density could be set to 10 percent of Einstein–de Sitter, about what was indicated by the relative peculiar velocities derived from the CfA redshift catalog discussed in Section 3.5.3. And this mass density was not seriously out of line with the baryon density indicated by light element abundances (Table 4.2). The primeval entropy per baryon would be a random function of position, with the power spectrum and amplitude adjusted so that the baryonic stellar halos of large galaxies would have formed early, at redshift $z \sim 10$ to 20. This seemed to be in line with the observations of fully developed quasars at high redshifts presumed to be hosted by massive galaxies. The model could be adjusted to be reasonably consistent with the bounds we had then on the CMB anisotropy. It did not require hypothetical nonbaryonic dark matter, a positive feature, but it did require ad hoc adjustment of initial conditions to allow galaxy assembly on relatively small scales and consistency with the upper bounds on CMB anisotropy on larger scales.

I conclude that up to the mid-1990s the two pictures—isocurvature initial conditions tuned to fit the observations and adiabatic scale-invariant initial conditions motivated by simplicity and inflation—showed roughly comparable promise. And it was not difficult to invent still other viable models (e.g., Peebles 1999). The abrupt change in the situation is the topic of Chapter 9.

5.2.7 BOTTOM-UP OR TOP-DOWN STRUCTURE FORMATION

Did cosmic structure grow in a hierarchical manner, smaller mass concentrations forming earlier and gravitationally gathering into the later assembly of more massive concentrations and on up? Or was there a first generation of massive concentrations that fragmented into young galaxies? In the early 1980s, both pictures—bottom-up and top-down—were supported by good arguments and challenged by others. Measurements of the galaxy low-order N-point correlation functions (which are discussed in Section 2.5) show that, on relatively small scales, the galaxy distribution is well approximated by a scale-invariant clustering hierarchy. That naturally invited the idea that structure was assembled in a hierarchical fashion, bottom-up. But the viscous dissipation of acoustic waves represented by equation (5.32) would have tended to suppress primeval fluctuations in the baryon distribution on smaller scales. This invited consideration of top-down formation. The apparent detection of a neutrino rest mass of a few tens of electron volts, which is discussed in Section 7.1.1, added considerable interest in the top-down picture, because these thermally produced neutrinos would have had velocities large enough that they would have smoothed the mass distribution out to mass scales far larger than that typical of galaxies.

The evidence that the galaxy space distribution on scales $r \lesssim 10$ Mpc is a good approximation to a scale-invariant fractal, with fractal dimension $D = 1.23$, is discussed in Section 2.6. It is illustrated in Soneira and Peebles (1978) by the comparison in their Figures 3 and 7 of the Lick galaxy map and a map of fractals, cut off at about 20 Mpc, constructed to match the shapes of the observed galaxy 2- to 4-point functions. Peebles (1974b) and Davis, Groth, and Peebles (1977) argued that this agrees with structure formation that grew out of adiabatic scale-invariant initial conditions that gravity gathered in a hierarchical assembly. There were earlier thoughts along these lines, as in Layzer (1954) and Peebles (1965).

Saslaw (1972) introduced an approach to a quantitative theory of the bottom-up picture by the methods developed for analysis of properties of nonideal gasses. Davis and Peebles (1977) aimed to use these methods to show how gravity in the expanding universe could produce a scale-invariant mass distribution that might account for the observed nearly scale-invariant clustering pattern of the galaxies. But numerical simulations of structure formation showed that the approximations Davis and Peebles introduced to make the analysis tractable fail at mass density contrasts of observational interest.

Efstathiou et al. (1988, 726) concluded from their numerical simulations that "It is paradoxical that this apparent absence of any characteristic scale is, in fact, inconsistent with scale-free gravitational clustering, even though it has often been used to argue in favour of such a process."

The galaxy two-point correlation function is remarkably close to a simple power law at separations ranging from about 10 kpc to about 10 Mpc. In numerical simulations of the evolution of the mass distribution in the established ΛCDM cosmology, the mass correlation function is quite different from this power law. Yet the simulations show formation of mass density peaks that look like good approximations to galaxies, and the space distributions of these peaks are a good fit to the statistics of the space distribution of real galaxies. The standard and accepted thinking is that we must learn to live with this curious situation in the bottom-up picture, though it does weaken the case for a simple gravitational instability picture.

There were serious arguments in the 1970s through the early 1980s for structure formation in the other direction: top-down. Under the assumption of adiabatic initial conditions considered in Section 5.2.6, small-scale fluctuations in the baryon distribution are dissipated by photon diffusion at the rate approximated by equations (5.31) and (5.32). Silk (1967, 1968) worked out the consequence; the effect is termed "Silk damping." Assuming all matter is baryonic (as seemed natural and not even worth mentioning then), Silk estimated that the primeval mass density fluctuations would be dissipated on scales smaller than about $M_S \sim 10^{11}$ to $10^{12}\ M_\odot$. Unless dissipation had been countered by really large primeval density fluctuations on smaller scales, or there had been a primeval isocurvature component, this would mean that the mass distribution at the start of structure formation had been smoothed on scales smaller than M_S. The first generation of nonlinear departures from homogeneity would have collapsed roughly to sheets of compressed and maybe shocked baryons in concentrations with masses of roughly M_S. Lynden-Bell (1962) showed this behavior, collapse to a sheet, in a homogeneous rotating ellipsoid. Zel'dovich (1970) showed that it follows in the initial stages of collapse of a mass distribution that is smoothed at some mass scale.

These considerations suggested to Zel'dovich (1970) and Sunyaev and Zel'dovich (1972) that the first generation of bound objects would have tended to be flattened; they termed them "pancakes." Their estimate of the characteristic mass set by the viscous smoothing is

$$M_S \sim 10^{12} \text{ to } 10^{14}\ M_\odot, \tag{5.53}$$

even larger than Silk had proposed. The introduction of the first idea for nonbaryonic dark matter, the hot dark matter discussed in Section 7.1, led to consideration of an even larger value of this characteristic mass. But for the present purpose, the point to note is that this value of M_S is well above the mass in the luminous parts of large galaxies, such as the Milky Way. Zel'dovich and Sunyaev accordingly proposed a top-down picture: the first generation of nonlinear mass concentrations would have been protoclusters that collapsed to pancakes that fragmented to form galaxies.

Zel'dovich and Sunyaev had a historical precedent. In *The Realm of the Nebulae*, Hubble (1936, 81) remarked on indications of

> the density of the cluster diminishing as the most frequent type advances along the sequence of classification [from early to late] ... with the dominance of late types among isolated nebulae in the general field ... [which] suggest as a possibility that nebulae may originate in clusters and that the disintegration of clusters may populate the general field.

This overview of the observations has not changed: redder, more gas-poor, early-type elliptical and lenticular galaxies are more common in the denser parts of clusters of galaxies; and bluer, more gas-rich, star-forming, late-type spiral and irregular galaxies are more common elsewhere. (It had been suspected that early-type galaxies evolve into late-type, hence the names, but a direction of evolution this way or the other is no longer considered likely.)

Another consideration, by de Vaucouleurs (1960, 585), is:

> A number of recent investigations ... have sharply brought out the conflict between masses of groups and clusters of galaxies estimated from mass-luminosity ratios (i.e., from rotation), on the one hand, and from velocity dispersion through the virial theorem, on the other ... [suggesting] that most groups and loose clusters of galaxies, especially those rich in spirals, are apparently unstable and may evaporate with lifetimes of the order of a few billion years, as predicted by Ambartzumian.

This mass conflict certainly was serious, and in the 1970s, de Vaucouleurs's point could have been taken to be in line with the Zel'dovich and Sunyaev picture of fragmenting pancakes, though I have found no mention of it. The conflict has been resolved in a different way by the hypothesis that the masses of galaxies and groups and clusters of galaxies are dominated by nonbaryonic dark matter. But as it happens, the considerations of what came to be termed "nonbaryonic hot dark matter" led to renewed interest in the top-down picture of structure formation. This is reviewed in Section 7.1.

There are two serious problems with the pancake picture. First, the redshift pattern around the relatively nearby Virgo cluster of galaxies indicates that the galaxies around this cluster, out to the distance of the Local Group, are drifting toward the cluster, as if gravitationally attracted by its mass. That is, this cluster seems to be growing by gathering together galaxies that have already formed and relaxed to mature-looking systems. The evidence is reviewed in

Section 3.6.4, on page 86. Recall also the Regős and Geller (1989) demonstration that galaxies are flowing toward the mass concentrations in other clusters of galaxies. These phenomena certainly look like bottom-up formation.

The second problem with the top-down pancake picture is that most galaxies are not in clusters. The early generation of protoclusters would have to have cast away most of their mass in fragments that became the common galaxies outside clusters. Zel'dovich (1978, 416) recognized this: "The walls must fragment into separate galaxies and clusters of galaxies." But in the gravitational instability picture, gravity is supposed to have gathered these large concentrations of mass. It is in the nature of gravity to continue gathering, not to start casting away. Another way to put it is that this version of the top-down scenario predicts a much greater mass clustering length than is observed in the galaxy distribution, and it predicts that galaxies are younger than clusters, contrary to the evidence. The straightforward reading is that since the time of galaxy formation, the galaxies have been gathering into groups and clusters and superclusters of galaxies, in the bottom-up picture.

5.3 Concluding Remarks

The benefits and confusions that may grow out of interactions of intuition and experience, and theory and observation, are well illustrated in this chapter. Turbulence is observed on scales ranging from everyday life to galaxies. The idea that structure in our universe grew out of primeval turbulence, or some other sort of chaos, thus may seem philosophically satisfying, as it did in earlier creation stories. An example of the appeal of fossil turbulent eddies as protogalaxies is seen in von Weizsäcker's (1951a) remark quoted on page 213. Another example from the same conference is Werner Heisenberg's (1951, 199) comment:

> wouldn't you say that if we believe in the expanding universe (I know that some of us do not but that is another matter), then we should also assume that there is an enormous energy in this primary cosmic gas which expands? Now, if there is this enormous kinetic energy of the gaseous masses, I suppose there must be turbulence, because the turbulent motion is the normal motion of the gas, whereas laminar flow is extremely exceptional.

In the laboratory, the production of laminar flow indeed requires careful arrangement, as Heisenberg says. And his point may be taken to include the fact that the expanding universe requires exceptional initial conditions: very close to exact primeval homogeneity expanding at very close to gravitational escape velocity. But given that, and general relativity, the enormous kinetic energy of matter in the expanding universe is not exponentially unstable to the development of turbulence. Intuition is valuable, but here is an example of intuition wrongly overshadowing analysis. This is not a criticism of the pioneers who were working with far less data and far sparser traditions of analysis compared to what came later in the history of cosmology. It is offered as a lesson, perhaps even a hope, that we may find that we still are fooling ourselves about some aspects of our standard world picture.

It would have been elegant to find that our universe began with perfect symmetry—homogeneous with no primeval dents in spacetime—and that the galaxies grew out of a spontaneous breaking of homogeneity. Perhaps even more elegant would be exact primeval homogeneity with zero entropy, but that is convincingly ruled out by the thermal CMB spectrum. Homogeneity breaking in a hot big bang may be modeled along the lines of perfectly good physics, as in the picture of cosmic field structures such as domain walls or cosmic strings. But it seems clear that nature did not choose this elegant way to produce the world for observers such as us. Copeland and Kibble (2009) emphasize that there still is ample motivation from fundamental physics to look for effects of cosmic field defects or structures that the universe may have acquired as it expanded and cooled, albeit now as a subdominant process in galaxy formation. The art of searches for signatures of cosmic strings in CMB sky maps is discussed by the Planck Collaboration (2014a). A quarter of a century ago, Vilenkin (1981c), Hogan and Rees (1984), and others proposed that flailing cosmic strings could produce detectible gravitational waves. The web page of the NANOGrav Physics Frontiers Center illustrates the ongoing search for this phenomenon.

A similar line of thought contemplated the breaking of primeval symmetry by an autocatalytic process driven by pure gravity or maybe the energy released in the conversion of light elements to heavier ones. Autocatalytic behavior is real: consider a fire. The possibility of this behavior in cosmology had to be considered, but it seems clear that it is not a significant factor in structure assembly in our expanding universe.

Explosions are observed to rearrange the baryons in galaxies, and numerical simulations display the importance of explosions in helping suppress the tendency of baryons to pile up in the centers of the many galaxies that are better described as pure disks. But the evidence from these simulations and the considerations presented in Chapters 6 and 7, along with the cosmological tests reviewed in Chapters 8 and 9, is that gravity is the prime mover for cosmic structure formation out of primeval dents in spacetime.

Community thinking in the 1980s was informed by tradition: gravity has figured in the thinking about galaxies for a long time, and explorations along this line naturally continued. There were variants of gravitational instability to consider: adiabatic or isocurvature, bottom-up or top-down. Early ideas about the consequences of cosmological inflation were influential, reinforcing the community preference for bottom-up formation from Gaussian nearly scale-invariant initial conditions. The idea of nonbaryonic dark matter in the CDM model discussed in Section 8.2 offered an attractive framework for more detailed and quantitative analyses of structure formation by gravity out of initial conditions that seemed reasonable in the inflation picture of the very early universe. But before considering this, we must review in the next two chapters how the community became persuaded that we might do well to add this hypothetical nonbaryonic matter to ideas about how cosmic structure formed.

CHAPTER SIX

Subluminal Mass

THE PRESENCE OF mass around galaxies in excess of what would be expected from the observed starlight has been known for many years, by many names, including missing mass, or hidden, unseen, invisible, nonluminous, underluminous, or subluminal mass, and eventually, dark matter. King's (1977, 7) comment on the situation is that

> The most serious problem in extragalactic astronomy today is the notorious "missing mass." (I know that Jerry Ostriker derides the term; he tells us, quite correctly, that it is not missing at all, since we know from its gravitation that it must be there. But this is just a difference of taste. I prefer to call it "missing" because it certainly is missing from our understanding.)

The astronomical phenomena reviewed in this chapter will be said to indicate the presence of "subluminal matter," a term that is intended to be descriptive and minimally pejorative. In this history, the term "dark matter," or DM, is reserved for the nonbaryonic mass component in the ΛCDM cosmology that was established at the turn of the century, when subluminal matter became interpreted as dark matter.

The presence of significant mass in subluminal matter was first suggested in the 1930s by the surprisingly large velocities of galaxies in clusters of galaxies. Either the clusters are flying apart, or they are held together by the gravitational attraction of mass outside the luminous parts of the cluster members, or maybe there is something wrong with our standard physics. By the mid-1970s, there was reasonably clear evidence for the presence of this subluminal mass in the outer parts of galaxies as well as in clusters of galaxies, assuming standard physics. At the time, particle physicists happened to be growing interested in the thought that there may be stable nonbaryonic matter that would not interact much with ordinary matter and radiation, apart from gravity. The possible connection to the astrononers' subluminal matter was recognized in the 1970s, but not widely advertised, perhaps because it was difficult to think of positive things to say about it. That changed in the early 1980s, when cosmologists adopted the particle physicists' hypothetical nonbaryonic matter as a way to reconcile the distinctly clumpy distribution of matter in galaxies with the smooth sea of cosmic microwave radiation, the CMB. This was a step toward the cosmology established at the turn of the century.

The subject of this chapter is the history of discovery of astronomical evidence of subluminal matter in large clusters of galaxies (Section 6.1), in groups of a few or just two galaxies that are close enough that they seem likely to be gravitationally bound (Section 6.2), and in individual spiral galaxies. There must be enough mass in spirals to account for the circular velocities of disk stars (Section 6.3), and the mass rotationally supported in the disk must be large enough that gravity can form spiral arms, but this mass component cannot be so large that the spiral arms grow to destroy the observed nearly circular motions in the disk (Section 6.4). These conditions require that most of the mass in a spiral galaxy is in a stable subluminal massive halo draped around the outskirts of the luminous parts of the galaxy.

6.1 *Clusters of Galaxies*

The evidence of a lack of understanding of masses on large scales first appeared in the 1930s with the difficulty of understanding what is holding together the great concentrations of galaxies known as rich clusters. The Swiss-American physicist and astronomer Fritz Zwicky pointed this out. Zwicky's (1933) analysis used the virial theorem,[1] the relation between the velocities of the galaxies in a cluster and the mass needed for gravity to hold the cluster together, which he approximated as

$$\frac{3}{2} M \langle v_{\parallel}^2 \rangle \simeq \frac{3}{10} \frac{GM^2}{R}. \tag{6.1}$$

Here M is the cluster mass; R is a measure of its radius; and $\langle v_{\parallel}^2 \rangle$ is the mean of the squares of the line-of-sight velocities $v_{\parallel}$ of the galaxies within the cluster, derived from the galaxy Doppler shifts relative to the mean for the cluster.

1. The virial theorem in Newtonian mechanics is that the time average of the kinetic energy of a gravitationally bound and slowly evolving system is half the time average of the negative of the gravitational potential energy:

$$K = \frac{U}{2}, \text{ where } K = \sum_i \frac{m_i \vec{v}_i^2}{2}, \ U = \sum_{i<j} \frac{G m_i m_j}{|\vec{r}_i - r_j|}.$$

Zwicky took $U = 3GM^2/(5R)$ for a homogeneous sphere of mass M and radius R, a reasonable approximation, given his modest redshift data. The factor of three on the left-hand side of equation (6.1) adds the two components of velocity dipersion transverse to the line of sight.

Equation (6.1) and similar approximations had been used to find the masses of galaxies from the velocities of the stars within them. Zwicky estimated the mass of the relatively nearby and large Coma Cluster (named for its position in the constellation Coma Berenices) by multiplying his count of galaxies in the cluster by an estimate of the typical mass of a galaxy. That total mass in equation (6.1) indicated that the velocity dispersion would be expected to be $\langle v_{\parallel}^2 \rangle \simeq 80\ \mathrm{km\ s^{-1}}$. But Zwicky (1933) had radial velocities of galaxies in the cluster from measurements of the redshifts of their spectra. This measured galaxy velocity dispersion is $\langle v_{\parallel}^2 \rangle \sim 1000\ \mathrm{km\ s^{-1}}$, far larger than expected from the sum of the masses of the galaxies in the cluster. Zwicky concluded that (in the translation here and below by Andernach, and Zwicky 2017, 5), "If this should be verified, it would lead to the surprising result that dark matter exists in much greater density than luminous matter."

Zwicky had only eight galaxy redshifts to estimate the velocity dispersion, he used a crude approximation to the distribution of mass in the cluster, and he had only rough estimates of galaxy masses. But all the approximations were sensible, and he correctly identified the phenomenon: If Newtonian physics is to be trusted, and the cluster is not flying apart, then the mass of this cluster is far greater than the mass in its observed stars.

Zwicky used a serious underestimate of the extragalactic distance scale, but it did not affect the ratio of the cluster mass to the sum of the luminous galaxy masses, because both scale with distance in the same way.[2] The underestimate did made the galaxies seem to have curiously large masses for their luminosities compared to ratios of mass to luminosity for known kinds of stars. Sidney van den Bergh (1999, 657) remarked that "It is of interest to note that Hubble's prestige was so great that none of the early authors thought of reducing Hubble's constant as a way of lowering their mass-to-light ratios."

In his second paper on this issue, Zwicky (1937a) presented a map of angular positions of galaxies in and near the Coma Custer. It looks much like the modern version: a prominent dense core is surrounded by a distribution of galaxies that fans out to the field with no discernible cluster edge. If the galaxies were not gravitationally bound to the cluster, then the dense core and the outskirts of this cluster would dissolve at an expansion speed of about $1000\ \mathrm{km\ s^{-1}}$. Zwicky's (1937a, 227) conclusion from the smooth and regular appearance of the galaxy distribution in the cluster was that "it is probably

2. The mass of a gravitationally bound object with radius r and internal velocity v is $M \sim v^2 r/G$ (equation (6.1)). The value of r for a galaxy or a cluster of galaxies is the product θD of the angular size θ and the distance D. The distance reckoned from the redshift cz using Hubble's law is $D = czH_0^{-1}$. The considerable overestimate of Hubble's constant H_0 thus underestimated r and hence M. But the measures of the cluster mass and the sum of the galaxy masses are underestimated by the same factor, so the ratio of the two is independent of H_0. The luminosity scales as D^2, so the ratio of mass to luminosity scales as $M/L \propto D^{-1} \propto H_0$.

legitimate to assume that clusters of nebulae such as the Coma cluster ... are mechanically stationary systems," albeit with surprisingly large masses relative to the sum of masses of the galaxies in the cluster.

Zwicky (1937a, 235) cautioned that

> On the other hand, the virial theorem can hardly be used with much confidence in cases such as the Virgo cluster and the Pisces cluster.[9] These clusters are much more open and asymmetrical than the Coma cluster and their boundaries are thus far ill defined. Accurate values of the gravitational potentials in these clusters are difficult to determine.

The footnote is a reference to Zwicky's (1937b) paper, "On a New Cluster of Nebulae in Pisces." It is now known as a part of the Perseus-Pisces Supercluster.

Despite Zwicky's caution, Sinclair Smith (1936) had already estimated the mass of the Virgo Cluster. He used an approximation similar to equation (6.1) (with cluster mass somewhere between $M \sim v^2 R/G$, as suggested by circular orbits, and $M \sim v^2 R/2G$, a measure of escape velocity). Smith's cluster mass estimate is some two orders of magnitude larger than the sum of Hubble's estimates of the galaxy masses in this cluster. Smith concluded that this subluminal mass possibly represents "internebular material, either uniformly distributed or in the form of great clouds of low luminosity surrounding the nebulae." Smith (1936) referred to Zwicky (1933); and Zwicky (1937a) referred to Smith (1936).

In a survey of increasing evidence for the presence of subluminal matter, Martin Schwarzschild (1954) referred to Smith (1936) and Zwicky (1937a). Schwarzschild's application of the virial theorem to the Coma Cluster used a more detailed representation of the spatial distribution of the galaxies (by counts of galaxies in strips as a function of perpendicular distance from the cluster center). He had 22 galaxy redshifts, from which he deleted the five largest deviations from the mean to reduce the possible effect of background and foreground galaxies seen in projection onto the cluster. He noted that this might eliminate some true cluster members, meaning that his cluster mass might be an underestimate. His final estimate of the root-mean-square line-of-sight velocity dispersion is $\langle v_\parallel^2 \rangle = 825$ km s^{-1}. This is reasonably close to Zwicky's estimate and close to later measurements of the central velocity dispersion based on many more cluster members: Sohn et al. (2017) report $\langle v_\parallel^2 \rangle = 947 \pm 31$ km s^{-1}.

With Zwicky, Schwarzschild concluded that the mass required to hold the Coma Cluster together is much larger than the mass present in the luminous parts of the galaxies. Schwarzschild suggested the excess mass might be contributed by stars of very low mass and so exceedingly low luminosities, or by the remnants of stars that have exhausted their supplies of nuclear fuel. Others returned to this thought, as in White and Rees (1978).

Schwarzschild (1954, 280) repeated Zwicky's assessment that "the assumption of approximate stationariness may be questionable for a loose and extended system like the Virgo cluster but appears fairly safe for a compact system like the Coma cluster." This assessment certainly agrees with the regular and compact appearance of the distribution of galaxies in Zwicky's (1937a) map of the Coma Cluster. The sentiment is repeated by Neyman, Page, and Scott (1961) in their review of a Conference on the Instability of Systems of Galaxies (Santa Barbara, California, August 10–12, 1961) and by Margaret Burbidge and Geoffrey Burbidge (1975), in their review of *The Masses of Galaxies*. (This 1975 paper is marked as received much earlier, February 1969, and it contains no references dated later than 1968. We may take it as their assessment of the situation in the late 1960s.)

Karachentsev (1966) presented measures of the masses needed for gravitational confinement of nine clusters of galaxies, to which he could add six mass estimates from the literature. The masses are consistently larger than the sum of the masses of the galaxies they contain, by about the same factor as for the Coma and Virgo clusters. Karachentsev (1966, 48) was willing to "discuss the possibilities of explaining the large values of f [the cluster mass-to-light ratios] by the presence of unobserved matter in the systems," but he noted that the "recently detected explosive processes in the nucleus of M 82 harmonize well with the general idea of instability in systems of galaxies."

The second idea is in line with thinking by de Vaucouleurs (1960) and others, as we have noted. The first idea, that clusters of galaxies are in close-to-stable gravitationally bound states, even the poor ones like Virgo, was established in the 1970s through the discovery that clusters are X-ray sources (Gursky et al. 1971). The Mitchell et al. (1976) detection of an emission line at 7 keV indicated transitions in iron ionized down to one or two electrons. This argued for X-ray emission by thermal bremsstrahlung in a hot intracluster plasma (that is, radiation by plasma electrons accelerated by the electric fields of the ions). Lea (1977) summarized the evidence for the plasma interpretation, in particular the consistency of the plasma temperature derived from the X-ray spectrum with the temperature required for plasma pressure support in the gravitational field indicated by the velocity dispersion of the galaxies.

The evidence for pressure support is illustrated in Figure 6.1 from Mushotzky et al. (1978). The dot-dashed line has the slope of the relation between the intracluster plasma temperature and the galaxy velocity dispersion near the cluster centers to be expected if plasma and galaxies have about

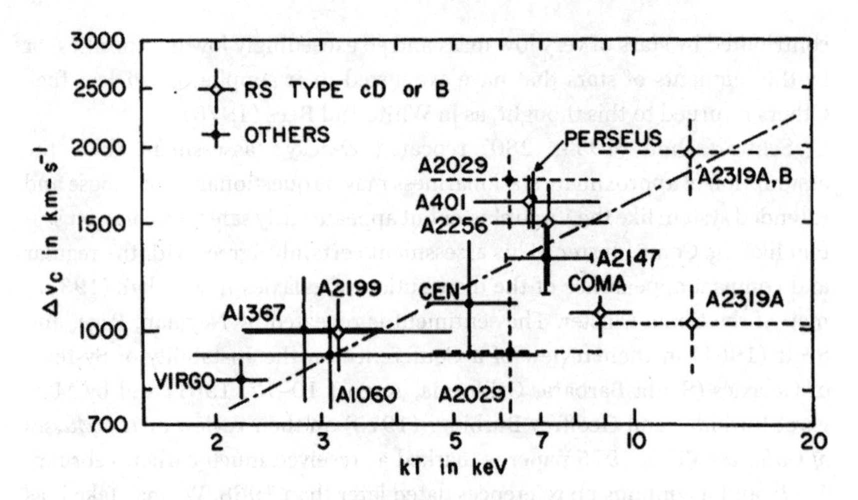

FIGURE 6.1. Cluster plasma temperatures and galaxy velocity dispersions. From Mushotzky et al. (1978). © AAS. Reproduced with permission.

the same space distribution and both are at close to dynamical equilibrium in the gravitational potential well of the cluster.[3] Within the uncertainties, the measurements of cluster plasma temperatures and cluster galaxy velocity dispersions agree with this relation. The stability of the Virgo Cluster had been questioned, but it is close to the expected pressure balance, at the lower left-hand edge of the figure.

The scatter of measurements in Figure 6.1 is large, in part because estimates of the galaxy velocity dispersions require separation of true cluster members from galaxies that are seen as close only in projection on the sky. A later version of Figure 6.1 by Lubin and Bahcall (1993) shows smaller scatter in a larger sample of clusters.

More evidence of subluminal matter in rich clusters came from the demonstration that luminous arcs, streaks of light across the faces of clusters of galaxies, are strongly distorted images of background galaxies that have been gravitationally lensed by a smooth distribution of cluster mass in subluminal matter (Hammer 1987; Soucail et al. 1988). The evidence of subluminal matter in the Virgo Cluster and others by the thermal Sunyaev-Zel'dovich suppression of the CMB temperature, as in the Planck Collaboration (2016) observations, is also compelling, but this was established after the community had generally come to accept the case for nonbaryonic dark matter in clusters.

3. The one-dimensional mass-weighted rms velocity dispersion averaged over electrons and ions in plasma at temperature T is $v_{\rm p} = \sqrt{k_{\rm B}T/m}$, where $k_{\rm B}$ is Boltzmann's constant, and m is the mean mass per particle in the plasma. If cluster plasma and the galaxies are near statistical equilibrium with the same space distribution, this should agree with the galaxy line-of-sight velocity dispersion: $v_{\rm g} = v_{\rm p}$. The dot-dash line in the figure has the slope of this relation, but the normalization is a little off.

6.2 Groups of Galaxies

Four examples illustrate the development of thinking about subluminal mass in groups of a few large galaxies. The first originated with Slipher (1917), who found that his measurements of redshifts of galaxies showed that most are positive, as if the galaxies were moving away from us, but that the nearest large spiral galaxy, the Andromeda Nebula M 31, has a negative redshift; this galaxy appears to be approaching us. (The velocity of approach relative to the solar system is about 300 km s^{-1}. After correction for our motion within our galaxy, M 31 is found to be approaching our galaxy at a little more than 100 km s^{-1}.) Kahn and Woltjer (1959, 705) wrote that

> It is well known that the density of galaxies in our neighborhood is larger than the average. This has led to the theory than our Galaxy is a member of the so-called Local Group, a group similar to the fairly numerous small groups of galaxies found all through extragalactic space. It seems reasonable to assume that most of these groups are systems of negative energy, i.e., that they are held together by gravitational forces.

From the condition that M 31 and the Milky Way are coming together after completing the better part of one orbit in 15 billion years, Kahn and Woltjer calculated that the mass of the Local Group is an order of magnitude larger than the mass in the luminous parts of the two galaxies. The argument is sensible and careful, and it proves to be accurate. But we should bear in mind Geoffrey Burbidge's (1975, L8) question: "And what of the possibility that our Galaxy and M31 are simply field galaxies, perhaps in the supercluster, but making a chance passage?" and Oort's (1958, 2) comment (continued from Section 5.2.2): "The character of a galaxy would appear to be determined by the measure of orderliness of the large-scale currents in the clump of the universe from which it contracted."

At the time it would have been fair to consider with Burbidge whether Oort's large-scale currents could have led galaxies to move about after formation, and if so whether M 31 may only be passing by us, close just as we flourish. Slipher's observation certainly suggests that the mass of the Local Group is much larger than what is seen in the stars, but in the 1970s, that interpretation could be questioned.

The second example is illustrated by Stephan's Quintet, a compact group of galaxies. Burbidge and Burbidge (1961, 244, 245) found that four of the group members have similar redshifts, as expected if they are in a physical group, but that the fifth member, NGC 7320, has redshift some 5000 km s^{-1} less than the mean of the other four. Their assessment was that "The chance that NGC 7320 is a foreground galaxy is very small [but that on] the other hand, it appears inherently improbable that NGC 7320 is a physical member of the quintet and is literally exploding away from the other members."

The explosion picture was to be considered, but Margaret Burbidge and Sargent (1971, 364) remarked that

> examination of the location of Stephan's Quintet on the Palomar 48-inch Sky Survey Atlas reveals that it lies quite near NGC 7331, a large, well-studied Sb galaxy. ... The velocities of NGC 7320, the discrepant member of Stephan's Quintet, and NGC 7331 are so similar as to suggest that the two galaxies may be physically related

and closer than the other galaxies. Kent (1981) confirmed this interpretation. His measurements of distances of the five galaxies, based on the correlation between the luminosity of a galaxy and its internal velocity dispersion, indicated that NGC 7320 is 11 Mpc from us, and the other four are at about 65 Mpc. With these distances r, the redshifts z of the five galaxies are reasonably close to what would be expected from Hubble's law, $v = cz = H_0 r$. The redshift of NGC 7320 is low, because it is well in front of the other four galaxies in Stephan's Quintet. The a priori probability that this would happen is small, but there are many knots of galaxies to look at.

A third example is the VV 172 group. It appears on the sky as five galaxies with quite similar angular sizes arranged in close to a straight line with nearly uniform spacings at separations close to the galaxy angular diameters. But Sargent (1968) found that the redshift of one of the galaxies is greater than the others by 21,000 km s^{-1}. Burbidge and Sargent (1971, 362) argued that

> There are three possibilities for explaining this strongly discordant velocity: either this is a case of the chance coincidence of a background object with a higher intrinsic luminosity, fitting into a convenient gap between the neighboring galaxies in the chain, as seen projected on the celestial sphere, or there is a non-velocity component in the redshift of the discordant object, or, thirdly, this object is being ejected with explosive violence out of the group.

The fourth example is from the study of relative motions of pairs of galaxies in samples taken to be gravitationally bound, because the two are close together in the sky and have roughly similar redshifts. Faber and Gallagher (1979, 166) assessed the attempts to measure galaxy masses from relative velocities of binary galaxies, with due attention to systematic errors. Their conclusion is that "In summary, for spatial separations greater than 100 kpc, the binary data indicate $M/L_B \simeq 35 - 50$, where the higher value applies if the orbits have moderate eccentricity."

This estimate of the mass-to-light ratio M/L_B (measured at the wavelength band noted in footnote 13 in Section 3.6.4) is an order of magnitude larger than the means of ratios of mass relative to luminosity in populations of stars in our neighborhood. That is, the dynamical analyses of binary galaxies pointed to the presence of subluminal mass. Faber and Gallagher presented all due caution about the difficulty of rejecting pairs of galaxies that are well separated along the line of sight but happen to appear close together in the sky: Erroneous inclusions would cause overestimates of relative velocities and masses of binaries; erroneous rejections of true bound pairs with large relative velocities would underestimate masses. And the curious situations in Stephan's Quintet and VV 172 cannot have inspired confidence in estimates of masses of individual groups in the 1970s.

In a study designed to address the problem of completeness and rejection of interlopers in samples of binary galaxies, Zaritsky and White (1994) analyzed a sample of positions and velocities of low-luminosity satellites of large spirals similar to the Milky Way that have no large nearby neighbors. Their data selection allowed good controls for their analysis, which indicated that the typical mass within $r = 200$ kpc of a large spiral is about $2 \times 10^{12} M_{\odot}$. This is an order of magnitude greater than the mass within the luminous parts of the galaxy. It is a characteristic measure of mass that has changed little since then.

At the time of writing, community interest in odd arrangements such as VV 172 has largely faded, possibly wrongly. Interest has turned to patterns in the positions and motions of dwarf satellites of large galaxies (e.g., Ibata et al. 2013; Tully et al. 2015). Interpretation in either situation is difficult, because our eyes are very effective at finding patterns, even if only apparent in truly random arrangements. And as these examples show astronomy offers an abundance of situations to observe and discern curious arrangements. Faber and Gallagher (1979) brought order to the situation by their assembly of data on binary systems, and later studies by Zaritsky and White (1994) made a reasonably clear case for massive subluminal halos around large spiral galaxies.

6.3 *Galaxy Rotation Curves*

Evidence of the presence of subluminal matter in the outer parts of spiral galaxies follows from measurements of the streaming speeds of stars and gas moving around the galaxies in the disks in roughly circular orbits. The centrifugal acceleration of this motion must be balanced on average by the gravitational attraction of the mass in the galaxy. The evidence emerging in the 1970s was that the mass in a typical spiral galaxy must be more broadly distributed than the starlight, meaning that the total mass must exceed the nominal estimates of the mass in stars. To keep the review of the growth of this evidence to a manageable length, I follow the stories of just four galaxies: M 31, NGC 3115, M 300, and NGC 2403. They offer a reasonably fair sample of how advances in technology improved observations that were driven in part by defined goals and in part by the simple desire to make the best possible measurements. And they illustrate how this research helped interest cosmologists in the idea that there is subluminal mass that might be nonbaryonic, perhaps with more conviction than astronomers might have considered appropriate at the time.

6.3.1 THE ANDROMEDA NEBULA

Figure 6.2 illustrates progress in the discovery of evidence of subluminal matter around galaxies from measurements of the velocity $v_c(r)$ of circular motion around the galaxy as a function of distance r from the center, known as the rotation curve. The figure shows observations of the Andromeda Nebula, M 31. It is the nearest large galaxy outside our Milky Way, a close to edge-on spiral that spans more than 4° across the sky. It offers the best chance to measure streaming velocities of stars and gas at many positions across the face of a galaxy.

The panels in Figure 6.2 are from the literature. I have taken the liberty of distorting them to approximate a common scale of radial velocity and angular position, and I have rotated or reflected them so the radial velocity increases upward as the angular position increases to the right in the figure, toward the northeast in the sky. The result is messy, but I think it offers a useful illustration of an important point: Recognition of the presence of subluminal matter on the outskirts of galaxies was a process that took decades.

Panel (a), at the top of Figure 6.2, is from Horace Babcock (1939). The smaller circles near the center are measurements of redshifts of stellar absorption lines in the high-surface-brightness starlight near the center of the galaxy. The four larger filled circles are Babcock's measurements of redshifts of emission lines from the plasma around massive young stars that are hot enough to have ionized the nearby interstellar medium. They were known as emission line regions, now termed H II regions. In a modern photograph of the galaxy, the H II regions are red from the prominent red H-α Balmer line of atomic hydrogen at wavelength 6600 Å (660 nm, 0.33 μ). Babcock's plates were sensitive only in the blue; he largely detected the (doublet) emission line of singly ionized oxygen at λ3727. The four larger open circles in Panel (a) show the same measurements reflected to the other side of the galaxy. They make the point that Babcock's four redshifts of emission line regions are consistent with a nearly circularly symmetric rotation curve, meaning the velocity v_c of rotation depends only on the distance r from the center. But that does not add much to the evidence that the rotation curve really is close to axially symmetric.

Babcock's (1939, 50) larger point is that

> the very great mass calculated in the preceding section for the outer parts of the spiral on the basis of the unexpectedly large circular velocities of these parts [and] the great range in the calculated ratio

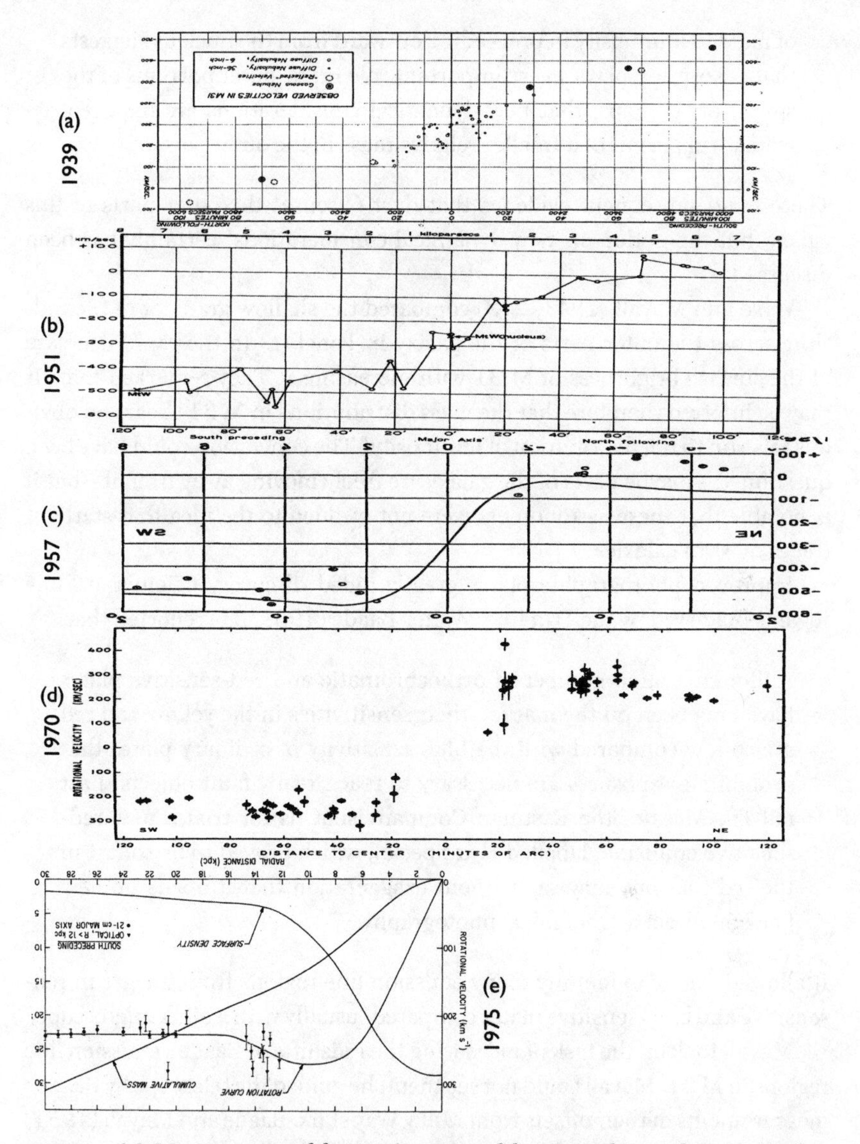

FIGURE 6.2. Measurements of the rotation curve of the nearest large galaxy, M 31. The panels have been rotated or reflected to bring them to a common orientation and angular scale. The sources from top to bottom are Babcock (1939); Mayall (1951); van de Hulst, Raimond, and van Woerden (1957); Rubin and Ford (1970); and Roberts and Whitehurst (1975). Panel (a) courtesy of the University of California Observatories. Panel (b) courtesy of the University of Michigan Press. Panels (d) and (e) © AAS. Reproduced with permission.

> of mass to luminosity in proceeding outward from the nucleus suggests that absorption plays a very important role in the outer portions of the spiral, or, perhaps, that new dynamical considerations are required, which will permit of a smaller relative mass in the outer parts.

There is no subsequent evidence that dust obscures the outer parts of this galaxy, but the notion of "new dynamical considerations" certainly has been discussed.

Wyse and Mayall (1941, 273) compared the shallow gradient of the redshift across the outer parts of the galaxy in Panel (a) to the rapid decrease of the surface brightness of M 31 with increasing *r*. They remarked that "It may be inferred therefore that the mass distribution [in M 31] bears no obvious relation to the distribution of luminosity." The conclusion could have been questioned—maybe parts of the galaxy are freely moving away from it—but it is notable that these astronomers were not wedded to the idea that starlight traces mass in galaxies.

Improved photographic plates greatly aided discovery of emission line regions made red by the H-α line. Walter Baade (1939, 31) reported that

> Although a large number of orthochromatic and red-sensitive plates have long been on the market, their sensitivities in the yellow and red are so low compared with the blue sensitivity of ordinary plates that prohibitive exposures are necessary to reach really faint objects. Last fall Dr. Mees of the Eastman Company sent us for trial a new red-sensitive emulsion, labelled H-α Special, which proved to be so fast in the red that one may say without exaggeration that it opens up new fields in direct astronomical photography.

It allowed Baade to identify many emission line regions from images in red-sensitive and blue-sensitive plates compared "usually with a blink microscope."

Mayall took on the task of measuring the redshifts of Baade's emission line regions in M 31. Mayall could not see them; he guided the telescope by Baade's measurements of their offsets from Milky Way stars. Baade and Mayall (1951) presented first results from this program at the Symposium on Cosmical Gas Dynamics in Paris in 1949.

Panel (b) in Figure 6.2, from Mayall (1951), shows more results from this program. He cautioned that the redshift measurements are not very precise, but note that the mean shapes of Panels (a) and (b) look similar. Mayall's (1951, 20) assessment of the shape of the rotation curve is that

> At nuclear distances greater than 65′ to 70′, to nearly 100′ north-following and 115′ south-preceding, there is good evidence of a decrease in rotational velocity with increasing nuclear distance. In other words, the "turn-around" points apparently occur at nuclear distances of 65′

> to 70′, where the main body of the spiral as shown on most ordinary photographs appears to end.

The "turn-around" would be expected if the mass were concentrated within about 70 arcmin from the center of the galaxy so that, at larger distances, the rotation curve reaches the Kepler relation $v_c(r) \propto r^{-1/2}$. Later observations have conditioned us to expect that rotation curves are close to flat well beyond the concentrations of starlight in spiral galaxies, which makes it difficult to see indications of Mayall's turn-around. We may suppose that astronomers then were conditioned to see in the outer parts of galaxies the Kepler rotation curve familiar from the solar system, and that is what Mayall saw.

Schwarzschild's (1954, 276) analysis of Mayall's (1951) measurements led him to "conclude that the best available observations of the rotational velocity in the Andromeda nebula are not discordant with the assumption of equal mass and light distribution."

The illustration of this conclusion in Schwarzschild's Figure 1 shows a broad scatter of redshifts, and indeed, Mayall had warned that the measurements are not very precise. I do not understand the systematic difference of measured redshifts on the two sides of the galaxy in Schwarzschild's figure and cannot assess the measurement of the distribution of starlight. But we see a useful caution from a capable astronomer.

Panel (c) in Figure 6.2 shows results from another landmark advance in technology, the ability to detect the 21-cm emission line of atomic hydrogen with a radio telescope capable of resolving positions along the long axis of M 31. This rotation curve was obtained using the 25-meter telescope at Dwingeloo, the Netherlands (van de Hulst, Raimond, and van Woerden 1957). Hugo van Woerden (in a personal communication, 2018) recalls this development:

> The M31 project was strongly led by Henk van de Hulst. Ernst Raimond and I were the first two observers with the new Dwingeloo 25-meter dish and the 21-cm line receiver. Between us, we did almost all the observing during the first run (26 days, 10 Oct–5 Nov 1956). In fact, Ernst did almost all the measurements of M31 (at night!); I ran several other projects. Henk had proposed the M31 program, after his 1957 prognosis (written in 1955) in IAU Symposium 4, and had a major share in developing the novel observing procedure with comparison fields. In January 1957, Ernst was enlisted by the army, and I took over the observations, with the help of some junior students. I had a major share in the reduction of observations of M31, but the analysis, model fitting etc. were in the hands of Henk, with some student help.

The solid curve in Panel (c) is the rotation curve $v_c(r)$ fitted to the 21-cm Dwingeloo measurements. I am indebted to Hugo van Woerden for advising me that this curve was given the slight maximum and minimum at about

30 arc min from the center because there is a similar feature in Schmidt's (1956) rotation curve for the Milky Way. But this feature agrees with the later measurements in Panels (d) and (e), along with the flat outer parts of the curve. The data points in Panel (c) are from Mayall's (1951) optical measurements plotted in a slightly different way from Panel (b).

The Dwingeloo measurement inspired another important measurement. Maarten Schmidt (1957) computed the mass distribution in a close-to-flat disk that produces the rotation curve plotted as the solid line in Panel (c). Schmidt (1957, 19) points out that Schwarzschild (1954)

> has suggested that the mass-luminosity ratio in M31 might be constant throughout the nebula. Our results tend to confirm this, although nothing as yet can be stated about the ratio in the innermost and the outermost parts of the nebula. ... Accurately calibrated observations of the surface brightness in different colours over the whole nebula are urgently needed.

Gérard de Vaucouleurs reported measurements of the surface brightness across M 31 "undertaken at the suggestion of Dr. M. Schmidt" to compare to the mass distribution indicated by the Dwingeloo 21-cm measurements. De Vaucouleurs (1958b) concluded that the ratio of mass to B-band luminosity in M 31 increases from $M/L_{\rm B} \simeq 8$ near the center to $M/L_{\rm B} \simeq 70$ at 2° from the center, close to the edge of the measured rotation curves in Figure 6.2. At this radius, the measured surface brightness of M 31 is only about 3 percent of the sky. Apart from warnings about the possible effects of obscuration by dust in M 31 and the possibility of different $M/L_{\rm B}$ values in disk and spheroid components, de Vaucouleurs did not comment on his evidence of a strikingly large value of $M/L_{\rm B}$ near the edge of the detectible luminous part of the galaxy. But it was characteristic of him not to offer comments about interpretation in a paper reporting measurements.

The uncertainty in the accuracy of the measurements of the outer part of the rotation curve makes it difficult to judge whether the large mass-to-light ratio in the outer parts of M 31 really was well established by 1958. But a simpler and more secure number is the ratio of Schmidt's total mass to de Vaucouleurs's total luminosity, $M/L_{\rm B} = 24$. This assumes the distance to M 31 is 630 kpc. At the more recent measurement, 780 kpc, the ratio is reduced to $M/L_{\rm B} = 19$. It still is well above what one would expect from the observed population of stars in the Milky Way.

Panel (d) in Figure 6.2 shows results from another technical advance. The Carnegie Institution of Washington, in the east and west coast branches, had encouraged Kent Ford to look into the use of image tube intensifiers made by RCA in Lancaster, PA, to improve quantum efficiency over the use of photographic plates alone. Ford (1968) reviews the active state of this art at the Carnegie Institution and other observatories. He and Vera Rubin at the

Department of Terrestrial Magnetism, in the east coast branch, used their image tube spectrograph to get greatly improved measurements of spectra of astronomical objects, among them redshifts of emission line regions in M 31. (A photograph of Rubin and Ford is shown in plate VI.)

In her autobiography, Rubin (2011) recalls how she and Ford learned of Baade's identification of 688 emission line regions in M 31. Mayall had used these identifications for the measurements in Panel (b). The finding list greatly aided the Rubin and Ford goal of measuring redshifts of emission lines across the face of the galaxy. And they had the image intensifier tube that reduced exposure times from photographic: "20 h to less than 1.5 h." Their rotation curve in Panel (d) is from Rubin and Ford (1970).

Rubin and Ford (1970, 381) reported that

> Mayall observed twenty-seven emission regions near the major axis; we have observed seventeen of these regions. The velocity agreement is not good. For ten regions near the NE major axis, the average difference ΔV (our value less Mayall's) is 97 km sec^{-1}; for seven regions near the SW major axis, $\Delta V = 31$ km sec^{-1}. However, Mayall (private communication) has cautioned against putting too much weight on his velocities, because of his long exposures and low dispersions. We have therefore not made use of these early values.

Baade and Mayall (1951, 171) had warned about this earlier: "When results from several measurements of the same object differ by as much as 100 km/s, as sometimes happens, there is little encouragement for refined analysis." And van de Hulst, Raimond, and van Woerden (1957, 13) wrote that "The agreement between optical [Mayall 1951] and radio data in Figure 12 is rather poor and the systematic deviation of nearly 100 km/sec in the NE half is disconcerting."

These are laudable examples of checking. And as we have noted, the large measurement uncertainties may have contributed to Schwarzschild's conclusion that he saw no significant evidence that the distributions of mass and starlight are different. But the trend of Mayall's measurements with position across the galaxy looks broadly similar to the more precise later results.

Rubin and Ford (1970) computed the ratio of integrated mass to integrated light within a given distance from the center of the galaxy, using de Vaucouleurs's (1958b) surface brightness measurements. (They did not estimate local values of M/L.) They found that $\int M / \int L$ (in their notation) increases from $1.0 \pm 1\, M_{\odot}/L_{\odot}$ at $r = 3$ kpc to 13 ± 0.7 at $r = 24$ kpc. The trend of increasing M/L with increasing distance from the center of this galaxy certainly was not new—Babcock (1939) suggested it—but here the trend is far better established. It is typical of Rubin and Ford that they did not speculate about the meaning of the increasing value of $\int M / \int L$ with increasing radius.

Roberts and Whitehurst (1975) used the "resurfaced 300-foot (91 m) telescope" in West Virginia for 21-cm measurements of the redshifts of the outer parts of the southwest side of M 31.[4] The radius in Panel (e) used an M 31 distance of 690 kpc, close to the present estimate. I regret that distorting their figure to the same scale and orientation as the others has a particularly ugly effect.

The Roberts and Whitehurst (1975) measurements are plotted as the nine filled circles to the left in Panel (e) in Figure 6.2. The velocities plotted as triangles are from the Rubin and Ford (1970) optical measurements, folded under the assumption of axial symmetry. The three 21-cm points at $r \approx 21$ kpc trend to slightly larger redshifts than the three Rubin and Ford optical measurements, but they are not seriously inconsistent within the measurement uncertainties. The six new data points at $r = 24$ to 30 kpc indicate the rotation curve is quite close to flat, out to these larger radii. Roberts and Whitehurst concluded that the local mass-to-light ratio reaches $M/L \simeq 200$ at $r \simeq 25$ kpc, and reaches $M/L \simeq 600$ at $r = 30$ kpc if you trust a "linear extrapolation" of de Vaucouleurs's (1958b) measurements of the distribution of starlight.

At the M 31 distance of 780 kpc, the Rubin and Ford (1970) measurements reach about 25 kpc radius, Roberts and Whitehurst (1975) reach 30 kpc, and the later study by Chemin, Carignan, and Foster (2009) reaches 40 kpc. This galaxy has an unusually luminous classical stellar bulge; in the Chemin et al. analysis, the stellar bulge makes the dominant contribution to the gravitational acceleration within ~ 3 kpc radius. They find that at $r \sim 10$ to 40 kpc, the mass in the disk contributes roughly 20 percent of the acceleration and almost all of the rest is in the subluminal halo.

The Andromeda Nebula, M 31, is an example of a spiral galaxy with a close to flat rotation curve and mass more broadly spread than starlight, on the usual assumption of standard gravity physics. Mayall (1951; Panel (b) in Figure 6.2) had evidence of the flat curve, but his caution about large measurement uncertainties may have led Schwarzschild (1954) to doubt its significance. Van de Hulst, Raimond, and van Woerden (1957; Panel (c)) confirmed Mayall's results, within reasonably measurement uncertainties, and Schmidt's (1957) mass with de Vaucouleurs's (1958b) luminosity showed the broader spread of mass relative to light. Rubin and Ford (1970; Panel (d)), Roberts and Whitehurst (1975; Panel (e)), and Chemin, Carignan, and Foster (2009) each

4. Roberts and Whitehurst (1975, 327) wrote that "Satisfactory measurements could not be made at large distances along the north-east major axis because the rotation of M 31 combined with its systemic velocity result in radial velocities which are similar to foreground galactic hydrogen resulting in serious velocity confusion." I am grateful to Hugo van Woerden for explaining that the Dwingeloo measurements on the northeast side of Panel (c) were obtained by differencing the spectrum on the galaxy from the spectra just above and below it, thus suppressing the foreground.

did still better. There is similar evidence for other spiral galaxies. But let us consider next a possible counterexample, the galaxy NGC 3115.

6.3.2 NGC 3115

The early-type S0 galaxy NGC 3115 shares with spirals a flattened shape with near rotational support, and it shares with ellipticals only modest amounts of dust and gas and young stars. It is 10 Mpc from us. It is not close to other large galaxies, which is unusual, because galaxies such as this one tend to be in dense regions where we can imagine that collisions or plasma ram pressure stripped away the gas and dust. And the history of thinking about its mass distribution is unusual, too.

Milton Humason made the first estimate of the rotation curve of this galaxy. It is presented in the Annual Report of the Director of the Mount Wilson Observatory, 1936–1937 (Adams and Seares 1937, 31):

> Humason has measured spectrographic rotations in ... NGC 3115, where the luminosity falls smoothly from the nucleus outward to undefined boundaries, the rotation follows the linear law
>
> $$r = 9.8x + 640 \text{ km/sec},$$
>
> where x, the distance from the nucleus along the major axis, is expressed in seconds of arc ... out to $x = \pm 45''$.

This is the rotation curve of matter rotating with uniform angular velocity. Oort (1940, 274) concluded from Humason's estimate that "the distribution of mass in the system appears to bear almost no relation to that of light." We see again that a most capable astronomer, Jan Oort, was willing to consider the possibility that mass and light are differently distributed in a galaxy.

Williams (1975) remeasured the rotation curve from redshifts of many stellar absorption line spectra across NGC 3115, "at the suggestion of J. H. Oort." Outside a steep gradient across the center of the galaxy, Williams's mean curve is reasonably close to flat, albeit with considerable scatter in these challenging measurements. It is quite different from what had been been accepted earlier for this galaxy.

Rubin and colleagues greatly reduced the scatter by using their Carnegie image tube spectrograph at the 4-m Mayall telescope at Kitt Peak, Arizona. Rubin, Peterson, and Ford (1976) presented the rotation curve shown in Panel (a) in Figure 6.3 as the abstract for a paper to be presented at a meeting of the American Astronomical Society, but the figure left no room in the abstract for an explanation. Ford, Peterson, and Rubin (1976) showed the same measurement in the *Carnegie Institution Year Book*. They explained there that they aligned the spectrograph slit along the close to edge-on disk of this galaxy. The trace of each stellar spectral line thus appears in the dispersed

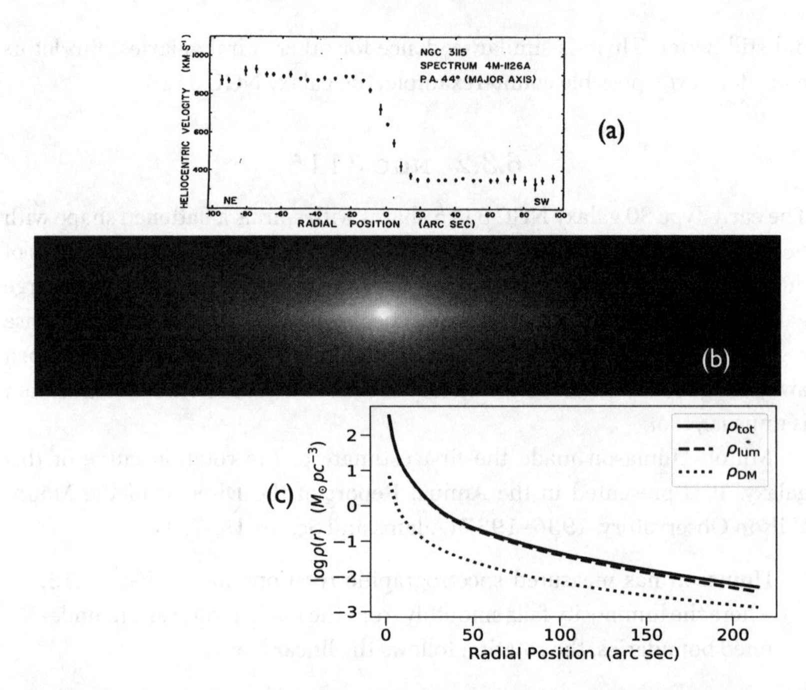

FIGURE 6.3. Observations of the galaxy NGC 3115. Panel (a) is the Rubin, Peterson, and Ford (1976) optical measurement of the rotation curve. Panel (b) is a 2.2μ 2MASS image. Panel (c) is Michele Cappellari's (personal communication, 2017) distributions of luminosity and mass. The angular scales are close to the same in the three panels. Panel (a) © AAS. Reproduced with permission. Panel (c) courtesy of Michele Cappellari.

image as a good approximation to the shape of the rotation curve. The result in Panel (a) resembles the rotation curve of M 31 in Figure 6.2, but here we see a strikingly pronounced gradient across the center between the outer quite close to flat parts. Rubin, Peterson, and Ford (1980) presented an analysis of their rotation curve measurement using optical surface brightness data largely from Strom et al. (1977). The Rubin, Peterson, and Ford (1980, 53) conclusion is that "A simple mass model suggests that the mass-to-luminosity ratio is constant from 1 to 5 kpc from the nucleus."

Panel (b) in Figure 6.3 is a 2MASS image of the galaxy at a wavelength ($\lambda \approx 2\mu$) that gives a reasonable representation of the distribution of mass in a normal mix of stars. The size of the image is adjusted to appear at about the same scale as the rotation curve in Panel (a).

Michele Cappellari (personal communications 2017) kindly made Panel (c) for me. It shows the run of mass density as a function of radius in the plane and the decomposition into the stellar and subluminal components. This is based on the dynamical modeling of the two-dimensional measurements of stellar kinematics and surface brightness in Cappellari et al. (2015). The great advances in detector efficiency and data rate allowed detailed measurements of stellar streaming motions to twice the radius Rubin et al. explored. The result shows little evidence of a subluminal massive halo out to six effective radii, that is, six times the radius that contains half the luminosity.

We have become conditioned to expect that large galaxies formed within subluminal massive halos. But the galaxy NGC 3115, which had yielded a false signal of such a halo, proves to show little evidence of subluminal matter. This phenomenon has not attracted much notice and may only be the result of a seriously unusual history of this galaxy. But it shows that observations of galaxies still can surprise us and challenge received knowledge.

6.3.3 NGC 300

Freeman (1970) presented the second reasonably clear example of a spiral galaxy in which starlight is not a good tracer of mass. Freeman's paper is celebrated for the demonstration that the distribution of starlight in many spiral galaxies can be expressed as the sum of an exponential disk and a spheroidal stellar bulge that resembles an elliptical galaxy. Also to be celebrated is his calculation of the gravitational acceleration in the plane of a flat axisymmetric exponential disk of mass and its application to NGC 300, a pure disk galaxy with a measured 21-cm rotation curve. The distance to this galaxy is about 2 Mpc, putting it just outside the nominal edge of the Local Group.

Freeman remarked that the surface brightness $\mu(r)$ of the stellar disk of a spiral galaxy (measured normal to the disk) often is well approximated as an exponential function of distance r from the center:

$$\mu(r) \propto e^{-ar}. \tag{6.2}$$

The constant a^{-1} is a characteristic disk radius. Freeman computed the gravitational acceleration of this distribution of mass in a pure disk, one that has no bulge component, and from that the predicted rotation curve $v_c(r)$. He found that the peak value of $v_c(r)$ in this exponential mass distribution is at radius

$$R_T = 2.151/a. \tag{6.3}$$

Freeman (1970) remarked that the nearby spirals NGC 300 and M 33 have little noticeable bulges and close-to-exponential disks, so their rotation curves might be compared to what would be expected from his computation, under the assumption that starlight traces mass. His clear case is NGC 300. Its measured characteristic optical disk scale length is $a^{-1} = 2.9$ arc min. If starlight traced mass in this disk, the circular velocity would reach peak value at radius $R_T = 2.151a^{-1} \simeq 6$ arc min. The 21-cm rotation curve of NGC 300 was measured by Shobbrook and Robinson (1967) using the 210-foot radio telescope at Parkes, Australia. The peak of this measurement of the rotation curve is at radius no less than 15 arc min, more than twice the 6 arc min expected if light traced mass. Freeman's (1970, 828) conclusion is that

> The HI rotation curve has $V_{\max}$ at $R \approx 15'$, which also happens to be the photometric outer edge of the system. If the HI rotation curve is correct, then there must be undetected matter beyond the optical extent of NGC 300; its mass must be at least of the same order as the mass of the detected galaxy.

Freeman was appropriately cautious about the accuracy of the NGC 300 rotation curve measurement. But Carignan and Freeman (1985) and Kent (1987) confirm that the 21-cm rotation curve is dominated by subluminal mass outside about 5 kpc.

Freeman (1970) did not cry up his evidence of subluminal matter; it is presented in an appendix in his paper. He did not refer to the emerging evidence of subluminal matter in M 31. The Rubin and Ford (1970) measurement of this rotation curve in Panel (d) in Figure 6.2 appeared in the same year, yet the earliest reference by Rubin and colleagues to Freeman's evidence of subluminal matter that I can find appeared a decade later (Burstein et al. 1982). This line of research was building a key part of the evidence that there was a hot big bang, but in the early 1970s, it was not yet a highly visible branch of science.

6.3.4 NGC 2403

The galaxy NGC 2403 is another pure-disk spiral; it is not far outside the Local Group at 3 kpc distance. Seth Shostak's (1972) PhD thesis at the California Institute of Technology, Pasadena, includes a rotation curve of this galaxy from his measurements of redshifts of the 21-cm line with the two 90-foot antenna interferometer in Owens Valley, California. His result is quite similar to Figure 6.2 for M 31: A steep gradient across the center joins close to flat outer parts with no indication of the turndown to be expected if the observations were reaching the edge of the mass distribution. Shostak (1972, 200) concluded that, since his observations were reaching the edge of luminous parts of the galaxy, his measured rotation curve "requires the presence of underluminous material in the outer regions of the galaxy." He concluded that the mass-to-light ratio increases from $M/L \sim 3$ near the center to $M/L \sim 15$ near the edge of the starlight.

Shostak did not have a measurement of the distribution of starlight in this disk galaxy. He used de Vaucouleurs's (1959) measurement of the characteristic radius a^{-1} (equation (6.2)) for the exponential distribution of starlight in M 33. From the relative physical sizes of the luminous parts of NGC 2403 and M 33, Shostak estimated that NGC 2403 has characteristic radius $a^{-1} \sim$ 2.7 arc min. Now to depart from his approach, and follow Freeman (1970), let us note from equation (6.3) that if the mass in NGC 2403 were distributed like the light in a thin rotationally supported exponential disk, then the rotation curve would be expected to peak at angular distance 6 arc min from the center of the galaxy. But Shostak's rotation curve is flat from 5 to 15 arc min.

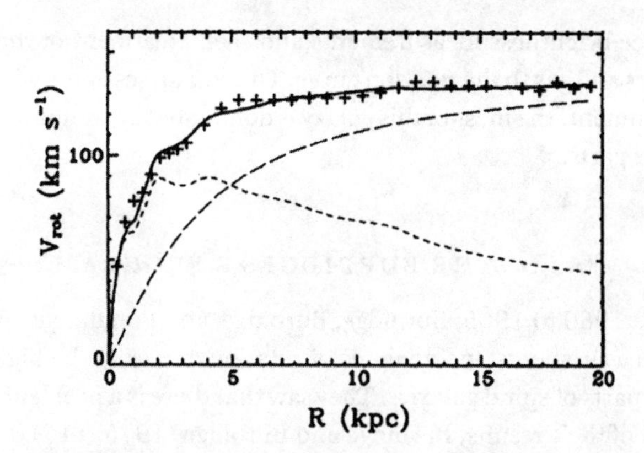

FIGURE 6.4. Kent's (1987) analysis of the distributions of mass in the galaxy NGC 2403. The solid curve is the model fit to the rotation curve measurements (marked as plus signs). The short-dash curve is the model contribution to the rotation curve by the mass in stars in the disk, and the long-dash curve is the contribution by mass in a spherical subluminal halo, to be added in quadrature. © AAS. Reproduced with permission.

As for NGC 300, it appears that the mass in NGC 2403 is not distributed like the starlight. He pointed to other examples of flat rotation curves: M 31 (Section 6.3.1), NGC 6574 (Demoulin and Chan 1969), M 101 (Rogstad 1971) and M 33.

In the publication based on his thesis, Shostak (1973) refrained from mentioning his discussion of the distribution of light. The argument was sensible, but speculative and perhaps best not published outside his thesis. But he again stated the essential point (Shostak 1973, 411): "The rotational velocities in NGC 2403 reach a maximum at 9 kpc, thereafter remaining relatively constant, thus suggesting that the bulk of this galaxy's mass lies outside its optical radius."

Figure 6.4 shows Kent's (1987) later measurements and analyses of the mass distribution in NGC 2403. His surface brightness measurements confirm Shostak's expectation that the surface brightness run is close to exponential. The plus signs in the figure show the rotation curve from 21-cm measurements by Shostak (1973) and Begeman (1987), the latter from the Westerbork Synthesis Radio Telescope in the Netherlands. The short-dashed curve is Kent's estimate of the contribution to the circular velocity by stars, assuming a constant mass-to-light ratio with the value of M/L adjusted to fit the innermost parts at $r \lesssim 2$ kpc without significant contribution from a subluminal halo. The long-dashed curve is the contribution from a massive spherical halo that completes the fit to the measurements. Begeman (1987) obtained a comparable result for NGC 2403 from the Westerbork 21-cm measurements and the optical surface brightness measurements by Wevers, van der Kruit, and Allen (1986). They used digital scans of 48-inch Palomar Schmidt Telescope plate images of the galaxy. And van Albada and Sancisi (1986) showed the Westerbork rotation curve for NGC 2403 with the Wevers et al. surface brightness measurements and their treatment of the contributions of stars and gas to the rotation curve. These analyses agree with Shostak's (1972) argument: the mass of this galaxy is dominated by subluminous matter in the outer parts.

6.3.5 THE BURBIDGES'S PROGRAM

In the years 1960 to 1965, Burbidge, Burbidge, and Prendergast presented a series of measurements and analyses of rotation curves in the higher surface brightness parts of spiral galaxies. They saw that there is a problem with mass. In a review of their results, Burbidge and Burbidge (1975, 116) report that

> it is clear that the constituents in elliptical galaxies, and in some spirals, must differ very greatly from the types of stellar population found in the solar neighborhood and in star clusters in our own Galaxy. They must differ in the sense that there is a large population of objects which contribute a great deal of mass but which contribute very little light. ... Since there is not thought to be a large amount of diffuse matter in galaxies with high mass-to-light ratios, this mass is thought to be in the form of very small mass stars (red dwarfs) or in the form of highly evolved stars (white dwarfs or neutron stars) or collapsed mass.

The evidence noted on page 243 is that this was written in about 1968. It is a good assessment of the situation at that time.

It was in the Burbidges's tradition to confine attention to the parts of galaxy rotation curves that could be well measured and not to speculate on what might be present farther out. A later example in this tradition is the Demoulin and Chan (1969) measurements of redshifts of emission lines that show the spiral galaxy NGC 6574 has a strikingly flat rotation curve. Demoulin and Chan did not comment on the curious absence of an approach to the Kepler circular velocity $v_c \propto r^{-1/2}$ one might have expected to see near the edge of the luminous part of this galaxy, if the mass were concentrated with the starlight.

6.3.6 CHALLENGES

Rogstad and Shostak (1972, 320) reported the similar rotation curves in NGC 2403, M 33, IC 342, M 101, and NGC 6946. Their conclusion is that

> we confirm here the requirement for low-luminosity material in the outer regions of these galaxies ($M/L \sim 20$), assuming exponentially decreasing surface luminosities (Freeman 1970). ... Therefore, estimates of the total masses to infinite radius represent a dubious extrapolation of the data.

Geoffrey Burbidge played an important role by expressing doubts about the evidence of massive halos around galaxies, as in Burbidge (1975). My impression is that he objected to the rush to judgment, what he termed the "bandwagon effect." His caution makes sense: Hopeful extrapolation beyond the direct evidence can be misleading. But extrapolation in the manner of Freeman and Shostak can be productive. One of the arts of science is to probe the boundaries between empty and productive speculation.

The word "bandwagon" rightly connotes the tendency to pay attention to ideas that are attracting attention, but it should not be taken to mean that there was broad interest in the full range of subluminal mass phenomena. For example, the Faber and Gallagher (1979) review of subluminal matter, which was important in drawing attention to the issues, focused more on systems of galaxies. They presented Bosma's (1978) extended 21-cm rotation curves of spiral galaxies, but they did not mention the rotation curve of M 31 (Figure 6.2) or the evidence from Freeman (1970) and Shostak (1972) of considerable subluminal mass in the outskirts of the nearby and well-examined galaxies NGC 300 and NGC 2403 (Sections 6.3.3 and 6.3.4).

Flat rotation curves, in which the circular velocity $v_c(r)$ is close to constant as the radius r extends to the faint, barely detected edges of the galaxy, are usually considered characteristic of subluminal matter. There are exceptions; a likely example is the flattened S0 galaxy NGC 3115 (Section 6.3.2). The nearby spiral galaxy M 81 looks reasonably normal, though it has an unusually large stellar bulge. But Rots (1974) found that its 21-cm rotation curve rises to a maximum at 250 km s^{-1} at 3 to 7 kpc radius, drops to 200 km s^{-1} at 15 kpc from the center, and past that the observations indicate serious departure from axial symmetry. There is no indication that M/L systematically increases with radius within 15 kpc, the apparently axisymmetric part of this galaxy (Rots 1974, Fig. 16; Roberts and Rots 1973). In short, we should be aware that there are galaxies that depart from the apparently normal order of things, it must be presumed because of untoward pasts.

At a discussion at an International Astronomical Union (IAU) meeting, Agris Kalnajs (1983) showed that the measured rotation curve of one of the galaxies Rogstad and Shostak mentioned, M 33 (= NGC 598), can be fitted by appropriate choice of the stellar mass-to-light ratio in the disk without the hypothetical subluminal matter. Kalnajs presented this and three other examples in which the measured rotation curves are flat or still rising in the outer parts of the optical images, yet are well fitted by the stellar mass with reasonable-looking values of M/L. It is recorded that during Kalnajs's (1983, 88) comments, "The audience becomes restless and the massive halo enthusiasts slowly regain their composure."

When classical stellar bulges are present in spiral galaxies, it adds the complication that the measured orbital acceleration in the disk is to be fitted to the sum of the gravitational accelerations of the mass in stars and gas in the disk, the mass in stars in a spheroidal component observed as a classical bulge and extending to a stellar halo, and the mass that may be present in a subluminal halo that is not even directly observed. The freedom to adjust the two mass-to-light ratios of the stellar components, disk and bulge, allows considerable uncertainty in the derived amount and distribution of the mass in subluminal matter. One of Kalnajs's (1983) four examples requires these two components, stellar bulge and disk. Kent (1986) presented more such examples. His digital (CCD) measurements of the surface brightness distributions in galaxies for which Rubin and colleagues had measured optical rotation curves show that the stellar luminosity distributions in a disk and in a classical stellar bulge—with the two free choices of M/L for disk and bulge, but with M/L in each component independent of radius—offer adequate fits to many optical rotation curve measurements with not unreasonable choices of the two mass-to-light ratios and without subluminal mass.

The challenges by Kalnajs and Kent were resolved in many cases by 21-cm measurements that allow tracing rotation curves to larger radii. For example, Kalnajs showed that the stellar distribution in M 33 could fit the rotation curve measurements to $r = 6$ kpc radius, but Corbelli and Salucci (2000) and López Fune, Salucci, and Corbelli (2017) present 21-cm rotation curve measurements to $r = 23$ kpc that make a clear case for an extended subluminal halo that dominates the rotation curve beyond about 5 kpc. And, although Kent (1986) could fit optical rotation curves to starlight in disk and bulge, Kent's (1987) results of fitting his photometry to 21-cm measurements that trace rotation curves to larger radii generally require the postulate of subluminal halos. Albert Bosma (1978, 1981a,b) showed dramatic examples of the power of 21-cm measurements with the Westerbork Synthesis Radio Telescope. The measurements demonstrate the presence of subluminal matter in the outer parts of late-type galaxies that in some cases is seen to be far more broadly spread than the extent of the luminous parts of the galaxies.

Athanassoula, Bosma, and Papaioannou (1987, 23) showed how the disk-bulge-halo degeneracy is further reduced by an important consideration from the issue of disk stability discussed in Section 6.4. The mass assigned to the subluminal matter must leave enough mass supported by rotation in the disk of a spiral galaxy so that gravity in the disk can produce spiral arms, but not so much mass that the arms grow unacceptably large and destroy near-circular rotation. Their conclusion from this consideration is that "We suggest that all galaxies need a halo to fit their rotation curve, and that those cases which show otherwise are due to the artificially short radius range of the rotation data."

A later check of the massive halo picture came from the effect of the gravitational deflection of light from more distant galaxies by the mass concentration around a large galaxy in the foreground. The gravitational deflection distorts images of the background galaxies; the systematic distortion, averaged over many galaxies, yields a measure of the mean distribution of mass around a galaxy. An early result from the Sloan Digital Sky Survey was the confirmation that the mean mass within 100 kpc of a large galaxy averages about ten times the mass within 10 kpc (Fischer et al. 2000), in a subluminal massive halo.

All these phenomena do not require nonbaryonic dark matter. Burbidge and Burbidge (1975), Roberts (1976), White and Rees (1978), and others made the sensible point that the subluminal mass could be dwarf stars or remnants of early generations of massive stars. These objects would have to become increasingly abundant relative to normal stars with increasing distance from the center of a galaxy, but that certainly might have happened as stars formed under different conditions in different parts of the galaxy. There is an important constraint, however, that if the subluminal mass were remnants of observed types of stars, or of dwarf stars that formed with the observed types, one would expect that the subluminal mass is streaming in the disk along with the observed stars. This is ruled out by the consideration of disk stability to be considered in Section 6.4.

All these analyses assume standard gravity physics. In Zwicky's (1937a, 228) second paper on the mass of the Coma Cluster, he remarked that the cluster mass problem depends on "the assumption that Newton's inverse square law accurately describes the gravitational interactions among nebulae." This caution was not often considered. For example, Rees (1984, 339) states that "the evidence for hidden mass which I summarized in Section 1—the evidence that M/L increases as we consider scales from 10 kpc out to a few Mpc, is incontrovertible in general terms, even though particular details are all subject to debate."

The statement that the evidence is incontrovertible agrees with the implicit nonempirical community decision to accept application of the inverse square law of gravity on scales extending to 1 Mpc, even though this is an extrapolation of some ten orders of magnitude beyond the tests on length scales ranging from the laboratory to the solar system. The decision was sensible and appropriate, but at the time, it was also appropriate to consider the possibility of alternative laws of gravitation on the scales of galaxies to clusters of galaxies and on to the scale of the observable universe.

Finzi (1963) proposed modifying the inverse square law for the acceleration of gravity to $g \propto r^{-3/2}$ at separations r greater than roughly 0.5 kpc. This would make it easier to account for the rotation curves of spiral galaxies without the postulate of subluminal matter. But Milgrom (1983) offered the most widely discussed and useful variant of this approach. He proposed that the gravitational acceleration a of a test particle at distance r from a compact mass M is given by the equation

$$GM/r^2 = a\mu(a/a_0), \quad \mu \rightarrow 1 \text{ at } a \gg a_0, \quad \mu \rightarrow a/a_0 \text{ at } a \ll a_0. \tag{6.4}$$

At accelerations well above the constant a_0 this is the usual Newtonian expression, $a = GM/r^2$. At $a \ll a_0$, the acceleration is smaller than Newtonian,

$a=(GMa_0)^{1/2}/r$, so a test particle orbiting M at distance r and speed v has orbital acceleration $v^2/r=(GMa_0)^{1/2}/r$. In this low-acceleration limit, the orbital velocity is

$$v_c=(GMa_0)^{1/4}. \tag{6.5}$$

Milgrom's prescription for modified Newtonian dynamics, or MOND, is a little more complicated than Finzi's, but it was shrewdly chosen. The rotation curve $v_c(r)$ at low accelerations is predicted to be flat, independent of radius, as observed in the outer parts of many galaxies, without the hypothetical subluminal matter. The circular velocity is predicted to scale as the fourth root of the mass. This is the Faber-Jackson (1976) relation between luminosity and stellar velocity dispersion in elliptical galaxies, $L\propto\sigma^{\gamma_e}$, $\gamma_e\simeq 4$, and the Tully-Fisher (1977) relation between luminosity and circular velocity in spirals, $L\propto v_c^{\gamma_s}$, $\gamma_s\simeq 4$. It continues to offer a good fit to observations of spiral galaxies on length scales from about 1 to 30 kpc (e.g., Lelli, McGaugh, and Schombert 2017).

On the length scales of cosmology, $c/H_0\sim 4000$ Mpc, the demanding tests to be reviewed in Chapter 9 make a compelling case that general relativity with the hypothetical nonbaryonic dark matter is a good approximation to reality. If this is accepted, as most have done, why is Milgrom's alternative theory so successful on the scale of galaxies? The community assessment is that this is an accident of the complexity of the application of standard physics to galaxy formation. Deciding whether we have adequate physics for analyses of the structures of galaxies, including curious examples such as NGC 3115 (Figure 6.3), or whether we have missed something interesting, calls for more data analyzed in better ways, as usual. Meanwhile the community decision is appropriate: work with standard physics and the hypothetical subluminal/nonbaryonic matter applied to a cosmology that fits demanding tests—until or unless we run into trouble.

Productive lines of research have been founded on our conditioned feeling that a simple phenomenological regularity may point to the operation of an underlying physical principle. Consider the cosmologies that grew out of Hubble's and Humason's observations of the wonderfully simple linear relation between galaxy redshifts and distances. But community opinion is that the wonderfully simple flat rotation curves of many spiral galaxies have to be transient accidents of the way matter of the various kinds was spun up as it settled into the inner parts of galaxies. I accept this as a useful working hypothesis, with reservations mostly kept to myself.

The same somewhat awkward philosophy might have to be applied to the curious phenomenon illustrated by the rotation curve of NGC 2403 in Figure 6.4. It appears that within 2 kpc distance from the center of this galaxy, the mass in stars makes the larger contribution to the gravitational acceleration, while at larger distances, the mass in subluminal matter makes the larger contribution. But the rotation curve shows no feature at the transition from the baryon-dominated to subluminal matter–dominated sources of gravity. This is a common phenomenon among nearby galaxes. Van Albada and Sancisi (1986) and Kent (1987) termed this curious behavior a "conspiracy."

6.4 Stabilizing Spiral Galaxies

As the evidence for subluminal mass from spiral galaxy rotation curve measurements was growing tighter in the early 1970s, evidence for the presence of subluminal mass in spirals came from a different and initially quite independent line of research, on the gravitational formation of the spiral arms that are such a distinctive feature of normal rotationally-supported galaxies.

Spiral arms of galaxies display a complex interplay of processes fed by intersecting streams of interstellar gas and plasma that can trigger star formation. This produces regions seen as blue from the light of the massive luminous hot young stars. Other regions are seen as red from the H-α recombination radiation by plasma ionized by stars. Threading through these regions are elegant streams of dust. But it was and is natural to think that these phenomena are driven by the simpler process of the gravitational gathering of mass along the spiral arms. In early studies of this gravitational process, it also was natural to suppose that most of the mass is in the observed disk and in the central stellar bulge if the galaxy has one. The bulge would be supported against the pull of gravity by nearly random stellar motions: a "hot" component. The disk would be supported by nearly circular streaming motion: a "cool" component with velocity dispersion small compared to the streaming motion.

A celebrated analysis by Alar Toomre (1964) showed that, if we can ignore the gravitational acceleration of the bulge, and if the circular streaming motion in the disk is quite smooth and uniform, then the disk is gravitationally unstable against the growth of departures from circular symmetry of mass and motion, with the characteristic growth time set by the orbital period. Toomre's condition sets the velocity dispersion required to suppress this instability on length scales small compared to the size of the galaxy. Toomre found reasonable consistency of his bound for stability with the observed velocity dispersions of stars in the solar neighborhood, which was encouraging. But this analysis applies to small-scale disturbances.

The formidable challenge of analytic assessment of the global stability or instability of a self-gravitating, rotationally supported disk of matter is made even more difficult by the need to consider the variety of possible forms of rotation curves observed in galaxies, the effect of the mass concentrated in stellar bulges, which can have quite different luminosities in different galaxies, and the effects of random motions that may fill single-particle phase space in many different ways. The situation changed in the 1960s with the introduction of digital computers that have the speed and memory needed for useful numerical simulations of the gravitational motions of stars in the disk of a galaxy. With computation came the ability to make figures to illustrate the behavior of large numbers of particles. Such images can be influential.

An notable early example is seen in Frank Hohl's (1970, 2) NASA Technical Report:

> The Control Data 6600 computer system at the Langley Research Center was used to integrate the equations of motion for each particle for systems containing from 50 000 to 200 000 particles. An initially cold balanced disk was found to be violently unstable.

These are impressively large numbers for computers at the time. They required approximation schemes to find useful estimates of the gravitational accelerations of particles given their positions, here in two dimensions. Hohl argued from his tests that his approximations for gravity are not responsible for the unhappy endings of his model galaxies.

The initial particle velocities in these simulations are a smooth circular motion in dynamical balance with the gravity of the mass in the particles, and in some trials, an added random velocity component to satisfy Toomre's (1964) criterion for suppression of small-scale instability. Other reports of results from this program (Hohl and Hockney 1969; Hockney and Hohl 1969; and Hohl 1971) confirm that stability is improved by assigning the particles this initial velocity dispersion, but that after a few orbits the disk nevertheless degenerates to a diffuse form largely supported by random motions, quite unlike what the stars and gas are observed to be doing in a normal spiral galaxy.

The example in Figure 6.5, from Hohl (1971), shows the evolution of a disk of particles in which the initial mean streaming motion balanced by gravity has uniform angular velocity, and the dispersion around the mean satisfies Toomre's condition. The time unit is the orbit period. After two orbits, the mass distribution has developed what might seem to be a promising approach to a grand design spiral, but by four orbits, the disk has broken up into support by random motions in the plane. Hohl (1970) reported numerical experiments with an initially exponential mass distribution, as observed for the distribution of stars in the galaxies NGC 300 and NGC 2403 (Sections 6.3.3 and 6.3.4), though we have seen that it is not close to the mass distribution derived from the rotation curves. Again, after one or two orbits, these simulation seriously depart from uniform streaming motion in a disk.

Miller and Prendergast (1968) also were exploring how to make the computation of gravitational accelerations fast enough for numerical simulations of the motions of stars in a spiral galaxy with interesting numbers of particles. This may have been a natural extension of the Burbidge, Burbidge, and Prendergast program of measurement and analysis of spiral galaxy rotation curves. Miller, Prendergast, and Quirk (1970) reported that, in simulations using their method of computation of the gravitational acceleration, an initially

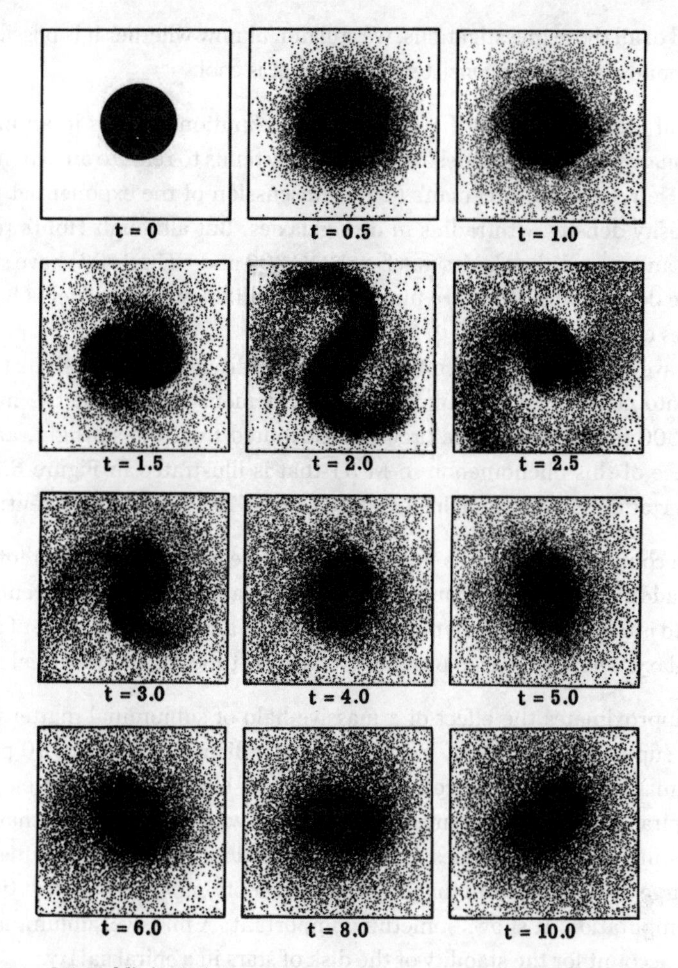

FIGURE 6.5. Hohl's (1971) numerical simulation of the evolution from an initially uniformly rotating disk of gravitating particles with initial velocity dispersion given by Toomre's condition for stability on small scales. © AAS. Reproduced with permission.

cold flow in a disk galaxy grows unreasonably hot. This is the same outcome as Hohl's, with an independent, though again approximate, scheme for computing accelerations.

Miller, Prendergast, and Quirk considered remedying the instability by "cooling" the particles, as may happen when "gas" particles collide and lose relative velocity while conserving momentum, perhaps as they are turning into stars. This approach has not proved to be of lasting interest, but it exemplifies the seriousness of the disk instability problem. Thus Miller's (1971, 89) examination of the situation led him to the conclusion that

> The search for reasons why the computer models [of disk galaxies] might be so hot has led through some interesting investigations. At the

end of all these investigations, we still don't know whether it is possible to build a static self-consistent model that is cool.

Hohl (1971) reported that the radial distribution of mass in an initially rotationally supported model for a disk galaxy tends to relax to an exponential form. He referred to Freeman's (1970) discussion of the exponential run of luminosity density with radius in disk galaxies. But although Hohl's relaxed model and pure disk galaxies (such as NGC 300 and NGC 2403) have similar surface density runs, they are fundamentally different: the model is hot, the galaxies cold.

I have found no evidence that these early discussions of disk instability took into consideration Freeman's (1970) demonstration that the mass in NGC 300 is distinctly more broadly distributed than the starlight, and the evidence of this phenomenon in M 31 that is illustrated in Figure 6.2. But they certainly could have helped inspire Hohl's (1970, 41) experiment:

> the computer model was modified to include a fixed central potential in addition to the self-consistent disk population of stars. The central field is taken to represent the halo population and the central core of the Galaxy, which appear to produce a relatively time-independent field.

This approximates the effect of a massive halo of subluminal matter that is stably supported by a nearly isotropic velocity distribution. With 10 percent of the mass in stars and the rest in the rigid halo, Hohl found the development of a spiral pattern in the particle distribution with fine structure that looks interestingly like real galaxies, and he found that it evolved without unacceptably large departures from the initial circular motion for the 8.5 orbit times of the computation. It shows something important: A massive subluminal halo might account for the stability of the disk of stars in a spiral galaxy.

The analytic approach to the study of disk stability was reaching a similar picture. The Kalnajs (1972) analysis indicates that a disk bound by gravity and supported by circular motion is unstable, just as the simulations had shown. Kalnajs considered the addition of random radial and angular velocities in the plane of the disk, which, if large enough, can suppress the instability, though this does not seem to be a promising model for the motions of stars in a normal spiral galaxy.[5] But Kalnajs (1972, 72) also argued that the instability is remedied by

> a simple yet useful modification of the disk problem to which our results can be applied. We may suppose that the disk is embedded in a uniform-density halo which supplies a fraction of the equilibrium force field, yet is rigid enough that it does not participate in the oscillations of the disk.

5. As Kalnajs concluded, galaxy disks would be stable if most of the mass is in the disk, subluminous, and in nearly random orbits confined to the plane of the disk, rather than in

Although Hohl and Kalnajs recognized from their simulations and analyses that a massive halo can stabilize the disk of a spiral galaxy, I have found no evidence in the literature that it occurred to them that their stabilizing hot component might be the subluminal mass that van de Hulst, Raimond, and van Woerden (1957) and Rubin and Ford (1970) inferred to be present on the outskirts of the spiral M 31, and Freeman (1970) inferred to be present in the outskirts of the pure disk galaxy NGC 300 (Sections 6.3.1 and 6.3.3). That connection was made in the literature by Ostriker and Peebles (1973).

The Ostriker and Peebles N-body simulations were motivated by Ostriker's experience with analyses of the stability of rotating stars (Ostriker and Bodenheimer 1973). It led him to suspect that the disk of a galaxy bound by the gravity of the mass rotating in the disk is not stable. The Ostriker and Peebles numerical simulations to check this in three dimensions have initially thin disks with initial mass within radius r that scales with radius as $M(<r) \propto r$. This produces the flat rotation curve that was already becoming familiar in the observations. Random velocities added to the streaming motions satisfy Toomre's (1964) condition for small-scale stability. The numerical method grew out of simulations of the growth of a cluster of galaxies in an expanding universe (Peebles 1970), done while I was visiting Los Alamos National Laboratory in Los Alamos, New Mexico. (It is a center for nuclear weapons research, and I was an immigrant alien, yet I was allowed to use one of their computers while supervised by an employee, who usually sat and read a book.) The gravitational acceleration was computed by direct sums over particles. My colleague E. J. (Ed) Groth devised the numerical methods that made this sum more efficient for what computers could do then. Even with improved efficiency, this computation of accelerations by direct sums limited the particle numbers to $N = 150$ to 500. But the simulations demonstrate pronounced disk instability if the mass that is holding the galaxy together is in nearly circular motion in the disk, and they demonstrate near stability if most of the mass is taken from the disk and placed in a subluminal halo modeled as a fixed central force (which, in a more realistic model, would be supported by a nearly isotropic distribution of halo particle orbits).

The major step in Ostriker and Peebles (1973) is the argument for the connection between the subluminal mass needed to stabilize disk galaxies and the subluminal mass needed to account for their rotation curves, along with the motions of galaxies in groups and clusters. Credibility of the simulations may

---

a more nearly spherical halo. In the standard picture, the dissipative settling of baryonic mass in the growing disk of a galaxy would gravitationally draw in dissipationless subluminal matter, but it would not concentrate the subluminal matter in the disk. This means evidence of subluminal matter concentrated in the disk would be interesting. The massive disk model is tested by comparing estimates of the mass in stars and diffuse baryons in the nearby plane of the Milky Way to the mass in the plane required to balance the vertical distributions of star positions and velocities. This is known as the Oort (1932) problem or limit. Courteau et al. (2014) conclude that this test is not yet secure.

have been aided by the computation of accelerations by simple sums over particles, because that method eliminated concern about artifacts introduced by faster methods of modeling accelerations. It was encouraging that these simulations agree with Ostriker's condition based on the analysis of the threshold of instability of rotating stars. It was important that the argument from stability agreed with the growing evidence from rotation curves for subluminal mass in and around spiral galaxies. And it must be added that the high visibility of research in cosmology and astronomy at Princeton University likely helped draw attention to the argument that most of the mass of the universe is subluminal matter draped around the galaxies.

The thinking in the mid-1970s, among those who were thinking about it, is illustrated by Bardeen's (1975, 317) statement:

> All lines of theoretical evidence lead to the same conclusion. Any disk which remotely resembles the disk of a spiral galaxy as represented by the neighborhood of the Sun or as contemplated in density wave theories of spiral structure will be globally unstable unless the disk contains only a rather small fraction of the total mass within its outer radius.

Alar Toomre's (1977a, 469) assessment is slightly different: "A lot of 'heat' is clearly needed, but must it be hidden in faint, massive halos, as Ostriker and Peebles suggested? Or might some very hot inner disks or 'spheroidal components' suffice already?"

The "heat," perhaps large departures from nearly uniform streaming motion in a subluminal disk or bulge, would aid stability of the observed stellar disk. An illustration of what grew out of this line of thinking concerning the classical bulge of stars rising out of the disk near the center in some spiral galaxies is shown in Figure 6.6. Panel (a) is Kent's (1987) analysis of the contributions to the rotation curve of the Andromeda Nebula (M31) by the stellar bulge, the disk of stars and gas, and the subluminal halo. Panel (b) shows the distributions of these components in a model that Sellwood and Evans (2001) show is stable against radial perturbations and large-scale departures from axial symmetry, demonstrated in both perturbation theory and numerical simulations. The model grew out of ideas in Toomre's student Tomas Zang's (1976) doctoral dissertation. The Sellwood and Evans model was not meant to be compared to M 31, but it looks similar. (The more recent decomposition of components by Chemin, Carignan, and Foster (2009) places less mass in the bulge and more in the subluminal halo, but the general idea remains.) In the Sellwood and Evans (2001) example, the "hot" stellar bulge aids stability, as does the external matter that approximates an extended subluminal halo.

The demonstration of disk stability remains a challenging problem. The galaxy NGC 2403 does not seem to have much of a stellar bulge. Kent's (1987) decomposition in Figure 6.4 on page 259 shows a reasonable match of the gravitational acceleration (indicated by the measured rotation curve in the

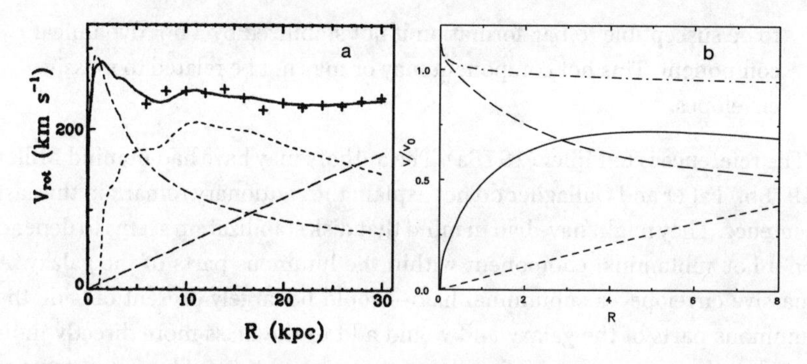

FIGURE 6.6. Panel (a) shows Kent's (1987) decomposition of the distributions of mass in the Andromeda Nebula. It may be compared to the model galaxy in Panel (b) that Sellwood and Evans (2001) show is stable. The contributions to the rotation curve are the stellar bulge (the long-short dashed curve in Panel (a) and the long-dashed curve in (b)) the disk (the short-dashed curve in (a) and the solid curve in (b)) and the subluminal halo (the straight dashed lines in both panels). © AAS. Reproduced with permission. I am grateful to Jerry Sellwood for giving me the figure in the form shown here.

inner parts of the galaxy) to the gravity of the mass distribution in the disk (traced by the starlight). This also is true of the more recent analysis by Fraternali et al. (2002, Fig. 10), assuming the disk mass-to-light ratio is in the range of 1.4 to 2.3. That is, there does not seem to be much room for a stabilizing inner massive stellar bulge in this galaxy. The evidence from the rotation curve is that this galaxy, along with other spirals that show little indication of a classical stellar bulge, has an extended massive subluminal halo that we must trust somehow serves to stabilize the disk all the way into a compact stellar nucleus. Jerry Sellwood's (private communication, 2019) assessment of the theoretical situation at the time of writing is that

> Our current understanding of the bar stability of disks is still far from complete. If halo mass, or pressure support, is indeed the solution in some cases, we now know that the coupling between the disk and halo requires a great deal more pressure supported matter, or "heat," than Ostriker and Peebles (1973) suggested.

The evidence for subluminal massive halos based on the issue of disk stability was not widely advertised in the years around 1980. I take as illustration the discussion of this line of evidence in the careful review of the subluminal mass situation by Faber and Gallagher (1979, 182). Their discussion of disk stability is largely confined to the statement that

> In addition to the dynamical evidence, there are other indirect indications of dark material in galaxies. The most important of these are the stability analyses of cold, self-gravitating axisymmetric disks (e.g., Ostriker & Peebles 1973, Hohl 1976, Miller 1978), which show them

> to be susceptible to bar-formation if not stabilized by a hot dynamical component. This hot component may or may not be related to massive envelopes.

(The reference is to Miller (1978a). The authors may have had in mind Miller 1978b.) Faber and Gallagher do not explain the cautionary remark in this last sentence. They might have had in mind that disk stabilization seems to depend on a hot subluminal component within the luminous parts of the galaxy. A massive envelope—a subluminal halo—would be largely present outside the luminous parts of the galaxy and would add to the mass more directly indicated by the rotation curve of a galaxy or relative motions of bound systems of galaxies.

Along with the stability of isolated disks, there is the issue of what happens as disks are violently disrupted when spiral galaxies merge. There are merger remnants characterized by tails and clouds of stars and gas that rise well away from central regions and look compact but quite disturbed compared to normal spiral galaxies. The analyses by Toomre and Toomre (1972) argued for the interpretation, now well accepted, that these objects are remnants of mergers of spiral galaxies. Toomre (1977b) stressed an interesting issue in this picture. The slow coalescence of the compact central regions to be expected if the bulk of the mass is in these compact parts would tend to present us with merger remnants with two nuclei, but this is not commonly observed. The central luminous parts of two merging spirals must have decelerated quickly to have been able to coalesce while the expanding tails are still nearby. Toomre (1977b, 415) points out that

> In principle at least, one can always embed them, prior to any fateful encounter, within some appreciably larger and more massive systems like the much-discussed extensive halos. Such outer parts would by definition interpenetrate and even splatter nicely as those visible disks only graze one another.

Barnes (1988) eventually illustrated this thought by numerical simulations of merging disk galaxies that show how the massive halos can absorb the energy and angular momentum of relative motions of the compact parts of the merging galaxies. That enables the coexistence of wonderfully long tidal tails with already merged central regions. It gives a good account of the observations.

6.5 *Recognizing Subluminal Matter*

Ostriker, Peebles, and Yahil (1974), and Einasto, Kaasik, and Saar (1974), independently presented overviews of the evidence for the presence of subluminal matter on scales from galaxies to clusters of galaxies. The latter paper does not mention disk stability; the Ostriker and Peebles (1973) paper may not even have reached them in the Soviet Union. Both groups concluded that the mean mass density is likely about 20 percent of the Einstein–de Sitter value, that is, $\Omega_m \simeq 0.2$ (which, it will be recalled, is independent of the value of Hubble's constant). This is not far from what was established later from broader evidence. The two 1974 papers did not add much to the combined expertise on the many aspects of subluminal matter, but they brought the evidence together in clear ways that attracted attention.

The serious case reviewed in these two papers for the presence of subluminal mass draped around galaxies includes its roles in helping to account for (1) the large apparent masses of groups and clusters of galaxies, (2) the rotation curves of spiral galaxies, (3) the stability of the disks of spiral galaxies, and (4) the properties of merging spirals. In the early 1970s, point (1) was recognized as motivation for the thought that neutrinos have nonzero rest mass and the mass in neutrinos might hold the Coma Cluster together. Point (2) was the more commonly mentioned evidence for subluminal matter in the early 1980s, as we moved toward the notion of nonbaryonic dark matter. Point (3) was one of the Ostriker, Peebles, and Yahil (1974) arguments for subluminal matter in a cosmologically interesting amount. It later helped motivate the addition of nonbaryonic matter to cosmology, but otherwise was less widely noticed at the time; perhaps a matter of fashion. Point (4) was even less well appreciated. Note that the direct illustration from Barnes's (1988) simulations appeared well after the community had come to accept the notion of nonbaryonic dark matter and its possibly deep significance. Eventualities set timings, but we can see that the growing pressure from the many developments reviewed in this section forced community interest and acceptance of subluminal matter. The added pressure from theoretical developments is reviewed in Chapter 8.

Fritz Zwicky's two early papers on the large ratio of mass to starlight in the Coma Cluster of galaxies were not often mentioned in print until the 1990s. This is illustrated by the NASA Astrophysics Data System counts of citations of the paper Zwicky (1937a), shown in Figure 6.7. The trend is similar for Zwicky (1933). (Recall that ADS citations are not complete, but this trend is real.) We can attribute the abrupt change in citation rate to the revolutionary advances in cosmology to be discussed in Chapter 9. It became fashionable to cite Zwicky on the Coma Cluster.

Figure 6.8 is another illustration of the history of thinking about issues of astronomical masses, here the observation that the apparent value of the mass of an object relative to its luminosity is larger for larger objects. Panel (a) shows Karachentsev's (1966) estimates of the mass-to-light ratio $f = M/L$ as a function of sizes of the systems as represented by their luminosities. Karachentsev collected estimates of masses of individual galaxies; masses from relative redshifts of pairs and triplets of galaxies close together in the sky and presumed to be gravitationally bound; and estimates of masses for a considerable

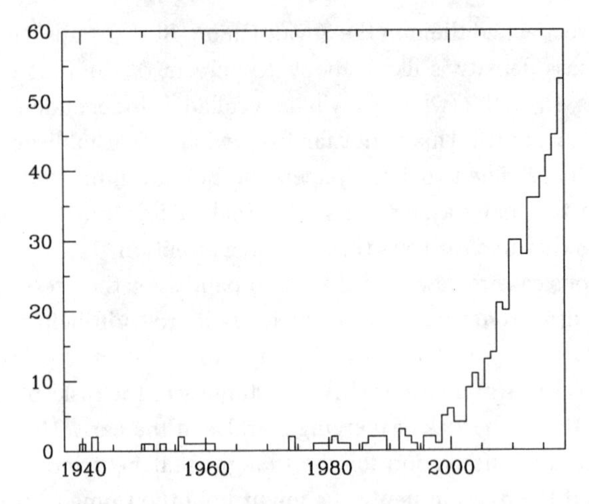

FIGURE 6.7. Citations per year to Zwicky (1937a).

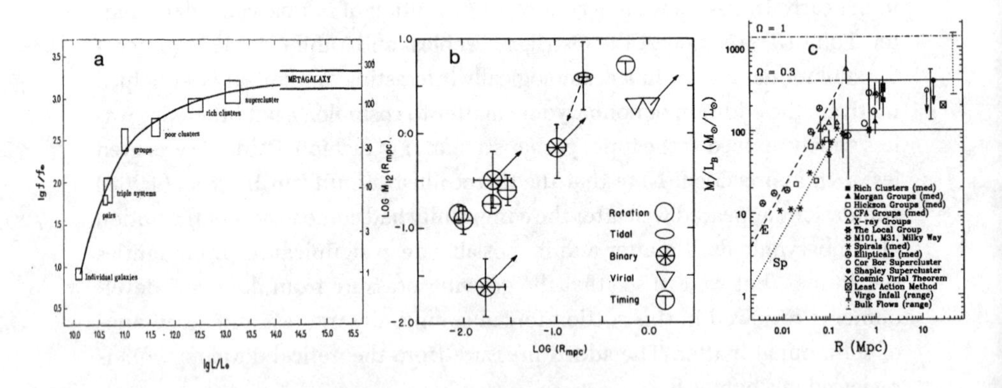

FIGURE 6.8. Finding the scaling of mass with size: Karachentsev (1966); Ostriker, Peebles, and Yahil (1974); and Bahcall, Lubin, and Dorman (1995). Panel (a) Reproduced by permission from Springer Nature. Panels (b) and (c) © AAS. Reproduced with permission.

number of groups and clusters from their internal spreads of redshifts, including Zwicky's case (discussed in Section 6.1). Not all these data are reliable, and not all agreed that many of these systems are gravitationally bound. But in 1966, Karachentsev saw the right picture: if standard gravity physics is to be trusted, and groups and clusters of galaxies are gravitationally bound, then most of the mass of the universe is in concentrations spread more broadly than the concentrations of starlight. Accordingly, a fair sample of all the mass would be detected only by the internal dynamics of sufficiently large systems. Data similar to Panel (a) in Figure 6.8 were the subject of the conference reviewed by Neyman, Page, and Scott (1961). We see that in the 1960s, the phenomenon of broadly spread subluminal mass was known and discussed in print by some.

But the community by and large was not prepared to consider it. Perhaps that was in part because it was difficult to know what to make of the phenomenon.

Panel (b), from Ostriker, Peebles, and Yahil (1974), shows estimates of the typical mass $M(<R)$ contained within distance R of a large spiral galaxy similar to the Milky Way. It reaches mass in excess of $10^{12}M_{\odot}$ at radii of a few hundred kiloparsecs, well above the mass in stars observed at radius ~ 10 kpc. Einasto, Kaasik, and Saar (1974) presented similar arguments. At the time, others were considering the thought (reviewed in Chapter 7) that there may be considerable mass in neutrinos of the known kinds with rest mass of a few tens of electron volts, and that this mass might be holding the Coma Cluster together. Later in the decade, ideas turned to a new variety of neutrinos that have a much larger rest mass; it became the prototype for the dark matter in the established cosmology. But in that community in the 1970s, not much notice was taken of the broad variety of evidence of subluminal matter illustrated in Figure 6.8 and reviewed at length in this chapter.

Panel (c) in Figure 6.8, from Bahcall, Lubin, and Dorman (1995), shows the mass-to-light ratios of galaxies and systems of galaxies as a function of length scale. We see the considerable improvement in the data. By this time, the community was well aware of the subluminal mass phenomena, and interest had turned to whether the mass might amount to the Einstein–de Sitter value, though we see in Panel (c) that the evidence of an approach to sufficiently large masses at $R \gtrsim 300$ kpc did not look promising. This history is discussed in Sections 3.6.3–3.6.5.

Lawrence Aller's (1995, 6) recollection of thinking in the 1940s is that

> a $\langle M/L \rangle$ ratio of ~ 14 for M31 suggests a large component of "dim" or invisible material. ... Folks worried about these matters at the time [the early 1940s] but few of the worries were published as it was considered bad taste to publish idle speculations. We, then of the younger generation, were particularly vulnerable on this issue.

In Morton Roberts's (2008, 287–288) review of this history, he concludes that

> The concept of dark matter surrounding galaxies was firmly established because of two important events: (1) the ability to extend rotation curves well beyond the optical boundary of a galaxy and (2) at this same time the need developed for a component in a spiral galaxy's make up that would stabilize its disk; theoreticians wanted it! What better way than with a previously unrecognized halo of dark matter? ... Let me conclude with a set of related questions, perhaps for the sociologist as well as the astronomer; I've described a well defined, well documented list of discoveries on the route to dark matter. What took us so long to accept it? How does it differ from the instant acceptance of the extragalactic nature of the nebulae after Hubble's announcement

> of Cepheids in M31? And how does it differ from the very rapid acceptance of dark energy?

Aller and Roberts do not mention Zwicky; they were involved in other aspects of the subluminal mass problem. Roberts raises an issue of sociology. I suspect that few were citing Zwicky's (1933) and (1937a) papers because few had been citing them and few recognized that there was good reason to look into these old papers.

So who discovered subluminal matter? Zwicky found the first well advertised evidence in the 1930s. His two papers were remembered through the 1970s, as we see in Figure 6.8, though not by many who felt they were worth citing, as we see in Figure 6.7. And to many others, Zwicky's effect seemed distinctly unlikely, as we see from the 1961 *Conference on the Instability of Systems of Galaxies* discussed on pages 243 and 274. Astronomers, like Tantalus, may look but not touch (outside the solar system). Inferences from astronomical observations usually are quite indirect and are made persuasive only by checks of consistency of independent lines of evidence. And it does help to have a theoretical pedigree.

Apart from Zwicky's first steps, it is meaningless to ask who discovered subluminal matter; this was a process of recognition growing out of the considerable variety of methods of observation and analysis of the great clusters of galaxies, groups of galaxies, and the rotation curves and stability of spiral galaxies. The community by and large became interested in subluminal matter when this considerable variety of observational evidence gained sanction from the theory that promoted subluminal matter to nonbaryonic dark matter in the early 1980s. The theoretical developments (reviewed in Chapter 8) did not change the observations, of course, but that is the nature of community assessments.

6.6 *What Is the Nature of the Subluminal Matter?*

The case from astronomy for subluminal matter did not require that the subluminal matter be exotic. Roberts and Whitehurst (1975, 244) concluded that their measurements of the rotation curve of M 31 shown in Figure 6.2 require

> significant mass at large R to keep V_c from decreasing. This mass is not visible in blue light at the ~1 percent of sky brightness level, yet the required mass in the form of the most common type of star in the solar neighborhood, dwarf M-type stars, will satisfy the upper limits on brightness required from photometry.

Martin Rees (1977, 348) expressed similar thoughts:

> By $z = 100$ as much as 90 percent of the primordial material could have condensed into stars. These stars would originally be grouped in units

> smaller than galaxies, but clustering would develop on progressively larger scales. This material would constitute the halos of galaxies.

White and Rees (1978, 342) enlarged on this:

> Of the many possibilities, the most plausible candidates are low-mass stars, burnt-out remnants of high-mass stars, or the remnants of supermassive stars, any of which might have formed soon after the primordial plasma recombined [but] nothing in our discussion would change if the dark mass consisted of, for instance, massive neutrinos, or black holes which formed before recombination.

The thought that the subluminal mass may be black holes, perhaps produced in the very early universe during first-order phase transitions, occurred to others (e.g., Crawford and Schramm 1982; Carr, Bond, and Arnett 1984; Lacey and Ostriker 1985).

Thoughts about subluminal matter such as stars or planets or black holes were tested by observations of gravitational microlensing: a compact mass that is close enough to the line of sight from a distant compact source of light, such as a star or quasar, acts as a gravitational lens that magnifies the solid angle of the source. The magnification increases the received radiation energy flux density. Such a microlensing event would be transient, with time scale set by the mass of the object and its transverse motion relative to the source.[6] Bohdan Paczyński (1986) pointed out that if the massive halo of our Milky Way galaxy consisted of subluminal objects, such as dwarf stars, planets, stellar remnants, or stellar-mass black holes, they would produce a microlensing event along a given line of sight with probability $P \sim 10^{-6}$. Paczyński suggested that it might be feasible to check this by monitoring the brightnesses of several million stars. A valuable by-product for astronomers would be the identification of many variable stars. This inspired the Optical Gravitational Lensing Experiment (OGLE) search for microlensing events by objects in the Milky Way, then the Expérience de Recherche d'Objets Sombres (EROS) and MACHO projects. Regarding the last, Kim Griest (1991, 412) pointed out that

> The nature of the dark matter known to exist in galactic halos is unclear. It could consist of elementary particles such as axions, light neutrinos, or members of the weakly interacting, massive particle (WIMP) class (such as the lightest supersymmetric particle). It could also consist of massive astrophysical objects such as brown dwarfs, Jupiters, or black

6. Refsdal (1964) and Liebes (1964) showed that stellar masses can produce gravitational microlensing. Press and Gunn (1973) pointed out that the probability that a compact luminous object at the Hubble distance is microlensed is on the order of the mass density parameter Ω in such lensing objects. This is independent of the lens masses, as one might expect: what dimensionless number other than Ω might be important in determining the probability?

hole remnants of an early generation of stars. (As a major alternative to WIMPs, this latter class should surely be collectively called massive astrophysical compact halo objects [MACHOs].)

Microlensing events are distinguished from variable and eclipsing stars by the well-characterized time-symmetric microlensing light curve. It is independent of wavelength and defined by the MACHO mass, the impact parameter, the relative transverse velocity of MACHO and source, and the distances of source and gravitational lens. The summary results in Alcock et al. (2000 and 2001) are that up to about 20 percent of the mass of the Milky Way halo might consist of MACHOs with masses in the range of 0.15 to $0.9M_\odot$, and that the full halo mass cannot be in MACHOs in the range of masses of 30 to $10^{-7}M_\odot$.

Ivan King at the May 1977 Yale Conference on The Evolution of Galaxies and Stellar Populations offers a summary challenge (King 1977, 9): "Can we really claim to know anything about the nature of the universe if we don't know the properties, or even the nature, of 90 percent of its material?"

What might this material be? There was a hint. Measured light-element abundances interpreted in the hot big bang theory indicate that the mass density in baryons is $\Omega_{\rm baryon} \lesssim 0.05$ (Table 4.2). In the collections of evidence for the presence of subluminal mass by Ostriker, Peebles, and Yahil (1974), and Einasto, Kaasik, and Saar (1974), the conclusions were that the total mass density is larger, $\Omega_{\rm mass} \simeq 0.2$. This larger value is in line with the considerable variety of measures of the mass density summarized in Table 3.2. The hint was that, if we have the observations and physics about right, most of the mass cannot have taken part in the element-building reactions in the hot big bang. Maybe $\Omega_{\rm baryon}$ is less than $\Omega_{\rm mass}$ because the subluminal matter is not baryonic. Maybe it is neutrinos with rest mass on the order of 30 eV. This thought was in the literature by 1974; Gott et al. (1974), who assembled the constraints in Figure 3.2 on page 76, mention it. That is, when King posed his question at the 1977 Yale conference, the ingredients for at least a partial answer to his question were being assembled. But it took another half decade to put this concept together in a picture that looked promising. This is the subject of the next chapters.

CHAPTER SEVEN

Nonbaryonic Dark Matter

IN THE YEARS around 1980, the main contribution of particle physics to cosmology was its offer of candidates for the nature of the astronomers' subluminal mass. The main contribution of cosmology to particle physics was its offer of bounds on the properties of real or conjectured particles and fields from the condition that their mean mass density not exceed the limit from the relativistic cosmological model with reasonable values of the expansion time and deceleration parameter.

The condition on the mean mass density was first used to establish an upper bound on the rest masses of the neutrinos in the electron and muon lepton families. In the hot big bang model, the mean number density of neutrinos relict from the hot big bang was set by relaxation to thermal equilibrium in the early universe. This number density requires that their masses not exceed $m_\nu \sim 30$ eV, give or take a factor of two or three.

The contribution of particle physics to cosmology in the early 1970s was the idea that Zwicky's (1933 and 1937a) subluminal mass in the Coma Cluster may be relict thermal neutrinos with rest mass $m_\nu \sim 30$ eV. Later in the decade, it was realized that a hypothetical new neutrino with the usual weak interaction but rest mass of about 3 GeV would have been largely annihilated as the universe expanded and cooled, leaving a residual number density that would amount to the mean mass density that cosmology favors. These GeV neutrinos were seen to be an even more promising candidate for the subluminal matter.

Neutrinos in the smaller of the allowed mass windows would have a large velocity dispersion in the early universe; they came to be termed hot dark matter or HDM (though the reduction of momentum as the universe expanded would have made them slowly moving in the present epoch). The hypothetical much more massive neutrino, $m_\nu \sim 3$ GeV, would have negligible primeval velocity dispersion; it was termed cold dark matter or CDM. The axion field added to the standard model for particle physics, or else particles remnant from broken supersymmetry, could equally well serve as CDM. One might even imagine that the CDM is sea of primeval black holes, though the MACHO gravitational lensing searches discussed on page 277 seriously bound acceptable black hole masses. The idea of a third class termed warm dark matter or WDM arose from the early thought that particles remnant from supersymmetry breaking might have primeval velocity dispersions that define a characteristic mass, set by how far free streaming would carry them, that might be comparable to the mass of a large galaxy. This certainly would be interesting for astrophysics, but no evidence of such a characteristic mass has emerged so far.

The terminology—HDM, WDM, and CDM—first appeared in print in the Proceedings of the Fourth Workshop on Grand Unification, Philadelphia, 1983. Primack and Blumenthal (1983, 265) wrote that "We are grateful to Dick Bond for proposing this apt terminology." In the following paper in the Proceedings, Szalay and Bond (1983) discussed this terminology, and it is further considered in Bond et al. (1984a,b). In a private communication, Bond recalls that "the HDM, WDM, CDM names began in a colloquium I gave in 1981 at The University of Chicago, with Chandrasekhar in the front row—and he stayed to the end of my talk."

7.1 *Hot Dark Matter*

The number density n_γ of CMB photons at temperature T remnant from the hot big bang, and the remnant number density n_ν of neutrinos in two spin states in one family, are

$$n_\gamma = 2\int \frac{d^3p}{(2\pi\hbar)^3}\frac{1}{e^{pc/k_BT}-1}, \qquad \langle n_\nu\rangle = 2\int \frac{d^3p}{(2\pi\hbar)^3}\frac{1}{e^{pc/k_BT_\nu}+1}. \qquad (7.1)$$

The second expression assumes that the neutrinos were relativistic when last thermally coupled to the radiation, at $T \gtrsim 10^{10}$ K. Annihilation of the thermal electron-positron pairs adds entropy to the radiation, making the radiation temperature after annihilation larger than the effective neutrino temperature by the factor $T_\gamma = (11/4)^{1/4} T_\nu$. These expressions work out to present mean number densities of

$$n_\gamma = 420 \text{ cm}^{-3}, \qquad \langle n_\nu\rangle = 113 \text{ cm}^{-3}, \qquad (7.2)$$

at the present radiation temperature $T_f = 2.725$ K. At neutrino rest mass m_ν, the mean mass density $m_\nu n_\nu$ in this neutrino family is

$$\Omega_\nu h^2 = 0.3\, m_{30}, \text{ where } m_\nu = 30\, m_{30} \text{ eV}. \qquad (7.3)$$

The second expression defines m_{30}.

Gershtein and Zel'dovich (1966) pointed out that the neutrino rest mass m_ν is constrained by the condition that the present mean mass density of the neutrinos relict from the hot big bang, $\rho_\nu = m_\nu n_\nu$, must not exceed that of a cosmological model with acceptable expansion time and deceleration parameter (of course assuming that the standard physics of the hot big bang cosmology is a useful approximation). With a conservative bound on the mass density the cosmology would allow, Gershtein and Zel'dovich concluded that the rest mass of each neutrino family cannot exceed $m_\nu \sim 400$ eV. Their mass density allowance was overly generous, but they made the point: cosmology offers a significant constraint on particle physics.

Marx and Szalay (1972) and Cowsik and McClelland (1972) refined the calculation. Their bounds are

$$m_\nu < 130 \text{ eV, and } m_\nu < 8 \text{ eV,} \tag{7.4}$$

respectively. The spread is a fair indication of the uncertainties in cosmological parameters.

Ramanath Cowsik (in a personal communication, 2016) recalls how he

> became aware of the dark-matter issue. Rood and King over the years of effort that culminated in 1972 measured the velocities of more than 200 galaxies in the Coma Cluster, and precisely determined the level of the virial discrepancy, originally discovered by Zwicky. This observational confirmation of the existence of unseen matter prompted me to think that if it dominated large clusters, such unseen matter should also dominate the dynamics of the universe as well. It also occurred to me that the matter was not just "unseen" but "unseeable," that is, weakly interacting particles. By then I had read "Physical Cosmology." This resulted in the papers of 1972 and 1973, at which time neutrinos were the only particles that had been discovered, and could play the role as dark matter.

Alexander Szalay (in a personal communication, 2019) recalls that

> there was nobody that time doing cosmology in Hungary, and I felt very much alone, but I found your "Physical Cosmology" and the Weinberg book in the library of the National Academy of Sciences and these taught me about what is modern cosmology. I started working on cosmological neutrinos in my undergraduate thesis, extending the ideas of Gershtein and Zel'dovich (1966). This led to the Neutrino'72 paper Marx and Szalay (1972). Then in my PhD at Eötvös University, Budapest (Szalay 1974) I extended the calculations to include dynamics, the approximate computation of the damping mass (Szalay and Marx 1976). In those years this was considered a rather esoteric idea, so I started to work with the Zeldovich group in Moscow in more mainstream cosmology, on "pancake theory." In 1980 I was a student at a cosmology summer school in Warsaw, Poland, when the news about the Lyubimov experiment arrived. The Director of the School, Marek Demianski remembered my thesis work and asked me to improvise two

> lectures on cosmological neutrinos. Joe Silk was a lecturer at the School, and after my talk he invited me to come to Berkeley, and also recommended me for an invited talk at the Texas Symposium that December, in Baltimore. I arrived to Berkeley, following the conference, just in time for the Christmas Party in the department, where I met Dick Bond for the first time. We became instant friends and many decades of collaborations followed. I still remember that remarkable year of 1980 as the turning point when Big Bang relics became the leading candidates for dark matter, and a key part of physical cosmology.

Szalay told Richard Feynman about his constraint on the neutrino mass when Feynman attended the conference Neutrino '72 in Hungary in the Summer of 1972. Szalay kindly allowed me to show in Figure 7.1 the illustration of Feynman's reaction to cosmology's contribution to particle physics.

Cowsik and McClelland (1973), Szalay (1974), and Szalay and Marx (1974, 1976) took note of the possibility that the mass density in neutrinos remnant from the hot big bang might close the universe. The phrase may refer to thoughts reviewed in Section 3.5 about a philosophically attractive universe, or it may only be a convenient benchmark for a cosmologically interesting mass density. The important step in these papers is the conjecture that nonbaryonic matter—neutrinos with nonzero rest mass—might be the astronomers' subluminal matter. All these papers mention the situation in the Coma Cluster discussed in Section 6.1, the system for which Zwicky (1933 and 1937a) first demonstrated evidence of subluminal mass (Section 6.1). None mention the later evidence of subluminal matter on smaller scales reviewed in Chapter 6. But the transformative idea about the nature of the subluminal mass was introduced. It happened as the evidence for subluminal mass on a broad range of scales was improving and becoming more widely noticed, the notice driven in part by the transformative idea, of course.

Cowsik and McClelland (1973, 8) state that "though the idea is undoubtedly not new, it does not appear to have been presented in published form before." We know now that similar thoughts about subluminal mass were arising in Hungary at essentially the same time. Apart from that, I have not encountered any evidence to contradict this statement. The dark matter these authors had in mind proves not to be the dominant kind, but their thinking was an important advance, and their physical reasoning is of lasting interest.

By paying attention to orders of magnitude, we can see the nature of the physical considerations involved in the idea of neutrinos with nonzero rest mass in a cluster of galaxies or draped around a single field galaxy. Consider a gravitationally bound and stable clump of neutrinos with radius R and mass M, perhaps containing a trace of baryons to approximate a galaxy. The neutrino rest mass is m_ν. The characteristic velocity v_ν of the neutrinos gravitationally bound in this clump satisfies $v_\nu^2 \sim GM/R$, in an approximation to

Feynman's Preface

These are the lectures in physics that I gave last year and the year before to the freshman and sophomore classes at Caltech. The lectures are, of course, not verbatim—they have been edited, sometimes extensively and sometimes less so. The lectures form only part of the complete course. The whole group of 180 students gathered in a big lecture room twice a week to hear these lectures and then they broke up into small groups of 15 to 20 students in recitation sections under the guidance of a teaching assistant. In addition, there was a laboratory session once a week.

The special problem we tried to get at with these lectures was to maintain the interest of the very enthusiastic and rather smart students coming out of the high schools and into Caltech. They have heard a lot about how interesting and exciting physics is—the theory of relativity, quantum mechanics, and other modern ideas. By the end of two years of our previous course, many would be very discouraged because there were really very few grand, new, modern ideas presented to them. They were made to study inclined planes, electrostatics, and so forth, and after two years it was quite stultifying. The problem was whether or not we could make a course which would save the more advanced and excited student by maintaining his enthusiasm.

The lectures here are not in any way meant to be a survey course, but are very serious. I thought to address them to the most intelligent in the class and to make sure, if possible, that even the most intelligent student was unable to completely encompass everything that was in the lectures—by putting in suggestions of applications of the ideas and concepts in various directions outside the main line of attack. For this reason, though, I tried very hard to make all the statements as accurate as possible, to point out in every case where the equations and ideas fitted into the body of physics, and how—when they learned more—things would be modified. I also felt that for such students it is important to indicate what it is that they should—if they are sufficiently clever—be able to understand by deduction from what has been said before, and what is being put in as something new. When new ideas came in, I would try either to deduce them if they were deducible, or to explain that it *was* a new idea which hadn't any basis in terms of things they had already learned and which was not supposed to be provable—but was just added in.

At the start of these lectures, I assumed that the students knew something when they came out of high school—such things as geometrical optics, simple chemistry ideas, and so on. I also didn't see that there was any reason to make the lectures

3

Dear Alex,
Thank you for your hospitality and help in Balatonfüred and Debrecen. I won't forget my visit. I was surprised to learn that neutrino mass could be limited by thoughts of cosmology!
Richard Feynman

FIGURE 7.1. Richard Feynman's note to Alex Szalay, Hungary, 1972 (by permission from Alex Szalay; The Feynman Lectures © 1963, California Institute of Technology).

the virial theorem (equation (6.1)). The neutrinos are nonrelativistic, so the characteristic neutrino momentum is $p_\nu \sim m_\nu v_\nu$. The maximum allowed number density of neutrinos in the system is, in order of magnitude, $n_\nu \lesssim p_\nu^3/\hbar^3$, which follows from the Heisenberg exclusion principle, $\delta p \delta x \geq 2\pi\hbar$, or from equation (7.1). Cowsik and McClelland (1973) based their estimate of the neutrino rest mass on the assumption that the cluster contains a degenerate sea of neutrinos, a very safe bound. Tremaine and Gunn (1979) pointed out that neutrinos are effectively collisionless after the break from thermal coupling in the early universe, so the density in single-particle phase space is conserved along the path of a neutrino. This is the Liouville theorem in classical mechanics. The bound on the density in phase space of one spin state of one kind of neutrino is thus

$$\frac{1}{(2\pi\hbar)^3(e^{pc/k_B T_\nu}+1)} \leq \frac{1}{2(2\pi\hbar)^3}. \tag{7.5}$$

Gravitational collapse mixes streams with different densities in phase space; Tremaine and Gunn (1979) put it that the phase space density distribution becomes "frothy." This reduces the mean density in phase space, which increases the neutrino mass m_ν needed to fit a massive subluminal halo. Now the mass of the system is the product of the neutrino number density, the volume of the system, and the neutrino rest mass: $M \sim n_\nu R^3 m_\nu$. The result of collecting and rearranging these expressions, a good exercise for the student, is

$$m_\nu^8 \gtrsim \frac{\hbar^6}{MR^3G^3}, \qquad m_\nu^4 \gtrsim \frac{\hbar^3}{GR^2 v_\nu}. \tag{7.6}$$

The first expression fixes the order of magnitude of the minimum neutrino mass m_ν, given the system radius R and the mass M (assumed to be dominated by neutrinos).

Cowsik and McClelland (1973) derived the first expression in equation (7.6), with an estimate of the numerical prefactor, and applied it to the mass distribution in the Coma Cluster of galaxies. They concluded that (Cowsik and McClelland 1973, 10) "we see that neutrinos, should they have a rest mass of a few eV/c^2, could close the Universe and account for the gravitational binding of the Coma cluster." Szalay (1974), Szalay and Marx (1976), and Schramm and Steigman (1981) arrived at similar conclusions in increasingly detailed studies.

The concept of nonbaryonic dark matter has proved to be of lasting interest. But this candidate for dark matter, HDM, is seriously challenged by its prediction of pancake structure formation. The problems this raises are discussed on pages 235 and 236 in Section 5.2.7 and reviewed on page 288 in Section 7.1.1 below. But let us consider here yet another issue.

Tremaine and Gunn (1979) explored conditions under which neutrinos with rest mass consistent with an acceptable cosmic mean mass density could account for the dark matter draped as a halo around a large galaxy such as M 31 (Figure 6.2). In a more detailed follow up of the Cowsik and McClelland (1973) discussion, Tremaine and Gunn improved the numerical prefactor in the second of the expressions in equation (7.6), took the neutrino velocity to be $v_\nu \sim 150$ km s^{-1} in a dark matter halo typical of a large galaxy, and took the radius to be $R \sim 20$ kpc, which might be typical of the extent of a subluminal or dark matter halo. This yielded their lower bound on the neutrino rest mass, $m_\nu \gtrsim 20$ eV, required to allow a massive halo of neutrinos to be as compact as those observed around large galaxies. Tremaine and Gunn estimated that the upper bound on the neutrino mass allowed by the cosmic mean mass density is

$m_\nu \lesssim 2.5h^2$ eV. This is well below their lower bound for an acceptable massive neutrino halo of a large galaxy. They concluded that neutrinos with nonzero rest mass are not promising candidates for subluminal matter.

This analysis could be debated, because Tremaine and Gunn (1979) used a low estimate for the mean mass density, $\Omega_m \sim 0.05$. At $\Omega_m \sim 0.2$, as Ostriker, Peebles, and Yahil (1974) and Einasto, Kaasik, and Saar (1974) had argued, the upper bound on m_ν would be multiplied by a factor of four, which might allow the neutrinos to fit around large galaxies. In a reassessment of the argument, Gunn (1982) concluded that the Tremaine and Gunn estimate of the mean mass density likely is too small but that their conclusion may still hold. But Aaronson (1983) and Lin and Faber (1983) found that the Tremaine and Gunn constraint applied to the dark matter in dwarf spheroidal galaxies requires neutrino mass on the order of 500 eV, or $\Omega_\nu h^2 \sim 5$. Even within the uncertain state of measurements of the cosmological parameters, this was unacceptable (though one might imagine that the subluminal mass is dwarf stars in dwarf galaxies and neutrinos in large ones).

7.1.1 APPARENT DETECTION OF A NEUTRINO REST MASS

During these considerations of whether neutrinos with rest mass on the order of 30 eV could be the subluminal mass in halos of galaxies, Lubimov et al. (1980) announced a possible laboratory detection of a nonzero neutrino rest mass of about this amount. That naturally raised great interest in the HDM model and sped recognition of its challenges.

The laboratory search for nonzero electron antineutrino rest mass was based on the measurement of the shape of the decay electron energy spectrum of tritium. The energy taken up by a neutrino rest mass would truncate the high-energy tail of the decay electrons. The Lubimov et al. experiment indicated the neutrino mass is in the range

$$14 \leq m_\nu \leq 46 \text{ eV at 99 percent confidence.} \tag{7.7}$$

Later measurements removed the evidence of this detection,[1] but at the time, the announcement was greeted with the interest to be expected from the prospect of something important. A family of neutrinos with rest mass in the range of the Lubimov et al. experiment, thermally produced in the hot big

1. Measurements that removed the lower bound on m_ν and reduced the upper bound include $m_\nu < 18$ eV at 95 percent confidence by Fritschi et al. (1986), and $m_\nu < 9.3$ eV at 95 percent confidence by Robertson et al. (1991). In their discussion of structure formation with HDM, Bond, Efstathiou, and Silk (1980) cited evidence for nonzero neutrino rest masses from Reines, Sobel, and Pasierb (1980), who announced tentative detection of neutrino oscillation with $m_1^2 - m_2^2 \sim 1$ eV^2. Neutrino oscillations have been detected, but perhaps not in this experiment.

bang, would contribute a cosmologically interesting mass density, maybe even the Einstein–de Sitter value $\Omega_m = 1$ (in equations (7.3) and (7.7)). Zel'dovich and Sunyaev (1980, 249) wrote that

> Our understanding of the universe as a whole would be fundamentally altered if it should be proved that neutrinos have a nonzero rest mass. In fact, the latest measurements announced by Lyubimov et al.[1,2] suggest that electronic antineutrinos do have a rest energy $m_\nu c^2 \simeq 30$ eV, corresponding to a rest mass $m_\nu \simeq 5 \cdot 10^{-32}$ g.

Reference 1 is to Lubimov et al. (1980). Reference 2 is to the same authors, *Soviet Journal of Nuclear Physics* 32 1980, in press, but the paper is not in volume 32 or 33 of this journal.

Bond, Efstathiou, and Silk (1980) and Doroshkevich et al. (1981) discussed two important implications of this HDM model for the subluminal matter. First, electromagnetic radiation would slip freely through the growing clustering of the dominant mass in neutrinos. This means the disturbance to the radiation by the formation of cosmic structure would be much less than if the mass were dominated by a baryonic plasma that would be dragged by the radiation. The effect is discussed further in Section 7.2. Second, the free streaming of the thermally formed neutrinos while they were relativistic would smooth the neutrino space distribution, setting a mass scale by the comoving distance the neutrinos would have streamed as they cooled. That would define a characteristic mass for cosmic structure, a very interesting signature to look for. Szalay and Marx (1976) seem to have been the first to have introduced this line of thinking for the HDM model.

Zel'dovich and Sunyaev had already pointed out that the dissipation of acoustic oscillations of an initially adiabatic plasma-radiation fluid would define a characteristic mass at the smoothing scale, and that the first generation of mass concentrations might be expected to form at this mass in pancake-like collapses (Section 5.2.7; Zel'dovich 1970, Sunyaev and Zel'dovich 1972.) The free streaming of HDM would smooth this nonbaryonic component of the primeval mass distribution to even larger scales, so if the dominant mass were these neutrinos, it would mean that the pancakes are more massive than Zel'dovich and Sunyaev previously had in mind. Estimates of the new characteristic mass defined by the smoothing by free streaming were presented by Bisnovatyi-Kogan and Novikov (1980); Bond, Efstathiou, and Silk (1980); and Doroshkevich et al. (1981), all of whom referred to the apparent laboratory detection of the neutrino rest mass. Doroshkevich et al. (1980); Sato and Takahara (1980); and Wasserman (1981) did not mention the detection but considered similar values of the neutrino rest mass, on the order of 25 eV. They arrived at similar characteristic masses. We see that interested members of the community soon were aware of an important implication of a neutrino rest mass $m_\nu \sim 30$ eV. The first generation of gravitationally formed mass concentrations would be expected to be much more massive than individual galaxies.

Before considering the problem with this picture, let us review the physical processes that fix the smoothing mass in the primeval HDM picture. This is another exercise in orders of magnitude, in the fashion of equation (7.6). Neutrinos with standard interactions and masses of a few tens of electron volts would have been moving at relativistic speeds when they broke thermal contact with the radiation and electron-positron pairs in the early universe. Thus they would have been endowed with typical momentum $p_\nu \sim k_B T_\nu / c$ when they began to move with negligible interaction with anything but gravity. This would make the mass density in relativistic neutrinos comparable to the mass density in radiation (though a little lower, because the occupation numbers tend to be smaller and, as indicated on page 280, T_ν is lower than the radiation temperature, but we are considering orders of magnitude). When the temperature fell to $k_B T_\nu \sim m_\nu c^2$, the mass density would have become dominated by the neutrinos as they became nonrelativistic and the photon energies continued to decrease as their wavelengths continued to increase. Now consider that at cosmic time t, neutrinos streaming about with speed $\sim v(t)$ would smooth primeval departures from a homogeneous neutrino distribution on the length scale $\lambda_S \sim v(t)t$. Since the mean mass density at t is $\rho(t) \sim (Gt^2)^{-1}$, the smoothing mass at time t is

$$M_S(t) \sim \rho(t)\lambda_S^3 \sim \frac{v(t)^3}{G^{3/2}\rho(t)^{1/2}}. \tag{7.8}$$

While the neutrinos are relativistic, $v(t) \sim c$, the smoothing mass M_S increases, because $\rho(t)$ is decreasing. The neutrinos become nonrelativistic at temperature $\sim m_\nu c^2 / k_B$. After that, the mass density is dominated by nonrelativistic neutrinos, so $\rho(t) \propto a(t)^{-3}$, and the neutrino streaming velocities vary as $v(t) \propto a(t)^{-1}$, causing this characteristic mass to decrease as $M_S \propto a(t)^{-3/2}$. This leaves the sea of neutrinos smoothed to the maximum value of M_S reached when the neutrinos became nonrelativistic. Collecting all this, we see that the present value of the smoothing mass is, in order of magnitude

$$M_S \sim \left(\frac{\hbar c}{G}\right)^{3/2} m_\nu^{-2} \sim 10^{15}\, m_{30}^{-2}\, M_\odot. \tag{7.9}$$

(This is the cube of the Planck mass, $(\hbar c/G)^{1/2} \sim 10^{-5}$ g, divided by the square of the neutrino mass.)

More serious computations of M_S in the early 1980s yield similar numerical results. For example, Bond, Efstathiou, and Silk (1980) present M_S at four times this value. It is a considerable increase in mass from the earlier idea of a smoothing length set by the pressure and viscosity of the primeval radiation and plasma acting as a fluid (equation (5.53)).

The baryon distribution would not be smoothed by free streaming of the neutrinos. But since the baryon mass density is supposed to be subdominant, the gravitational effect of baryons would tend not to be significant until the neutrinos started to collapse to self-gravitating concentrations that gathered up the baryons (as discussed Section 5.1).

Challenges to the pancake picture are discussed on page 235 in Section 5.2.7. These problems were made manifest by numerical simulations of the evolution of the mass distribution in a neutrino-dominated universe with primeval adiabatic initial conditions, where the coherence length prior to galaxy formation is set as in equation (7.9). Melott et al. (1983) concluded from their numerical studies that this model requires galaxy formation at unrealistically low redshifts. White, Frenk, and Davis (1983, L1) explored how parameters might be adjusted so that the large coherence length in the neutrino picture might be reconciled with the observations of galaxy positions and relative motions. They concluded that "We find this length to be too large to be consistent with the observed clustering scale of galaxies if other cosmological parameters are to remain within their accepted ranges." This conclusion is illustrated in Hut and White (1984), who compare the observed distribution of nearby galaxies with an example of the quite different distribution expected in the neutrino model.

The direct way to see the problem is to recall that if the dominant mass is in the neutrinos, then the first generation of gravitationally formed mass concentrations would have had typical mass $\sim M_S$ (equation (7.9)) comparable to that of a rich cluster of galaxies. Since rich clusters are rare, most of these mass concentrations would have to have scattered their mass in fragments that became the common galaxies well outside clusters. But gravity does not operate that way: gravity gathers mass and galaxies.

The announcement of an actual laboratory detection of a neutrino rest mass called for a close study of the HDM model and its pancake picture for cosmic structure formation. The laboratory evidence of detection soon faded, but not before it inspired careful studies of the nonlinear gravitational and hydrodynamical collapse of a pancake of baryons and nonbaryonic dark matter (e.g., Doroshkevich et al. 1983; Shapiro, Struck-Marcell, and Melott 1983; Bond et al. 1984a,b). This research was illuminating but not very closely guided by the phenomenology of the galaxy distribution.

A way out to be considered was that the primeval adiabatic departures from homogeneity on relatively small mass scales may have been large enough to overcome the suppression due to the free streaming of the HDM. My attempt at this in Peebles (1982a) did not look promising.

Another way out is to postulate that the baryons at high redshift were in clumps tight enough to have formed the central baryon-dominated parts of the galaxies. Then as the universe expanded and the HDM cooled gravity would drape the neutrinos around the baryon clumps, maybe producing the observed dark matter halos around galaxies. This picture was not much discussed, perhaps because, as Bond, Efstathiou, and Silk (1980, page 1983) put it, "Such a theory seems unattractive (galaxies exist because galaxies have always existed)." Later thoughts about isocurvature versions of this picture are reviewed in Section 8.4.

It is an interesting empirical consideration that pancake collapse occurring at about the present epoch would produce patterns similar to the large-scale voids and frothy clustering observed in the galaxy spatial distribution (Zel'dovich, Einasto, and Shandarin 1982). This picture taken literally certainly would not do, because the galaxies had to have broken away from the general expansion well before the present epoch. The point is much easier to make now, of course, well after these developments. Thus Silk's (1982) careful assessment of galaxy formation theory offers challenges to the pancake scenario but does not reject it. But we see that the quantitative assessments from numerical simulations by Melott et al. (1983) and White, Frenk, and Davis (1983) clearly rule out the adiabatic HDM pancake picture.

Blumenthal, Pagels, and Primack (1982), and Bond, Szalay, and Turner (1982) considered warm dark matter, maybe a hypothetical weakly interacting particle remnant from supersymmetry that has a larger rest mass, perhaps 1 keV. The particle number density relict from the hot early universe would have to be well below that of the neutrinos, but that might be because these particles had been thermally produced and then decoupled in the very early universe, when the entropy was shared among many more species of particles. The earlier decoupling and larger particle mass might reduce the troublesomely large mass of the first generation in equation (7.9) to perhaps $10^{12}M_\odot$, about the mass of a large galaxy with its dark matter halo. Blumenthal et al. considered this mass to be "more natural" for a scenario for cosmic structure formation, and Bond et al. pointed out that it could allow hierarchical bottom-up structure formation starting from large galaxies, leaving the far more numerous lower-mass dwarfs to be formed by fragmentation of large ones. There is evidence of fragmentation in tidal disruptions of galaxies, but this does not offer a natural account of the large numbers of dwarfs that are not now near any large galaxy. These dwarfs seem to point to bottom-up formation from a smaller mass. More considerations of how the HDM model might be modified to meet its challenges are reviewed in Section 8.4.

7.2 *Cold Dark Matter*

The prototype for the nonbaryonic cold dark matter in the standard cosmology that was established a quarter century later came from particle physics in five essentially simultaneous and independent papers published in 1977. They exhibit at most modest awareness of the considerable range of evidence from astronomy for the subluminal mass that became known as dark matter. But it was soon realized that this kind of nonbaryonic matter is more promising than HDM.

7.2.1 WHAT HAPPENED IN 1977

The first idea from particle physics for what became the nonbaryonic cold dark matter, or CDM, of cosmology was a hypothetical neutrino with rest mass $m_L \sim 3$ GeV and the low-energy V-A interaction and Fermi coupling constant of the weak interaction of the known lepton families. This still is an acceptable candidate for the nonbaryonic dark matter in the established ΛCDM cosmology, if we allow a modest adjustment of the strength of the weak interaction for this particle so as to fit the absence of its direct laboratory detection so far. The story of how this idea was introduced is worth recalling as an example of the capricious ways of progress in the natural sciences.

The assumption of the standard weak interaction fixes the low-energy rate constant σv for the creation and annihilation reactions:

$$\nu_L + \nu_L \leftrightarrow e^+ + e^-, \ \mu^+ + \mu^-, \text{ etc.}, \tag{7.10}$$

for the hypothetical ν_L neutrino with mass m_L. In the early universe, at temperatures well above $m_L c^2/k_B$, these massive neutrinos would have been moving at relativistic speeds in thermal equilibrium with the radiation and other particles, and their number density would have been given by the second part of equation (7.1). These neutrinos would have remained in statistical equilibrium of creation and annihilation as the temperature fell below $m_L c^2/k_B$ and the neutrino number density decreased relative to the photon number density, roughly by the exponential factor $\exp[-m_L c^2/k_B T] \ll 1$ in equation (7.1). At mass $m_L \sim 3$ GeV, the remnant number density that escaped annihilation would amount to an interesting mass density for cosmology.[2]

This candidate for the astronomers' subluminal mass is known as a weakly interacting massive particle, or WIMP, and the fact that a neutral particle with the standard weak interaction and a mass similar to that of the nucleons has

2. Recall that the density parameter for HDM is $\Omega_\nu h^2 = 0.3\, m_{30}$ for neutrino mass $m_\nu = 30\, m_{30}$ eV (equation (7.3)). These low-mass neutrinos would have decoupled from the thermal sea while relativistic. If the mass of the new neutrino were 3 GeV, eight orders of magnitude larger, annihilation before their decoupling must have reduced the number density by the factor $\sim 10^8 \sim \exp[m_L c^2/k_B T_L]$ to get the wanted residual mass density. This puts the wanted decoupling temperature at $T_L \simeq 0.05\, m_L c^2/k_B$. The temperature T_L fixes the wanted expansion time t_L at decoupling in terms of m_L. Lee and Weinberg (1977a) put the weak interaction rate coefficient for the sum of the reactions in equation (7.10), with an allowance for 14 annihilation channels, at $\sigma v \simeq 3 \times 10^{-27} m_L^2$ cm^{3}s^{-1}, with m_L expressed in GeV. The expansion time t_L and remnant neutrino number density n_L at thermal decoupling, when annihilation of the heavy neutrinos effectively ended, satisfy the condition $\sigma v n_L t_L \sim 1$. Gamow (1948a) used this condition when he introduced nuclear physics to cosmology (equation (4.18)). Putting this together with the expansion time in the early universe, $t_L = 3.2 \times 10^{19} (T_0/T_L)^2$ s, with $T_0 = 2.725$ K, gives density parameter $\Omega_L \simeq 4 m_L^2$, with the mass m_L in GeV. Finer calculation gets to $m_L \simeq 3$ GeV for Ω_L close to unity.

Table 7.1. The 1977 Introduction of the CDM Prototype

Paper	Date received	Date published
Hut (1977)	April 25	July 18, 1977
Lee and Weinberg (1977a)	May 13	July 25, 1977
Sato and Kobayashi (1977)	May 23	December 1, 1977
Dicus, Kolb, and Teplitz (1977)	May 31	July 25, 1977
Vysotskii, Dolgov, and Zel'dovich (1977)	June 30	August 5, 1977

an interesting mass density is known as the WIMP miracle.[3] The miracle is somewhat diminished by the freedom to adjust m_L to get the wanted mass density, of course

The evidence I have found is that the five sets of authors of the papers listed in Table 7.1 independently arrived at the idea of this new neutrino, or WIMP, thermally produced in the early universe and with a mass well above the upper bounds on the known three types. The dates of publication in the last column of the table show the usual broad scatter of times between submission and publication. The second column lists the dates received by the journals as marked in the papers; they span just two months. All five papers use the constraint on the cosmic mean mass density from the relativistic Friedman-Lemaître cosmology with a reasonably acceptable expansion time or deceleration parameter. All refer to the prior suggestion that the astronomers' subluminal mass in the Coma Cluster of galaxies may be one of the known neutrinos with rest mass on the order of tens of electron volts. Hut (1977); Sato and Kobayashi (1977); and Vysotskii, Dolgov, and Zel'dovich (1977) cite Szalay and Marx (1976); Lee and Weinberg (1977a) cite Cowsik and McClelland (1973); and Dicus, Kolb, and Teplitz (1977) cite Szalay and Marx (1974). But there is no suggestion in any of the five papers that their new neutrino, with rest mass m_L comparable to that of the nucleons, might be the astronomers' subluminal mass, replacing HDM.

We have some indications of their thinking. Piet Hut (in a personal communication, 2018), whose paper was submitted first, recalls that

> As an undergraduate in astrophysics I stumbled upon some writings by a fellow Dutchman, Tjeerd de Graaf, in the form of typed lecture notes about nucleosynthesis during the first few minutes after the Big Bang, well before Weinberg's book with that title came out. Seeing how quantitative we can be about such a remote history of our Universe was an important inspiration for me to study the Big Bang. I started my PhD research in December 1976, when I joined Tini Veltman's group

3. Kolb and Turner (1990) state that the term "WIMP" originated in research at the University of Chicago and Fermilab.

> in theoretical physics in Utrecht. Soon after I arrived, I mentioned to Veltman my interest in the physics of nucleosynthesis, and the possibility to use the very early Universe in addition to what we can learn about elementary particles from accelerators. Veltman then suggested to look at the properties of neutrinos, and pretty soon I managed to derive the results that I published in my paper "Limits on masses and number of neutral weakly interacting particles." The limits on the masses and number of light neutrinos were easiest to derive. It was only after I had already written a draft for the preprint, together with a figure of the dependence of the number of degrees of freedom on neutrino mass, that I suddenly wondered what would happen at much larger masses. And as soon as I asked myself that question, it again was not that difficult to derive the answers that I then plotted in that figure. As often in research, asking a new question is the real bottleneck, and more of a challenge than answering the question, once asked. When I later saw the paper by Lee and Weinberg, I sent a copy of my paper to Steven Weinberg. He sent me in response a short hand-written letter, in the days before email, telling me that it was clear that they and I had reached the same conclusion. Receiving that letter was one of the most encouraging highlights in my student life.

Hut adds, "no, I did not think about missing mass; that was something I became aware of only later."

Sato and Kobayashi (1977) did not derive the mass of a stable heavy neutrino with the remnant mass density allowed by cosmology, but the concept of such a particle is there. The delay in publication allowed Sato and Kobayashi to add in proof references to the papers by Lee and Weinberg (1977a) and Dicus, Kolb, and Teplitz (1977). Dicus, Kolb, and Teplitz cited Lee and Weinberg (1977a), yet their paper is published in *Physical Review Letters* immediately after the Lee and Weinberg paper. Kolb (in a personal communication, 2017) explains how this came about:

> We did a crude freeze-out calculation and first calculated the contribution to Omega from neutrinos and the corresponding limit on the neutrino lifetime. We saw that neutrinos of mass of a few GeV could be dark matter. We were working in a particle theory group; Dicus and Teplitz were particle phenomenologists. I was a graduate student just looking for something to do. We didn't have much knowledge about cosmology. We were writing up the paper when we heard of the Lee-Weinberg paper. We decided not to compete with Lee and Weinberg, but to write a companion paper concentrating on the lifetime limits.

This qualifies as independent recognition of WIMPs. Sato and Kobayashi also reported considerations of the effect of neutrino decay, as in $\nu_L \rightarrow \nu_e + \gamma$, with due attention to the observability of decay products. The paper by Goldman and Stephenson (1977), on massive neutrino decay, might have been added to the table, but the idea of an interesting remnant mass density is not mentioned.

By 1977 there was a considerable astronomical literature on the phenomena of masses around galaxies and in groups and clusters of galaxies. It is reviewed in Chapter 6. I had discussed much of this in my book, *Physical Cosmology* (Peebles 1971a). Weinberg (1972) also reviewed the mass problem in his book, *Gravitation and Cosmology,* but in less detail, in some 5 of 600 pages of text. That is about proportionate to Weinberg's recollections (in a personal communication, 2017) of his impression of the subluminal mass issue at the time of the Lee-Weinberg paper:

> in 1977 I did know about the need for dark matter in clusters of galaxies, so why did Lee and I in our paper only mention the possibility that particles like heavy neutrinos might close the universe? I think it is because at that time the cosmological argument [from the constraint on the mean mass density] for what we now call cold dark matter seemed stronger to me than the argument from clusters of galaxies. The cosmic mass density contributed by galaxies alone would give a deceleration parameter of only a few percent, whereas red shift surveys were reporting values of order one. On the other hand, I did not know whether the results given by the virial theorem for clusters of galaxies could be trusted, because the clusters might not be gravitationally bound.

We see that, despite limited communication between the particle physics and astronomy communities, the latter had a good case for the presence of subluminal matter around galaxies, and the former had a good candidate for the nature of this matter.

Thoughts of a GeV-mass neutrino may have been encouraged by the emerging evidence of the tau lepton family with the charged tau mass approaching 2 GeV. Sato and Kobayashi (1977); Dicus, Kolb, and Teplitz (1977); and Vysotskii, Dolgov, and Zel'dovich (1977) mentioned the indications of a massive weakly interacting particle first announced by Perl et al. (1975). And Lee and Weinberg (1977a) referred to Lee and Weinberg (1977b), who referred to Perl et al. These were events in electron-positron collisions at the Stanford Linear Accelerator announced as "Evidence for Anomalous Lepton Production in $e^+ - e^-$ Annihilation." By 1977, Martin Perl and colleagues had concluded that they had detected a new lepton family with charged τ lepton mass 1.90 ± 0.10 GeV and the associated ν_τ neutrino mass less than about 0.6 GeV (Perl et al. 1977). This ν_τ would not do for dark matter unless the annihilation cross sections for the reactions in equation (7.10) were a little stronger than expected from the electron and muon families, but maybe the hint for other massive particles was there.

In a personal communication about this book, Hut wrote that "I'm sure I would have mentioned it [the tau family] in the text of the paper, had I known about it, since that would have been a welcome piece of extra observational constraint." But we can imagine that others of the papers in Table 7.1 mentioned the process of discovery of the τ lepton and its neutrino, because it offered a precedent for a still more massive neutrino. Reinforcing this thought is the comment in the paper by Gunn et al. (1978, 1016), who referred to the Lee and Weinberg (1977a) paper (with Ben Lee a coauthor of both), that

> It might be noted in passing, to bolster the reader's confidence in the seriousness and reality of leptons heavier than the muon, that experiments at the e^+-e^- colliding beam machine SPEAR have accumulated substantial evidence for the existence of a charged heavy lepton, called τ^-, with a mass of 1.9 GeV (Perl et al. 1975 and 1976).

It might also be noted that ideas about the substructures of neutrons and protons—Feynman's partons and Gell-Mann's quarks—may have invited some to think about other kinds of GeV-mass particles. In correspondence, Kolb recalled yet another hint, the "high-y anomaly" in scattering of neutrinos by atomic nuclei. It is mentioned in Lee and Weinberg (1977b). The anomaly was withdrawn, but it seems that for some, it served as an aid to thoughts of a new GeV-mass neutrino.

I venture to add a modest proposal about the five-paper coincidence. Physicists share information, by research papers and conference proceedings, as in Hut's example of de Graaf's typed lecture notes, and I imagine often and most effectively by conversations. This may include explanations of specific ideas, but I have in mind thoughts that may not be very well worked out and are somehow communicated while not explicitly stated—ideas that are "in the air." My experience suggests the effect is real, but I must leave informed assessment to those who have given far more thought to how we all communicate.

We might pause to take note of Lee and Weinberg's (1977a, 167) conclusion that "if a stable heavy neutral lepton were discovered with a mass of order 1–15 GeV, the gravitational field of these heavy neutrinos would provide a plausible mechanism for closing the universe." The notion of closing the universe is not uncommon in this history. The thought that HDM might do it was explored in the early 1970s (as discussed in Section 7.1), and Gott et al. (1974) concluded it could not be done by baryons (page 78 in Section 3.6.3).

So how did five groups at so close to the same time arrive at the idea of a particle that proved to have the right properties for cold dark matter, the CDM, in the established cosmology? We must consider first simple coincidence. But in retrospect, at least, it does seem natural to have considered the idea of a new neutrino with the standard weak interaction and a mass large enough that the thermally produced neutrinos in the early universe largely annihilated before decoupling but left interesting remnants. Maybe the idea also seemed natural in 1977. And maybe the idea somehow was "in the air."

7.2.2 THE SITUATION IN THE EARLY 1980S

The Lee and Weinberg (1977a) paper prompted explorations of astrophysical effects of a new heavy stable neutrino, first by Gunn et al. (1978) and Steigman et al. (1978), with Gary Steigman a coauthor of both papers. In the former, we find the comment (Gunn et al. 1978, 1023) that "Heavy noninteracting neutral particles present since the early universe, constrained by the arguments above to cluster with the matter, could not be better as stuff to constitute the dynamical missing mass," and in the latter (Steigman et al. 1978, 1060) that "heavy neutrinos are an ideal material from which to form the "missing mass" in clusters of galaxies and galactic halos."

Gunn et al. (1978) pointed out that gravity would be expected to cause the nonbaryonic dark matter to settle into gravitationally bound mass concentrations, without dissipation, of course. Diffuse baryons, whether gas or plasma, might be expected to settle with dissipation to the centers of the concentrations of nonbaryonic matter before complete conversion to stars. The result would be a halo of nonbaryonic dark matter, perhaps WIMPs, draped around the outskirts of a galaxy of stars. It could account for the evidence discussed in Sections 6.3 and 6.4, from disk galaxy rotation curves and stability, that points to mass in excess of what is observed in stars present on the outskirts of galaxies. And it would help explain why the disk stars and gas in spiral galaxies are supported by rotation, as discussed on page 218. White and Rees (1978) were pursuing similar thoughts about how galaxies formed, but they felt the more plausible idea is that the subluminal matter is the remnants of early generations of stars. Thus I conclude that Gunn et al. (1978) are to be credited for the introduction of the now quite firmly established idea that the dark matter around galaxies is nonbaryonic: CDM, or HDM that is not too warm.

Not long after the Gunn et al. paper, Tremaine and Gunn (1979, 407) analyzed the phase-space constraint on neutrinos as HDM draped around galaxies (discussed on page 284 in Section 7.1). They remark that "Lee and Weinberg's hypothetical heavy leptons (mass $\sim$ 1 GeV) are not ruled out by this argument." This is because, for a given spread of velocities in a galaxy, the bound on the neutrino density in phase space allows a much larger number density of the much more massive WIMPs that Gunn et al. (1978) were considering (as one sees from equation (7.6)).

At the time of writing the standard and accepted theory, ΛCDM, postulates the existence of something that acts like a gas of stable particles that have negligible interaction apart from gravity, have negligible primeval particle velocity dispersion, and behave as if they had particle masses in the considerable range of values that allows us to treat the CDM as a gas. The 3-GeV neutrinos, or WIMPs, would do, provided we allow a little adjustment of the strength of interaction with the lower-mass leptons. A sea of black holes, perhaps remnant from first-order phase transitions in the early universe, would do as well, provided that the black hole masses are small enough not to damage galaxies and, from a later test, small enough to have escaped detection by the gravitational lensing probe discussed on page 277. Early thinking along this line is reviewed on page 77.

The CDM might be the axion field of particle physics. The early discussions by Abbott and Sikivie (1983) and Dine and Fishler (1983) considered the constraint from cosmology on the axion mean mass density, as had been done earlier for the two kinds of neutrinos, hot and cold. But at about the same time, Preskill, Wise, and Wilczek (1983, 131) announced that "we are led to entertain the possibility that axions make up a significant part of the dark matter of the universe." Ipser and Sikivie (1983, 925, 927) argued that

> The evidence[1] that individual galaxies possess massive dark halos with masses exceeding that of the luminous galactic matter by a factor ~ 10 has generated extensive investigation into the makeup, origin, and influence of such halos ... the axion fits the bill remarkably well.

The reference (1) is to the discussions of the mass density in Ostriker, Peebles, and Yahil (1974) and Einasto, Kaasik, and Saar (1974). We see that it had not taken particle physicists long to become aware of some interesting and suggestive astronomical phenomena.

Another candidate for CDM is a stable supersymmetric particle remnant from the breaking of primeval supersymmetry. Cabibbo, Farrar, and Maiani (1981) proposed that the subluminal mass in the galaxy might be photinos. Pagels and Primack (1982, 224) wrote that "Although a neutrino-dominated universe is becoming increasingly attractive to cosmologists, it is worth considering the possibility that much of the missing mass is gravitinos." Joel Primack's recollection (in a personal communication, 2018) is that

> I think Pagels & Primack 1982 was the first paper that suggested that the lightest super-partner particle, stable because of conserved R-parity, was a natural candidate for the dark matter. ... I had been very impressed with the evidence for dark matter in Faber and Gallagher's 1979 *Annual Reviews* article and in conversations with Sandy [Faber]. So to answer your question, it was research motivated by particle physics that led us to propose the lightest super-partner particle as a candidate for dark matter.

Thoughts that a gravitino with mass on the order of 1 keV might act as warm dark matter (WDM) that defines an interesting Jeans mass, maybe that of a large galaxy, were explored by Pagels and Primack (1982); Blumenthal,

Pagels, and Primack (1982); and Bond, Szalay, and Turner (1982). Ideas about WDM have continued to attract attention, but so far there has not been enough evidence to warrant adjustment of the standard ΛCDM cosmology.

Ellis et al. (1984) presented a broad survey of the possibilities for subluminal matter offered by the breaking of supersymmetry, in the paper "Supersymmetric Relics from the Big Bang." John Ellis (personal communication, 2016) recalls that at the time he was

> among those who took astrophysics more seriously and followed it more closely than most particle physicists, writing papers on a variety of astroparticle physics topics starting even before the 1980s. I do not remember exactly when I personally became convinced about dark matter, but it must have been in the early 1980s.

Another of the authors of this paper, Keith Olive (personal communication, 2017), recalls that

> I certainly did take it [subluminal mass] seriously ... I became convinced for the need for nonbaryonic dark matter in a paper with Dennis Hegyi in 1983 entitled "Can galactic halos be made of baryons?"

The abstract of the Hegyi and Olive (1983, 28) paper is worth considering:

> Several arguments are presented indicating that the apparently nonluminous matter forming massive halos of spiral galaxies is not baryonic. There are difficulties with a halo dominated by gas, snowballs, dust and rocks, jupiters, low mass stars, dead stars and neutron stars. Also, halos may not be composed of black holes unless they are either extremely efficiently accreting or primordial. Consequently, it appears that a significant fraction of the universe may be in the form of massive neutrinos, gravitinos, monopoles, etc.

This is a healthy inspection of the astronomical evidence in the context of ideas about what might be suggested by particle physics.

7.2.3 THE SEARCH FOR DARK MATTER DETECTION

The announcement of an apparent laboratory detection of the electron-type neutrino rest mass certainly added to the interest in the HDM discussed in Section 7.1. The idea of the much greater rest masses of WIMPs, or what might be expected of relict supersymmetric partners, encouraged thoughts of detection in more massive dark matter particles. It might be indirect, by detection of radiative decay or annihilation of dark matter in astronomical objects, or direct, by laboratory detection of interactions of dark matter with baryons. I take it as an example of the fascination of an interesting and seriously challenging but perhaps not quite impossible measurement that experiments to detect this hypothetical nonbaryonic matter began before there was much empirical evidence that there really is nonbaryonic matter to be detected. The eventual identification of the natures of the dominant forms of nonbaryonic matter will be a brilliant discovery that will advance thinking in fundamental physics and cosmology. But since the growth of cosmology so far has not been seriously influenced by the searches for dark matter detection, this review is confined to starting ideas.

Following the Ipser and Sikivie (1983) proposal that the CDM might be axions, Sikivie (1983) explored the idea of a laboratory detection of a sea of axion dark matter by its coupling to the electromagnetic field. In an early discussion of the idea that the dark matter may be a remnant from supersymmetry, Cabibbo Farrar, and Maiani (1981, 155) declared that "Light photinos could provide the missing mass in the galaxy and give rise to an observable UV background."

In the first discussion of the astrophysics of WIMPs, the particle physicists' 3-GeV neutrino, Gunn et al. (1978, 1030) argued that

> The production rate of γ-rays via annihilation in structures like galactic halos now is interestingly high, even though the annihilation times are very long. Our picture of the annihilation processes is very crude, however; and until a better model is forthcoming, we regard the predictions as tantalizing: the prospect of detecting the "missing" matter through its fundamental properties is a very exciting one.

Stecker (1978); de Rújula and Glashow (1980); Sciama (1984); and Silk and Srednicki (1984) developed thoughts about detection of photons or antiprotons from dark matter radiative decay or annihilation. Ideas about the role of decaying dark matter in cosmic structure formation are considered in Section 8.4.2.

If WIMPs exist and interact with baryons, perhaps with the strength of the known neutrinos, they could be trapped by scattering in stars. Gunn et al. (1978, 1027) pointed out that this could have observable effects. It

> can change both radiative transport and luminosity estimates of stars, and also influence the nucleosynthetic stellar evolution. The points here are that part of the radiation pressure necessary to hold a star up against its gravity could then arise in part from the neutrino annihilation photons, thereby relieving the burden on the core of the star to produce, by conventional nucleosynthesis, the major part of the photon budget. This, in turn, means that the core can be cooler than under conventional scenarios of element production, which in turn means that reaction rates proceed more slowly.

Steigman et al. (1978, 1051) argued that in stars,

> the effect of even a small contribution of heavy neutrinos would be enormous. These particles would provide alternative sources of gravitation, energy transport, and luminosity, and could drastically modify stellar structure... the predicted solar neutrino flux (normal, massless neutrinos) measureable by current detectors would be substantially reduced in agreement with the observations.

Krauss et al. (1985) and others pursued this thought about how to reconcile the theory and measurements of the solar neutrino flux. But the resolution of the smaller-than-predicted rate of solar neutrino detections was later seen to be that the known neutrinos have rest masses, likely well below 1 eV, that allow oscillation among flavor states of solar neutrinos.

Silk, Olive, and Srednicki (1985) and Krauss, Srednicki, and Wilczek (1986) introduced the idea that CDM particles trapped by scattering in the Sun or Earth's interior, if sufficiently concentrated, may produce a detectable flux of annihilation neutrinos. Annihilation photons from CDM gravitationally concentrated in the dense central parts of galaxies may be observable. And effects of CDM on the structures of planets and stars must be subtle, because none has been noticed, but the thought certainly is interesting.

If the subluminal matter were WIMPs (massive neutrinos that have close to the standard weak interaction), they would be scattered by atomic nuclei through the weak neutral current. This would mean that dark matter in the galaxy would on occasion scatter off atomic nuclei in baryonic material in the laboratory, which might be detected by the effects of the momentum and energy deposited by the recoil. Drukier and Stodolsky (1984) discussed how this might be used to detect the known electron-type neutrinos produced by thermonuclear reactions in the Sun. They took note of the much greater difficulty of detecting subluminal matter, if HDM, but did not discuss the more favorable possibility of detecting WIMPs with their greater rest mass. Goodman and Witten (1985); Cabrera, Krauss, and Wilczek (1985); and Wasserman (1986) took this step. Drukier, Freese, and Spergel (1986, 3495) added the consideration that "Earth's motion around the Sun can produce a significant annual modulation in the signal," a potentially important signature.

Early results from the laboratory search for direct detection of effects of scattering of dark matter particles were reported in Ahlen et al. (1987) and Caldwell et al. (1988). The great progress in the technology of detection of nonbaryonic matter in the laboratory, and detection of what nonbaryonic matter may be doing in astronomical objects, is a heroic enterprise that calls for its own history. As the time of writing, the dreams of nonbaryonic dark matter detection, or of phenomena that might point us to more interesting physics in the dark sector, remain unrequited. That does not argue against the postulate of nonbaryonic dark matter, of course, or against Wilkins Micawber's philosophy: Perhaps something will turn up.

CHAPTER EIGHT

The Age of Abundance of Cosmological Models

THINKING IN THE cosmology community in the early 1980s and later was informed by two old ideas, that our universe is homogeneous on average and is usefully described by general relativity, and by two new ones. The first of these is the cosmological inflation picture discussed in Section 3.5.2. Some took the picture to be too elegant to be wrong, and in this sense, it might be compared to the community faith in general relativity, but with the difference that inflation is a framework in which one may place a considerable variety of theories. The second is the new idea of nonbaryonic CDM discussed in Chapter 7, which opened the possibility of more interesting cosmologies beginning with the CDM model. The motivation for this model, and the list of what it assumes, are discussed in Sections 8.1 and 8.2.

The CDM cosmological model owed its initial popularity to simplicity, which allowed the analytic and numerical explorations of cosmic structure formation reviewed in Sections 8.2 and 8.3. But the model is flawed by its awkwardly large mass density. That could be adjusted, of course: maybe the nonbaryonic matter is warm, or decaying, or self-interacting, or something completely different; maybe the initial conditions suggested by inflation are to be adjusted; or maybe Einstein's cosmological constant should be reconsidered. This is the subject of Section 8.4.

My thinking about cosmological models in the early 1980s was influenced by results from the search for departures from an exactly homogeneous sea of cosmic microwave radiation. Two groups—Fabbri et al. (1980) and Boughn, Cheng, and Wilkinson (1981)—announced evidence of detection of anisotropy at $\delta T/T \sim 1 \times 10^{-4}$ on large angular scales. I took this seriously—the authors of the second paper were colleagues—and crafted a model that fit the measurements by application of the Sachs-Wolfe relation (to be discussed below) to an ansatz about initial conditions (Peebles 1981b). I think this model is reasonably simple, even elegant, but the details are irrelevant, because Boughn et al. withdrew the detection, and Fixsen, Cheng, and Wilkinson (1983) placed a new, tighter bound on the anisotropy. I knew about the tighter bound when I introduced the CDM model before publication of the Fixsen, Cheng, and Wilkinson paper. Again, they were colleagues. My paper, Peebles (1982b), on the CDM cosmological model, was meant to save the gravitational instability picture from the tighter anisotropy bound.

8.1 Why Is the CMB So Smooth?

In the 1970s, improving upper bounds on the CMB anisotropy were revealing the striking difference between the distinctly clumpy distribution of matter in galaxies and the much smoother distribution of the sea of microwave radiation. I recall informal discussions of the possibility that this situation may falsify the gravitational instability picture for cosmic structure formation. There were declarations of this thought in print. The analysis of the effect of the gravitational growth of the clustering of mass on the CMB by Silk and Wilson (1981) led them to conclude that "already we believe it possible to assert that adiabatic fluctuations in the standard model are untenable for any combination of n and Ω_0." This standard model is a universe of radiation and baryons with mass density parameter Ω_0 (Ω_m in equation (3.13)) and the initial condition of primeval adiabatic mass fluctuations with a power-law power spectrum with spectral index n (in equations (5.49) and (8.7)).

But a new thought was discussed on page 232 in Section 5.3.6. In the HDM picture, the neutrinos would freely slip through the radiation, minimizing the disturbance to the radiation needed for the gravitational assembly of galaxies and their clustering. Doroshkevich et al. (1981, 37) put it that

> The fluctuations of neutrinos that have a mass larger than M_ν have an uninterrupted growth. Due to their coupling to photons, the fluctuations of baryons smaller than the horizon can start growing only after recombination. Starting from the same initial amplitude at the moment of recombination, the neutrinos have much larger fluctuation amplitudes than the baryons and photons of the background radiation. At the recombination, the baryon Jeans mass drops very quickly to the value $10^5 M_\odot$. The growth of baryon fluctuations is accelerated by the large inhomogeneities formed in the neutrino density until the same amplitude is reached.... This process thus provides large fluctuations in baryons and neutrinos and small $\delta T/T$ in the photon background.

Here M_ν is the mass in equation (7.9) of the first generation of mass fluctuations to break away from the expansion of the universe, set by the free streaming of neutrinos in the HDM picture. The resulting top-down structure formation is seriously problematic, as discussed in Sections 5.2.7 and 7.1.1.

But the broader point is that nonbaryonic dark matter would slip through the radiation.

I introduced the CDM cosmological model in Peebles (1982b) in response to the Fixsen, Cheng, and Wilkinson (1983) improvement in the bound on the CMB anisotropy. This model replaces HDM with the nonbaryonic CDM reviewed in Section 7.2. It was meant to be a counterexample to the apparent challenge to the gravitational instability picture from the tight upper bound on the CMB anisotropy. I was not aware of it at the time, but as just noted, the starting idea had already been introduced: The CMB would be minimally disturbed by the growing clustering of nonbaryonic dark matter, whether HDM or CDM, leaving the more modest disturbances to the CMB caused by gravity and the smaller mass density in baryons.

8.2 The Counterexample: CDM

A counterexample to save the phenomenon is best kept simple. Thus I used the Einstein–de Sitter parameters $\Lambda = 0$ and $\Omega_m = 1$, even though the measured galaxy relative velocity dispersion in the CfA redshift survey had already convinced me that the mean mass density likely is lower than this. (This is entered in row 13 in Table 3.2. It and other considerations that in 1982 convinced me the mass density likely is less than Einstein–de Sitter were published the following year, in Davis and Peebles 1983a,b.) For simplicity, I took the nonbaryonic DM to be initially cold. The massive neutrino—the WIMP—introduced in 1977, would do. The warm DM that Pagels and Primack (1982) and Blumenthal, Pagels, and Primack (1982) had introduced would have done equally well, but cold is simpler. The model assumes primeval adiabatic Gaussian scale-invariant initial conditions. This follows from a simple interpretation of the recently introduced concept of cosmological inflation. But again, I was more influenced by the argument from simplicity for these initial conditions independently suggested by Harrison (1970) and Peebles and Yu (1970), and developed by Zel'dovich (1972). My computation of the mass-fluctuation power spectrum took account of radiation and DM, but it ignored the modest mass in baryons. That made the computation easier and again simplified the model by eliminating a parameter: $\Omega_{\rm baryon}$ need only be much less than unity.[1]

1. The baryons are needed for the acoustic oscillations, illustrated in Figure 5.2, that are important in later cosmological tests, but not for the CMB anisotropy on large angular scales that worried us in the early 1980s. It might be noted that Peebles (1981a) computed the matter power spectrum for adiabatic initial conditions in linear perturbation theory, taking account of radiation and baryons (Figure 5.3); Peebles (1982a,b) computed the spectrum for radiation and CDM, ignoring baryons and massless neutrinos; and Blumenthal et al. (1984) and Vittorio and Silk (1984) reported computations for the full case of radiation, CDM, baryons, and massless neutrinos. Bond and Efstathiou (1987)

The computation of the predicted anisotropy of the microwave background radiation and the spatial distribution of the matter in this CDM model must follow evolution from conditions at high redshift, when baryons, CDM, neutrinos, and radiation are assumed to have the same space distribution, through the early stage of evolution in which baryons and radiation behave as an approximation to a viscous fluid, then through decoupling as radiation starts to diffuse through the baryons and then breaks free as the plasma combines. Subsequently the radiation propagates through almost purely neutral baryons and the slightly dented spacetime, while the baryonic matter joins the growing clustering of the cold dark matter.

These are a lot details to consider. But the important point is that there is not the complexity of seriously nonlinear processes, such as turbulence or star formation, which are impossible to analyze from first principles. In the CDM model and its variants, the departures from homogeneity on large scales are small and computable in perturbation theory, and the predicted evolution on the smaller scales of clusters of galaxies can be modeled to reasonable accuracy by numerical simulations. This, along with serious advances in the observations, made possible demanding cosmological tests.

The form for the power spectrum of the mass distribution well after decoupling, introduced in Peebles (1982b), is

$$P(k) = \frac{Ak}{(1+\alpha k+\beta k^2)^2}, \quad \alpha = 6h^{-2}\ \mathrm{Mpc}, \ \beta = 2.65h^{-4}\ \mathrm{Mpc}^2. \tag{8.1}$$

It has the primeval scale-invariant shape $P \propto k$ on large length scales, small k. It swings to $P \propto k^{-3}$ on smaller scales, larger k. The origin of this small-scale behavior follows by the argument in footnote 13 on page 231. The computation, adapted from Peebles (1982a), takes account of the tight coupling of radiation and plasma as the mode wavelengths start to oscillate, but it does not take into account the effect of the mode oscillations, an approximation that was good enough for the immediate need. Successive improvements to equation (8.1), taking account of the mass in the baryons and the effect of massless neutrinos, are in Bond and Efstathiou (1984); Davis et al. (1985); Bardeen et al. (1986); and Efstathiou, Bond, and White (1992).

Computing the predicted large-scale CMB anisotropy in this CDM model requires the normalization A in equation (8.1). I used results from the program of measurements of the galaxy two-point correlation function summarized in Peebles (1980). This statistic is usefully summarized as the standard deviation (root-mean-square value) of the count of galaxies in randomly placed spheres of a given size. The convenient choice of the sphere radius was $8h^{-1}$ Mpc, about the smallest at which the growth of the fractional fluctuations

---

demonstrated the baryon acoustic oscillations caused by the interaction of plasma and radiation in the CDM model.

in the mass density $\delta M/M$ might be reasonably well approximated by linear perturbation theory, and about the largest for which the measurements of the two-point function we had then could be trusted. The normalization in Peebles (1982b) assumes galaxies trace mass on this scale:

$$\sigma_8(\text{mass}) = \frac{\delta M}{M} \approx \sigma_8(\text{galaxies}) = \frac{\delta N}{N} \approx 1 \text{ at } R = 8h^{-1} \text{ Mpc}. \tag{8.2}$$

The numerical value for the radius is rounded from equation (2.18). Later discussions that take account of the idea that $\sigma_8(\text{mass})$ may be significantly less than $\sigma_8(\text{galaxies})$ are discussed in Section 8.4.

Sachs and Wolfe (1967) derived the wanted relation between the large-scale mass distribution and the CMB anisotropy in an Einstein–de Sitter universe with adiabatic initial conditions. The form of the Sachs-Wolfe relation I used, from Peebles (1980, equation (93.25)), relates the angular distribution of the CMB to the Fourier transform of the mass distribution.[2] In the linear perturbation theory used in this computation, the second moments of the mass determine the second moments of the CMB. The latter are usefully expressed in terms of the spherical harmonic expansion of the CMB temperature as a function of angular position across the sky.[3] The result in the 1982 CDM model is[4]

$$\frac{T(\theta,\phi)}{\langle T\rangle} = 1 + \sum a_l^m Y_l^m(\theta,\phi), \quad \langle |a_l^m|^2\rangle^{1/2} = 3.5\times 10^{-6}\left(\frac{6}{l(l+1)}\right)^{1/2}. \tag{8.3}$$

The CMB temperature anisotropy spectrum $\langle |a_l^m|^2\rangle$ in this equation is the Sachs-Wolfe gravitational effect on the angular distribution of the CMB by primeval adiabatic departures from homogeneity in the CDM model with primeval power spectrum $P_k \propto k$ normalized to the present universe by

2. I take this opportunity to apologize for not thinking to transfer the reference to Sachs and Wolfe (1967) in Peebles (1980) to Peebles (1982b).

3. The use of spherical harmonics to represent the angular distributions across the sky of galaxies and clusters of galaxies was introduced in Yu and Peebles (1969). Peebles (1982b) introduced its application beyond dipole to the variation of the CMB temperature across the sky. Note here that at $l > 0$, the real and imaginary parts of the spherical harmonic Y_l^m have zeros spaced at minimum separation π/l radians in the azimuthal and polar directions. The problem of converging lines of zeros toward the poles is solved, because Y_l^m is close to zero at polar angle $\theta \lesssim m/l$. Note also that the ensemble average value $\langle |a_l^m|^2\rangle$ for a statistically isotropic process is independent of m.

4. Turner, Wilczek, and Zee (1983) independently set down elements of the CDM model, but their inadequate estimate of the predicted CMB anisotropy led them to conclude that the model could not accommodate scale-invariant primeval conditions. Abbott and Wise (1984) independently found that the CMB anisotropy spectrum scales as $\langle |a_l^m|^2\rangle \propto [l(l+1)]^{-1}$ for scale-invariant initial conditions, by a method different from but physically equivalent to Peebles (1982b). Abbott and Wise did not consider the normalization. Bond and Efstathiou (1987) wrote down the scaling of the angular power spectrum $\langle |a_l^m|^2\rangle$ with degree l for the generalization of the primeval power spectrum to $P_k \propto k^n$.

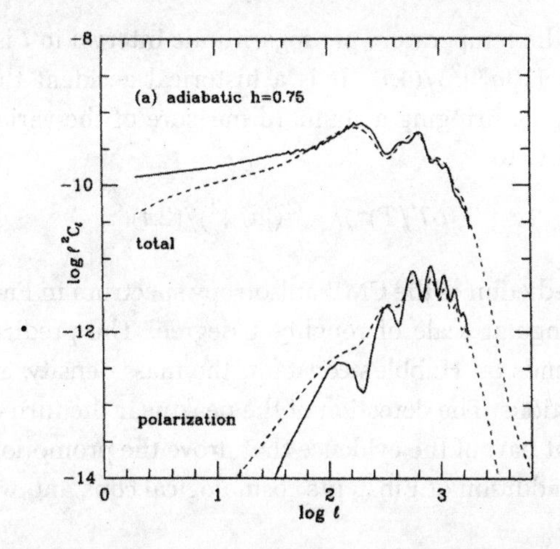

FIGURE 8.1. The Bond and Efstathiou (1987) computation of the CMB temperature and polarization angular power spectrum in the CDM model. By permission of Oxford University Press on behalf of the Royal Astronomical Society.

the assumption that galaxies trace mass. It is independent of the Hubble parameter h. And since the CMB anisotropy in this model is well below the upper bounds on anisotropy we had at the time, it was sufficient for a counterexample.

The quadrupole $l = 2$ anisotropy in this first computation can be compared to the Bennett et al. (2003) first-year quadrupole measurement,

$$\begin{aligned} \langle |a_2^m|^2 \rangle^{1/2} &= 3.5 \times 10^{-6} \quad \text{Peebles (1982}b\text{)}, \\ &= 5 \pm 1 \times 10^{-6} \quad \text{Spergel et al. (2003)}, \end{aligned} \tag{8.4}$$

which is reasonably close.

Computation of the CMB anisotropy on smaller angular angular scales must take account of the effect of the acoustic oscillations of the plasma-radiation fluid prior to decoupling. That was discussed in more primitive notation (and neglecting the nonbaryonic dark matter that had not yet been invented) in Peebles and Yu (1970). Bond and Efstathiou (1987) computed the CDM model prediction of the CMB angular power spectrum $\langle |a_l^m|^2 \rangle^{1/2}$ as a function of the spherical harmonic degree to larger l, beyond the quadrupole. Their result is shown in Figure 8.1. The dashed curves in the figure are from a convenient approximation to the computation of the solid curves.

The measure of the CMB anisotropy on the vertical axis of Figure 8.1, $l^2 C_l = l^2 \langle |a_l^m|^2 \rangle$, is motivated by the following consideration. Since $\langle |a_l^m|^2 \rangle$ is independent of m, the mean square value of the departure of the CDM temperature from isotropy is, in the spherical harmonic expansion in equation (8.3), with the convention that the integral of $|Y_l^m|^2$ over the sphere is unity,

$$\langle (\delta T/T)^2 \rangle = \sum_l (2l+1) \langle |a_l^m|^2 \rangle / (4\pi). \tag{8.5}$$

The variance of the temperature per logarithmic interval in l is well approximated by $l(2l+1)\langle|a_l^m|^2\rangle/(4\pi)$. It is a historical accident that $2l+1$ was replaced by $2(l+1)$, bringing a standard measure of the variance per logarithmic interval of l to

$$\langle(\delta T/T)^2\rangle_l = l^2\langle|a_l^m|^2\rangle/(2\pi). \tag{8.6}$$

The CMB prediction of the CMB anisotropy spectrum in Figure 8.1 peaks at $l \sim 200$, an angular scale of roughly 1 degree. The predicted value of l at the peak depends on Hubble's constant, the mass density, and the curvature of space sections. The detection of the peak near the turn of the century was an important part of the evidence that drove the promotion of the CDM model, with the addition of Einstein's cosmological constant, to the standard cosmology.

In the 1980s, it was not at all clear that the CDM model is a useful approximation,[5] or if so, that the angular power spectrum is the best way to compare the CMB anisotropy theory and measurements. Thus Bond and Efstathiou (1987) and Efstathiou and Bond (1987) considered isocurvature as well as adiabatic primeval conditions, and they computed the CMB two-point correlation functions as well as power spectra.[6]

When I introduced the CDM model in Peebles (1982b), I considered it an example of what might have happened, likely to be one of many to be explored. I did not imagine that it might so readily grow into a convincing picture of the early universe. But I might have expected that the very simplicity of the CDM model would draw interest. That was aided by a common interpretation of the inflation picture (Section 3.5.2), which would have the flat space sections and initially adiabatic, Gaussian, and scale-invariant initial conditions assumed in the CDM model. But I was surprised at how seriously the model was taken, and was uneasy about it, because I saw no reason to be confident that nature

5. Simpson and Hime (1989) found indications of detection of a new neutrino species with mass $m_\nu = 17$ keV that mixes in the electron-type interaction. It led Bond and Efstathiou (1991) and others to consider how this neutrino would affect the predicted distributions of matter and radiation. Bond and Efstathiou found tentative advantages of adding to the cosmological model this neutrino with lifetime postulated to be about a year, but interest in this possibility for a slightly more complicated dark sector faded.

6. The correlation function and angular power spectrum are mathematically equivalent, but their utility in practice may be quite different. I learned this from the book by Blackman and Tukey (1959), *The Measurement of Power Spectra from the Point of View of Communications Engineering*. Blackman and Tukey showed that power spectra are more informative than correlation functions for many applications. That led me to use power spectra based on spherical harmonic expansion rather than Fourier expansion in our first analyses of the spatial distributions of extragalactic objects. But N-point correlation functions usually prove to be the more useful statistics in this situation. The angular distribution of the CMB is best characterized by power spectra and higher moments, as Blackman and Tukey recommended, in spherical harmonic expansion.

shared our ideas of simplicity. It had not taken me long to think up this model and compute the CMB anisotropy in equation (8.4), and I could see how to set up other models, maybe not quite as simple, that could equally well fit the observational constraints (which were not all that tight then). I continued to invent such models until the late 1990s, when the CMB anisotropy measurements started to reveal the anisotropy peak predicted in the CDM model and illustrated in Figure 8.1. That went a long way toward persuading me that nature may have taken our simplest way, apart from the curious presence of Einstein's cosmological constant, Λ, and the hypothetical nonbaryonic dark matter. The many tests since then continue to agree with the CDM model with the addition of Λ. It is a remarkable advance, although at the time of writing, the natures of Λ and CDM remain unknown.

8.3 CDM and Structure Formation

A viable cosmology must offer a platform for an acceptable analysis of how the galaxies formed in all their rich phenomenology. This is a cosmological test, but one that is difficult to assess, because the galaxies formed by complex nonlinear processes, at the heart of which are the still very poorly characterized properties of star formation. It means that studies of galaxy formation must rely on prescriptions to be explored by analytic and numerical methods and adjusted to fit the observations.

Before the revolution in cosmology, some studies of cosmic structure formation tested cosmological models; the prime example is the analyses of clusters of galaxies that helped lead us to a low-density universe (summarized in categories B_1 and B_2 in Table 3.3). That test was persuasive, because the gravitational assembly and evolution of clusters seems to be simple enough to be modeled in numerical simulations sufficiently accurately to serve as a guide to the nature of the cosmology.

The more common tradition has been to explore how a picture for structure formation may be framed to fit a given cosmological model. An early example is the study of the gravitational origin of the rotation of galaxies (Section 5.2.3). Still earlier is the Eggen, Lynden-Bell and Sandage (1962) picture of the formation of our Milky Way galaxy (Section 5.2.3) that Partridge and Peebles (1967) placed in the context of the gravitational instability of the expanding big bang cosmological model, along the lines reviewed in Section 5.1.[7] Elliptical galaxies contain little gas and few young stars, which might suggest they formed

7. The Partridge and Peebles estimate of the spectrum one might expect of a young galaxy included the prominent atomic hydrogen Ly-α resonance emission line. We proposed that this line could be a good marker for the search for distant young galaxies, seen as they were in the past because of the light-travel time, and at redshifts large enough to bring the Ly-α line into the optical. I don't remember worrying that these resonance photons have a large cross section for scattering by hydrogen atoms, so they may be absorbed by dust as

by the gravitational gathering of matter that had already been largely converted to stars. Larson (1969) and Gott (1975) placed exploration of this mode of formation of an elliptical galaxy in the context of an expanding universe.

Galaxies are much more complicated than clusters of galaxies. Gas and plasma in a newly forming galaxy can be expected to gain internal energy from shocks and turbulence driven by gravitational collapse along with radiation and winds from massive young stars, lose energy by thermal bremsstrahlung emission by plasma and by radiative decay of collisionally excited atoms, and be rearranged by winds and explosions. This would happen in the messy early concentrations of mass that gravity may have gathered. In a series of papers, Richard Larson pioneered exploration of the budget for loss and gain of energy in galaxy formation and evolution (e.g., Larson 1969, 1976 and 1983). Spitzer (1956) offered the thought that the collapse of a protogalaxy might have left thermally supported coronae of plasma in the outer parts at densities low enough that the energy dissipation times exceed the Hubble expansion time. But plasma at the mean baryon mass density at our position in the Milky Way, $n \sim 1$ proton cm^{-3}, with the plasma temperature $T \sim 10^7$ K that can be gravitationally confined by the mass of the galaxy, has a cooling time $\sim 10^7$ years, shorter than the collapse time $\sim 10^8$ years. Thus the radiative loss of energy by plasma in a young galaxy can be quite significant. Binney (1977), Rees and Ostriker (1977), and Silk (1977) introduced considerations of how the relative rates of cooling and free gravitational collapse in a protogalaxy can determine the nature of its evolution.

Gunn et al. (1978) and White and Rees (1978) introduced considerations of the role that a subluminal massive halo might play in the formation of the luminous parts of a galaxy (as discussed on page 218). The CDM cosmological model introduced in Peebles (1982b) offered a more definite basis for exploration of how gravity might grow concentrations of nonbaryonic matter, in which the baryons might dissipatively settle to form a galaxy within a massive dark matter halo. Early explorations are in Peebles (1984a), and in greater detail, Blumenthal et al. (1984) and Bardeen et al. (1986). The Blumenthal et al. considerations of how the CDM model can accommodate the rich phenomenology of cosmic structure on the broad range of scales from galaxies to superclusters of galaxies continue to be widely cited.

In the 1970s, applications of statistical measures of the space distribution and motions of the galaxies offered the prospect of a cosmological test: compare the observed measures of the galaxy spatial distribution to what would be expected from the gravitational build-up of mass clustering in an expanding universe. Ed Groth and I spent a lot of time exploring numerical N-body simulations of the evolution of the mass distribution in an expanding model universe. We meant to compare the mass correlation functions in the

---

the photons diffuse around a newly forming galaxy with a halo of atomic hydrogen. But the Ly-α line proves to be a useful marker for young galaxies.

simulations to the measurements we were making of the galaxy low-order position and velocity correlation functions, as discussed in Davis, Groth, and Peebles (1977). With scale-invariant initial conditions in the scale-invariant Einstein–de Sitter model, the mass distribution in pure gravity N-body simulations will relax to a scaling solution for the mass correlation functions, after enough time has elapsed to allow transients to die down. We could not find this scaling behavior, and what we had did not seem to be close to convergence to scaling. Our few publications include a conference proceeding early on (Peebles 1973b) and a conference abstract (Groth and Peebles 1975) presented at about the time we were giving up.

Others took up the challenge of simulating the evolution of the distributions of mass and galaxies in an expanding universe; examples include Press and Schechter (1974); Doroshkevich and Shandarin (1976); Aarseth, Gott, and Turner (1979); Efstathiou, Fall, and Hogan (1979); Miller and Smith (1981); Centrella and Melott (1983); Melott et al. (1983); Miller (1983); Kauffmann and White (1992); and Governato et al. (2010). The results from a lot of work show that the parameters meant to describe what the baryons are doing can be adjusted to fit the observations, including the properties of the galaxies and their space distribution and motions, in impressive detail. The ΛCDM cosmology passes this test. But the complexity makes it difficult to assess the weight of this case from the study of cosmic structure formation.

Davis et al. (1985), in a celebrated paper known as DEFW, presented early numerical simulations of the evolution of structure in the CDM model (before it became the ΛCDM cosmology). This is along the lines of Groth and Peebles (1975) but with the far more capable numerical computers of a decade later. It was attractive to turn to the CDM model and away from simulations of structure formation in a universe of baryons, because baryons behave in the complex ways Richard Larson had been exploring. In the CDM model, the dominant mass interacts only with gravity, so one might hope to arrive at a useful first approximation to cosmic evolution by ignoring the complications of the baryons. It is easy to set up numerical simulations of the evolution of the distribution of a gas of particles that move under the influence of gravity alone. Accurate computation of the gravitational accelerations given the particle positions is time-consuming, but one can work on that. And simulations of the CDM model are convenient, because initial conditions that are close to scale invariant grow into a first generation of nonlinear structures that form across a broad range of mass scales at about the same time, reducing the span of cosmic time called for in simulations.[8] Adding to all this was the motivation

8. Recall the discussion in footnote 13 in Section 5.2.6: Under scale-invariant initial conditions, the mass fluctuation power spectrum on the scale of galaxies approaches $P(k) \propto k^{-3}$ prior to the onset of formation of nonlinear mass concentrations, meaning the amplitude of the mass fluctuations scales only as the logarithm of the length scale.

from the community interest in the CDM model on the basis of simplicity, inflation, and maybe promise.

DEFW introduced a change of thinking: identify galaxies with the mass concentrations in the simulation, and compare the galaxy space distribution to the distribution of these mass density peaks, not to the measures of the mass distribution. Carlos Frenk (in a personal communication, 2018) recalls that

> We started working on what became DEFW towards the end of 1982 when the department in Berkeley had just bought a then amazingly powerful new computer, a VAX 780.... In the first part of the DEFW project we assumed that the distribution of galaxies traced the distribution of mass and we were upset to find that, with this assumption, an open model with $\Omega_m \sim 0.2$ gave an acceptable match to the CfA survey while the much more appealing flat $\Omega_m = 1$ model did not. The problem with the latter, as you know, was that matching the observed galaxy clustering amplitude required a very large amplitude of mass fluctuations and thus rms pairwise peculiar velocities much larger than those that you and Marc [Davis] had measured in the CfA redshift survey. As consolation we ran a flat model with a cosmological constant which looked as good as the open model from the point of view of clustering and peculiar velocities and at least had the virtue of having a flat geometry, even if at the expense of the then highly unattractive inclusion of Λ. In 1984, with the simulations in hand and seeing what Nick [Kaiser] had come up with to explain the large clustering amplitude of galaxy clusters, we figured out that if we extended his ideas to galaxies and deployed the same "high-peak" trick to them, then the required mass fluctuations would be reduced by b^2 and we could then reconcile the $\Omega_m = 1$ CDM model with the CfA galaxy two-point correlation function and the measured rms velocities. This is how the idea of "biased galaxy formation" came about but we soon convinced ourselves that, regardless of the desire to have $\Omega_m = 1$, the idea that galaxies formed in high peaks made physical sense and would be relevant for any value of Ω_m.

The early DEFW numerical simulations were pure dark matter, and the peaks were identified in the initial conditions. A satisfactory study of how the galaxies formed must take account of the complex behavior of the baryons and identify the model galaxies as they formed. But DEFW set the direction of thinking.

The long history of ideas about how the galaxies formed includes the early debates reviewed in Chapter 5 on whether gravity can assemble the mass concentration of a galaxy in an expanding universe, and if so, whether gravity can cause the galaxy to rotate. The behavior of a concentration of baryons with the dimensions of a galaxy is complex. The idea that most of the mass of a galaxy is CDM allowed an easy entry to the problem: Ignore the baryons at first, and then introduce analyses of the behavior of the baryons in increasing detail guided by what is found to work. It is difficult to argue that the results added much weight to the cosmological tests, but they did help keep the community interested in the CDM model and its variants in the 1990s.

8.4 Variations on the Theme

The CDM cosmology in Peebles (1982b) assumes the Einstein–de Sitter model with $\Lambda = 0$ and $\Omega_m = 1$. That was for simplicity, and despite two reservations. First, the evidence seemed to me to be that the mass density is less than Einstein–de Sitter. Second, I was not at all confident that simplicity is a reliable guide to a better cosmology. I was right about the first, wrong about the second. But arriving at this second conclusion required sifting through the ideas to be reviewed this section. For this purpose, let us follow the standard practice of naming the 1982 model sCDM, for standard, though it never was that.

Warm dark matter, WDM, was introduced in the same year as sCDM, by Pagels and Primack (1982) and Blumenthal, Pagels, and Primack (1982). They suggested that a characteristic mass set by WDM streaming might account for the masses of large galaxies. The idea has not had much effect so far, but the idea, with a smaller characteristic mass, continues to be discussed. Variants of what might be termed sWDM could have been considered along with the variants of sCDM to be discussed here, but I have found little evidence that this has happened.

Recall the growing list of measures summarized in Table 3.2 and Figure 3.5 that indicate the cosmic mean mass density is about a third of the Einstein–de Sitter value. The evidence includes dynamical measures of the mass clustered with galaxies on scales from ~ 0.3 Mpc to ~ 10 Mpc; the cluster mass function, evolution, spatial correlation function, and baryon mass fraction; and the positive spatial correlation of galaxy positions extending to ~ 50 Mpc (this last assuming initial conditions are adiabatic and scale-invariant, as in the sCDM model). But this low mass density was counter to the common opinion in the years around 1990 that the mass density surely is Einstein–de Sitter: $\Omega_m = 1$. That would have to mean that these mass density measurements are all biased low (Section 3.5.3). Was it reasonable to have assumed that biasing had such a systematically similar effect on estimates of Ω_m on scales ranging from ~ 0.3 Mpc to ~ 30 Mpc? Or was it more reasonable to take it that the mass density likely is only a third of the sCDM model?

The issue grew more interesting with the Smoot et al. (1992) announcement of detection of the CMB anisotropy by the differential microwave radiometers (DMR) experiment on the NASA Cosmic Background Explorer satellite, COBE. The background radiation temperature anisotropy in the first-year data is $\delta T/T = 1.1 \times 10^{-6}$ measured on an angular scale $\theta \sim 10°$. The detection of such a small departure from isotropy is impressive, though later measurements have done even better. The detection was exciting, because the degree of anisotopy is about what one would expect in the sCDM cosmology and its variants. It was important, because it is a measure of the departure from a homogeneous mass distribution (depending on the initial conditions, of course, usually assumed to be adiabatic). And it was a new constraint on cosmological models.

Bunn and White (1997) reported that the sCDM model normalized to the COBE anisotropy in the 4-year CMB measurements (Bennett et al. 1996) requires the mass fluctuation amplitude (defined in equation (8.2)) to be the equivalent of $\sigma_8 = 1.22$. But they pointed out that the Viana and Liddle (1996, Fig. 3) condition to fit the abundance of rich clusters of galaxies requires $\sigma_8 \simeq (0.6 \pm 0.1)\Omega_m{}^{-0.4}$. Since $\Omega_m = 1$ in sCDM, this is a considerable discrepancy that might be added to the other evidence that Ω_m is well below unity. But there were other ideas to consider.

8.4.1 TCDM

Bunn and White (1997, 20) concluded that the

> COBE-normalized "standard" CDM [that is, sCDM] . . . predicts significantly too much small-scale power and is therefore ruled out. However, any of several slight changes to the model can easily resolve this inconsistency. Perhaps the simplest solution is a slight tilt to the power spectrum. Inflationary models typically predict spectral indices slightly less than unity, and a value of n of 0.8 or even less is quite natural in such models.

This simple solution became known as the tilted cold dark matter or TCDM model. The only difference from sCDM is the primeval mass fluctuation power spectrum, tilted from scale-invariance to

$$P_k \propto k^n, \quad n < 1. \tag{8.7}$$

To see how this remedies some problems with sCDM, note that the angular resolution of the COBE measurement, $\theta \sim 10°$, subtends a comoving length scale of roughly 500 Mpc at high redshift, when expanded to the present epoch. Under adiabatic initial conditions, the CMB anisotropy translates to the departures from a homogeneous mass distribution on this scale. The departures from homogeneity on much smaller scales, about 10 Mpc, are relevant for the gravitational formation of clusters. The sCDM overprediction of fluctuation power on the scales of clusters thus is remedied by tilting the primeval power spectrum. As White, Efstathiou, and Frenk (1993) had anticipated, this resolves the problem with clusters. It does leave the evidence that the mean mass density is less than Einstein–de Sitter, from the array of measures summarized in Figure 3.5, but a common feeling at the time was that this might be dealt with separately.

The attention to inflation is worth noting. It is natural to expect a slowing of the expansion rate during inflation, which would tilt initial conditions from scale invariance in the direction wanted for the TCDM model. But the value of n is not predicted by inflation; its implementation can be chosen to make n larger or smaller than unity. This flexibility is illustrated by the titles of two papers: "Designing Density Fluctuation Spectra in Inflation" (Salopek, Bond, and Bardeen 1989), and "Arbitrariness of Inflationary Fluctuation Spectra" (Hodges and Blumenthal 1990). A broad variety of initial conditions is afforded by the choice of the potential $V(\phi)$ as a function of the scalar field ϕ in a single-field inflation model. Two scalar fields offer still greater variety. Thus one response to the failure of sCDM was to adjust the model for inflation, as in the Bunn and White tilt.

One can postulate the tilt in equation (8.7) or offer a model. If the expansion parameter during inflation is a power law in physical time, $a \propto t^p$, it produces the power law initial condition in equation (8.7) with $n = (p-3)/(p-1)$ (Lucchin and Matarrese 1985; Liddle, Lyth, and Sutherland 1992). In another approach, Freese, Frieman, and Olinto (1990) introduced a single-field model they termed "natural inflation," because their field potential is an arguably natural form in a grand unified model for particle physics. The primeval mass fluctuations in natural inflation are close to equation (8.7), but for some, the pedigree is better. Cen et al. (1992); and Adams et al. (1993) concluded that TCDM and the almost-equivalent natural inflation model look promising but require work, including significant galaxy position biasing to get around the dynamical evidence for $\Omega_m \sim 0.3$: perhaps a trace of HDM, perhaps a cosmological constant.

8.4.2 DDM AND MDM

In the decaying dark matter (DDM) model, the dark matter has been decaying, perhaps into other kinds of nonbaryonic matter, perhaps into something that interacts with baryonic matter in an observable way. Early examples are the Dicus, Kolb, and Teplitz (1977 and 1978) and Sato and Kobayashi (1977) discussions of massive neutrinos with possibly interesting radiative decay lifetimes.

Davis et al. (1981) proposed that the early universe may have contained both HDM, the thermally produced sea of neutrinos with rest mass of a few tens of electron volts discussed in Section 7.1, and WIMPS, another neutrino family with the much larger rest mass discussed in Section 7.2. In this model, the more massive neutrinos are assumed to have decayed after their gravity had served to amplify the growth of structure on the scale of galaxies, leaving the HDM to provide most of the present Einstein–de Sitter mass density. Free streaming may have left the HDM close enough to smooth on a scale of about

10 Mpc (as in the HDM cosmology in Section 7.1.1) that this mass component would not have been detected in dynamical probes on smaller scales. That is, the mass in and around galaxies could be small enough to account for their small relative velocity dispersion. The theme was pursued in early papers by Doroshkevich and Khlopov (1984); Fukugita and Yanagida (1984); Gelmini, Schramm, and Valle (1984); Hut and White (1984); and Olive, Seckel, and Vishniac (1985). Turner, Steigman, and Krauss (1984) proposed a variant: After the massive dark matter particles had gravitationally boosted structure formation, their decay might have produced a sea of relativistic nonbaryonic matter. The dominant mass density in this relativistic dark matter would be close to smooth across the Hubble length, leaving subdominant mass clustered in and around the galaxies. The galaxies would have the observed small relative velocities driven by their small contribution to the mean mass density. The effect would be roughly similar to that of a cosmological constant, except that the cosmic expansion time would be considerably shorter, a serious but at the time perhaps not fatal problem.

The MDM model (also known as mixed dark matter; cold plus hot dark matter, as in CHDM; and two-component dark matter) places the stable CDM largely in massive halos around the galaxies, while the greater mass density in the more smoothly distributed HDM brings the total to the Einstein–de Sitter value. This is the Davis et al. (1981) two-component scheme but without decay of any of the nonbaryonic matter. The model received considerable attention; early examples are Fang, Li, and Xiang (1984); Shafi and Stecker (1984); Umemura and Ikeuchi (1985); and Valdarnini and Bonometto (1985).

8.4.3 ΛCDM AND τCDM

Models with $\Omega_m = 1$ are challenged by the considerable variety of evidence summarized in Figure 3.5 that the mass density is well below unity, $\Omega_m \sim 0.3$. The ΛCDM model accepts this and postulates that Einstein's cosmological constant Λ serves to keep space sections flat. The move in this direction was foreshadowed by Gunn and Tinsley (1975), who marshaled empirical indications of "An Accelerating Universe." It was first proposed in the context of CDM by Peebles (1984b); Rees (1984); Turner, Steigman, and Krauss (1984); and Kofman and Starobinsky (1985). It is a simpler, more direct way than biasing to interpret the considerable variety of probes of the mass density discussed in Section 3.6.5 while preserving the flat space sections indicated by inflation (and, arguably, the elegance of this spacetime geometry).

The τCDM model for the power spectrum of the mass distribution is designed to allow examination of effects of adjustments of cosmological parameters in a variety of models, including ΛCDM. Efstathiou, Bond, and White (1992) present a generalization of equation (8.1) for the mass fluctuation power spectrum:

Plate I

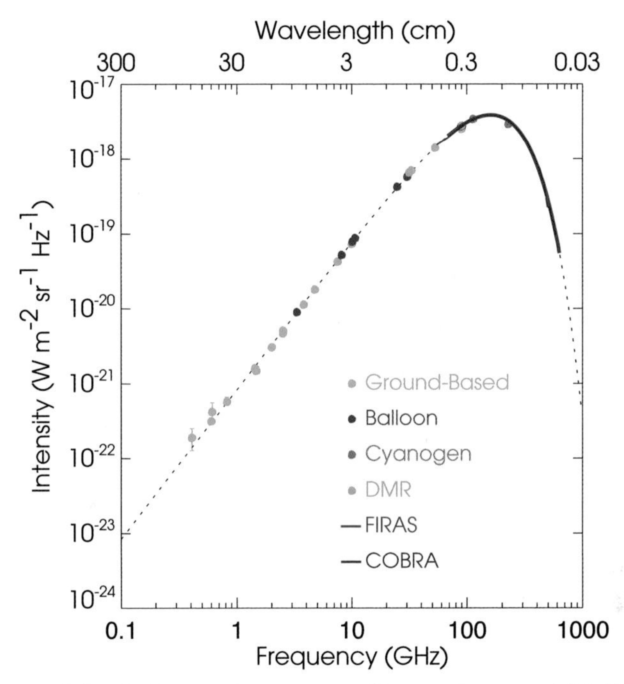

Measurements of the CMB intensity spectrum discussed on page 173 (Alan Kogut, Goddard Space Flight Center).

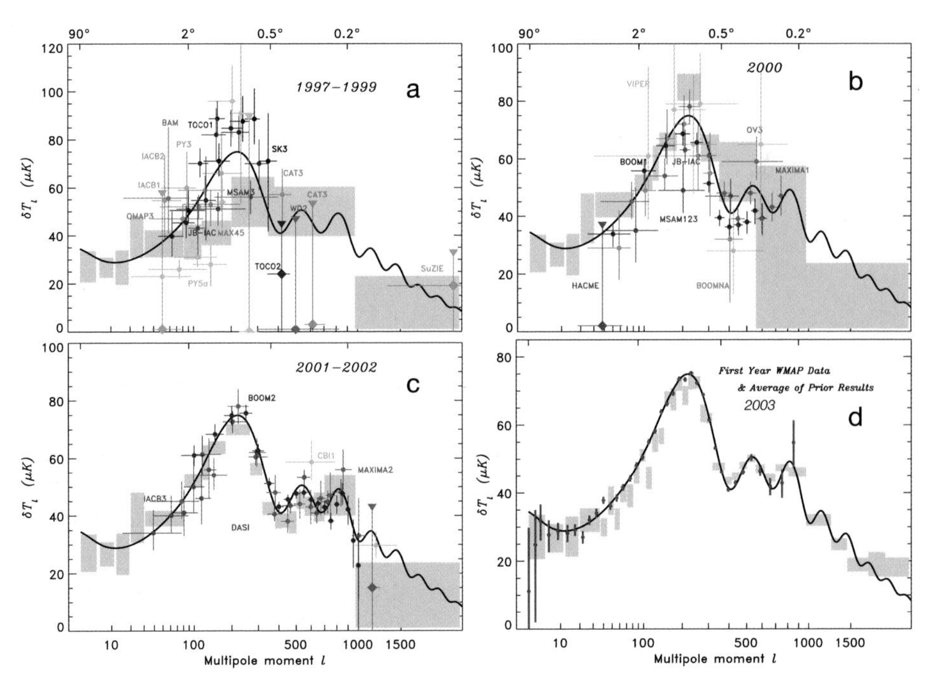

Progress in measuring the CMB temperature anisotropy power spectrum discussed on page 333 (Peebles, Page, and Partridge 2009). Reproduced with permission of Cambridge University Press through PLSclear.

Plate II

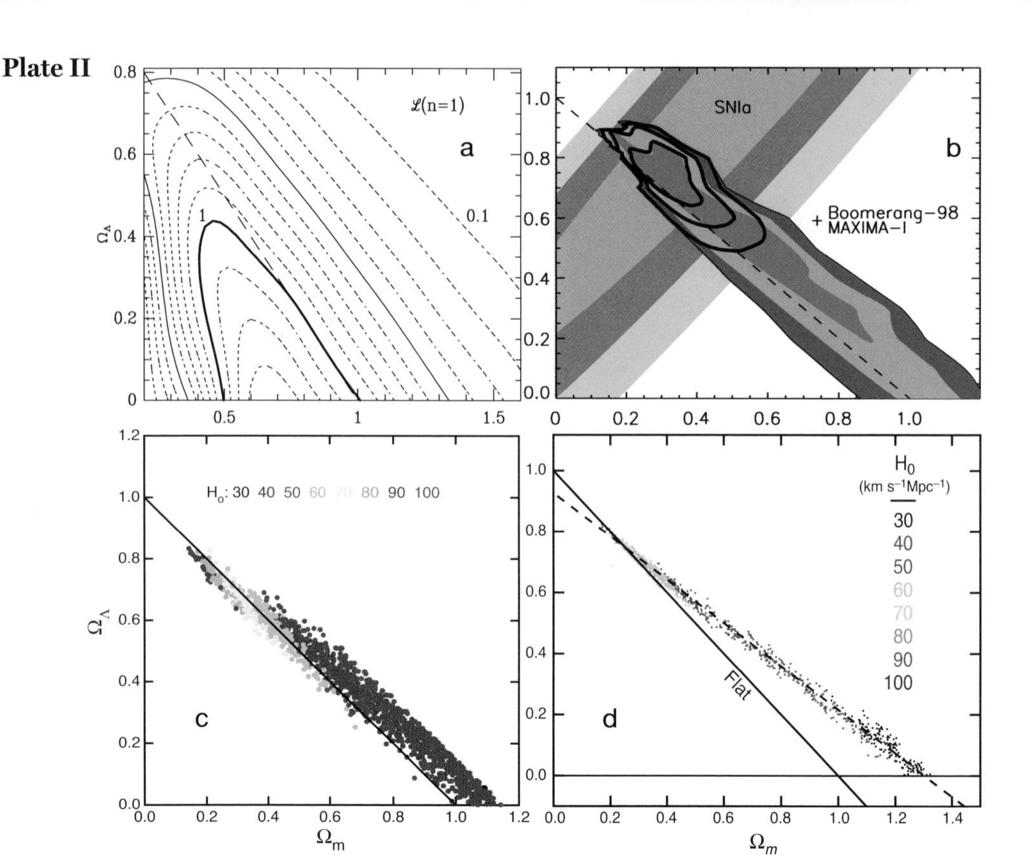

Progress in constraining parameters in the ΛCDM model by CMB anisotropy measurements (page 334). Panel (a) illustrates the situation in 1997 (Bunn and White 1997); (b) the situation in 2001 (Jaffe et al. 2001); (c) in 2002 pre-WMAP (Wright 2019); and (d) in 2007 (Spergel, Bean, Doré et al. 2007).

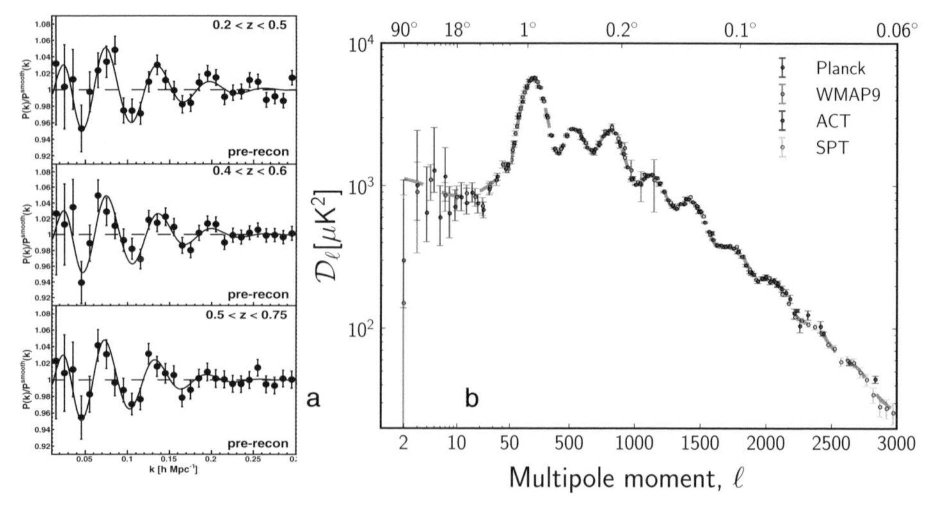

Signatures of primeval acoustic waves in the coupled plasma and radiation in the early universe (page 336). Panel (a) is the power spectrum of the spatial distribution of the galaxies, from Beutler et al. (2017). Panel (b) is the power spectrum of the angular distribution of the cosmic microwave background radiation, from the Planck Collaboration (2014b).

Plate III

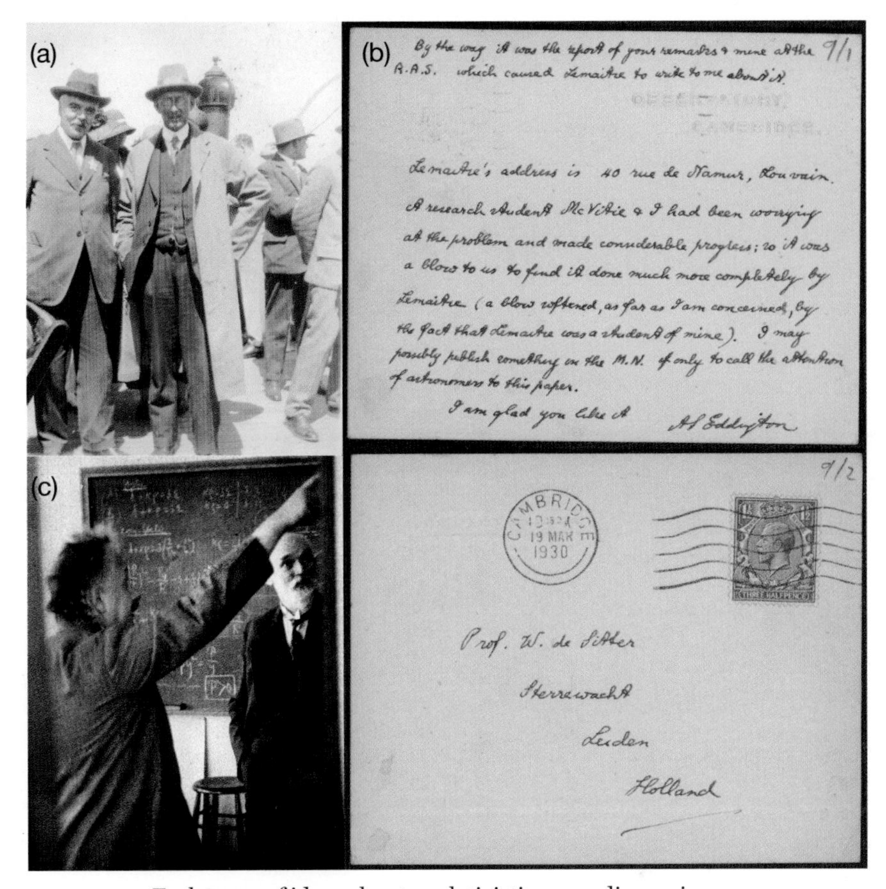

By the way it was the report of your remarks & mine at the R.A.S. which caused Lemaître to write to me about it.

Lemaître's address is 40 rue de Namur, Louvain.

A research student McVittie & I had been worrying at the problem and made considerable progress; so it was a blow to us to find it done much more completely by Lemaître (a blow softened, as far as I am concerned, by the fact that Lemaître was a student of mine). I may possibly publish something in the M.N. if only to call the attention of astronomers to this paper.

I am glad you like it

A S Eddington

9/1

9/2

CAMBRIDGE
19 MAR
1930

THREE HALFPENCE

Prof. W. de Sitter
Sterrewacht
Leiden
Holland

Exchanges of ideas about a relativistic expanding universe.

Frank Watson Dyson, an astronomer experienced in solar observations, and Eddington to the right in Panel (a), led observations of the solar eclipse of 1919 to measure displacements of images of stars as the Sun passed near them in the sky. The results favored the relativistic prediction, and were widely celebrated, but the judgment in retrospect is that the measurements are not very secure and added little to the test from the successful relativistic correction to the orbit of the planet Mercury.

The postcard to de Sitter gives Lemaître's address in Belgium; Eddington had made Lemaître famous (Section 3.1). The photograph of Einstein and de Sitter in Panel (c) was taken in 1932, the year of publication of their argument that one might, for the moment, ignore space curvature (and the cosmological constant) in the cosmological model that became the favorite in the years around 1990.

Plate IV

Participants at the 1953 Symposium on Astrophysics at the University of Michigan include George Gamow and Walter Baade (Gamow is the taller of the two), front center of this group photograph. Baade had identified regions of emission of the H-α line of atomic hydrogen in the galaxy M 31. Vera Rubin, standing between Gamow and Baade two rows back, and pictured also in Plate VI, used Baade's finding list to measure redshifts of these regions. Section 6.3.1 reviews this serious addition to the case that M 31 has a subluminal massive halo.

Geoffrey Burbidge, holding sunglasses, is on the far left one row from the back, and Margaret Burbidge is beside him. They were authorities on galaxies in the years around 1970. Geoff later recognized that there is more helium than seemed likely to have been produced in stars. Gamow mentioned at this conference his theory of helium production in the early universe. But the two did not make the connection.

Donald Osterbrock and Irene Osterbrock are in the back row at the far left. Gamow's ideas impressed Don. He and Jack Robertson were among the first to propose in print that the helium abundance is large and may be a remnant from the early universe (Section 4.3.1).

Alan Sandage, one row in front of and between the Burbidges, pioneered the redshift-magnitude test discussed in Section 9.1. Beverly Oke, to our right and in front of Sandage, applied the test with James Gunn.

Nelson Limber, behind and just to our right of Geoff Burbidge, and Vera Rubin, presented early applications of the two-point measure of the galaxy distribution (Section 2.5).

Lawrence Aller is in the back row to the right of Baade. His recollections of early evidence of subluminal matter are on page 275.

Nancy Boggess, fourth to the right of Rubin, also appears in Plate VIII.

Photo courtesy Ed Spiegel.

Plate V

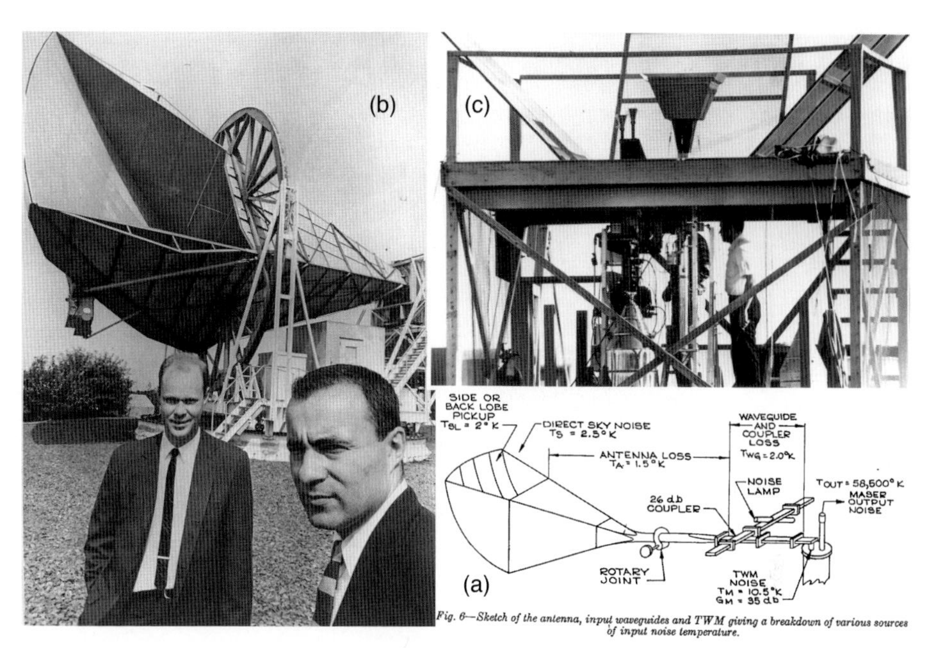

Identifying the sea of thermal radiation.

Panel (a) shows the noise budget in a Bell Telephone Laboratories microwave communications experiment, balanced by assigning 2 K to radiation from the ground (DeGrasse et al. 1959b). But engineers knew that likely is an overestimate; there was a problem with excess noise. Panel (b) shows Robert Wilson, left, and Arno Penzias. Their careful examinations failed to identify a local source of this excess noise in the Bell system in the background. Panel (c) shows the Princeton experiment, Wilkinson holding a screwdriver, and Roll standing behind the Dicke radiometer. Section 4.4 reviews how news of the Princeton search for thermal radiation from the early universe led to the realization that the excess noise in the Bell systems is the radiation the Princeton group was looking for.

Plate VI

Vera Rubin and Kent Ford.

Kent Ford's image intensifier made feasible Vera Rubin's astronomical observations in their studies of internal motions of galaxies. Rubin is examining the photographic plate taken from the envelope in her left hand. This was a standard astronomers' pose, no longer seen much. Their measurement of the rotation curve of M 31 is shown in Panel (d) in Figure 6.2.

Photograph taken in 1984 at the Department of Terrestrial Magnetism of the Carnegie Institute of Washington, presented Courtesy Carnegie Institution for Science, Administrative Archives.

Wagoner, Fowler, and Hoyle, with Donald D. Clayton left to right, 1967.

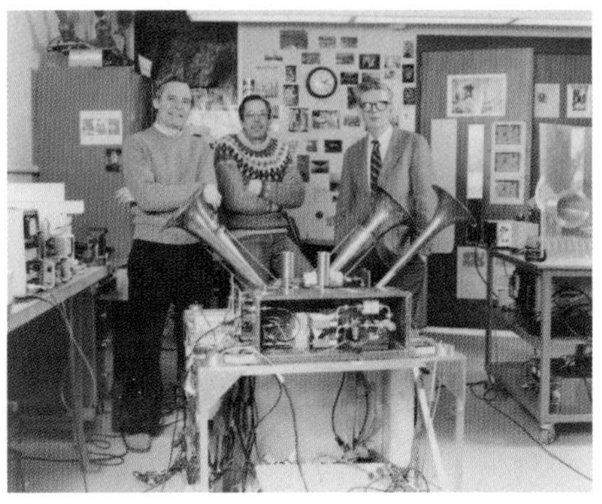

Wilkinson, Peebles, and Dicke left to right, in the late 1970s.

Dicke's invitation to think about implications of the Roll and Wilkinson search for radiation left from a hot big bang led me to reinvent Gamow's scenario for the formation of helium. At the same time, Hoyle's recognition that the helium abundance is larger than would be expected from stars led him and Tayler to point out that the helium could be from Gamow's hot big bang. Recognition of the microwave sea that might be from the big bang led to more detailed computations of element formation in a hot big bang, by Wagoner, Fowler, and Hoyle, in the bottom left photograph, and by me, to the right (Sections 4.3.3 and 4.6).

Meanwhile, Wilkinson was leading the search for anisotropy of the radiation, the start of a long trail to the demanding tests reviewed in Section 9.2. A balloon carried the instrument in front of us above the disturbing radiation from the atmosphere. It probed the anisotropy by switching between pairs of horn antennae pointing to different parts of the sky.

Plate VII

Cosmology in the Soviet Union.

Panel (a): George Marx, Yakov Zel'dovich, and Igor Novikov (left to right) in Hungary in about 1970; (b): Jaan Einasto, left, and Andrei Doroshkevich at a Conference in Tallinn in 1977; and (c): Nikolay Shakura, left, and Rashid Sunyaev in the 1970s.

Zel'dovich with Semen Solomonovich Gershtein pointed out that neutrino rest masses are bounded by the condition that their mean mass density not exceed what the relativistic cosmology would allow. Marx with Alex Szalay turned that thought into the first model for nonbaryonic dark matter (Section 7.1). Doroshkevich and Novikov pointed out that the Bell microwave communications experiments test Gamow's hot big bang theory (Section 4.4.3). Einasto, Kaasik, and Saar pointed to the variety of evidence that the mean mass density is well above what is in the luminous parts of galaxies: dark matter (Section 6.5). Shakura and Sunyaev explored the physics of accretion of matter into a black hole from a surrounding accretion disk. Sunyaev and Zel'dovich introduced the effect of scattering of the CMB by hot electrons in clusters of galaxies on the CMB intensity spectrum (Section 4.1). Section 10.4 offers observations of the conditions for this productive research in cosmology in the Soviet Union.

Panel (a) courtesy of Igor Novikov; (b) courtesy of Jaan Einasto; (c) courtesy Rashid Sunyaev.

Plate VIII

COBRA rocket experiment.

COBE satellite experiment.

A pressing question through the 1980s was whether the sea of microwave radiation is in near thermal equilibrium. Serious departures from a homogeneous and isotropic universe would present us with a mix of radiation temperatures determined by the variety of local expansion histories, a nonthermal situation. Large energy release in relativistic collapses or early generations of stars also could seriously disturb the radiation. The measurements from the COBE and COBRA groups presented in 1990 independently showed that the intensity spectrum is very close to thermal (Section 4.5.4). On the whole, our universe has evolved in a simple way.

Members left to right of the COBRA team: Ed Wishnow, Mark Halpern, and Herbert Gush. The COBE science working group, left to right: Ed Cheng, David Wilkinson, Rick Shafer, Tom Murdock, Steve Meyer, Charles Bennett, Nancy Boggess, Mike Janssen, Bob Silverberg, Sam Gulkis, John Mather, Harvey Moseley, Phil Lubin, Ned Wright, Mike Hauser, George Smoot, Rai Weiss, and Tom Kelsall.

Left: University of British Columbia Archives, Photo by Martin Dee [UBC 44.1/531]. Right: Photo courtesy John Mather.

$$P(k) = \frac{Bk}{\{1+[6.4q+(3.0q)^{3/2}+(1.7q)^2]^\nu\}^{2/\nu}}, \quad q = \frac{k}{\Gamma}, \quad \nu = 1.13. \qquad (8.8)$$

The unit of the wavenumber k is $1h\ \mathrm{Mpc}^{-1}$. The amplitude B and the number Γ are free parameters, and in ΛCDM, we have $\Gamma = \Omega_m h$. Efstathiou et al. give other expressions for Γ for other models. I take the wonderfully detailed equation (8.8) to be a good illustration of the ingenuity devoted to the search for a believable picture of the behavior of the universe, from galaxies to the largest observable scales.

The ΛCDM model is counter to the long tradition of nonempirical arguments against Λ reviewed in Section 3.5.1, but it proves to pass the revolutionary advances in the cosmological tests at the turn of the century. Before discussing how this came about, we should consider still more ideas about how to reach a well-established cosmology.

8.4.4 OTHER THOUGHTS

Gott et al. (1974) discussed the case for a low-density cosmological model with open space sections. In the 1970s, this model was not considered unreasonable. The later change of thinking was largely inspired by the interpretation of cosmological inflation in the early 1980s, which held that the universe is close to homogeneous because the great expansion during inflation swept away space curvature gradients and with them the curvature of space sections. Even though Abbott and Schaefer (1986) shared the common opinion that a cosmology with open space sections is contrary to inflation, they felt it is worth considering anyway. And Ellis, Lyth, and Mijić (1991) felt they could craft an acceptable model for inflation in an open universe.

The open CDM (OCDM) variant of sCDM accepts the evidence for low mass density, $\Omega_m \sim 0.3$, and assumes open space sections with $\Lambda = 0$. Through the mid-1990s, this model could be adjusted to reasonably acceptable fits to the constraints (e.g., Wilson 1983; Blumenthal, Dekel, and Primack 1988; Kamionkowski et al. 1994; Ratra and Peebles 1995). It was convincingly falsified at the turn of the century.

The primeval isocurvature baryon model, PIB, introduced in Peebles (1987a,b) is even more iconoclastic; it was introduced as a counterexample to the idea that the observations require nonbaryonic matter. The model assumes primeval isocurvature conditions, meaning that a clumpy distribution of the baryons in the early universe is compensated by tiny perturbations to the radiation that serve to eliminate primeval spacetime curvature fluctuations. The matter density, all baryons, is chosen to be consistent with the big bang nucleosynthesis (BBNS) constraint (Section 4.6), and with $\Lambda = 0$, this requires open space sections. The power spectrum of the primeval baryon space distribution is a power law, and the power law index and amplitude are free parameters. In the mid-1990s, the model was considered to be seriously constrained but perhaps not ruled out (e.g., Efstathiou and Bond 1987; Cen, Ostriker, and Peebles 1993). Hu, Bunn, and Sugiyama (1995, L62) conclude that

> At present, none of the simplest models for structure formation fares well in comparison with the combined observations of the CMB and large-scale structure; it is, therefore, perhaps unwise to dismiss this scenario [PIB] as entirely unviable.

I introduced a more elaborate isocurvature model (Peebles 1999) just in time for it to be falsified, along with PIB, in a particularly manifest way by the detection of the acoustic oscillations illustrated in Figure 5.2 and shown in Plates I and II.

Sahni, Feldman, and Stebbins (1992) reconsidered Lemaître's (1931d) proposal that the expanding universe passed through the near-static hovering phase discussed in Section 3.6. Lemaître made great contributions to cosmology, but this one has not proved to be promising. Messina et al. (1992) discussed the interesting effect on structure formation in the sCDM cosmology by seriously large positive or negative skewness of the primeval mass density fluctuations. The large mass density in sCDM is problematic, of course, but at the time, one could have considered ΛCDM with skewness. That line of thought was ruled out at the turn of the century by the close-to-Gaussian CMB anisotropy. Bartlett et al. (1995) pointed out that many of the challenges to the sCDM cosmology would be relieved if the extragalactic distance scale were about twice the astronomers' measurements. But the low dynamical estimates of Ω_m in Table 3.2 are not sensitive to the distance scale, and it certainly was difficult to imagine how the astronomers' distance calibrations could have been so far off. The idea had to be considered and seen to be seriously challenged.

We see in this review the broad range of inventive ideas put forth in the search for clues to a better cosmological model. Motivations surely differed, but I think a common feeling was excitement at the thought that we may be approaching constraints tight enough to make the case that a particular cosmology is a reasonably convincing approximation to what actually happened. Let us consider now examples of the evolution of thoughts about how close we may have been to this end game in the 1990s.

8.5 How Might It All Fit Together?

Dekel, Burstein, and White (1997, 176) argue that

> the order by which more specific models should be considered against observations, are guided by the principle of Occam's Razor, *i.e.*, by simplicity and robustness to initial conditions. The caveat is that different researchers might disagree on the evaluation of "simplicity".

> It is commonly assumed that the simplest model is the Einstein–de Sitter model, $\Omega_m = 1$ and $\Omega_\Lambda = 0$. One property that makes it robust is the fact that Ω_m remains constant at all times with no need for fine tuning at the initial conditions (the "coincidence" argument [2]).
>
> The most natural extension according to the generic model of inflation is a flat universe, $\Omega_{tot} = 1$, where Ω_m can be smaller than unity but only at the expense of a nonzero cosmological constant.
>
> These simple models could serve as useful references, and even guide the interpretation of the results, but they should not bias the measurements.

These cautious statements give a sensible picture of the state of the art in the mid-1990s.

The reference [2] in these statements is to Bondi (1960), who pointed out that if space curvature and Λ vanish, then the density parameter for the total mass is $\Omega_m = 1$, independent of time. As discussed in Section 3.5.1, this would mean we need not have flourished at some particular epoch during the course of evolution of the universe. The community also was quite aware of the problem that a value of Λ acceptable for cosmology appears to be distinctly unsuitable from the point of view of quantum physics, for the reasons considered on page 59. Kofman, Gnedin, and Bahcall (1993, 8) put it that "from both cosmology and fundamental physics, there is a major difference whether the cosmological constant is exactly zero or very small (in Planck units)." Preference for the former is seen in the comment in the paper Davis, Efstathiou et al. (1992) given in Section 3.5.1. Marc Davis, whose many contributions to this subject are discussed at length in this history, recalls (in a personal communication, 2018) that

> I remember when we got together in Cambridge, and we were very reluctant to allow Omega less than 1. It just wasn't done, the title of this paper says it all: "The end of cold dark matter?"

Another caveat is understood, of course: nature must agree. Marc Davis, again in a personal communication (2018), said that

> If we had been willing to accept more than one parameter in our models, yes we could have made more progress. But we absolutely had to be forced, with data unambiguously pointing at Lambda.

Taylor and Rowan-Robinson (1992, 396) presented a comparison of the observational constraints on seven variants of sCDM, including ΛCDM. Their conclusion is that

> We find only one completely satisfactory model, in which the Universe has density $\Omega = 1$, with 69% in the form of cold dark matter, 30%

provided by hot dark matter in the form of a stable neutrino with mass 7.5 eV, and 1% baryonic.

This is MDM, the mix of CDM and HDM discussed in Section 8.4.2, with $\Lambda = 0$. Numerical simulations of cosmic structure formation led Davis, Summers, and Schlegel (1992, 359) to a similar conclusion:

> The MDM model thus seems to resolve a long-standing problem of large-scale structure, namely the disparate estimates of Ω on small and large scales. Velocity fields are reduced on small scales and increased on large scales, increasing the Mach number of the cosmic velocity field.[31] This with its other successes in matching large-scale structure in the Universe[8,32] makes the model worthy of serious consideration.

The references are to Ostriker and Suto (1990), who characterized the small dispersion around the mean galaxy flow as a large cosmic Mach number; van Dalen and Schaefer (1992); and Taylor and Rowan-Robinson (1992). The last two papers also present arguments for the MDM model. The paper by Klypin et al. (1993, 1) presents a similar assessment:

> C+HDM looks promising as a model of structure formation. The presence of a hot component requires the introduction of a *single* additional parameter beyond standard CDM—the light neutrino mass or, equivalently, Ω_ν—and allows the model to fit essentially all the available cosmological data remarkably well. The τ neutrino is predicted to have a mass of about 7 eV, compatible with the MSW explanation of the solar neutrino data together with a long-popular particle physics model.

The title of a later review of the situation by Primack (1997) is "The Best Theory of Cosmic Structure Formation is Cold + Hot Dark Matter." In this cosmological model, the CDM component would be more strongly clustered around individual galaxies, and the HDM component more broadly spread, because the neutrinos were streaming about at high speeds at high redshift. This is in the direction of reconciling the small mass density inferred from observations at relatively small scales and the COBE normalization on large scales with the Einstein–de Sitter mass density $\Omega_m = 1$ assumed in these models. But of course, there still was the problem with the considerable variety of evidence in Table 3.2 for a lower mass density.

Primack's (1997) paper is the last carefully organized argument for MDM that I have found; interest in the model had quite abruptly faded. I have not been able to find a direct demonstration at that time that MDM had been falsified, or looked likely to be. And in that year, Perlmutter et al. (1997) presented evidence from measurements of the supernova redshift-magnitude relation that Ω_m is close to unity. It was withdrawn, but at the time, reconciling the apparently large value of Ω_m with the smaller mass densities indicated by the probes of the mean mass density summarized in Figure 3.5 called for some special arrangement. The MDM model seems to have been worth considering. But instead, the community was growing attached to the simplicity of adding Einstein's cosmological model Λ to keep space sections flat while lowering the mean mass density from sCDM to agree with the evidence for that, despite the unlikely quantum physics. Later evidence is that we must indeed learn to live with Λ, and that while neutrinos do have nonzero rest masses, they are well below those considered in the MDM models.

Arguments for ΛCDM include Efstathiou, Sutherland, and Maddox (1990, 705), who conclude from their measurement of the galaxy correlation function (Section 3.6.5) that

> the successes of the CDM theory can be retained and the new observations accommodated in a spatially flat cosmology in which as much as 80% of the critical density is provided by a positive cosmological constant, which is dynamically equivalent to endowing the vacuum with a non-zero energy density.

S. White et al. (1993, 432) came to this conclusion from their estimate of the cluster baryon mass fraction (Section 3.6.4). Their thought is that "The flat universe required by the inflation model can be rescued by a non-zero cosmological constant, a possibility which has other attractive features[41] but which still conflicts with dynamical evidence for large Ω_0." Reference 41 is to the Efstathiou, Sutherland, and Maddox (1990) evidence for mass density less than $\Omega_m = 1$ from the range of positive correlations of galaxy positions. The mention of evidence of a large mass density is not explained, but at the time, the result from the POTENT method was widely discussed. It argued for consistency with $\Omega_m = 1$ (Dekel et al. 1993), though we have seen that it is difficult to reconcile this large mass density with the bulk of the evidence illustrated in Figure 3.5. And the improvements in the data and methods of analysis of the mass density later that decade, in particular from method E in Table 3.3, led Willick et al. (1997) to conclude that the mean mass density likely is in the range $0.16 \lesssim \Omega_m \lesssim 0.34$ for reasonable values of Hubble's constant.

Recall now that at a given value of the mass density parameter less than unity, perhaps $\Omega_m \approx 0.3$, the OCDM model with $\Lambda = 0$ and open space sections and the ΛCDM model with flat space sections equally well fit mass density measurements on scales ~ 1 Mpc, clusters on scales ~ 10 Mpc, and the cutoff of correlated galaxy positions at ~ 100 Mpc. This is the geometrical degeneracy discussed in footnote 9.2 on page 334 in Chapter 9. Kofman, Gnedin, and Bahcall (1993, 2) argued that the degeneracy between OCDM and ΛCDM might be broken by the consideration that "a positive cosmological constant helps to overcome the (possible) 'age problem' of the universe;

i.e., the age of the oldest globular clusters ... is larger than the age of the universe for $\Omega = 1$ and $h \simeq 0.5$ ($t_0 = 13$ Gyr)." This issue also led Krauss and Turner (1995) and Chaboyer et al. (1996) to argue for ΛCDM and against OCDM. But the Kofman, Gnedin, and Bahcall (1993) qualifier "possible" is to be noted: A reliable value of the Hubble parameter h was difficult to establish.

Gott et al. (1974) assembled the parameter constraints shown in Figure 3.2 on page 76. Ostriker and Steinhardt (1995) updated the figure; they had tighter constraints on Hubble's constant and the expansion time, and they replaced the baryon mass density by the network of constraints on the total mass density reviewed in Section 3.6.4. The case for mass density parameter in the range $0.2 \lesssim \Omega_m \lesssim 0.4$ had grown reasonably persuasive, as they argue, though not yet generally accepted. Ostriker and Steinhardt proposed that the degeneracy between OCDM and ΛCDM is broken by the OCDM prediction of an unacceptably large CMB anisotropy. This is an important addition to the constraints, but at the time, the significance of the CMB anisotropy measurements could be debated.

The bottom figure in Plate I shows the progress of CMB anisotropy measurements compiled by Peebles, Page, and Partridge (2009).[9] Kamionkowski, et al. (1994, Fig. 2) concluded that at $\Omega_m = 0.3$ and $\Lambda = 0$, and with scale-invariant initial conditions, the COBE-normalized peak anisotropy is $\delta T_l = 55\mu$K at $l = 400$ This is allowed by the measurements up to the year 1999 in Plate 1. Ostriker and Steinhardt (1995) argued that since the mass fluctuation spectrum grows more slowly in OCDM than in ΛCDM, the primeval spectrum in OCDM should be tilted to $n = 1.15$, which would increase the power on small scales. They found that this with $\Omega_m = 0.375$ predicts $\delta T_l \approx 95\mu$K at $l = 400$, which is unacceptably large. This consideration can be compared to the Bunn and White (1997) analysis. They used the four-year COBE CMB anisotropy data. In their OCDM model with $n = 1$, the COBE normalization indicates $\sigma_8 = 0.64$. This is smaller than their estimate of what is needed to account for the abundance of rich clusters, $\sigma_8 = 0.87 \pm 0.14$ at $\Omega_m = 0.4$. Bunn and White pointed to tilting the other way, to $n = 0.8$ with $\Omega_m = 1$. This fits the cluster abundance but requires biasing on smaller scales.

We must consider two other issues. First, reionization following the dark ages has suppressed the CMB anisotropy by scattering by free electrons. The effect is now known to be modest, but Figure 2 in Kamionkowski, Spergel, and Sugiyama (1994) shows the considerable suppression of the anisotropy that could be contemplated in the mid-1990s. Second, the constraint from the

9. To convert the measure of anisotropy in Ostriker and Steinhardt (1995) and other papers at the time to the value of δT_l in micro Kelvins on the vertical axes in Plate I, take the square root of the anisotropy and multiply by 2.725×10^6.

CMB anisotropy assumes primeval adiabatic departures from homogeneity. The PIB primeval isocurvature model discussed on page 315 was thought to be viable until the late 1990s.

The nonempirical argument from inflation certainly favors ΛCDM over OCDM with its curved space sections, though we have seen that to some, the nonempirical consideration from quantum physics argued against ΛCDM. The empirical evidence supported ΛCDM over OCDM because it allows a larger expansion time that is an easier fit to stellar evolution ages, and perhaps a better fit to the CMB anisotropy. But in the mid-1990s, the interpretations of these measurements were not secure. Some argued that the addition of Λ could be avoided by turning to MDM or isocurvature models, and others pointed out that since the astronomers were finding estimates of Hubble's constant in the range 50 to 100 km s^{-1} Mpc^{-1}, it was not unreasonable to consider $H_0 \sim 40$ km s^{-1} Mpc^{-1}, which would help reconcile sCDM with the expansion time and CMB anisotropy. Others remained agnostic, of course.

The state of the empirical evidence changed at the end of the 1990s when the measurements of the CMB anisotropy were in better condition, and results were emerging from the programs to measure the cosmological redshift-magnitude relation. These are the two empirical lines of research that drove the revolutionary consolidation of evidence discussed in Chapter 9. Bahcall et al. (1999) conclude that this assembled evidence is converging to the ΛCDM cosmology. In their assessment of the evidence, Bond and Jaffe (1999, 61) took a more cautious line: The count of parameters in the cosmological models they were considering "is thus at least 17, and many more if we do not restrict the shape of $\mathcal{P}_\Phi(k)$ through theoretical considerations of what is 'likely' in inflation models."

Bond and Jaffe took account of the accumulated constraints from the COBE four-year CMB anisotropy data and emerging measurements on smaller angular scales, and they had preprints from the two teams measuring the redshift-magnitude relation. But Bond and Jaffe chose not to make a pronouncement on the most promising cosmological model; they instead concluded with analyses of the prospects for tighter tests from measurements in progress.

We see in this section the great energy devoted to ingenious model-building. The community in the 1980s and 1990s does not seem to have been seriously dismayed by the excess of ideas over observations, though that tended to be my feeling. Morale was maintained by the ongoing input from observational and experimental programs and by the perception that the models under discussion were not unpromising. Thoughts of elegance can inspire, too, and it was exciting to think that the right way forward may bring us to a well-founded cosmology that would be seen to be elegant in a sense to be discovered.

A natural tendency to be conservative may help explain why most ideas about a viable cosmology explored in the two decades leading up to the turn of the century are minimal departures from the original sCDM version, along with minimal variations from the early ideas about inflation. That approach could have missed the right direction forward, but the evidence to be summarized in the next chapter is that it did not.

CHAPTER NINE

The 1998–2003 Revolution

THE CHANGE IN the state of empirical cosmology in the five years from 1998 to 2003 was great enough to be termed a revolution. It was driven by two important programs, the measurements of the cosmological redshift-magnitude relation and of the pattern of the cosmic microwave background (CMB) radiation distribution across the sky. The two programs reached the precision needed for significant constraints on cosmological models at essentially the same time. Quick acceptance of their interpretation was driven by the impressive consistency of implications of these two quite different ways to look at the universe and, equally important, by the consistency with other lines of evidence gathered in the years of research before the revolution.

The measurements of the redshift-magnitude relation are reviewed in Section 9.1. The discovery that the CMB anisotropy pattern agrees with what is expected in the family of CDM models, and hence is capable of straightforward interpretation, is discussed in Section 9.2. Section 9.3 is a consideration of how the pieces of evidence, new and old, fitted together well enough that by 2003 the empirical revolution may be said to have been completed. Research in empirical cosmology went on, of course, with important new discoveries. But this history terminates at the end of the revolution when the ΛCDM cosmology had become an established branch of natural science. This chapter ends with the contemplation in Section 9.4 of the future of research on this subject.

9.1 The Redshift-Magnitude Test

Tolman (1934a) wrote out the relation of the redshift of an object, its intrinsic luminosity, and its observed energy flux density (albeit scattered among derivations in §§178 to 183). In astronomers' terms, this is the relation between the distance modulus, which is the difference between the apparent and absolute magnitudes of an object, and its redshift. Thus it also is known as the redshift-magnitude or z-m relation. Measurements of the redshifts z

and apparent magnitudes m of a class of objects can be an accurate measurement of the redshift-magnitude relation only if the objects can be brought close to a common intrinsic luminosity. If that could be assured, then the measured z-m relation would constrain the values of the cosmological parameters by application of Tolman's equations, always assuming the relativistic Friedman-Lemaître cosmological model.

In his book, *The Realm of the Nebulae*, Hubble (1936, 154–155) took note of the

> exceedingly rare supernovae which appear in any one nebula at average intervals of the order of 500 to 1,000 years. From the scanty data available, they are believed to reach, at maxima, fairly uniform intrinsic luminosity which is comparable with the average luminosity of the nebulae themselves. Supernovae can be detected at immense distances and, in principle, they are a criterion of distance about as reliable as that of the total luminosities of the nebulae. Actually, however, the maxima are so seldom observed and the [super]novae themselves are so rare that they contribute very little to the present problem.

Hubble was exploring the newly opened world outside the Milky Way, the realm of the nebulae, with the tools at hand. At the turn of the century, supernovae became a key tool in the exploration of how the realm of the nebulae came into existence.

Allan Sandage's (1961a) assessment of how the 200-inch telescope in the Palomar Mountain Range in southern California might be used to probe cosmic evolution led him to the conclusion that measurements of the z-m relation for the most luminous galaxies in rich clusters of galaxies seemed promising. Hubble and Humason had already extended the z-m relation for the more luminous cluster members to redshift $z \simeq 0.1$, 10 percent of the speed of light (Hubble 1936). Sandage proposed to continue this program to larger redshifts at greater distances, and Sandage and Hardy (1973) reached clusters at redshift $z \approx 0.4$. With careful attention to how the luminosities of the most prominent cluster members might depend on cluster properties, their formal estimate of the deceleration parameter defined in equation (3.23) is $q_0 \approx 1 \pm 1$. Gunn and Oke (1975) found a similar result.

Sandage (1961a) had cautioned that the corrections for evolution of the intrinsic luminosities of galaxies may be problematic. In her PhD dissertation, Beatrice Tinsley (1967) emphasized that previous estimates of the evolution of galaxy luminosities at best showed the need for closer analysis. She set out to improve the situation.

Loh and Spillar (1986b) found an interesting constraint on the cosmological parameters from a related measure, the count of galaxies as a function of redshift. Tolman (1934a) also wrote out this theory. Loh and Spillar had a large sample of photometric redshifts (by the methods reviewed on page 94 in

Section 3.6.4), and they had a test for galaxy evolution under the assumption that all galaxies evolve at the same fractional rate. This was a significant start to addressing an important problem, but now that the cosmological parameters are well established by other means, the Tinsley and Loh-Spillar programs are central to the study of how the galaxies formed and evolved.

Gustav Tammann proposed that the control of evolution of the right type of supernovae may be less problematic than it is for galaxies; supernovae may be more suitable for measuring the redshift-magnitude relation. Tammann pointed to a particular class, those characterized as Type I by the absence of hydrogen features in their spectra. The early conclusion in Tammann (1978, 325) is that

> SNe of Type I in E/S0 galaxies—and also those in spiral galaxies after reliable corrections for intrinsic absorption—are nearly standard candles [that] may be useful for the determination of q. Although SNeI are at maximum much fainter than first-ranked cluster galaxies, they may have the same or even smaller intrinsic magnitude dispersion ($\leq 0^m.3$) and be less affected by evolutionary effects over cosmological times than the latter.

Here q is the deceleration parameter q_0 defined in equation (3.23) on page 53, a first measure of the z-m relation.

Woosley and Weaver (1986) reviewed the accumulating evidence that Type I supernovae are to be separated into three classes. Type Ia were and are thought to be thermonuclear explosions of white dwarf stars, the energy coming from the conversion of the carbon and oxygen in the white dwarf to atomic nuclei near the peak of the nuclear binding energy curve near iron. The less-luminous types Ib and Ic were taken to be relativistic collapses of more massive stars that had lost their hydrogen envelopes by stellar winds, which would account for the absence of hydrogen features in their spectra. The restriction to Type Ia reduces the scatter of peak luminosities. The scatter is further reduced by taking account of the correlation between the SNeIa peak luminosity and the rate of decline of the luminosity after the peak. Pskovskii (1977) pointed out this effect. Phillips (1993) demonstrated how the assembled observations of this correlation may be used to reduce the scatter of SNeIa luminosities to a more nearly common value; it became known as the Phillips relationship. It could be explored in detail from the light curves—luminosity as a function of time—of SNeIa discovered in the systematic Calán/Tololo Supernova Survey at the University of Chile and the Cerro Tololo Inter-American Observatory (e.g., Hamuy et al. 1993, 1995 and 1996).

The Calán/Tololo Supernova Survey discovered these objects on photographic plates. The low quantum efficiency limited the distances to which supernovae could be detected to redshift $z \sim 0.1$, but the large plate area allowed sampling large numbers of galaxies. These data were an important baseline for the observations at higher redshifts that were made possible by the greater quantum efficiency of digital detectors. Stirling Colgate had placed early emphasis on this; Colgate, Moore, and Colburn (1975, 1429) report that

> We have coupled an ITT F4089 magnetically focused image intensifier to an RCA 4826 intensified silicon target vidicon. Our objective was to obtain an astronomical sensor that could be used for reliable and reproducible astronomical measurements for the computer-controlled automated supernova search and spectral measurements.

Colgate (1979) offered arguments following those of Tammann (1978) for the particular advantage of using SNeI to measure the z-m relation and pointed to a new way to detect them: the use of the charge coupled devices, or CCDs, of the kind planned for use in the Space Telescope.

The elements of CCD supernova detection would be to obtain digital images of the same piece of sky spaced at about one month intervals, which would fit the rise times of SNeI; adjust the images from successive observations, so the stars are closely aligned; subtract the photon counts in cells from successive exposures; identify positions of significant departures from the mean difference; reject those due to the motion of asteroids, variability of stars, and the like; and persuade colleagues who are observing at large telescopes to measure spectra of interesting candidates for supernovae. Robert Kirshner (in a personal communication, 2019) points out that

> the true observational problem is harder—the sky brightness varies from night to night, so you have to take that out, and the seeing varies from night to night, so the images are not the same size each night. In practice, we do the disgusting thing of blurring the better image to match the poorer one, then doing the subtraction.

And Hamuy et al. (1993, 2398) put it that

> Unfortunately, the appearance of a SN is not predictable. As a consequence of this we cannot schedule the followup observations *a priori*, and we generally have to rely on someone else's telescope time. This makes the execution of this project somewhat difficult.

Hansen, Nørgaard-Nielsen, and Jørgensen (1987) were among the first to use a CCD detector to find supernovae. After two years of observation, Nørgaard-Nielsen et al. (1989) reported discovery of a supernova with the light curve characteristic of an SNeIa with the time the scale stretched by the measured redshift, $z = 0.31$. It was a striking demonstration of the promise of this method to measure the redshift-magnitude relation, but a larger telescope and larger field of view were needed for better statistics.

In an exchange occasioned by an article in the magazine *Sky and Telescope* (Kahn 1987) on the supernova search project under way at the University of California, Berkeley, Colgate (1987, 230) wrote that

> The Berkeley supernova search has been a major stimulation. Without the affirmation of the automated search concept by Luis Alvarez, Richard Muller, Carl Pennypacker, and others, our work would have suffered and our momentum failed. Our many discussions have been mutually helpful. Astronomy needs several automated searches with modern equipment. Finding many supernovae early offers the best hope of understanding them in depth. Once we do, they may become the best standard candles for measuring the scale of our universe.

Muller and Pennypacker (1987, 230) wrote in response that "We feel that our supernova search at Berkeley is largely an outgrowth of the work begun by Stirling Colgate and his colleagues."

The UC Berkeley program of development of technology for an automated supernova search was in progress at the time of the *Sky and Telescope* article. It grew into the Berkeley Supernova Search, which became the Berkeley Supernova Cosmology Project. By 1990, the group had added many of its key players and was announcing detections of supernovae. The program is described by Perlmutter et al. (1990). Saul Perlmutter (2012) recalls that at about this time, he had become principal investigator. A decade later, the project had a clear and important detection of a departure of the redshift-magnitude relation from the power law observed at lower redshifts.

The development of the competing program, the High-Z Supernova Search, traces back to the early 1990s, when Brian Schmidt joined Robert Kirshner's supernova research group at Harvard University. Schmidt's doctoral dissertation with Kirshner is on "Type II Supernovae, Expanding Photospheres, and the Extragalactic Distance Scale." Schmidt (2012) recalls that in 1994 he joined Nicholas Suntzeff in collaboration with the Calán-Tololo Supernova Survey to form the High-Z Supernova Search program. Schmidt became principal investigator in 1996. Adam Riess joined Kirshner's group as a graduate student in 1993 and worked with Bill Press on calibration and application of the Phillips relationship in Type Ia supernovae. Press was a powerful ally, the lead author of the book *Numerical Recipes* (Press et al. 1992). This catalog of numerical methods with computer programs to implement them is valuable to many, including me. (I have given the reference to the FORTRAN version of the book, because FORTRAN is the only programing language I know.) The application to the Phillips relationship in Riess, Press, and Kirshner (1995 and 1996) shows Press's characteristic way of thinking.[1]

1. These papers use observations of supernovae to measure the extragalactic distance scale, and Hubble's constant. The light curve of the Type Ia supernova in the nearby galaxy

Riess (2012) recalls how he followed this path further in resolving the problem of correction for obscuration by dust in the supernova spectrum observed through the screen of interstellar dust in the host galaxy and the Milky Way, and for correction for the correlation of the intrinsic supernova spectrum with its luminosity.

There was a false start. Perlmutter et al. (1997) reported that their first measurements of seven supernovae at redshifts near $z \sim 0.4$ agree with the Einstein–de Sitter model: $\Lambda = 0$ with mass density $\Omega_{\rm m} = 1$. This was attractive from the point of view of elegance of the cosmological model, though reconciling it with other cosmological tests would have been an interesting challenge. But this preliminary indication was soon replaced.

The two groups presented measurements of the supernova redshift-magnitude relation with the precision and range of redshifts needed for a critical cosmological test: first the indication that the mass density is less than Einstein–de Sitter in Perlmutter et al. (1998) and Garnavich et al. (1998); then clear evidence of the signature of the cosmological constant Λ in Riess et al. (1998) and Perlmutter et al. (1999). The two groups reached these conclusions at effectively the same time. (In the above order, the papers were published in January 1998, February 1998, September 1998, and June 1999. This is a negligible spread of time compared to how long it took to get there.)

Both groups used the observations of SNeIa at redshifts $z \lesssim 0.1$ from the Calán/Tololo Supernova Survey as essential contributions to the baseline for interpretation of their observations at higher redshifts; each of the four papers referred to some combination of Hamuy et al. (1993, 1995, and 1996). The High-Z Supernova Search team roughly doubled the number of supernovae in their $z \lesssim 0.1$ baseline by adding the data from Riess et al. (1999). The Supernova Cosmology Project had about four times the number of supernovae at larger redshifts. But the Calán-Tololo/High-Z Supernova Search teams had superior experience in the art of astronomical photometry, including corrections for the effects of dust and intrinsic variability of the supernovae spectra, and they had generally smaller uncertainties in apparent magnitude measurements. The result is that the two teams had the similarly tight constraints on the z-m relation shown in Figure 9.1 and the cosmological parameters in Figure 9.2.

We see an analog to the measurements of the CMB intensity spectrum illustrated in Figure 4.6 on page 173. For both the spectrum and z-m measurements, many years of preparation by two independent groups led to the

---

NGC 5253 was well observed, and Riess, Press, and Kirshner (1995) showed how to take account of the Phillips relationship to find the ratio of distances of NGC 5253 to distances of SNeIa at larger redshifts. They had a Cepheid variable measure of the distance to NGC 5253. The SNeIa ratios of distances carry this Cepheid distance to galaxies far enough away that the corrections for their peculiar motions are unimportant. Research on establishing the extragalactic distance scale is an important part of the story not reviewed in this book.

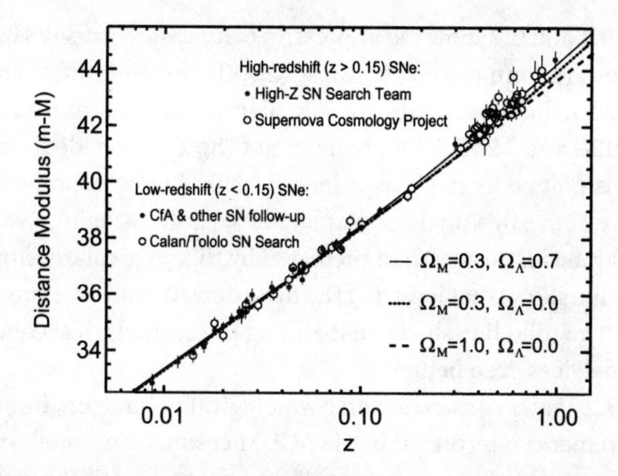

FIGURE 9.1. Measurements of the redshift-magnitude relation by the Supernova Cosmology Project and the High-Z SN Search Team (Riess 2000, Perlmutter and Schmidt 2003). This figure reproduced by permission from Springer Nature.

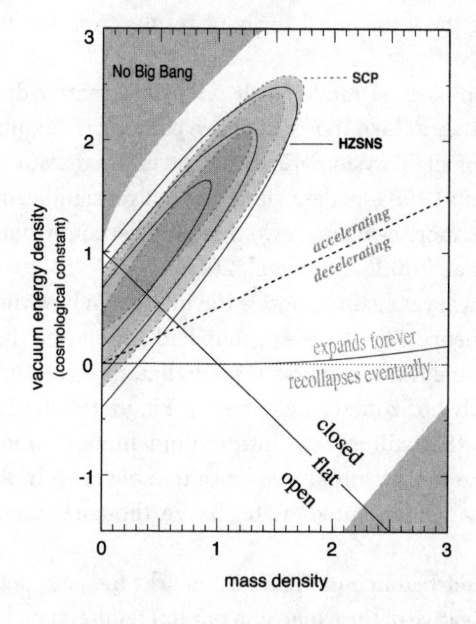

FIGURE 9.2. The constraints on cosmological parameters from the redshift-magnitude measurements (Riess 2000, Perlmutter and Schmidt 2003). This figure reproduced by permission from Springer Nature.

demonstration of an important result. In both cases, the result from one group would have been influential, but the essentially independent results presented so close in time made the evidence particularly persuasive. (Kirshner 2010 presents a careful list of elements common to both z-m groups, including data in the $z < 0.1$ baseline. But the supernova discoveries, measurements, and analyses are nearly independent.)

Figures 9.1 and 9.2 show the impressive consistency of results from the two groups. They are from reviews by Riess (2000) and Perlmutter and Schmidt (2003); the data in these figures are only slightly different from what was published in 1998 and 1999. The prediction of the Einstein–de Sitter model in Figure 9.1 is plotted as the line of long dashes. The measurements by both groups curve above this prediction at larger redshift, meaning that the supernovae are fainter in the sky than predicted by this particularly simple model. The dotted line shows that lowering the mass density while keeping $\Lambda = 0$ gives a better fit. The solid line shows that adding the cosmological constant in the ΛCDM model does even better.

Figure 9.2 shows constraints on cosmological parameters from the SNeIa *z-m* measurements interpreted in the ΛCDM cosmological model generalized to allow open or closed space sections. The horizontal axis is the matter density parameter Ω_m. The vertical axis is the parameter Ω_Λ that represents Einstein's cosmological constant Λ in the Friedman-Lemaître equations (3.3). The confidence contours are approximately 68, 95, and 99 percent. The solid and dashed contours show the consistency of results from the two independent experiments.

An open cosmological model with $\Lambda = 0$ and matter density parameter $\Omega_m \sim 0.3$ just touches the outer confidence contour in Figure 9.2. Given this and the chance of slight systematic error, the case against the OCDM model from the 1998 and 1999 *z-m* data alone was not yet significant. But by the end of the revolution, more *z-m* data gave tighter constraints that seriously falsify OCDM (Knop et al. 2003; Tonry et al. 2003).

We might pause yet again to notice the interest in how the world will end, if you trust the theory. The Berkeley group had announced that "Our eventual goal is to determine if the universe is open, flat, or closed" (Perlmutter et al. 1990). The nearly horizontal solid curve in Figure 9.2 marks the separation between models that will expand into the indefinitely remote future—a big freeze—and those that eventually collapse to a big crunch. By the end of the revolution, the evidence pointed to a big freeze, though that is an extrapolation too far.

The straight dashed line in Figure 9.2 marks the parameters at which the second time derivative of the expansion parameter $a(t)$ vanishes at the present epoch, meaning that the deceleration parameter q_0 in equation (3.23) vanishes, and the expansion rate now is changing from the earlier deceleration due to the attraction of matter to the acceleration driven by Λ. At mass density parameter $\Omega_m \lesssim 0.3$, these measurements would allow the coincidence that we flourish when cosmic acceleration vanishes, $q_0 = 0$. But later constraints put the transition to acceleration at about the time of formation of the solar system.

Section 3.5 reviews the arguments for the particular elegance of the Einstein–de Sitter cosmological model with $\Omega_m = 1$ and $\Omega_\Lambda = 0$. We see in

Figure 9.2 that the supernova measurements clearly are inconsistent with Einstein–de Sitter. There already was the reasonably persuasive case reviewed in Sections 3.6.4 and 8.4 that the mass density is well below Einstein–de Sitter, but more is better: We are reaching for a large extension of the established world picture.

The two groups were well aware of a longstanding issue: if the universe is evolving, surely its contents are evolving, too. How can we be confident that observations of objects nearby, at redshift z close to zero, and objects observed in the universe when it was younger, at $z \sim 1$, can be reduced to a common intrinsic luminosity? Both groups gave careful consideration to this question and argued that evolution is not likely to have corrupted their interpretation of the measurements. Particularly important is the check announced by Riess et al. (2004, 665), that "We have discovered 16 Type Ia supernovae (SNe Ia) with the Hubble Space Telescope (HST) and have used them to provide the first conclusive evidence for cosmic deceleration that preceded the current epoch of cosmic acceleration."

In the Friedman-Lemaître equation (3.3) with the cosmological parameters indicated by the redshift-magnitude measurements, the expansion of the universe is accelerating now but was decelerating, because of the larger mass density, at redshift $z > z_e \approx 0.67$ (equation (3.27)). Riess et al. (2004) presented Hubble Space Telescope observations of 15 SNeIa at redshifts z in the range $z_e < z < 1.6$, where the expansion is predicted to have been decelerating. The fit of their data to equation (3.26) indicates the signature of this transition, at $z_e = 0.46 \pm 0.13$. This is 1.6 standard deviations below the prediction, comfortably within the measurement uncertainty and consistent with the argument that the supernovae luminosities are well controlled. This result, along with the scrutiny of control of luminosity evolution by experts who had not taken part in the measurements, supported the community consensus for detection of cosmic acceleration and Einstein's cosmological constant.

But on the other hand the community is rightly conditioned to worry about evolution, and there is always the chance that nature has presented us with some other cause for the apparent detection of cosmic deceleration at $z > z_e$ and acceleration at lower redshifts. My impression is that in the absence of other evidence pointing to the ΛCDM model with $\Omega_m \approx 0.3$, the community would have separated into three groups. One would have emphasized that the measured z-m relation is consistent with the prediction of the 1948 steady-state cosmological model, within reasonable allowance for measurement errors. That group surely would have been small, however, despite the impressive success of this steady-state prediction. It is difficult to see how the 1948 steady-state model could be reconciled with the evidence of cosmic evolution, including the thermal spectrum of the CMB (Figure 4.7) and the observations of galaxies at high redshift that look systematically different from—younger than—nearby galaxies. A second, larger, group would have concluded that we must learn to live with Λ. I suspect that a third, maybe even larger, group would have agreed that the two supernovae teams and those outside the groups who checked on their arguments have made exemplary explorations for systematic errors caused by undetected evolution of the natures of the supernovae, or by the selection of different subsets of the supernovae at different redshifts, or by obscuring matter with different wavelength dependence at different redshifts. But the observers cannot have checked everything. I imagine this third group asking which is more reasonable: that Λ has this extraordinarily awkward value, or that there is some subtle systematic error in the measurements and Λ really vanishes? The division did not happen because the CMB anisotropy measurements that were being obtained at the same time as the redshift-magnitude observations produced an independent case for the low mass density ΛCDM cosmology, as reviewed in Section 9.2.

Cosmologists have been discussing the redshift-magnitude relation since the 1930s. At the turn of the century, two groups at last and in effect at the same time independently obtained consistent, seriously constraining measurements of the relation. This memorable achievement was recognized by the award of the 2011 Nobel Prize for Physics to the group leaders Saul Perlumtter (2012), Adam Riess (2012), and Brian Schmidt (2012).

9.2 The CMB Temperature Anisotropy

In the original sCDM cold dark matter model, and the variants discussed in Section 8.4, acoustic (pressure) oscillations of the plasma and radiation while coupled in the early universe leave distinctive patterns in the distributions of matter and the CMB radiation. In the Fourier expansion of the mass distribution, plane waves with amplitudes that are growing at decoupling of plasma and radiation end up with larger amplitudes than waves with amplitudes that are approaching zero at decoupling. The effect on the matter power spectrum—the square of the amplitudes in the Fourier expansion of the mass distribution—is the oscillating form illustrated in Figure 5.2 on page 206. The oscillations are less pronounced in the ΛCDM model, because only the baryons are coupled to the radiation in the early universe, but the effect is observed in the galaxy distribution and in the angular distribution of the CMB.

The effect of the plasma-radiation acoustic oscillations on the the radiation temperature observed as a function of angular position across the sky is seen in the analog of the Fourier power spectrum, the spherical harmonic expansion

$$T(\theta,\phi) = \sum_{l,m} a_l^m Y_l^m(\theta,\phi), \quad (\delta T_l)^2 \equiv \frac{l^2}{2\pi}\langle |a_l^m|^2 \rangle, \tag{9.1}$$

where $\langle |a_l^m|^2 \rangle$ is the average over m of the squares of the expansion coefficients a_l^m for given l (as in equations (8.3) and (8.6)).

The term $l=0$ in this expansion vanishes, because the mean has been subtracted; and the term $l=1$ is considered separately, because the evidence is that it is dominated by the dipole anisotropy produced by our motion through the sea of radiation. The anticipated form of the angular power spectrum in sCDM and related models at $l \geq 2$ is illustrated in Figure 8.1, from Bond and Efstathiou (1987). A decade later, as the two supernova groups were reaching a significant cosmological test from the measurement of the redshift-magnitude relation, the emerging evidence from the CMB measurements was that the angular spectrum may actually look like Figure 8.1. The discovery that there is an anisotropy peak in $(\delta T_l)^2$, and that it is present in the range of angular scales expected from the range of cosmological parameters then under discussion, was an encouraging hint that the CDM picture actually may be a useful approximation. It led me to abandon my search for alternatives to CDM.

The bottom figure in Plate I (in the color plate section) illustrates the progress in measuring the CMB temperature anisotropy spectrum. The panels were made by Lyman Page and published in Peebles, Page, and Partridge (2009), where the measurements summarized in these panels are reviewed. The horizontal axis in each panel is the spherical harmonic degree l. The vertical axis is δT_l in equation (9.1), but note that the vertical scales change as the measurements improve. The gray patches in each panel are weighted means from the accumulated measurements prior to the dates in the panel; the data points in each panel are the results published in the indicated year or range of years; and the solid curve in each panel is the prediction of the ΛCDM model with parameters fitted to the first-year anisotropy measurements by the Wilkinson Microwave Anisotropy Probe, WMAP.

An early goal of the anisotropy measurements was to check for the presence of the anisotropy peak predicted in sCDM and its variants. We see from the gray patches in Panel (a) in the bottom figure in Plate I that there was little evidence of this peak prior to 1997. One could argue that the peak is detected in the 1997–1999 data in Panel (a), but the case is clearer in Panel (b). The presence of the second and third of the predicted peaks is reasonably well demonstrated in Panel (c). This was a valuable addition to the case that sCDM and variants approach a useful approximation to reality. Main contributors to this advance are the MAXIMA (Millimeter-wave Anisotropy Experiment Imaging Array) measurements reported by Hanany et al. (2000) and the BOOMERanG (Balloon Observations Of Millimetric Extragalactic Radiation and Geophysics) measurements reported by Netterfield et al. (2002).

The WMAP first-year results in Panel (d), reported by Hinshaw et al. (2003), tightly trace out the spectrum over the first two peaks. This is even better evidence that the CDM picture is on about the right track. The large error flags at $l \lesssim 10$ in this panel show the scatter, or "cosmic variance," between the predicted ensemble average model value and the realization present in our CMB sky. The large error flags at $l \gtrsim 600$ are at the limit of the angular resolution of the instrument. The summary results from the Planck spacecraft mission (Planck Collaboration 2018) and advances in ground-based measurements 15 years later are much tighter and extend to much smaller angular scales. But the main elements of the picture have not changed much.

The top figure in Plate II illustrates progress in constraining the values of the mean mass density and Einstein's cosmological constant, represented by the parameters $\Omega_{\rm m}$ and Ω_Λ, respectively, and computed under the assumption of a relativistic CDM cosmological model with mass density dominated by CDM and Gaussian adiabatic scale-invariant initial conditions. The constraints show a distinct degeneracy: a far tighter bound on the sum $\Omega_{\rm m} + \Omega_\Lambda$ than on the difference $\Omega_{\rm m} - \Omega_\Lambda$.[2]

Panel (a) in the top figure in Plate II is from Bunn and White (1997). Lineweaver (1998); Efstathiou et al. (1999); and Tegmark and Zaldarriaga (2000) obtained similar results. The anisotropy data for these analyses are similar to what is shown in Panel (a) in the bottom figure in Plate I.

Panel (b) in the top figure in Plate II is from Jaffe et al. (2001). The blue region shows the range of parameter values allowed by CMB anisotropy measurements similar to the year 2000 data in Panel (b) in Plate I. The likelihood contours are roughly equivalent to 1, 2, and 3 standard deviations and are computed under the constraint that Hubble's constant and the baryon mass density cannot be quite outside conventional estimates. De Bernardis et al. (2000) present similar results. The bounds on the cosmological parameters from the CMB anisotropy data in Panels (a) and (b) run close to the dashed line at which the space curvature term Ω_k vanishes. It certainly was and is interesting that cosmological inflation as usually understood requires flat space sections, meaning $\Omega_k = 1 - \Omega_{\rm m} - \Omega_\Lambda = 0$.

The yellow and orange bands in Panel (b) show the range of parameters allowed by SNeIa z-m measurements, as in Figure 9.2. It was a notable advance to see that $\Omega_{\rm m}$ and Ω_Λ can be chosen to fit both of these very different probes of the universe. The degeneracy of constraints from the z-m

2. This is the geometrical degeneracy discussed by Bond, Efstathiou and Tegmark (1997); Hu, Sugiyama, and Silk (1997); and Zaldarriaga, Spergel and Seljak (1997). The present CMB radiation temperature is well measured, so the temperature at which the radiation allows the plasma to combine sets the redshift $z_{\rm dec} \sim 1200$ at decoupling of baryons and radiation. At decoupling, the effects of space curvature and Λ can be ignored as small compared to the densities of matter and radiation. The known values of the present CMB temperature and $z_{\rm dec}$, with $\Omega_b h^2$ from BBNS, for example, leaves three adjustable parameters, $\Omega_{\rm m}$, Ω_Λ, and h, to fix conditions at decoupling. Any combination of the three that yield the same matter density at decoupling and the same angular size distance back to decoupling produces close to the same angular power spectrum. (This is apart from the CMB anisotropy at angular scales larger than the Hubble length at decoupling.) Three free parameters to fit two constraints allows a considerable spread of values for one of the parameters.

measurements in Figure 9.2 runs in the perpendicular direction to the degeneracy from the CMB anisotropy, so the two together offer tight bounds on both Ω_m and Ω_Λ, provided you trust the theory.

Edward Wright[3] made Panel (c) in Plate II from the accumulated CMB anisotropy measurements in 2002, with the important contributions of the MAXIMA and BOOMERanG experiments, but before the WMAP data. Each dot is a trial set of parameters in the CDM model that yields a reasonable fit to the anisotropy measurements. The sensitivity to the value of Hubble's constant, H_0, is indicated by the colors of the dots that match the colors of the values of H_0 near the top of the panel. This is a helpful way to show how the model fit to the anisotropy depends on the extragalactic distance scale, the third parameter.

The constraints in Panel (c) in Plate II would allow the Einstein–de Sitter parameters $\Omega_m = 1$ and $\Omega_\Lambda = 0$ if Hubble's constant were $H_0 \approx 40$ km s^{-1} Mpc^{-1}. This is well below astronomers' estimates, but the situation had to be considered (Shanks 1985; Bartlett et al. 1995; M. White et al. 1995; Lineweaver et al. 1997). The WMAP first-year data shift the band of allowed parameters to exclude Einstein–de Sitter based on the anisotropy data alone, even allowing a distinctly awkward value of the Hubble parameter. This is shown in Figure 13 in Spergel et al. (2003). Wright's figure in Panel (d) is based on the WMAP three-year observations (Spergel et al. 2007, Figure 21). It shows impressively tight constraints.

The astronomical bounds on Hubble's constant, let us say, 60 to 80 km s^{-1} Mpc^{-1}, with the bounds from the CMB anisotropy in Panel (c), require $\Omega_m \sim 0.3$ and $\Omega_\Lambda \sim 0.7$. The former agrees with what is indicated by the array of probes of the mean mass density reviewed in Sections 3.6.3–3.6.5. The latter agrees with the redshift-magnitude measurements interpreted in the big bang cosmology.

9.3 What Happened at the Turn of the Century

In 1995, there was a modest and sensible empirical case for the ΛCDM model that assumes the CDM picture with $\Omega_m \sim 0.3$ and Einstein's cosmological constant to keep space sections flat. A modest case could also be made for OCDM, the open version with $\Lambda = 0$. Both would fit most of the evidence equally well, but the expansion time in OCDM seemed short compared to stellar evolution ages. The general theory of relativity predicts a longer expansion time in ΛCDM, which makes it comfortably consistent with stellar evolution. But this solves the age problem by introducing a parameter, Λ. How can we judge whether this is more than an adjustment of the theory to fit what is

3. A black and white version of Panel (c) is published in Wright (2004). I am grateful to Ned Wright for allowing me to use his original version.

observed, as in a just-so story? The answer must be to check for consistency with independent ways to constrain the parameters and otherwise test the theory without more adjustments. This became possible, in abundance, during the revolutionary advance of cosmology in the years around 2000.

Since the ΛCDM theory is incomplete, we need not pause in this review to debate whether modest discrepancies in the derived parameters in excess of nominal measurement uncertainties are to be attributed to problems with the theory or with the measurements. We are seeking to understand the nature of the case that the ΛCDM theory gives us a useful approximation to what actually happened as the universe expanded and cooled from a state that would have been radically different from what it is now. The example in the bottom figure in Plate II is the comparison of theory and measurements of the spatial power spectrum of the galaxy distribution and the angular power spectrum of the CMB radiation temperature as a function of position across the sky (the latter defined in equation (9.1)).

During the revolution, Percival et al. (2001) presented a reasonably clear detection of the effect of the acoustic (pressure) oscillations in the galaxy distribution. The effect is even clearer in Cole et al. (2005); and it is seen in still more detail in the Beutler, Seo, Ross et al. (2017) spectra in the three slices of redshift shown in Panel (a)[4] in the bottom figure in Plate II. The Eisenstein et al. (2005) demonstration of the acoustic oscillation effect in the Fourier transform of the spatial power spectrum—the galaxy two-point position correlation function—combines the row of peaks and valleys in the power spectrum into a single peak in the correlation function, which aids precision. But the spatial spectrum in Panel (a) offers a more intuitive comparison to the peaks and valleys in the angular spectrum in Panel (b).

Panel (b) in the bottom figure in Plate II is from the Planck Collaboration (2014b); it combines the Planck satellite measurements to 2012, the WMAP 9-year data that were important on larger angular scales, and on smaller scales the ground-based Atacama Cosmology Telescope and South Pole Telescope measurements.

Panel (a) is based on observations of galaxies at redshifts $z \lesssim 1$. Panel (b) is based on observations of radiation that propagated to us nearly freely since decoupling at redshift $z \sim 1000$. The data in the two panels were obtained by different methods of observation and analyses of different phenomena observed in the universe as it was at two different stages of its evolution. Beutler et al. (2017, 2424) conclude that "Our constraints for all redshift bins

4. The label "pre-recon" in the figure means that these spectra are based on the observed galaxy positions. Reconstruction adjusts positions to take account of the post-recombination displacements gravitationally driven by the mass distribution. This improves the signal at smaller scales, but for our purpose complicates the test. Other details of the data reduction that improve parameter constraints, a subtle business, need not be considered here.

Table 9.1. Checks of Cosmological Parameters[a]

	Matter density Ω_m	Baryon density $\Omega_{baryon}h^2$	Helium Y_p
$z \lesssim 0.1$	~ 0.3[b]	~ 0.015	0.23 ± 0.01
$z \sim 1$	0.26 ± 0.06	—	—
$z \sim 10^3$	0.28 ± 0.02[b]	0.022 ± 0.001	0.33 ± 0.08[c]
$z \sim 10^9$	—	0.021 ± 0.002	0.25 ± 0.01

[a] At the end of the revolution, and assuming flat space sections.
[b] Assumes $h \approx 0.7$.
[c] First detected later, in the WMAP 7-year observations.

are in good agreement with the Planck prediction within CDM." That is, within measurement uncertainties, the same ΛCDM theory and parameter values fits these two quite different ways to probe the universe. It is impressive that the theory passes this searching test.

The test displayed in Panel II is richer than the demonstration of consistency of derived parameters, because the theory also successfully accounts for the detailed shapes of the two spectra. But translating that into an effective number of independent checks is not very instructive. It is best left as a visual impression, but an important one to bear in mind.

Table 9.1 summarizes other tests by comparisons of derived parameter values. The first row is based on what might be termed "local observations," at distances small compared to the Hubble length. The rich history of measurements of the mean mass density along the various lines of evidence summarized in Table 3.3 and Figure 3.5 is illustrated by the number of results sampled in Table 3.2 on page 80. Understanding these results as they were being obtained was confused by the difficulty of assessing systematic errors, and in particular, the possible difference between the distributions of galaxies and mass that could reconcile some of the observations to the elegant Einstein–de Sitter model. But that is not true of others. My conclusion, explained in Section 3.6.4, is that the results from this wealth of research offer a sensible case that the cosmic mean mass density is about a third of the Einstein–de Sitter prediction.

This low mass density was not a welcome result in the 1990s, but we have two checks if we accept the argument from inflation that the universe is cosmologically flat. First, by the end of the revolution, measurements of the supernova redshift-magnitude relation, with the assumption of flat space sections, yielded the entry in the second row of the second column in Table 9.1 (Knop et al. 2003; Tonry et al. 2003). Second is the result of the fit of the ΛCDM model to the WMAP satellite first-year measurements of the CMB anisotropy, again under the assumption of flat space sections (Spergel et al. 2003). This tests our picture of what happened at $z \gtrsim 10^3$.

The prime point is not the precision of these three measures of the mass density, which is modest indeed for the astronomical measurements. It is the consistency from phenomena observed at three different stages in the evolution of the universe: locally, at redshifts $\lesssim 0.03$; the measurements of the redshift-magnitude relation to redshift $z \sim 1$; and the observations of the CMB anisotropy, which reflects what was happening at $z \gtrsim 1000$. These data were obtained and analyzed in quite different ways. Their consistency demonstrates that ΛCDM theory passes a demanding test, far beyond what could be done before the revolution.

The third column in Table 9.1 lists the baryon mass density. The history of estimates of this quantity is sampled in Table 4.2 on page 178. The measurements based on the pre-galactic deuterium abundance depend on the theory of light-element production in the early stages of expansion of the universe, at redshifts $z \sim 10^9$. The scatter in these measurements has been considerable, because the deuterium measurement is difficult. But by the end of the revolution, we see a stable case for the entry in the bottom row of Table 9.1 (Kirkman et al. 2003). The baryon mass density figures in a different way in the velocity of sound in the acoustic plasma-radiation waves that produced the statistical patterns in the observed distributions of matter and radiation illustrated in Plate II. The entry in the third row is from the WMAP first-year data, again under the assumption of flat space sections. We have yet a third measure from the cluster baryon mass fraction, under the reasonable-looking assumption that clusters of galaxies are large enough to have captured a fair sample of the cosmic mass fraction in baryons. In the 1990s, this assumption was used with the baryon mass density from BBNS to estimate the mean mass density (category B_2 in Table 3.3). But now we can turn it around: Given the present mean mass density, which seems well constrained, the cluster baryon mass fraction ~ 0.05 yields the baryon mass density in the first row. The estimate is crude but significant.

To recite again the prime point: Here are consistent measures of the baryon mass density based on the BBNS theory of thermonuclear reactions at redshift $z \sim 10^9$; the acoustic oscillation theory of what happened at $z \gtrsim 10^3$; and the observations of the baryon mass fraction in clusters of galaxies at $z \lesssim 0.1$.

The baryon mass density is bounded below by what is seen in stars: $\Omega_{\rm stars} \sim 0.003$ (Fukugita and Peebles 2004). Since this is much less than the other three measures, we must postulate that most of the baryons are now "hidden" in coronae of galaxies, plasma between the galaxies, low-mass stars, and maybe even smaller and fainter objects. This does not indicate a flaw in ΛCDM; nature is entitled to hide baryons. But probes for the hidden baryons will be followed with interest.

We have measures of the cosmic helium mass fraction Y_p from astronomical observations, the acoustic oscillation theory of the patterns in the distributions of galaxies and the CMB, and the BBNS theory of thermonuclear reactions in the early universe. Pagel's (2000), survey of the astronomical situation indicates the value of Y_p in the first row of Table 9.1. Sarkar's (1996) BBNS computation with current estimates of the baryon density is entered in the bottom row. At the turn of the century, the two were consistent. The effect of the mass in helium on the dynamics of the acoustic oscillation of the baryon-radiation fluid was detected after the revolution. For completeness, I enter in the third row and last column of Table 9.1 the first detection from the Komatsu et al. (2011) analysis of the Wilkinson Microwave Anisotropy Probe 7-year measurements. The Planck spacecraft mission placed a considerably tighter bound (see Table 4.1). The theory passes yet another serious test.

There is a long history of discussions of whether the cosmic expansion time that cosmology can allow is consistent with evolution ages found from geology and astronomy. The WMAP first-year measurements put the expansion time in the ΛCDM cosmology at a strikingly well-constrained value, $t_0 = 1.34 \pm 0.03 \times 10^{10}$ years (Spergel et al. 2003). This is not inconsistent with stellar evolution ages, a significant though not precise test. The age from cosmology proves to be a useful constraint on the study of star formation history. As both subjects develop, they may tighten to the point of checking the cosmology.

We have a serious check on time scales from the comparison of constraints on the value of Hubble's constant based on observations of the distances and redshifts of relatively nearby galaxies and based on the acoustic oscillation theory of the patterns in the distributions of galaxies and the CDM at $z \gtrsim 10^3$. At the end of the revolution, the two were reasonably consistent at $h \approx 0.7$, which was comforting. At present, authorities are divided on whether increased precision is revealing a significant difference between the two measurements, and if so, whether the difference reveals the need to adjust ΛCDM or correct some subtle systematic error in the astronomical measurement. But this issue is secondary to the prime point, that these two measures based on quite different phenomenology are quite similar.

Despite the hazard of loss of credibility by protesting too much, I emphasize again the broad variety of observations that probe the universe in such different ways and offer a story that is close enough to consistency to make the case for the ΛCDM theory about a compelling as it gets in natural science. This great extension of established physical science was made possible by impressive advances in technology applied with great effort. But it was also essential that the theoretical predictions are reliable, because they are computable from first principles or reasonably close to it. The tests do venture into the complexity of the behavior of baryons in clusters of galaxies, but that seems to be well enough understood for the situation at the turn of the century.

I expect more is to come from the galaxies that are observed in rich detail, which promises a broad probe of cosmology. Sorting through the complexities of the histories of the galaxies is a great challenge and opportunity.

The ΛCDM theory was assembled out of the simplest ideas that would allow a fit to the observations. It certainly would not be surprising to find that better observations require a better cosmology: perhaps an adjustment of the gravity theory, maybe a more interesting dark sector. But the modest sizes of real or apparent inconsistencies among the broad variety of ways to examine the universe lead me to expect that a better theory will predict a universe much like ΛCDM, because what is observed seems to be so much like ΛCDM. This is the best we could have hoped for, and a remarkable advance of science.

9.4 The Future of Physical Cosmology

It may be tempting to ask whether we have at last arrived at an empirically complete theory of the evolution of the universe since redshift $z \sim 10^{10}$, requiring only the addition of more decimal places to the measurements of the parameters of ΛCDM. My experience in a half century of research in this subject, with occasional time off, does not encourage me to attempt an answer. I have been repeatedly surprised by the advances in precision of the measurements. I do not recall ever pausing to wonder whether this at last is about as far as they can go, and it feels unnatural to ask myself that now.

Reluctance to attempt an answer is reinforced by cautionary examples from the history of science. In his book, *Light Waves and Their Uses*, Albert Michelson (1903, 23–24) wrote that

> The more important fundamental laws and facts of physical science have all been discovered, and these are now so firmly established that the possibility of their ever being supplanted in consequence of new discoveries is exceedingly remote.

The elegant experimental methods Michelson describes in this book indeed remain standard parts of physics, though advances in technology allow them to be done better. In the same sense, I expect that ΛCDM is not likely to be supplanted, apart from adjustments that are modest relative to what is now observed to accommodate tighter measurements and maybe the different predictions of a deeper theory. Something similar happened to the physical science Michelson was discussing.

In Michelson's (1903, 163) chapter on the fascinating array of experimental probes of the ether, he reports that

> The phenomenon of the aberration of the fixed stars can be accounted for on the hypothesis that the ether does not partake of the Earth's motion in its revolution about the sun. All experiments for testing this hypothesis have, however, given negative results, so that the theory may still be said to be in an unsatisfactory condition.

Thomson Lord Kelvin (1901, 7), in his essay "Nineteenth Century Clouds over the Dynamical Theory of Heat and Light," notes that Michelson's failure to detect motion through the ether might be resolved by the

> brilliant suggestion made independently by FltzGerald and by Lorentz of Leyden, to the effect that the motion of ether through matter may slightly alter its linear dimensions, according to which if the stone slab constituting the sole plate of Michelsen and Morley's apparatus has, in virtue of its motion through space occupied by ether, its lineal dimensions shortened one one-hundred-millionth in the direction of motion, the result of the experiment would not disprove the free motion of ether through space occupied by the earth.

The ether is Kelvin's Cloud I, a step to a revolution: special relativity. Kelvin's Cloud II is the phenomenology of heat capacities. He does not refer to Planck, but here, too, are steps to a revolution: quantum physics. At the time of writing, I take Cloud I over the cosmology of the twenty-first century to be the issue of what happened in the very early universe that replaces the extrapolation of ΛCDM back to a singularity. I take Cloud II to be the enigmatic simplicity of the dark sector of ΛCDM.

The cosmological inflation picture offers a framework for a resolution of Cloud I. We must pay attention to it, because inflation played an important part in forming the ideas that led to the ΛCDM model. Recall the hints from inflation for flat space sections and the close-to-adiabatic Gaussian power law of primeval departures from homogeneity with a slight and natural tilt from scale invariance. Human ingenuity and persistence being what they are, I count as a historical inevitability the construction of a logically complete theory connecting standard physics with ΛCDM, or some improved, empirically based cosmology, to the nature of the universe before the big bang. If this new and logically complete theory produces no new testable predictions, should it be taken to show us what really happened before the big bang? I do not hesitate to assert that we almost certainly know what really happened long ago and far away, back to when the light elements were forming and out to where we see the young galaxies. I would not say that about a logically complete theory of what happened before the big bang without tests from new predictions. Perhaps the logic and elegance of a new and complete theory will compel acceptance by generations to come. But a more happy possibility, to my way of thinking, is that Cloud I over ΛCDM will be found to offer hints to new cosmological tests and a better theory of how the world around us came to be the way it is.

Cloud II over cosmology has a different character from the clouds over twentieth-century physics. There is no empirical problem with the dark sector of ΛCDM, no hint to a better theory; the complaint is the nonempirical feeling that this large segment of the universe is not likely to be so exceedingly simple. Maybe in the excitement of the development of a cosmology that could be tested and even established, we have inadvertently put aside something interesting to be dealt with later, and then forgotten it. Maybe as the cosmological tests are tightened, they will at last reveal the failure of the extrapolation of standard physics. It happened with Kelvin's Clouds.

CHAPTER TEN

The Ways of Research

EACH BRANCH OF the natural sciences has its particular operating conditions, but there are commonalities. I have advertised cosmology as a worked example of how the reach of established science has grown, in this case in a way that is simple enough to be examined in some detail in the space of this book. I offer for consideration thoughts drawn from this story that may more broadly illustrate the nature of the enterprise of natural science.

10.1 Technology

It is obvious but must be stated that research in the natural sciences depends on technology that was developed largely for other purposes. Let us consider examples in the story of how modern cosmology grew.

Walter Baade (1939) reported that the Eastman Company had offered him red-sensitive photographic plates labeled "H-α Special." It allowed Baade to obtain images of galaxies on red- and blue-sensitive plates taken with the 100-inch telescope on Mount Wilson, and by blink comparisons find regions luminous in the red H-α emission line at 6600 Å that are so useful for measurements of Doppler shifts. The plate label might indicate this was an advance in technology for the purpose of astronomy, but I imagine the improvement of panchromatic film had broader commercial appeal.

Walter Baade's finding lists for emission line regions allowed Vera Rubin and Kent Ford to use their image intensifier to make precise measurements of the redshifts of this line in enough regions across the face of the nearby spiral galaxy M 31 for a reasonable estimate of the streaming motion of the stars and gas, and from that the mass needed to hold the galaxy together. With Gérard de Vaucouleurs's measurements of the distribution of stars, the streaming speeds gave evidence of a faint envelope of mass in the outskirts of this galaxy. The construction of radio telescopes with the angular resolution needed to trace Doppler shifts of the 21 cm emission line of atomic hydrogen across the faces of nearby galaxies gave similar indications. These were important steps in the chain of evidence that led to the nonbaryonic dark matter at the heart of the established cosmology. Fritz Zwicky discovered evidence of the subluminal mass in the Coma Cluster of galaxies in the 1930s. The presence of subluminal mass draped around galaxies was a realization that grew through decades of research aided by the successive advances in technology illustrated in Figure 6.2.

Technology enabled far more efficient measurements of galaxy redshifts. In the first statistical application of a dynamical measure of galaxy masses, Margaret Geller and I had a list of 527 measured redshifts of galaxies. Now there are measurements of redshifts and optical spectra of millions of galaxies. The technology that made this possible was not aimed at astronomy; it was adapted, in part for the purpose of obtaining enough data on galaxy positions and motions for a meaningful determination of the cosmic mean mass density. Large parts of this program depend on the high quantum efficiencies of digital photon detectors and the methods of data storage and analysis that can deal with the vast amounts of data these arrays of detectors produce. It made possible the great programs of measurements of the mean mass density in the 1990s reviewed in Sections 3.6.4 and 3.6.5.

10.2 Human Behavior

It also is obvious but should be stated that the ways of research in science are the ways people tend to operate in general. Some of us are capable of focusing great energy on a well-motivated and challenging problem, whether building a pyramid or developing the vast amount research summarized in Table 3.2 on the value of the cosmic mean mass density. An early motivation for the latter was to discover whether the density might be large enough to cause the universe to stop expanding, but by the 1990s, the focus had grown tighter: just measure the mass density. The results were scattered through the literature and not easily assessed, and they played a lesser role in the revolutionary developments in the years around 2000 than might have been expected. But this mass density program produced a valuable guide to the thinking in the late 1990s that was building up to the revolution, and it made a valuable contribution to the empirical case for the standard and accepted cosmology.

A working condition that may be particularly relevant for cosmology is the tendency to take a personal interest in the results: how did our universe begin, what is it like now, and how might it end? This is on scales of space and time exceedingly different from what we experience, but I expect it is adaptively advantageous to consider such things as how the world will end, even when there is no need to make preparations.

I count as an aspect of such modes of thought the tendency in the mid-1980s through the 1990s to expect that the large-scale mean mass density is such that the universe will continue to expand at escape velocity for a long time to come. Otherwise we flourish at a particular epoch, just as the rate of expansion departs from escape velocity. Such arguments from elegance can be productive: They led to serious consideration of the great extrapolation of general relativity to a homogeneous expanding universe. And such arguments can fail: They beclouded considerations of the mean mass density and Einstein's cosmological constant, though all that eventually was straightened out. Large failures, the likes of the extrapolation of classical physics to the scale of atoms, which forced the radical adjustment to the world view of quantum physics, have not been encountered in empirical cosmology so far. I join the ranks of those who hope that will change.

An interest in how the world might end many billions of years from now calls for another broadly useful trait: the optimistic trust in extrapolations from what we know. An example in cosmology was the general decision to stick to general relativity along with more firmly established physics. Maybe this was in part because general relativity theory is particularly elegant, but surely also in part because others had been doing so back to Einstein. We may term this decision an implicit and sensible nonempirical assessment of the situation, though I don't know whether many at the time would have put it that way.

The vast extrapolation from established physics to cosmology has also invited adventurous ideas, such as continual creation, white holes, superconducting magnetized cosmic strings, quantum spacetime foam, multiverses, and the hypothetical dark matter and dark energy, along with gravity physics and/or initial conditions contrived to employ or do away with such ideas.

So how did we arrive at a cosmology that most authorities agree is securely established? The way forward for the theory proved to be simple: extrapolate standard physics applied to a cosmological model chosen largely on the grounds of the simplest possible adjustments to observational constraints as they developed. This cosmology offers tests that bring theory and observation together by computations largely in perturbation theory, which we can trust. If there were a better but more subtle theory, our methods of research would not have been likely to find it, because most of us are content with ΛCDM. The pragmatic thinking here, as in all natural science, is that if there is a better theory, we will eventually realize it by the failure of what we have accepted. It happened in the atomic physics that drove us to quantum theory.

10.3 Roads Not Taken

Suppose Einstein had decided to become a musician. We would have had special relativity in flat spacetime; others were on the trail to it. But general relativity would have come later if at all, and we would not have had Einstein's influential proposal that the universe is homogeneous and isotropic in the average over local fluctuations. It would have allowed more room for the idea that a local explosion caused the galaxies to fly apart into empty space.

The explosion picture still would have been controversial—what do we make of the isotropy of Hubble's deep galaxy counts, and the large redshifts?—but worthy of debate.

If the early hints of homogeneity had failed to attract interest, then the growing use of microwaves for communication would have done it by forcing attention to the engineers' excess microwave noise. The manifest isotropy of this radiation—the absence of a change in the noise level as receivers moved across the sky—would have presented a conundrum if we had not already had the notion of a homogeneous expanding universe. Could the isotropy of the radiation really mean we are at the center of a spherically symmetric universe? Don't all those other galaxies look to be equally good homes for observers? It would seem more likely that observers on any of the many other galaxies would see what we see. But if we are in a homogeneous static sea of microwave radiation, then observers in most galaxies would be moving through the sea and observing the Doppler effect, which makes the radiation hotter in the direction of motion, colder in the opposite direction. We do not see a large anisotropy; why would we be so special?

It would be an interesting exercise for someone conversant with special relativity in this alternative world to show how conditions in flat spacetime can be set up so that observers in all galaxies see the same general recession of the other galaxies and the same isotropic sea of microwave radiation. We have the answer: the limit of the Robertson-Walker line element and the Friedman-Lemaître equations (3.1) and (3.3) at zero mass density and zero cosmological constant. A coordinate labeling for this situation gives the line element

$$ds^2 = dt^2 - (t/R)^2 \left[\frac{dx^2}{1 + (x/R)^2} + x^2(d\theta^2 + \cos^2\theta d\phi^2) \right], \tag{10.1}$$

where R is a constant. This is a coordinate labeling of the flat spacetime of special relativity. Let each of the galaxies be moving at the constant speed set by placing each at a fixed coordinate label $\vec{x}$. Set the radiation to be isotropic with the same spectrum everywhere on an initial hypersurface of fixed t. Observers in all the galaxies would see that the radiation remains isotropic with the same evolving spectrum as the galaxies move apart—we might say as the universe expands—at a fixed rate. And the observers would understand that if the spectrum is thermal then the radiation temperature is cooling as $T \propto t^{-1}$. It is a good exercise to show this using special relativity theory.

If the alternative world we are considering has our technology and our interest in its applications, then it seems inevitable that, soon after the excess microwave noise was recognized, it would have been found to have a close-to-thermal spectrum. Consider the illustrations in Figure 4.4, showing how quickly the first evidence of a thermal spectrum was obtained in our world once the presence of the radiation was recognized. It is a short step from that to notice that the universe would have been arbitrarily hot in the past, at $t \to 0$: a hot big bang.

Gamow's intuitive genius presented us with the idea of a hot big bang. He was thinking in terms of general relativity, but he remarked that his considerations work in the Newtonian limit if the radiation energy density is assigned a mass density to be added to the density of matter for the total value, ρ. Then Newton's equation for the rate of change of the radius $a(t)$ of an expanding shell of unit mass moving with the galaxies is

$$\frac{1}{2}\left(\frac{da}{dt}\right)^2 - \frac{4}{3}\pi G \rho a^2 = E, \tag{10.2}$$

where E is a constant. This is one of the Friedman-Lemaître equations without Einstein's cosmological constant Λ and with the usual constant R^{-2} replaced by the constant Newtonian energy E. The time derivative of equation (10.2) with the local energy equation for ρ in equation (3.4) yields

$$\frac{1}{a^2}\frac{d^2 a}{dt^2} = -\frac{4}{3}\pi G(\rho + 3p). \tag{10.3}$$

This is the other Friedman-Lemaître equation. It requires that pressure acts as an active gravitational mass density, which might have been unexpected, but Gamow could have lived with it, and so I expect would physicists in an alternative world without Einstein. If physicists in this alternative world knew nuclear physics, then equation (10.3) and the microwave radiation would be sufficient to work out light element abundances to compare to what is observed, assuming there are astronomers in the alternative world interested in looking into it.

The problem with this line of thought is that the expansion parameter $a = t/R$ in the flat spacetime line element in equation (10.1) disagrees with equation (10.3), given that we want a model universe with nonzero mass density. It calls for a bold thought: replace t/R in the line element (10.1) with the function $a(t)$ in equation (10.3). That produces a line element that describes curved spacetime. Again, we might suppose that imaginative physicists would recognize this and be willing to live with it. It provides a setup adequate for an analysis of the redshift-magnitude measurements, maybe without Λ, absent Einstein. But perhaps that would lead someone to invent Λ to fit the measurements.

Equation (10.3) would have to be generalized to take account of the observation that the mass distribution is clumpy. Without Einstein but with Newton, the generalization of the line element to describe the nonrelativistic behavior of matter seems reasonably straightforward; the generalization to the behavior of the radiation is more complicated. Maybe this alternative world ends up with gravity physics that resembles general relativity only in linear perturbation theory. Or maybe, as Feynman argues (in Feynman, Morínigo, and Wagner 1995), without Einstein, the path through field theory would take physicists in the alternative world to general relativity in a less intuitively demanding way. Einstein's genius in hitting on the theory of gravity and applying it to a homogeneous universe does not seem to have been needed, though it certainly gave our cosmology a powerful head start.

In our world, there was an alternative path forward after recognition of the thermal CMB. While some in the 1990s were pursuing the great goal of measuring the mean mass density, others were working on the longstanding issue of the consistency of evolution ages from geology and astronomy with what might fit cosmological models. It led to influential though not firmly based arguments for a positive value of Einstein's cosmological constant, wanted because this increases the expansion time. If this line of research had been pushed harder in the 1990s, if probes of time scales had been given the attention and resources at a level comparable to what was being devoted to the measurements of the mass density, might the time scale considerations have carried more weight in the establishment of the ΛCDM theory?

Is there another path to a different gravity theory that fits the evidence we have, or might have obtained by pursuing other lines of thought equally well or even better than ΛCDM? I expect it will not be known for sure unless evidence turns up that forces reconsideration of the present directions of thought. Before the revolution, I felt there was some chance of this happening to cosmology. It seems wildly unlikely now, but equally unlikely that ΛCDM will remain the standard model. We make progress by successive approximations.

10.4 The Social Construction of Science

It is sometimes said that the laws of physics were "there," waiting to be discovered. I would rather put it that we operate on the assumption that nature operates by rules we can discover, in successive approximations. But however it is put, the program of natural science certainly has been productive.

Cosmology is a social construction—what else? In the 1920s, the large-scale homogeneity of the universe was a nonempirical social construction, taken seriously by physicists who chose to think about it not from any significant evidence but rather from some combination of respect for Einstein's ideas and the convenience of one of the few analytic solutions to Einstein's general relativity field equation. In the 1930s, Hubble's deep galaxy counts and Hubble and Humason's deep redshift-magnitude measurements offered some support for homogeneity, but through the 1950s, a fractal universe was a viable option, with the elegant property that a mass distribution with fractal dimension $D = 1$ can be analyzed in Newtonian mechanics on all length scales (apart from the occasional black hole). But we have seen that the great interest in microwave communication made it difficult to have avoided discovering the microwave background radiation with its pretty clear hint of large-scale homogeneity. This is one of the observations that drove the concept of large-scale homogeneity from a nonempirical social construction to a well-tested part of our physical world picture.

Cosmology is an empirical construction, with data ranging from systematic programs of measurements designed for cosmology to unintended by-products, such as the excess noise in microwave radiometers (Section 4.4.2), the observation of absorption from the first excited level of the interstellar molecule CN (Section 4.4.1), and the curiously large abundance of helium (Section 4.3.1). Data and ideas about their interpretation can be misleading: Consider the elegance of a fractal universe and the observations that de Vaucouleurs (1970) could advance in its favor. Wrong ideas can be productive: The now well-supported theory of element formation in stars grew in part from the need to account for the presence of the chemical elements in the steady-state cosmology (Section 4.2.1).

Ideas are promoted from social constructions to elements of the generally accepted world picture for reasons that seem easy to sort out but difficult to encode in a general rule for admission to the canon of standard and accepted science. It is a collective subjective decision. There are some general guidelines, however. A theory might offer a testable prediction, something that did not figure in devising the theory. If the test is positive, it is a serious addition to the case for the theory. If it fails but can be remedied by adjusting the theory, the result is a modest addition to the case, because it shows that the theory at least can accommodate the facts. I take an extreme example of this from the situation in cosmology in the early 1960s, when I started looking into the subject. I disliked the scant empirical basis but found comfort in the fact that the estimate of Hubble's constant, H_0, gave a characteristic mass density, H_0^2/G, and a characteristic time, H_0^{-1}. In the early 1960s, these two quantities were not far from estimates of the mean mass density in galaxies and the ages of rocks and stars. It reassured me to see that the theory could not be all bad.

An early example of the transition to more demanding cosmological tests is shown in Figure 3.2 on page 76, from the year 1974. It shows rough consistency of more constraints than the number of free parameters that the authors were considering. The results certainly was encouraging, but I do not remember talk of it establishing cosmology. The tests completed in the half-decade around the year 2000, and reviewed in Section 9.3, are far tighter and more reliable and examine the universe from many more directions. They established ΛCDM to the satisfaction of most cosmologists by a weighing of the evidence that seems intuitively clear and sensible but not suggestive of a prescription for general use.

Cosmology shares with particle physics the possibility of computations from first principles or reasonably close to it. This enabled the remarkably demanding tests from measurements of the remnants of acoustic oscillations in the distributions of galaxies and the sea of microwave radiation. Cosmology shares with many branches of science the challenge of complexity. Observations of galaxies were important to the birth of modern cosmology, but their properties do not figure in the precision tests at the turn of the century and later, because the properties of galaxies are not at all easy to understand. I share the suspicion/hope that the puzzles of galaxy rotation curves and all the rest of the phenomenology of these objects still have something of value to teach us about cosmology.

The sociologists William Ogburn and Dorothy Thomas (1922), and Robert Merton (1961) gave the name "multiples" to the tendency in natural science for ideas and observations to appear from more than one apparently independent source. A familiar example is the independent and effectively simultaneous recognition of natural selection by Darwin and Wallace. Maybe some examples of multiples in cosmology are only coincidental, and some may be consequences of our tendency to present history as a sequence of developments that makes the process seem more orderly than it is. But maybe some of it hints at the sometimes subtle ways we communicate. It is remarkable that the five groups listed in Table 7.1 on page 291 introduced what became the prototype for the cold dark matter of the established ΛCDM theory. They seem to have been little influenced by one another or by the astronomers' broad range of evidence of subluminal mass. The same might be said of the particle physicists' explorations of baryon formation in the hot big bang just when it was needed for cosmological inflation. And I offer as a prime example the revival of research in cosmology in the two decades following the Second World War.

The end of the war made room for a burst of innovation in pure and applied sciences. Four memorable people active in physical cosmology in the postwar years are Robert Dicke, George Gamow, Fred Hoyle, and Yakov Zel'dovich. Gamow's book on nuclear physics in three successively larger editions, 1931, 1937, and 1949, show that he was well prepared for his postwar ideas about nuclear reactions in a hot early universe. Gamow offered interesting thoughts about cosmology before the war, but his great contributions to this subject were published in the year 1948. He was a brilliant intuitive physicist, but his imaginative mind was not suited for attention to pesky details or the creation of a productive research group. Gamow's part of the story could have been different if he had paid attention to someone like Joe Weber, who came out of the war with an understanding of microwave technology (page 151). If he had, I suspect we would have ended up about where we are, but earlier.

Hoyle paid close attention to observations in his pioneering theoretical examinations of the structure and evolution of stars and the chemical elements they produce. But my impression from our few conversations is that he felt that the distance and time scales of cosmology are so immense that we cannot rely on empiricism: Observations may falsify a cosmological model, but they cannot establish it; cosmology must rely on philosophical considerations. That was a reasonable assessment in the immediate post-war years, and the 1948 steady-state cosmology that he, Bondi, and Gold produced is a magnificent philosophical construction. But as the observational evidence grew far beyond anything he could have expected in the 1940s, or I had hoped for in the 1960s, Hoyle seemed to have been unable to let go of his guiding principle for cosmology, unable to accept that the observations were leading us to an empirically motivated and supported theory.

The greatest contributions to cosmology by Gamow and Hoyle were published in the same year, 1948: Hoyle with Bondi and Gold on the steady-state model, and Gamow on the hot big bang model. They surely acted independently, apart from sharing the creative energy released by the end of the war. Maybe this was a coincidence, or something "in the air." Hoyle gave the big bang model its name, but he was mocking the idea. Gamow (1954a) pointed to the problem of the age distribution of the galaxies in the 1948 steady-state model, but I have not found any other evidence that he paid much attention to it. And I have not encountered reason to think that Zel'dovich and Dicke took serious interest in the steady-state model.

Dicke and Zel'dovich came out of war research, Zel'dovich on nuclear weapons ending up at Arzamas-16, the Soviet Union's version of Los Alamos, and Dicke on radar and other microwave technology at the Radiation Laboratory at MIT in the USA. Both established productive research groups. Early members of Dicke's Gravity Research Group are shown in Plate VI. Members of Zel'dovich's research group with other important actors in the Soviet Union are pictured in Plate VII.

Zel'dovich was a theorist whose interests ranged from combustion to particle physics to the very early universe. I have seen no evidence that Zel'dovich questioned the application of general relativity, along with the rest of standard physics, to the scales of cosmology. Dicke was offended by the scant empirical tests of gravity physics and set out to improve them. He was interested in theory, provided it had some possible connection to what might be measured. I like to think that my empiricist philosophy is my own choice, but Dicke certainly reinforced it.

To the best of my knowledge, Zel'dovich and Dicke were not aware of each other when each decided to create a program of research in gravity physics and cosmology, yet another coincidence. Zel'dovich knew and was interested in what Gamow had been doing in cosmology, but at first felt that Gamow's hot big bang theory violates what was known about light element abundances. He promptly accepted the evidence for the hot big bang theory from detection of the microwave background radiation, and he made important contributions to its analysis. A notable example is the Sunyaev-Zel'dovich effect on the microwave intensity spectrum by the hot plasma in clusters of galaxies. Another from many is Zel'dovich's cosmological bound on neutrino masses that inspired Alex Szalay's thinking about hot dark matter, which is the precursor of the nonbaryonic matter of the standard ΛCDM theory.

Let us note also that when Zel'dovich's thought about neutrino rest masses led Szalay and Marx in Hungary to consider hot dark matter, Cowsik and McCelland were independently considering it in the United States. And when Doroshkevich and Novikov in Moscow were thinking about the cosmic radiation background and a possible thermal component at a few Kelvins, I was thinking along the same lines in Princeton (with the results in Figure 4.3 on page 152), again quite independently on both sides, as far as I can tell.

It will not have escaped the reader's attention that Zel'dovich and colleagues in the Soviet Union were making great contributions to cosmology despite serious challenges: delayed access to journals from outside the Soviet Union, restrictions on publication of their own research in the journals that people outside the Soviet Union were reading, and restrictions on travel outside the Soviet Union. The first satellite measurements of the cosmic microwave sea were by the Relikt mission in the Soviet Union. It was launched in 1983, 6 years before the USA COBE satellite, and it clearly mapped out the dipole anisotropy caused by our motion through the radiation. The collapse of the Soviet Union prevented the launch of the follow-up mission Relikt-2. We may suppose that a serious collapse of stock markets in the capitalist parts of the world could have prevented the launch of COBE, while a softer collapse of the Soviet Union could have allowed completion of the Relikt-2 mission. A lesson I draw is that pure curiosity-driven research can flourish in a tightly restrictive, authoritarian society, provided the maximal leaders approve. It may have helped that before turning to cosmology, Zel'dovich contributed to the success of the Soviet nuclear weapons program. He was made a Hero of Socialist Labor for this work.

Dicke liked astronomy; his early publications include the radial distribution of stars in a globular cluster (Dicke 1939) and a report of the detection of microwave radiation from the Sun and Moon (Dicke and Beringer 1946). After the war, he spent a productive decade of research on what might be termed "quantum optics." In the late 1950s, he decided to apply the great advances in the technology of laboratory physics to the neglected subject of empirical gravity physics: classical gravity experiments that could be done better and new ones that had become possible.

When I arrived at Princeton as a beginning graduate student in 1958, members of his Gravity Research Group already were working on elegant experiments probing gravity, and others were examining such arcane things (to me) as the dates and positions of early observations of solar eclipses, which test how the Moon and Sun have been moving over centuries. That was motivated by Dicke's fascination with the possibility that the strength of the gravitational interaction may be so small because it has been decreasing for a long time as the universe evolved. He felt the idea makes a lot of sense, as do those who now seek a more complete fundamental theory. But Dicke most liked looking into the many phenomena in the laboratory and to be found in all the other branches of natural science that might show signs of something interesting about gravity. His brilliant experimental methods grew into the monitoring of continental uplift following the last ice age; the seismographic network that monitors underground explosions; the laser tracking of the Moon that produces precision tests of gravity physics; and the microwave technology that has taught us so much about the structure and evolution of the universe (as reviewed in Peebles 2017). Dicke never abandoned interest in the search for new gravity physics, but he was quite content that the group he had formed pursued other varieties of observation and theory. His Gravity Research Group remains productive in its fourth generation.

In the early 1980s, Zel'dovich was an energetic advocate of the pancake picture of structure formation. Since it is so clearly wrong (as discussed on pages 235 and 236) why was the idea pursued in the Soviet Union? Andrey Kravtsov, in a personal communication (2018), recalls that it does not seem likely to have been a shortage of debate:

> what I witnessed during my studies in the early 1990s is that arguments and criticisms within these sciences were often very fierce. I pretty commonly witnessed someone standing up after a seminar or talk and severely questioning all of the foundations of presented research and even qualifications of speaker.

The Moscow cosmology group surely faced this wonderfully lively scene and profited from it. But to make progress, the group had to concentrate on the issues at hand, and tight concentration can obscure the larger picture. I expect that the conditions in the Soviet Union, seriously exacerbated by restrictions on travel, led to a scarcity of interactions with independent groups that better understood the astronomy and had the intellectual heft to break the concentration of the Moscow group and push it off dead center. This did happen, of course.

Few people were active in research in cosmology before the 1990s. I expect this contributed to the persistence of ideas in cosmology that would not pass pressing debate. There are many examples. Consider Heisenberg's judgment, expressed at the 1949 Paris meeting on "Motion of Gaseous Masses of Cosmical Dimensions." The fact that turbulent flow is the rule, laminar flow the artificial exception, led him to expect that the analog of laminar flow in a homogeneous expanding universe is a doubtful concept. Hubble flow may be said to be close to laminar—smooth and regular—and this indeed calls for exceedingly special initial conditions, but once the initial conditions are set up, Hubble flow, unlike fluid flow, is not exponentially unstable. (Lemaître's hovering model is an exception, to be sure.) Another example is Gamow's 1954 question to advocates of the steady-state cosmology: what do you make of the apparently modest range of ages of nearby galaxies, so unlike the broad spread of ages to be expected in the 1948 version of this cosmology? I do not understand why Gamow's question attracted so little attention. I take as an example along these lines from the 1980s the pancake picture for structure formation. More people were active in cosmology then, but cosmologists in the Soviet Union were not able to interact very closely with most of them, and we see that communication can be essential. I take as an example from the 1990s the scant influence of the rich suite of evidence that the cosmic mass density is only about a third of the critical value for expansion at escape velocity. The two barriers to communication here were that the evidence was too widely scattered to be readily assessed, and for many, the Einstein–de Sitter model was too elegant to be wrong. The willingness to fool ourselves surely is a basic element of human nature, maybe even adaptively useful. Maybe there are examples of this effect in the more heavily peopled branches of natural science. But we also see in the history of cosmology examples of the eventual detection and correction of such wrong turns.

As is the normal course of events in natural science, most research in cosmology aims to improve what Kuhn (1962) might have agreed to term the "normal science of cosmology," while some pursue ideas that are potentially transformative, inevitably with modest chances of success. Fred Hoyle followed the dream of a philosophically compelling cosmology. Bob Dicke dreamed of detection of the evolution of dimensionless parameters of physics. Both failed, and we celebrate their contributions to the advance of science. We celebrate Allan Sandage, who sought to measure the cosmological deceleration parameter by observations of galaxies, a seriously challenging approach; Stirling Colgate, who sought to do it by observations of supernovae with more efficient photon detectors, and failed; those who finally reached that goal; and those who obtained serious constraints on the deceleration parameter in a way that the pioneers could not have anticipated: close observation of the microwave sky. The ΛCDM theory is established in the sense that it passes an abundance of tests. But there are clouds over ΛCDM and dreams of completion that we may be sure will lead to something new and maybe transformative.

REFERENCES

Aaronson, M. 1983. Accurate Radial Velocities for Carbon Stars in Draco and Ursa Minor: The First Hint of a Dwarf Spheroidal Mass-to-Light Ratio. *Astrophysical Journal Letters* **266**: L11–L15 (285)

Aaronson, M., Huchra, J., Mould, J., et al. 1982. The Velocity Field in the Local Supercluster. *Astrophysical Journal* **258**: 64–76 (87,97)

Aarseth, S. J., Gott, J. R., III, and Turner, E. L. 1979. N-Body Simulations of Galaxy Clustering. I. Initial Conditions and Galaxy Collapse Times. *Astrophysical Journal* **228**: 664–683 (309)

Abbott, L. F., and Schaefer, R. K. 1986. A General, Gauge-Invariant Analysis of the Cosmic Microwave Anisotropy. *Astrophysical Journal* **308**: 546–562 (315)

Abbott, L. F., and Sikivie, P. 1983. A Cosmological Bound on the Invisible Axion. *Physics Letters B* **120**: 133–136 (296)

Abbott, L. F., and Wise, M. B. 1984. Large-Scale Anisotropy of the Microwave Background and the Amplitude of Energy Density Fluctuations in the Early Universe. *Astrophysical Journal Letters* **282**: L47–L50 (304)

Abel, T., Bryan, G. L., and Norman, M. L. 2002. The Formation of the First Star in the Universe. *Science* **295**: 93–98 (121)

Abell, G. O. 1958. The Distribution of Rich Clusters of Galaxies. *Astronomical Journal Supplement* **3**: 211–288 (18,32,85)

Adams, F. C., Bond, J. R., Freese, K., et al. 1993. Natural Inflation: Particle Physics Models, Power-Law Spectra for Large-Scale Structure, and Constraints from the Cosmic Background Explorer. *Physical Review D* **47**: 426–455 (313)

Adams, W. S. 1941. Some Results with the COUDÉ Spectrograph of the Mount Wilson Observatory. *Astrophysical Journal* **93**: 11–23 (153–156)

Adams, W. S., and Seares, F. H. 1937. Mount Wilson Observatory Annual Report **9**, 39 pp. (255)

Ahlen, S. P., Avignone, F. T., Brodzinski, R. L., et al. 1987. Limits on Cold Dark Matter Candidates from an Ultralow Background Germanium Spectrometer. *Physics Letters B* **195**: 603–608 (299)

Alam, S., Ata, M., Bailey, S., et al. 2017. The Clustering of Galaxies in the Completed SDSS-III Baryon Oscillation Spectroscopic Survey: Cosmological Analysis of the DR12 Galaxy Sample. *Monthly Notices of the Royal Astronomical Society* **470** 2617–2652 (112)

Albrecht, A., Coulson, D., Ferreira, P., and Magueijo, J. 1996. Causality, Randomness, and the Microwave Background. *Physical Review Letters* **76**: 1413–1416 (229)

Albrecht, A., and Steinhardt, P. J. 1982. Cosmology for Grand Unified Theories with Radiatively Induced Symmetry Breaking. *Physical Review Letters* **48**: 1220–1223 (62)

Albrecht, A., and Turok, N. 1985. Evolution of Cosmic Strings. *Physical Review Letters* **54**: 1868–1871 (228)

Alcock, C., Allsman, R. A., Alves, D. R., et al. 2000. The MACHO Project: Microlensing Results from 5.7 Years of Large Magellanic Cloud Observations. *Astrophysical Journal* **542**: 281–307 (278)

Alcock, C., Allsman, R. A., Alves, D. R., et al. 2001. MACHO Project Limits on Black Hole Dark Matter in the 1–30$M_{\odot}$ Range. *Astrophysical Journal Letters* **550**: L169–L172 (278)

Alcock, C., Fuller, G. M., and Mathews, G. J. 1987. The Quark-Hadron Phase Transition and Primordial Nucleosynthesis. *Astrophysical Journal* **320**: 439–447 (181)

Alfvén, H. 1965. Antimatter and the Development of the Metagalaxy. *Reviews of Modern Physics* **37**: 652–665 (34)

Aller, L. 1995. Early Clues to Abnormal Mass/Light Ratios in Galaxies: Messier 31 and Messier 33. In *Sources of Dark Matter in the Universe*, Singapore: World Scientific, ed. David B. Cline, pp. 3–8 (275)

Aller, L. H., and Menzel, D. H. 1945. Physical Processes in Gaseous Nebulae XVIII. The Chemical Composition of the Planetary Nebulae. *Astrophysical Journal* **102**: 239–263 (139,146)

Alpher, R. A. 1948a. On the Origin and Relative Abundance of the Elements. PhD thesis, The George Washington University (127,131–137,176,183)

Alpher, R. A. 1948b. A Neutron-Capture Theory of the Formation and Relative Abundance of the Elements. *Physical Review* **74**: 1577–1589 (131–134,136,183)

Alpher, R. A., Bethe, H., and Gamow, G. 1948. The Origin of Chemical Elements. *Physical Review* **73**: 803–804 (123,133,147,164,182)

Alpher, R. A., Follin, J. W., and Herman, R. C. 1953. Physical Conditions in the Initial Stages of the Expanding Universe. *Physical Review* **92**: 1347–1361 (137,147,164)

Alpher, R. A., and Herman, R. C. 1948a. Evolution of the Universe. *Nature* **162**: 774–775 (125,131,138)

Alpher, R. A., and Herman, R. C. 1948b. On the Relative Abundance of the Elements. *Physical Review* **74**: 1737–1743 (132)

Alpher, R. A., and Herman, R. C. 1949. Remarks on the Evolution of the Expanding Universe. *Physical Review* **75**: 1089–1095 (132)

Alpher, R. A., and Herman, R. C. 1950. Theory of the Origin and Relative Abundance Distribution of the Elements. *Reviews of Modern Physics* **22**: 153–212 (127,137,175)

Alpher, R. A., and Herman, R. C. 1953. The Origin and Abundance Distribution of the Elements. *Annual Review of Nuclear and Particle Science* **2**: 1–40 (150)

Alpher, R. A., and Herman, R. C. 1988. Reflections on Early Work on 'Big Bang' Cosmology. *Physics Today* **41**: 24–34 (127,130,152)

Alpher, R. A., and Herman, R. C. 2001. *Genesis of the Big Bang*. Oxford: Oxford University Press (123,198)

Andernach, H., and Zwicky, F. 2017. English and Spanish Translation of Zwicky's (1933) The Redshift of Extragalactic Nebulae. arXiv:1711.01693 (241)

Applegate, J. H., and Hogan, C. J. 1985. Relics of Cosmic Quark Condensation. *Physical Review D* **31**: 3037–3045 (181)

Athanassoula, E., Bosma, A., and Papaioannou, S. 1987. Halo Parameters of Spiral Galaxies. *Astronomy and Astrophysics* **179**: 23–40 (262)

Aver, E., Olive, K. A., and Skillman, E. D. 2015. The Effects of He I λ10830 on Helium Abundance Determinations. *Journal of Cosmology and Astroparticle Physics* **7**: 011, 23 pp. (176)

Baade, W. 1939. Stellar Photography in the Red Region of the Spectrum. *Publications of the American Astronomical Society* **9**: 31–32 (250,343)

Baade, W. 1952. Report of Meeting. *Transactions of the International Astronomical Union* **8**: 397 (73)

Baade, W., and Mayall, N. U. 1951. Distribution and Motions of Gaseous Masses in Spirals. In *Problems of Cosmical Aerodynamics; Proceedings of a Symposium on the Motion of Gaseous Masses of Cosmical Dimensions*, Paris, August 16–19, 1949. Dayton, OH: Central Air Document Offices, pp. 165–184 (250,253)

Babcock, H. W. 1939. The Rotation of the Andromeda Nebula. *Lick Observatory Bulletin* **498**: 41–51 (248,253)

Bahcall, N. A., and Cen, R. 1992. Galaxy Clusters and Cold Dark Matter: A Low-Density Unbiased Universe? *Astrophysical Journal Letters* **398**: L81–L84 (100,104)

Bahcall, N. A., Fan, X., and Cen, R. 1997. Constraining Ω with Cluster Evolution. *Astrophysical Journal Letters* **485**: L53–L56 (100)

Bahcall, N. A., Lubin, L. M., and Dorman, V. 1995. Where Is the Dark Matter? *Astrophysical Journal Letters* **447**: L81–L85 (107,275)

Bahcall, N. A., Ostriker, J. P., Perlmutter, S., and Steinhardt, P. J. 1999. The Cosmic Triangle: Revealing the State of the Universe. *Science* **284**: 1481–1488 (321)

Bahcall, N. A., and Soneira, R. M. 1983. The Spatial Correlation Function of Rich Clusters of Galaxies. *Astrophysical Journal* **270**: 20–38 (30)

Ballinger, W. E., Heavens, A. F., and Taylor, A. N. 1995. The Real-Space Power Spectrum of IRAS Galaxies on Large Scales and the Redshift Distortion. *Monthly Notices of the Royal Astronomical Society* **276**: L59–L63 (93)

Bardeen, J. M. 1968. Radiative Transfer in Perturbed Friedmann Universes. *Astronomical Journal Supplement* **73**: 164 (205)

Bardeen, J. M. 1975. Global Instabilities of Disks. In *Proceedings of IAU Symposium 69, Dynamics of Stellar Systems*, Besançon, France, September 9–13, 1974. Dordrecht: Reidel, pp. 297–320 (270)

Bardeen, J. M. 1980. Gauge-Invariant Cosmological Perturbations. *Physical Review D* **22**: 1882–1905 (199)

Bardeen, J. M. 1986. Galaxy Formation in an Omega = 1 Cold Dark Matter Universe. In *Inner Space/Outer Space: The Interface between Cosmology and Particle Physics*, pp. 212–217. Chicago: University of Chicago Press (64–67)

Bardeen, J. M., Bond, J. R., Kaiser, N., and Szalay, A. S. 1986. The Statistics of Peaks of Gaussian Random Fields. *Astrophysical Journal* **304**: 15–61 (303,308)

Bardeen, J. M., Steinhardt, P. J., and Turner, M. S. 1983. Spontaneous Creation of Almost Scale-Free Density Perturbations in an Inflationary Universe. *Physical Review D* **28**: 679–693 (62)

Barnes, J. E. 1988. Encounters of Disk/Halo Galaxies. *Astrophysical Journal* **331**: 699–717 (272)

Barrow, J. D. 2017. Some Generalities about Generality. In *The Philosophy of Cosmology*, Eds. K. Chamcham, J. Silk, J. D. Barrow, and S. Saunders. Cambridge: Cambridge University Press (212)

Barrow, J. D., and Matzner, R. A. 1977. The Homogeneity and Isotropy of the Universe. *Monthly Notices of the Royal Astronomical Society* **181**: 719–727 (211)

Bartlett, J. G., and Blanchard, A. 1996. The Significance of the Cosmic Virial Theorem. *Astronomy and Astrophysics* **307**: 1–7 (89)

Bartlett, J. G., Blanchard, A., Silk, J., and Turner, M. S. 1995. The Case for a Hubble Constant of 30 km s^{-1} Mpc^{-1}. *Science* **267**: 980–983 (316,335)

Bartlett, J. G., and Silk, J. 1993. Galaxy Clusters and the COBE Result. *Astrophysical Journal Letters* **407**: L45–L48 (100,104)

Bashinsky, S., and Bertschinger, E. 2002. Dynamics of Cosmological Perturbations in Position Space. *Physical Review D* **65** 123008, 19 pp. (105)

Baum, W. A. 1957. Photoelectric Determinations of Redshifts beyond 0.2 c. *Astronomical Journal* **62**: 6–7 (53,94)

Bean, A. J., Ellis, R. S., Shanks, T., Efstathiou, G., and Peterson, B. A. 1983. A Complete Galaxy Redshift Sample—I. The Peculiar Velocities between Galaxy Pairs and the Mean

Mass Density of the Universe. *Monthly Notices of the Royal Astronomical Society* **205**: 605–624 (89,92)

Becker, R. H., Fan, X., White, R. L., et al. 2001. Evidence for Reionization at $z=6$: Detection of a Gunn-Peterson Trough in a $z=6.28$ Quasar. *Astronomical Journal* **122**: 2850–2857 (204)

Beckman, J. E., Ade, P. A. R., Huizinga, J. S., Robson, E. I., and Vickers, D. G. 1972. Limits to the Sub-millimetre Isotropic Background. *Nature* **237**: 154–157 (171)

Begeman, K. G. 1987. HI Rotation Curves of Spiral Galaxies. PhD thesis, Kapteyn Institute, University of Groningen, the Netherlands (259)

Bennett, C. L., Banday, A. J., Górski, K. M., et al. 1996. Four-Year COBE DMR Cosmic Microwave Background Observations: Maps and Basic Results. *Astrophysical Journal Letters* **464**: L1–L4 (104,312)

Bennett, C. L., Halpern, M., Hinshaw, G., et al. 2003. First-Year Wilkinson Microwave Anisotropy Probe (WMAP) Observations: Preliminary Maps and Basic Results. *Astrophysical Journal Supplement* **148**: 1–27 (305)

Bennett, D. P., and Bouchet, F. R. 1988. Evidence for a Scaling Solution in Cosmic-String Evolution. *Physical Review Letters* **60** 257–260 (228)

Berlind, A. A., and Weinberg, D. H. 2002. The Halo Occupation Distribution: Toward an Empirical Determination of the Relation between Galaxies and Mass. *Astrophysical Journal* **575**: 587–616 (26)

Bernstein, G. M., Fischer, M. L., Richards, P. L., Peterson, J. B., and Timusk, T. 1990. A Measurement of the Spectrum of the Cosmic Background Radiation from 1 to 3 Millimeter Wavelength. *Astrophysical Journal* **362**: 107–113 (171)

Bertschinger, E. 1993. Galaxy Formation and Large-Scale Structure. *Annals of the New York Academy of Sciences* **688**: 297–310 (60)

Bertschinger, E., and Dekel, A. 1989. Recovering the Full Velocity and Density Fields from Large-Scale Redshift-Distance Samples. *Astrophysical Journal Letters* **336**: L5–L8 (102)

Bertschinger, E., Dekel, A., Faber, S. M., Dressler, A., and Burstein, D. 1990. Potential, Velocity, and Density Fields from Redshift-Distance Samples: Application: Cosmography within 6000 Kilometers per Second. *Astrophysical Journal* **364**: 370–395 (103)

Bethe, H. A. 1947. *Elementary Nuclear Theory*. New York: Wiley and Sons (125)

Beutler, F., Seo, H.-J., Ross, A. J., et al. 2017. The Clustering of Galaxies in the Completed SDSS-III Baryon Oscillation Spectroscopic Survey: Baryon Acoustic Oscillations in the Fourier Space. *Monthly Notices of the Royal Astronomical Society* **464**: 3409–3430 (336)

Binney, J. 1974. Galaxy Formation without Primordial Turbulence: Mechanisms for Generating Cosmic Vorticity. *Monthly Notices of the Royal Astronomical Society* **168**: 73–92 (220)

Binney, J. 1977. The Physics of Dissipational Galaxy Formation. *Astrophysical Journal* **215**: 483–491 (308)

Bisnovatyi-Kogan, G. S., and Novikov, I. D. 1980. Cosmology with a Nonzero Neutrino Rest Mass. *Astronomicheskii Zhurnal* **57**: 899–902. English translation in *Soviet Astronomy* **24**: 516–517 (286)

Blackman, R. B., and Tukey, J. 1959. *The Measurement of Power Spectra from the Point of View of Communications Engineering*. New York: Dover (306)

Blumenthal, G. R., Dekel, A., and Primack, J. R. 1988. Very Large Scale Structure in an Open Cosmology of Cold Dark Matter and Baryons. *Astrophysical Journal* **326**: 539–550 (98,315)

Blumenthal, G. R., Faber, S. M., Primack, J. R., and Rees, M. J. 1984. Formation of Galaxies and Large-Scale Structure with Cold Dark Matter. *Nature* **311**: 517–525 (302,308)

Blumenthal, G. R., Pagels, H., and Primack, J. R. 1982. Galaxy Formation by Dissipationless Particles Heavier Than Neutrinos. *Nature* **299**: 37–38 (289,297,302, 311,315)

Boesgaard, A. M., and Steigman, G. 1985. Big Bang Nucleosynthesis: Theories and Observations. *Annual Review of Astronomy and Astrophysics* **23**: 319–378 (176–179)

Bok, B. J. 1934. The Apparent Clustering of External Galaxies. *Harvard College Observatory Bulletin* **895**: 1–8 (19)

Bok, B. J. 1946. The Time-Scale of the Universe. *Monthly Notices of the Royal Astronomical Society* **106**: 61–75 (72)

Bond, J. R., Centrella, J., Szalay, A. S., and Wilson, J. R. 1984a. Dark Matter and Shocked Pancakes. In *Proceedings of the Third Moriond Astrophysics Meeting, La Plagne, March 1983*. Eds. Jean Audouze and Jean Tran Thanh Van. Dordrecht: Reidel, pp. 87–99 (280,288)

Bond, J. R., Centrella, J., Szalay, A. S., Wilson, J. R. 1984b. Cooling Pancakes. *Monthly Notices of the Royal Astronomical Society* **210**: 515–545 (280,288)

Bond, J. R., and Efstathiou, G. 1984. Cosmic Background Radiation Anisotropies in Universes Dominated by Nonbaryonic Dark Matter. *Astrophysical Journal Letters* **285**: L45–L48 (303)

Bond, J. R., and Efstathiou, G. 1987. The Statistics of Cosmic Background Radiation Fluctuations. *Monthly Notices of the Royal Astronomical Society* **226**: 655–687 (302–306,333)

Bond, J. R., and Efstathiou, G. 1991. The Formation of Cosmic Structure With a 17 keV Neutrino. *Physics Letters B* **265**: 245–250 (306)

Bond, J. R., Efstathiou, G., and Silk, J. 1980. Massive Neutrinos and the Large-Scale Structure of the Universe. *Physical Review Letters* **45**: 1980–1984 (285–289)

Bond, J. R., Efstathiou, G., and Tegmark, M. 1997. Forecasting Cosmic Parameter Errors from Microwave Background Anisotropy Experiments. *Monthly Notices of the Royal Astronomical Society* **291**: L33–L41 (334)

Bond, J. R., and Jaffe, A. H. 1999. Constraining Large-Scale Structure Theories with the Cosmic Background Radiation. *Philosophical Transactions of the Royal Society of London Series A* **357**: 57–75 (321)

Bond, J. R., Szalay, A. S., and Turner, M. S. 1982. Formation of Galaxies in a Gravitino-Dominated Universe. *Physical Review Letters* **48**: 1636–1639 (289,297)

Bondi, H. 1947. Spherically Symmetrical Models in General Relativity. *Monthly Notices of the Royal Astronomical Society* **107**: 410–425 (193)

Bondi, H. 1952. *Cosmology*. Cambridge: Cambridge University Press (7,22,31,54)

Bondi, H. 1960. *Cosmology*, second edition. Cambridge: Cambridge University Press (7,51,54–58,141,317)

Bondi, H., and Gold, T. 1948. The Steady-State Theory of the Expanding Universe. *Monthly Notices of the Royal Astronomical Society* **108**: 252–270 (50,51,54)

Bondi, H., Gold, T., and Sciama, D. W. 1954. A Note on the Reported Color-Index Effect of Distant Galaxies. *Astrophysical Journal* **120**: 597–599 (51)

Bonnor, W. B. 1957. Jeans' Formula for Gravitational Instability. *Monthly Notices of the Royal Astronomical Society* **117**: 104–116 (187)

Bortolot, V. J., Clauser, J. F., and Thaddeus, P. 1969. Upper Limits to the Intensity of Background Radiation at $\lambda = 1.32, 0.559$, and 0.359 mm. *Physical Review Letters* **22**: 307–310 (166)

Bosma, A. 1978. The Distribution and Kinematics of Neutral Hydrogen in Spiral Galaxies of Various Morphological Types. PhD thesis, University of Groningen, The Netherlands (261)

Bosma, A. 1981a. 21-cm Line Studies of Spiral Galaxies. I. Observations of the Galaxies NGC 5033, 3198, 5055, 2841, and 7331. *Astronomical Journal* **86**: 1791–1924 (262)

Bosma, A. 1981b. 21-cm Line Studies of Spiral Galaxies. II. The Distribution and Kinematics of Neutral Hydrogen in Spiral Galaxies of Various Morphological Types. *Astronomical Journal* **86**: 1825–1846 (262)

Boughn, S. P., Cheng, E. S., and Wilkinson, D. T. 1981. Dipole and Quadrupole Anisotropy of the 2.7 K Radiation. *Astrophysical Journal Letters* **243**: L113–L117 (227,300)

Boynton, P. E., Stokes, R. A., and Wilkinson, D. T. 1968. Primeval Fireball Intensity at $\lambda = 3.3$ mm. *Physical Review Letters* **21**: 462–465 (165)

Brans, C., and Dicke, R. H. 1961. Mach's Principle and a Relativistic Theory of Gravitation. *Physical Review* **124**: 925–935 (49,60)

Briel, U. G., Henry, J. P., and Böhringer, H. 1992. Observation of the Coma Cluster of Galaxies with ROSAT during the All-Sky Survey. *Astronomy and Astrophysics* **259**: L31–L34 (101)

Bunn, E. F., and White, M. 1997. The 4 Year COBE Normalization and Large-Scale Structure. *Astrophysical Journal* **480**: 6–21 (312,320,334)

Burbidge, E. M., and Burbidge, G. R. 1961. A Further Investigation of Stephan's Quintet. *Astrophysical Journal* **134**: 244–247 (245)

Burbidge, E. M., and Burbidge, G. R. 1975. The Masses of Galaxies. In *Galaxies and the Universe*, pp. 81–121. Eds. A. Sandage, M. Sandage, and J. Kristian. Chicago: University of Chicago Press (243,260,263)

Burbidge, E. M., Burbidge, G. R., Fowler, W. A., and Hoyle, F. 1957. Synthesis of the Elements in Stars. *Reviews of Modern Physics* **29**: 547–650 (129)

Burbidge, E. M., and Sargent, W. L. W. 1971. Velocity Dispersions and Discrepant Redshifts in Groups of Galaxies. In *Proceedings of a Study Week on Nuclei of Galaxies, Rome, April 1970*. Ed. D. J. K. O'Connell. Amsterdam: North Holland, pp. 351–386 (246)

Burbidge, G. R. 1958. Nuclear Energy Generation and Dissipation in Galaxies. *Publications of the Astronomical Society of the Pacific* **70**: 83–89 (140)

Burbidge, G. R. 1975. On the Masses and Relative Velocities of Galaxies. *Astrophysical Journal Letters* **196**: L7–L10 (245,261)

Burles, S., and Tytler, D. 1998. On the Measurements of D/H in QSO Absorption Systems. *Space Science Reviews* **84**: 65–75 (180)

Burstein, D., Rubin, V. C., Thonnard, N., and Ford, W. K., Jr. 1982. The Distribution of Mass in SC Galaxies. *Astrophysical Journal* **253**: 70–85 (258)

Cabibbo, N., Farrar, G. R., and Maiani, L. 1981. Massive Photinos: Unstable and Interesting. *Physics Letters B* **105**: 155–158 (296,298)

Cabrera, B., Krauss, L. M., and Wilczek, F. 1985. Bolometric Detection of Neutrinos. *Physical Review Letters* **55**: 25–28 (299)

Caldwell, D. O., Eisberg, R. M., Grumm, D. M., et al. 1988. Laboratory Limits on Galactic Cold Dark Matter. *Physical Review Letters* **61**: 510–513 (299)

Caldwell, R. R., Davé, R., and Steinhardt, P. J. 1998. Quintessential Cosmology. *Astrophysics and Space Science* **261**: 303–310 (60)

Cameron, A. G. W. 1957. *Stellar Evolution, Nuclear Astrophysics, and Nucleogenesis*. Chalk River, Ontario: Atomic Energy of Canada, CRL-41 (129)

Cappellari, M., Romanowsky, A. J., Brodie, J. P., et al. 2015. Small Scatter and Nearly Isothermal Mass Profiles to Four Half-Light Radii from Two-Dimensional Stellar Dynamics of Early-Type Galaxies. *Astrophysical Journal Letters* **804**: L21, 7 pp. (256)

Carignan, C., and Freeman, K. C. 1985. Basic Parameters of Dark Halos in Late-Type Spirals. *Astrophysical Journal* **294**: 494–501 (258)

Carlberg, R. G., Yee, H. K. C., Ellingson, E., et al. 1997. The Dynamical Equilibrium of Galaxy Clusters. *Astrophysical Journal Letters* **476**: L7–L10 (101,107)

Carpenter, E. F. 1938. Some Characteristics of Associated Galaxies. I. A Density Restriction in the Metagalaxy. *Astrophysical Journal* **88**: 344–355 (32)

Carr, B. J. 1975. The Primordial Black Hole Mass Spectrum. *Astrophysical Journal* **201**: 1–19 (77)

Carr, B. J. 1988. Submillimetre Excess. *Nature* **334**: 650–651 (170)

Carr, B. J., Bond, J. R., and Arnett, W. D. 1984. Cosmological Consequences of Population III Stars. *Astrophysical Journal* **277**: 445–469 (277)

Carter, B. 1974. Large Number Coincidences and the Anthropic Principle in Cosmology. In *Confrontation of Cosmological Theories with Observational Data*. Krakow, September 1993. Dordrecht: Reidel, pp. 291–298 (61)

Cen, R., and Ostriker, J. P. 1992. Galaxy Formation and Physical Bias. *Astrophysical Journal* **399**: L113–L116 (67)

Cen, R., Ostriker, J. P., and Peebles, P. J. E. 1993. A Hydrodynamic Approach to Cosmology: The Primeval Baryon Isocurvature Model. *Astrophysical Journal* **415**: 423–444 (316)

Centrella, J., and Melott, A. L. 1983. Three-Dimensional Simulation of Large-Scale Structure in the Universe. *Nature* **305**: 196–198 (309)

Chaboyer, B., Demarque, P., Kernan, P. J., and Krauss, L. M. 1996. A Lower Limit on the Age of the Universe. *Science* **271**: 957–961 (320)

Chan, K. L., and Jones, B. J. T. 1975. The Evolution of the Cosmic Radiation Spectrum Under the Influence of Turbulent Heating. I. Theory; II. Numerical Calculation and Application. *Astrophysical Journal* **200**: 454–470 (216)

Chandrasekhar, S., and Henrich, L. R. 1942. An Attempt to Interpret the Relative Abundances of the Elements and Their Isotopes. *Astrophysical Journal* **95**: 288–298 (122)

Chapline, G. F. 1975. Cosmological Effects of Primordial Black Holes. *Nature* **253**: 251–252 (77)

Charlier, C. V. L. 1922. How an Infinite World May Be Built Up. *Meddelanden fran Lunds Astronomiska Observatorium* Serie I, **98**: 1–37 (16,32)

Chemin, L., Carignan, C., and Foster, T. 2009. H I Kinematics and Dynamics of Messier 31. *Astrophysical Journal* **705**: 1395–1415 (254,270)

Chernin, A. D. 1972. Dynamic Motions in the Early Universe. *Astrophysical Letters* **10**: 125–128 (220)

Chibisov, G. V. 1972. Damping of Adiabatic Perturbations in an Expanding Universe. *Astronomicheskii Zhurnal* **49**: 74–84. English translation in *Soviet Astronomy—AJ* **16**: 56–63 (205)

Clayton, D. D. 1964. Chronology of the Galaxy. *Science* **143**: 1281–1286 (74)

Clayton, D. D., Fowler, W. A., Hull, T. E., and Zimmerman, B. A. 1961. Neutron Capture Chains in Heavy Element Synthesis. *Annals of Physics* **12**: 331–408 (182)

Code, A. D. 1959. Energy Distribution Curves of Galaxies. *Publications of the Astronomical Society of the Pacific* **71**: 118–125 (51)

Cole, S., Norberg, P., Baugh, C. M., et al. 2001. The 2dF Galaxy Redshift Survey: Near-Infrared Galaxy Luminosity Functions. *Monthly Notices of the Royal Astronomical Society* **326**: 255–273 (83)

Cole, S., Percival, W. J., Peacock, J. A., et al. 2005. The 2dF Galaxy Redshift Survey: Power-Spectrum Analysis of the Final Data Set and Cosmological Implications. *Monthly Notices of the Royal Astronomical Society* **362**: 505–534 (105,336)

Colgate, S. A. 1979. Supernovae as a Standard Candle for Cosmology. *Astrophysical Journal* **232**: 404–408 (326)

Colgate, S. A. 1987. Still Seeking Supernovae. *Sky and Telescope* **74**: 229–230 (327)

Colgate, S. A., Moore, E. P., and Colburn, J. 1975. SIT Vidicon with Magnetic Intensifier for Astronomical Use. *Applied Optics* **14**: 1429–1436 (326)

Cooke, R. J., Pettini, M., Nollett, K. M., and Jorgenson, R. 2016. The Primordial Deuterium Abundance of the Most Metal-Poor Damped Lyman-α System. *Astrophysical Journal* **830**: 148, 16 pp.(180)

Copeland, E. J., and Kibble, T. W. B. 2009. Cosmic Strings and Superstrings. *Proceedings of the Royal Society of London Series A* **466**: 623–657 (223,237)

Corbelli, E., and Salucci, P. 2000. The Extended Rotation Curve and the Dark Matter Halo of M33. *Monthly Notices of the Royal Astronomical Society* **311**: 441–447 (262)

Couchman, H. M. P., and Carlberg, R. G. 1992. Large-Scale Structure in a Low-Bias Universe. *Astrophysical Journal* **389**: 453–463 (67)

Courteau, S., Cappellari, M., de Jong, R. S., et al. 2014. Galaxy Masses. *Reviews of Modern Physics* **86**: 47–119 (9,269)

Cowie, L. L., and Hu, E. M. 1987. The Formation of Families of Twin Galaxies by String Loops. *Astrophysical Journal Letters* **318**: L33–L38 (228)

Cowsik, R., and McClelland, J. 1972. An Upper Limit on the Neutrino Rest Mass. *Physical Review Letters* **29**: 669–670 (281)

Cowsik, R., and McClelland, J. 1973. Gravity of Neutrinos of Nonzero Mass in Astrophysics. *Astrophysical Journal* **180**: 7–10 (77,282–284,291)

Crawford, M., and Schramm, D. N. 1982. Spontaneous Generation of Density Perturbations in the Early Universe. *Nature* **298**: 538–540 (277)

Crittenden, R. G., and Turok, N. 1995. Doppler Peaks from Cosmic Texture. *Physical Review Letters* **75**: 2642–2645 (229)

Crovini, L., and Galgani, L. 1984. On the Accuracy of the Experimental Proof of Planck's Radiation Law. *Nuovo Cimento Lettere* **39**: 210–214 (117)

Dallaporta, N., and Lucchin, F. 1972. On Galaxy Formation from Primeval Universal Turbulence. *Astronomy and Astrophysics* **19**: 123–134 (215)

Danese, L., Burigana, C., Toffolatti, L., de Zotti, G., and Franceschini, A. 1990. Theoretical Implications of the CMB Spectral Distortions. In *The Cosmic Microwave Background: 25 Years Later*, pp. 153–172. Eds. N. Mandolesi and N. Vittorio. Dordrecht, Netherlands: Kluwer Academic Publishers (170)

Dautcourt, G., and Wallis, G. 1968. The Cosmic Blackbody Radiation. *Fortschritte der Physik* **16**: 545–593 (164)

Davidson, W., and Narlikar, J. V. 1966. Cosmological Models and Their Observational Validation. *Reports on Progress in Physics* **29**: 539–622 (182)

Davis, M. 1987. Evidence for Dark Matter in Galactic Systems. In *Proceedings of IAU Symposium 117, Dark Matter in the Universe*, pp. 97–109 (64–67)

Davis, M., Efstathiou, G., Frenk, C. S., and White, S. D. M. 1985. The Evolution of Large-Scale Structure in a Universe Dominated by Cold Dark Matter. *Astrophysical Journal* **292**: 371–394 (67,97,191,303,309)

Davis, M., Efstathiou, G., Frenk, C. S., and White, S. D. M. 1992. The End of Cold Dark Matter? *Nature* **356**: 489–494 (59,317)

Davis, M., Geller, M. J., and Huchra, J. 1978. The Local Mean Mass Density of the Universe: New Methods for Studying Galaxy Clustering. *Astrophysical Journal* **221**: 1–18 (85,88)

Davis, M., Groth, E. J., and Peebles, P. J. E. 1977. Study of Galaxy Correlations: Evidence for the Gravitational Instability Picture in a Dense Universe. *Astrophysical Journal Letters* **212**: L107–L111 (233,309)

Davis, M., Huchra, J., Latham, D. W., and Tonry, J. 1982. A Survey of Galaxy Redshifts. II. The Large Scale Space Distribution. *Astrophysical Journal* **253**: 423–445 (68,89,97)

Davis, M., Lecar, M., Pryor, C., and Witten, E. 1981. The Formation of Galaxies from Massive Neutrinos. *Astrophysical Journal* **250**: 423–431 (313)

Davis, M., and Nusser, A. 2016. Re-examination of Large Scale Structure & Cosmic Flows. In *Proceedings of IAU Symposium 308, The Zel'dovich Universe: Genesis and Growth of the Cosmic Web*, Tallinn, Estunia, June 23–28. Cambridge: Cambridge University Press, pp. 310–317 (96)

Davis, M., Nusser, A., and Willick, J. A. 1996. Comparison of Velocity and Gravity Fields: The Mark III Tully-Fisher Catalog versus the IRAS 1.2 Jy Survey. *Astrophysical Journal* **473**: 22–42 (103)

Davis, M., and Peebles, P. J. E. 1977. On the Integration of the BBGKY Equations for the Development of Strongly Nonlinear Clustering in an Expanding Universe. *Astrophysical Journal Supplement* **34**: 425–450 (89,97,233)

Davis, M., and Peebles, P. J. E. 1983a. A Survey of Galaxy Redshifts. V. The Two-Point Position and Velocity Correlations. *Astrophysical Journal* **267**: 465–482 (29,68,91,97,302)

Davis, M., and Peebles, P. J. E. 1983b. Evidence for Local Anisotropy of the Hubble Flow. *Annual Review of Astronomy and Astrophysics* **21**: 109–130 (302)

Davis, M., Summers, F. J., and Schlegel, D. 1992. Large-Scale Structure in a Universe with Mixed Hot and Cold Dark Matter. *Nature* **359**: 393–396 (318)

Davis, M., Tonry, J., Huchra, J., and Latham, D. W. 1980. On the Virgo Supercluster and the Mean Mass Density of the Universe. *Astrophysical Journal Letters* **238**: L113–L116 (87)

Dawid, R. 2013. *String Theory and the Scientific Method.* Cambridge: Cambridge University Press (3)

Dawid, R. 2017. The Significance of Non-Empirical Confirmation in Fundamental Physics. arXiv:1702.01133 (3)

de Bernardis, P., Ade, P. A. R., Bock, J. J., et al. 2000. A Flat Universe from High-Resolution Maps of the Cosmic Microwave Background Radiation. *Nature* **404**: 955–959 (334)

DeGrasse, R. W., Hogg, D. C., Ohm, E. A., and Scovil, H. E. D. 1959a. Ultra-Low-Noise Measurements Using a Horn Reflector Antenna and a Traveling-Wave Maser. *Journal of Applied Physics* **30**: 2013 (156)

DeGrasse, R. W., Hogg, D. C., Ohm, E. A., and Scovil, H. E. D. 1959b. Ultra-Low-Noise Antenna and Receiver Combination for Satellite or Space Communication. *Proceedings of the National Electronics Conference* **15**: 371–379 (156,158)

Dekel, A., Bertschinger, E., Yahil, A., et al. 1993. *IRAS* Galaxies versus POTENT Mass: Density Fields, Biasing, and Ω. *Astrophysical Journal* **412**: 1–21 (103,319)

Dekel, A., Burstein, D., and White, S. D. M. 1997. Measuring Omega. In *Critical Dialogues in Cosmology, Proceedings of a Conference Held at Princeton in June 1996*. Ed. Neil Turok. Singapore: World Scientific, pp. 175–192. (316)

Demoulin, M.-H., and Chan, Y. W. T. 1969. Rotation and Mass of the Galaxy NGC 6574. *Astrophysical Journal* **156**: 501–508 (259)

de Rújula, A., and Glashow, S. L. 1980. Galactic Neutrinos and UV Astronomy. *Physical Review Letters* **45**: 942–944 (298) de Sitter, W. 1917a. On the Relativity of Inertia. Remarks Concerning Einstein's Latest Hypothesis. *Koninklijke Nederlandse Akademie van Wetenschappen Proceedings Series B Physical Sciences* **19**: 1217–1225 (14,17)

de Sitter, W. 1917b. Einstein's Theory of Gravitation and Its Astronomical Consequences. Third Paper. *Monthly Notices of the Royal Astronomical Society* **78**: 3–28 (17,38)

de Swart, J. G., Bertone, G., and van Dongen, J. 2017. How Dark Matter Came to Matter. *Nature Astronomy* **1** 0059, 8 pp. (9)

de Vaucouleurs, G. 1953. Evidence for a Local Supergalaxy. *Astronomical Journal* **58**: 30–32 (17)

de Vaucouleurs, G. 1958a. Further Evidence for a Local Super-cluster of Galaxies: Rotation and Expansion. *Astronomical Journal* **63**: 253–265 (17)

de Vaucouleurs, G. 1958b. Photoelectric Photometry of the Andromeda Nebula in the UBV System. *Astrophysical Journal* **128**: 465–488 (252–254)

de Vaucouleurs, G. 1959. Photoelectric Photometry of Messier 33 in the *U*, *B*, *V* System. *Astrophysical Journal* **130**: 728–738 (258)

de Vaucouleurs, G. 1960. The Apparent Density of Matter in Groups and Clusters of Galaxies. *Astrophysical Journal* **131**: 585–597 (235,243)

de Vaucouleurs, G. 1970. The Case for a Hierarchical Cosmology. *Science* **167**: 1203–1313 (32,349)

de Vaucouleurs, G., and Corwin, H. G., Jr. 1985. The Distance of the Hercules Supercluster from Supernovae and SBC Spirals, and the Hubble Constant. *Astrophysical Journal* **297**: 23–26 (74)

de Vaucouleurs, G., and de Vaucouleurs, A. 1964. *Reference Catalogue of Bright Galaxies*. Austin: University of Texas Press (88)

de Vaucouleurs, G., and Peters, W. L. 1968. Motion of the Sun with Respect to the Galaxies and the Kinematics of the Local Supercluster. *Nature* **220**: 868–874 (222)

Dicke, R. H. 1939. The Radial Distribution in Globular Clusters. *Astronomical Journal* **48**: 108–110 (352)

Dicke, R. H. 1961. Dirac's Cosmology and Mach's Principle. *Nature* **192**: 440–441 (60)

Dicke, R. H. 1968. Scalar-Tensor Gravitation and the Cosmic Fireball. *Astrophysical Journal* **152**: 1–24 (181)

Dicke, R. H. 1969. The Age of the Galaxy from the Decay of Uranium. *Astrophysical Journal* **155**: 123–134 (74)

Dicke, R. H. 1970. *Gravitation and the Universe*. Memoirs of the American Philosophical Society, Jayne Lectures for 1969. Philadelphia: American Philosophical Society (59)

Dicke, R. H., and Beringer, R. 1946. Microwave Radiation from the Sun and Moon. *Astrophysical Journal* **103**: 375 (352)

Dicke, R. H., Beringer, R., Kyhl, R. L., and Vane, A. B. 1946. Atmospheric Absorption Measurements with a Microwave Radiometer. *Physical Review* **70**: 340–348 (162)

Dicke, R. H., and Peebles, P. J. E. 1965. Gravitation and Space Science. *Space Science Reviews* **4**: 419–460 (144,163)

Dicke, R. H., and Peebles, P. J. E. 1979. The Big Bang Cosmology—Enigmas and Nostrums. In *General Relativity: An Einstein Centenary Survey*, Eds. S. W. Hawking and W. Israel. Cambridge: Cambridge University Press, pp. 504–517. (59,160)

Dicke, R. H., Peebles, P. J. E., Roll, P. G. and Wilkinson, D. T. 1965. Cosmic Black-Body Radiation. *Astrophysical Journal* **142**: 414–419 (162,164,183)

Dicus, D. A., Kolb, E. W., and Teplitz, V. L. 1977. Cosmological Upper Bound on Heavy-Neutrino Lifetimes. *Physical Review Letters* **39**: 168–171 (291–293,313)

Dicus, D. A., Kolb, E. W., and Teplitz, V. L. 1978. Cosmological Implications of Massive, Unstable Neutrinos. *Astrophysical Journal* **221**: 327–341 (313)

Dimopoulos, S., and Susskind, L. 1978. Baryon Number of the Universe. *Physical Review D* **18**: 4500–4509 (63)

Dine, M., and Fischler, W. 1983. The Not-So-Harmless Axion. *Physics Letters B* **120**: 137–141 (296)

Dingle, H. 1933a. Values of T_μ^ν and the Christoffel Symbols for a Line Element of Considerable Generality. *Proceedings of the National Academy of Sciences* **19**: 559–563 (193)

Dingle, H. 1933b. On Isotropic Models of the Universe, with Special Reference to the Stability of the Homogeneous and Static States. *Monthly Notices of the Royal Astronomical Society* **94**: 134–158 (193)

Dirac, P. A. M. 1938. A New Basis for Cosmology. *Proceedings of the Royal Society of London Series A* **165**: 199–208 (49)

Djorgovski, S., and Davis, M. 1987. Fundamental Properties of Elliptical Galaxies. *Astrophysical Journal* **313**: 59–68 (85)

Dmitriev, N. A., and Zel'dovich, Y. B. 1963. The Energy of Accidental Motions in the Expanding Universe. *Journal of Experimental and Theoretical Physics U.S.S.R.* **45**: 1150–1155. English translation in *Soviet Physics JETP* **18**: 793–796, 1964 (84)

Dolgov, A. D. 1983. An Attempt to Get Rid of the Cosmological Constant. In *The Very Early Universe, Proceedings of the Nuffield Workshop*, Cambridge, June 21–July 9, 1982. Eds. G. W. Gibbons and S. T. C. Siklos. Cambridge: Cambridge University Press, pp. 449–458 (60)

Doroshkevich, A. G. 1970. The Space Structure of Perturbations and the Origin of Rotation of Galaxies in the Theory of Fluctuation. *Astrofizika* **6**: 320–330 (218)

Doroshkevich, A. G., and Khlopov, M. I. 1984. Formation of Structure in a Universe with Unstable Neutrinos. *Monthly Notices of the Royal Astronomical Society* **211**: 277–282 (314)

Doroshkevich, A. G., Khlopov, M. I., Sunyaev, R. A., Szalay, A. S., and Zel'dovich, Ya. B. 1981. Cosmological Impact of the Neutrino Rest Mass. *Annals of the New York Academy of Sciences* **375**: 32–42 (286,301)

Doroshkevich, A. G., Kotok, E. V., Shandarin, S. F., and Sigov, I. S. 1983. Analysis of the Large-Scale Structure of the Universe. *Monthly Notices of the Royal Astronomical Society* **202**: 537–552 (288)

Doroshkevich, A. G., and Novikov, I. D. 1964. Mean Density of Radiation in the Metagalaxy and Certain Problems in Relativistic Cosmology. *Doklady Akademii Nauk SSSR* **154**: 809–811. English translation in *Soviet Physics Doklady* **9** 111–113 (152,159,161,165)

Doroshkevich, A. G., and Shandarin, S. F. 1976. On the Local Anisotropy of Expansion of the Universe. *Monthly Notices of the Royal Astronomical Society* **175**: 15P–18P (309)

Doroshkevich, A. G., Zel'dovich, Y. B., and Novikov, I. D. 1967. The Origin of Galaxies in an Expanding Universe. *Astronomicheskii Zhurnal* **44**: 295–303. English translation in *Soviet Astronomy—AJ* **11**: 233–239 (210,221)

Doroshkevich, A. G., Zel'dovich, Y. B., Sunyaev, R. A., and Khlopov, M. Y. 1980. Astrophysical Implications of the Neutrino Rest Mass. II. The Density Perturbation Spectrum and Small-Scale Fluctuations in the Microwave Background. *Pis'ma Astronomicheskii Zhurnal* **6**: 457–464. English translation in *Soviet Astronomy Letters* **6**: 252–256 (286)

Dressler, A., Faber, S. M., Burstein, D., et al. 1987. Spectroscopy and Photometry of Elliptical Galaxies: A Large-Scale Streaming Motion in the Local Universe. *Astrophysical Journal Letters* **313**: L37–L42 (85)

Drukier, A. K., Freese, K., and Spergel, D. N. 1986. Detecting Cold Dark-Matter Candidates. *Physical Review D* **33**: 3495–3508 (299)

Drukier, A., and Stodolsky, L. 1984. Principles and Applications of a Neutral-Current Detector for Neutrino Physics and Astronomy. *Physical Review D* **30**: 2295–2309 (299)

Durrer, R., Gangui, A., and Sakellariadou, M. 1996. Doppler Peaks in the Angular Power Spectrum of the Cosmic Microwave Background: A Fingerprint of Topological Defects. *Physical Review Letters* **76**: 579–582 (229)

Durrer, R., Kunz, M., and Melchiorri, A. 2002. Cosmic Structure Formation with Topological Defects. *Physics Reports* **364**: 1–81 (229)

Durrer, R., and Sakellariadou, M. 1997. Microwave Background Anisotropies from Scaling Seed Perturbations. *Physical Review D* **56**: 4480–4493 (229)

Eddington, A. S. 1923. *The Mathematical Theory of Relativity*. Cambridge: Cambridge University Press (38,78)

Eddington, A. S. 1930. On the Instability of Einstein's Spherical World. *Monthly Notices of the Royal Astronomical Society* **90**: 668–678 (71,188)

Eddington, A. S. 1936. *Relativity Theory of Protons and Electrons*. Cambridge: Cambridge University Press (49)

Efstathiou, G., and Bond, J. R. 1987. Microwave Anisotropy Constraints on Isocurvature Baryon Models. *Monthly Notices of the Royal Astronomical Society* **227**: 33P–38P (306,316)

Efstathiou, G., Bond, J. R., and White, S. D. M. 1992. COBE Background Radiation Anisotropies and Large-Scale Structure in the Universe. *Monthly Notices of the Royal Astronomical Society* **258**: 1P–6P (303,314)

Efstathiou, G., Bridle, S. L., Lasenby, A. N., Hobson, M. P., and Ellis, R. S. 1999. Constraints on Ω_Λ and Ω_m from Distant Type Ia Supernovae and Cosmic Microwave Background Anisotropies. *Monthly Notices of the Royal Astronomical Society* **303**: L47–L52 (334)

Efstathiou, G., Fall, S. M., and Hogan, C. 1979. Self-Similar Gravitational Clustering. *Monthly Notices of the Royal Astronomical Society* **189**: 203–220 (309)

Efstathiou, G., Frenk, C. S., White, S. D. M., and Davis, M. 1988. Gravitational Clustering from Scale-Free Initial Conditions. *Monthly Notices of the Royal Astronomical Society* **235**: 715–748 (234)

Efstathiou, G., and Jones, B. J. T. 1979. The Rotation of Galaxies: Numerical Investigations of the Tidal Torque Theory. *Monthly Notices of the Royal Astronomical Society* **186**: 133–144 (218)

Efstathiou, G., and Jones, B. J. T. 1980. Angular Momentum and the Formation of Galaxies by Gravitational Instability. *Comments on Astrophysics* **8**: 169–176 (218)

Efstathiou, G., Sutherland, W. J., and Maddox, S. J. 1990. The Cosmological Constant and Cold Dark Matter. *Nature* **348**: 705–707 (30,99,319)

Eggen, O. J., Lynden-Bell, D., and Sandage, A. R. 1962. Evidence from the Motions of Old Stars That the Galaxy Collapsed. *Astrophysical Journal* **136**: 748–766 (219,307)

Einasto, J., Kaasik, A., and Saar, E. 1974. Dynamic Evidence on Massive Coronas of Galaxies. *Nature* **250**: 309–310 (272,275–296)

Einstein, A. 1917. Kosmologische Betrachtungen zur allgemeinen Relativitätstheorie. *Sitzungsberichte der Königlich Preußischen Akademie der Wissenschaften*, Berlin, pp. 142–152 (6,12,13,38,45,78,187)

Einstein, A. 1923. *The Meaning of Relativity*. Princeton, NJ: Princeton University Press (13)

Einstein, A. 1931. Zum kosmologischen Problem der allgemeinen Relativitätstheorie. *Sitzungsberichte der Preußischen Akademie der Wissenschaften* Berlin, pp. 236–237 (39)

Einstein, A. 1945. *The Meaning of Relativity*, second edition. Princeton, NJ: Princeton University Press (57)

Einstein, A., and de Sitter, W. 1932. On the Relation between the Expansion and the Mean Density of the Universe. *Proceedings of the National Academy of Sciences* **18**: 213–214 (56,71,78)

Eisenstein, D. J., Zehavi, I., Hogg, D. W., et al. 2005. Detection of the Baryon Acoustic Peak in the Large-Scale Correlation Function of SDSS Luminous Red Galaxies. *Astrophysical Journal* **633**: 560–574 (105,207,336)

Eke, V. R., Cole, S., Frenk, C. S., and Henry, J. P. 1998. Measuring Ω_0 Using Cluster Evolution. *Monthly Notices of the Royal Astronomical Society* **298**: 1145–1158 (100)

Eke, V. R., Cole, S., Frenk, C. S., and Navarro, J. F. 1996. Cluster Correlation Functions in *N*-Body Simulations. *Monthly Notices of the Royal Astronomical Society* **281**: 703–715 (104)

Ellis, G. F. R., Lyth, D. H., and Mijić, M. B. 1991. Inflationary Models with $\Omega \neq 1$. *Physics Letters B* **271**: 52–60 (315)

Ellis, J., Hagelin, J. S., Nanopoulos, D. V., Olive, K., and Srednicki, M. 1984. Supersymmetric Relics from the Big Bang. *Nuclear Physics B* **238**: 453–476 (297)

Everett, H. 1957. "Relative State" Formulation of Quantum Mechanics. *Reviews of Modern Physics* **29**: 454–462 (61)

Evrard, A. E. 1989. Biased Cold Dark Matter Theory: Trouble from Rich Clusters? *Astrophysical Journal Letters* **341**: L71–L74 (100)

Fabbri, R., Guidi, I., Melchiorri, F., and Natale, V. 1980. Measurement of the Cosmic-Background Large-Scale Anisotropy in the Millimetric Region. *Physical Review Letters* **44**: 1563–1566 (227,300)

Faber, S. M. 1993. What I Learned This Week in Paris (About Cosmic Velocity Fields). In *Cosmic Velocity Fields, Proceedings of the 9th IAP Astrophysics Meeting*. Eds. François R. Bouchet and Marc Lachièze-Rey. Gif-sur-Yvette: Editions Frontieres, pp. 485–496. (107)

Faber, S. M., and Gallagher, J. S. 1979. Masses and Mass-to-Light Ratios of Galaxies. *Annual Review of Astronomy and Astrophysics* **17**: 135–187 (246,261,271,296)

Faber, S. M., and Jackson, R. E. 1976. Velocity Dispersions and Mass-to-Light Ratios for Elliptical Galaxies. *Astrophysical Journal* **204**: 668–683 (264)

Fall, S. M. 1975. The Scale of Galaxy Clustering and the Mean Matter Density of the Universe. *Monthly Notices of the Royal Astronomical Society* **172**: 23p–26p (84,106)

Fall, S. M. 1979. Dissipation, Merging and the Rotation of Galaxies. *Nature* **281**: 200–202 (218)

Fall, S. M., and Efstathiou, G. 1980. Formation and Rotation of Disc Galaxies with Haloes. *Monthly Notices of the Royal Astronomical Society* **193**: 189–206 (218)

Fall, S. M., and Romanowsky, A. J. 2018. Angular Momentum and Galaxy Formation Revisited: Scaling Relations for Disks and Bulges. *Astrophysical Journal* **868** article id. 133, 13 pp. (220)

Fang, L. Z., Li, S. X., and Xiang, S. P. 1984. Clustering in a Two-Component Universe. *Astronomy and Astrophysics* **140**: 77–81 (314)

Faulkner, J. 2009. The Day Fred Hoyle Thought He Had Disproved the Big Bang Theory. In *Finding the Big Bang*, Eds. P. J. E. Peebles, Page, and Partridge. Cambridge: Cambridge University Press, pp. 244–258 (138)

Feldman, H., Juszkiewicz, R., Ferreira, P., et al. 2003. An Estimate of Ω_m without Conventional Priors. *Astrophysical Journal Letters* **596**: L131–L134 (104)

Feynman, R. P., Morínigo, F. B., and Wagner, W. G. 1995. *Feynman Lectures on Gravitation*. Reading, MA: Addison-Wesley (348)

Field, G. B. 1965. Thermal Instability. *Astrophysical Journal* **142**: 531–567 (185)

Field, G. B. 1971. Instability and Waves Driven by Radiation in Interstellar Space and in Cosmological Models. *Astrophysical Journal* **165**: 29–40 (205)

Field, G. B., Herbig, G. H., and Hitchcock, J. 1966. Radiation Temperature of Space at λ2.6 mm. *Astronomical Journal* **71**: 161 (156)

Finzi, A. 1963. On the Validity of Newton's Law at a Long Distance. *Monthly Notices of the Royal Astronomical Society* **127**: 21–30 (263)

Fischer, P., McKay, T. A., Sheldon, E., et al. 2000. Weak Lensing with Sloan Digital Sky Survey Commissioning Data: The Galaxy-Mass Correlation Function to $1h^{-1}$ Mpc. *Astronomical Journal* **120**: 1198–1208 (263)

Fisher, K. B., Davis, M., Strauss, M. A., et al. 1994. Clustering in the 1.2-Jy IRAS Galaxy Redshift Survey—II. Redshift Distortions and $\xi(r_{\rm p}, \pi)$. *Monthly Notices of the Royal Astronomical Society* **267**: 927–948 (92)

Fisher, K. B., Huchra, J. P., Strauss, M. A., et al. 1995. The IRAS 1.2 Jy Survey: Redshift Data. *Astrophysical Journal Supplement* **100**: 69–103 (92)

Fixsen, D. J., Cheng, E. S., and Wilkinson, D. T. 1983. Large-Scale Anisotropy in the 2.7-K Radiation with a Balloon-Borne Maser Radiometer at 24.5 GHz. *Physical Review Letters* **50**: 620–622 (222,301)

Ford, W. K., Jr. 1968. Electronic Image Intensification. *Annual Review of Astronomy and Astrophysics* **6**: 1–12 (252)

Ford, W. K., Jr., Peterson, C. J., and Rubin, V. C. 1976. The Rotation Curve of the E7/S0 Galaxy NGC 3115. *Carnegie Institution Year Book* **75**: 124–125 (255)

Fraternali, F., van Moorsel, G., Sancisi, R., and Oosterloo, T. 2002. Deep H I Survey of the Spiral Galaxy NGC 2403. *Astronomical Journal* **123**: 3124–3140 (271)

Freedman, W. L., Madore, B. F., Gibson, B. K., et al. 2001. Final Results from the Hubble Space Telescope Key Project to Measure the Hubble Constant. *Astrophysical Journal* **553**: 47–72 (48,74)

Freeman, K. C. 1970. On the Disks of Spiral and S0 Galaxies. *Astrophysical Journal* **160**: 811–830 (257–261,268)

Freese, K., Frieman, J. A., and Olinto, A. V. 1990. Natural Inflation with Pseudo Nambu-Goldstone Bosons. *Physical Review Letters* **65**: 3233–3236 (313)

Friedman, A. 1922. Über die Krümmung des Raumes. *Zeitschrift für Physik* **10**: 377–386 (38)

Friedman, A. 1924. Über die Möglichkeit einer Welt mit Konstanter Negativer Krümmung des Raumes. *Zeitschrift für Physik* **21**: 326–332 (38–40)

Fritschi, M., Holzschuh, E., Kündig, W., et al. 1986. An Upper Limit for the Mass of $\bar{\nu}_e$ from Tritium β-Decay. *Physics Letters B* **173**: 485–489 (285)

Fry, J. N., and Peebles, P. J. E. 1978. Statistical Analysis of Catalogs of Extragalactic Objects. IX. The Four-Point Galaxy Correlation Function. *Astrophysical Journal* **221**: 19–33 (31)

Fukugita, M., Futamase, T., Kasai, M., and Turner, E. L. 1992. Statistical Properties of Gravitational Lenses with a Nonzero Cosmological Constant. *Astrophysical Journal* **393**: 3–21 (95)

Fukugita, M., and Peebles, P. J. E. 2004. The Cosmic Energy Inventory. *Astrophysical Journal* **616**: 643–668 (338)

Fukugita, M., Takahara, F., Yamashita, K., and Yoshii, Y. 1990. Test for the Cosmological Constant with the Number Count of Faint Galaxies. *Astrophysical Journal Letters* **361**: L1–L4 (95)

Fukugita, M., and Yanagida, T. 1984. Constraints on the Mass of Muon Neutrinos and Their Possible Role in the Galaxy Formation. *Physics Letters B* **144**: 386–390 (314)

Gamow, G. 1946. Expanding Universe and the Origin of Elements. *Physical Review* **70**: 572–573 (122,214)

Gamow, G. 1948a. The Origin of Elements and the Separation of Galaxies. *Physical Review* **74**: 505–506 (119,123,125,130–132,136–138,154,164,176,182,192,196, 201,209,290)

Gamow, G. 1948b. The Evolution of the Universe. *Nature* **162**: 680–682 (125,131,138, 139,154,164,176,182,201)

Gamow, G. 1949. On Relativistic Cosmogony. *Reviews of Modern Physics* **21**: 367–373 (127,132,136,142,159,175,183)

Gamow, G. 1950. Half an Hour of Creation. *Physics Today* **3**: 16–21 (151)

Gamow, G. 1952a. *The Creation of the Universe*. New York: Viking (151)

Gamow, G. 1952b. The Role of Turbulence in the Evolution of the Universe. *Physical Review* **86**:251 (213)

Gamow, G. 1953a. In *Proceedings of the Michigan Symposium on Astrophysics*, June 29–July 24, 29 pp. (129,138,140–142)

Gamow, G. 1953b. Expanding Universe and the Origin of Galaxies. *Danske Matematisk-fysiske Meddelelser* **27**, number 10, 15 pp. (138)

Gamow, G. 1954a. On the Steady-State Theory of the Universe. *Astronomical Journal* **59**: 200 (54,111,140,351,353)

Gamow, G. 1954b. On the Formation of Protogalaxies in the Turbulent Primordial Gas. *Proceedings of the National Academy of Science* **40**: 480–484 (214)

Gamow, G.1954c. Modern Cosmology. *Scientific American* **190** number 3: 54–63 (129)

Gamow, G. 1956a. The Physics of the Expanding Universe. *Vistas in Astronomy* **2**: 1726–1732 (138)

Gamow, G.1956b. The Evolutionary Universe. *Scientific American* **195**, number 3: 136–156 (130)

Gamow, G., and Critchfield, C. L. 1949. *Theory of Atomic Nucleus and Nuclear Energy-Sources*. Oxford: Clarenden Press (123–126,133,139,154,162,182)

Gamow, G., and Hynek, J. A. 1945. A New Theory by C. F. von Weizsäcker of the Origin of the Planetary System. *Astrophysical Journal* **101**: 249–254 (213)

Gamow, G., and Teller, E. 1939. The Expanding Universe and the Origin of the Great Nebulae. *Nature* **143**: 116–117 (139,198,201,209,214)

Garnavich, P. M., Kirshner, R. P., Challis, P., et al. 1998. Constraints on Cosmological Models from Hubble Space Telescope Observations of High-z Supernovae. *Astrophysical Journal Letters* **493**: L53–L57 (328)

Geller, M. J., and Peebles, P. J. E. 1973. Statistical Application of the Virial Theorem to Nearby Groups of Galaxies. *Astrophysical Journal* **184**: 329–342 (84,88,93)

Gelmini, G., Schramm, D. N., and Valle, J. W. F. 1984. Majorons: A Simultaneous Solution to the Large and Small Scale Dark Matter Problems. *Physics Letters B* **146**: 311–317 (314)

Gershtein, S. S., and Zel'dovich, Y. B. 1966. Rest Mass of Muonic Neutrino and Cosmology. *ZhETF Pis'ma* **4**: 174–177. English translation in *JETP Letters* **4**: 120–122 (280)

Giacconi, R., Gursky, H., Paolini, F. R., and Rossi, B. B. 1962. Evidence for X Rays from Sources Outside the Solar System. *Physical Review Letters* **9**: 439–443 (21)

Gisler, G. R., Harrison, E. R., and Rees, M. J. 1974. Variations in the Primordial Helium Abundance. *Monthly Notices of the Royal Astronomical Society* **166**: 663–672 (181)

Goenner H. 2001. Weyl's Contributions to Cosmology. In *Hermann Weyl's Raum-Zeit-Materie and a General Introduction to His Scientific Work.* Ed. E. Scholz. Basel: Birkhäuser, DMV Seminar **30** pp. 105–137 (38)

Gold, T., and Hoyle, F. 1959. Cosmic Rays and Radio Waves as Manifestations of a Hot Uuniverse. In *Paris Symposium on Radio Astronomy*, July 30, 1958. Ed. R. Bracewell, pp. 583–588 (184)

Goldman, T., and Stephenson, G. J., Jr. 1977. Limits on the Mass of the Muon Neutrino in the Absence of Muon-Lepton-Number Conservation. *Physical Review D* **16**: 2256–2259 (293)

Goodman, M. W., and Witten, E. 1985. Detectability of Certain Dark-Matter Candidates. *Physical Review D* **31**: 3059–3063 (299)

Gott, J. R., III. 1975. On the Formation of Elliptical Galaxies. *Astrophysical Journal* **201**: 296–310 (308)

Gott, J. R., III. 1982. Creation of Open Universes from de Sitter Space. *Nature* **29**: 304–306 (63)

Gott, J. R., III, Gunn, J. E., Schramm, D. N., and Tinsley, B. M. 1974. An Unbound Universe? *Astrophysical Journal* **194**: 543–553 (75–78,111,179,278,294,315,320)

Gott, J. R., III, Gunn, J. E., Schramm, D. N., and Tinsley, B. M. 1976. Will the Universe Expand Forever? *Scientific American* **234**: 62–72 (78)

Gott, J. R., III, and Turner, E. L. 1976. The Mean Luminosity and Mass Densities in the Universe. *Astrophysical Journal* **209**: 1–5 (83,88,101,106)

Gould, R. J. 1967. Origin of Cosmic X Rays. *American Journal of Physics* **35**: 376–393 (21)

Gould, R. J., and Burbidge, G. R. 1963. X-Rays from the Galactic Center, External Galaxies, and the Intergalactic Medium. *Astrophysical Journal* **138**: 969–977 (184)

Governato, F., Brook, C., Mayer, L., et al. 2010. Bulgeless Dwarf Galaxies and Dark Matter Cores from Supernova-Driven Outflows. *Nature* **463**: 203–206 (309)

Griest, K. 1991. Galactic Microlensing as a Method of Detecting Massive Compact Halo Objects. *Astrophysical Journal* **366**: 412–421 (277)

Groth, E. J., Juszkiewicz, R., and Ostriker, J. P. 1989. An Estimate of the Velocity Correlation Tensor: Cosmological Implications. *Astrophysical Journal* **346**: 558–565 (97)

Groth, E. J., and Peebles, P. J. E. 1975. *N*-Body Studies of the Clustering of Galaxies. *Bulletin of the American Astronomical Society* **7**: 425 (309)

Groth, E. J., and Peebles, P. J. E. 1977. Statistical Analysis of Catalogs of Extragalactic Objects. VII. Two- and Three-Point Correlation Functions for the High-Resolution Shane-Wirtanen Catalog of Galaxies. *Astrophysical Journal* **217**: 385–405 (28–31,99)

Gunn, J. E. 1982. Some Remarks on Phase-Density Constraints on the Masses of Massive Neutrinos. In *Astrophysical Cosmology, Proceedings of the Study Week on Cosmology and Fundamental Physics*, Vatican City State, September 28–October 2, 1981. Vatican City State: Pontifica Academia Scientiarum, pp. 557–562 (285)

Gunn, J. E. 1987. Conference Summary. In *Proceedings of IAU Symposium 117, Dark Matter in the Universe*, pp. 537–549 (64)

Gunn, J. E., Lee, B. W., Lerche, I., Schramm, D. N., and Steigman, G. 1978. Some Astrophysical Consequences of the Existence of a Heavy Stable Neutral Lepton. *Astrophysical Journal* **223**: 1015–1031 (218–220,294,298,308)

Gunn, J. E., and Oke, J. B. 1975. Spectrophotometry of Faint Cluster Galaxies and the Hubble Diagram: An Approach to Cosmology. *Astrophysical Journal* **195**: 255–268 (77,324)

Gunn, J. E., and Peterson, B. A. 1965. On the Density of Neutral Hydrogen in Intergalactic Space. *Astrophysical Journal* **142**: 1633–1641 (203)

Gunn, J. E., and Tinsley, B. M. 1975. An Accelerating Universe. *Nature* **257**: 454–457 (314)

Gursky, H., Kellogg, E., Murray, S., et al. 1971. A Strong X-Ray Source in the Coma Cluster Observed by *UHURU*. *Astrophysical Journal* **167**: L81–L84 (243)

Gush, H. P. 1974. An Attempt to Measure the Far Infrared Spectrum of the Cosmic Background Radiation. *Canadian Journal of Physics* **52**: 554–561 (171)

Gush, H. P. 1981. Rocket Measurement of the Cosmic Background Submillimeter Spectrum. *Physical Review Letters* **47**: 745–748 (172)

Gush, H. P., Halpern, M., and Wishnow, E. H. 1990. Rocket Measurement of the Cosmic-Background-Radiation mm-Wave Spectrum. *Physical Review Letters* **65**: 537–540 (172)

Guth, A. H. 1981. Inflationary Universe: A Possible Solution to the Horizon and Flatness Problems. *Physical Review D* **23**: 347–356 (62)

Guth, A. H. 1984. The New Inflationary Universe. *Annals of the New York Academy of Sciences* **422**: 1–14 (65)

Guth, A. H. 1991. Fundamental Arguments for Inflation. In *Observational Tests of Cosmological Inflation*, Durham, U.K., December 10–14, 1990. Eds. T. Shanks, A. J. Banday, R. S. Ellis, et al. NATO ASI series **348**: 1–21 (64)

Guth, A. H., and Pi, S.-Y. 1982. Fluctuations in the New Inflationary Universe. *Physical Review Letters* **49**: 1110–1113 (62)

Guyot, M., and Zel'dovich, Y. B. 1970. Gravitational Instability of a Two-Component Fluid: Matter and Radiation. *Astronomy and Astrophysics* **9**: 227–231 (201)

Halley, E. 1720. Of the Infinity of the Sphere of Fix'd Stars. *Philosophical Transactions of the Royal Society of London Series I* **31**: 22–24 (186)

Hamilton, A. J. S. 1993. Ω from the Anisotropy of the Redshift Correlation Function in the IRAS 2 Jansky Survey. *Astrophysical Journal Letters* **406**: L47–L50 (92)

Hamilton, A. J. S., Tegmark, M., and Padmanabhan, N. 2000. Linear Redshift Distortions and Power in the IRAS Point Source Catalog Redshift Survey. *Monthly Notices of the Royal Astronomical Society* **317**: L23–L27 (93,96)

Hammer, F. 1987. A Gravitational Lensing Model of the Strange Ring-Like Structure in A370. In *Proceedings of the Third IAP Workshop*, Paris, France, June 29–July 3. Gif-sur-Yuette, France: Editions Frontieres, pp. 467–473 (244)

Hamuy, M., Maza, J., Phillips, M. M., et al. 1993. The 1990 Calán/Tololo Supernova Search. *Astronomical Journal* **106**: 2392–2407 (325–328)

Hamuy, M., Phillips, M. M., Maza, J., et al. 1995. A Hubble Diagram of Distant Type Ia Supernovae. *Astronomical Journal* **109**: 1–13 (325,328)

Hamuy, M., Phillips, M. M., Suntzeff, N. B., et al. 1996. The Absolute Luminosities of the Calán/Tololo Type Ia Supernovae. *Astronomical Journal* **112**: 2391–2397 (325,328)

Hanany, S., Ade, P., Balbi, A., et al. 2000. MAXIMA-1: A Measurement of the Cosmic Microwave Background Anisotropy on Angular Scales of 10′–5°. *Astrophysical Journal Letters* **545**: L5–L9 (333)

Hansen, L., Nørgaard-Nielsen, H. U., and Jørgensen, H. E. 1987. Search for Supernovae in Distant Clusters of Galaxies. *ESO Messenger* **47**: 46–49 (326)

Harrison, E. R. 1970. Fluctuations at the Threshold of Classical Cosmology. *Physical Review D* **1**: 2726–2730 (231,302)

Harwit, M. 1961. Can Gravitational Forces Alone Account for Galaxy Formation in a Steady-State Universe? *Monthly Notices of the Royal Astronomical Society* **122**: 47–50 (185)

Hauser, M. G., and Dwek, E. 2001. The Cosmic Infrared Background: Measurements and Implications. *Annual Review of Astronomy and Astrophysics* **39**: 249–307 (119)

Hawking, S. W. 1966. Perturbations of an Expanding Universe. *Astrophysical Journal* **145**: 544–554 (190)

Hawking, S. W. 1971. Gravitationally Collapsed Objects of Very Low Mass. *Monthly Notices of the Royal Astronomical Society* **152**: 75–78 (77)

Hawking, S. W. 1982. The Development of Irregularities in a Single Bubble Inflationary Universe. *Physics Letters B* **115**: 295–297 (62)

Hawkins, E., Maddox, S., Cole, S., et al. 2003. The 2dF Galaxy Redshift Survey: Correlation Functions, Peculiar Velocities and the Matter Density of the Universe. *Monthly Notices of the Royal Astronomical Society* **346**: 78–96 (93)

Hayashi, C. 1950. Proton-Neutron Concentration Ratio in the Expanding Universe at the Stages Preceding the Formation of the Elements. *Progress of Theoretical Physics* **5**: 224–235 (128,138,147–151)

Hegyi, D. J., and Olive, K. A. 1983. Can Galactic Halos Be Made of Baryons? *Physics Letters B* **126**: 28–32 (297)

Heisenberg, W. 1951. Discussion. In *Proceedings of the Symposium on the Motions of Gaseous Masses of Cosmical Dimensions*, Paris, August 16–19, 1949, p. 199 (217,236,353)

Herzberg, G. 1950, *Molecular Spectra and Molecular Structure I. Spectra of Diatomic Molecules.* Second edition. New York: Van Nostrand (153)

Hinshaw, G., Spergel, D. N., Verde, L., et al. 2003. First-Year Wilkinson Microwave Anisotropy Probe (WMAP) Observations: The Angular Power Spectrum. *Astrophysical Journal Supplement* **148**: 135–139 (333)

Hockney, R. W., and Hohl, F. 1969. Effects of Velocity Dispersion on the Evolution of a Disk of Stars. *Astronomical Journal* **74**: 1102–1124 (266)

Hodges, H. M., and Blumenthal, G. R. 1990. Arbitrariness of Inflationary Fluctuation Spectra. *Physical Review D* **42**: 3329–3333 (313)

Hogan, C. J., and Rees, M. J. 1984. Gravitational Interactions of Cosmic Strings. *Nature* **311**: 109–114 (237)

Hogg, D. C., and Semplak, R. A. 1961. The Effect of Rain and Water Vapor on Sky Noise at Centimeter Wavelengths. *Bell System Technical Journal* **40**: 1331–1348 (163)

Hohl, F. 1970. Dynamical Evolution of Disk Galaxies. NASA Technical Report, NASA-TR R-343, 108 pp. (266,268)

Hohl, F. 1971. Numerical Experiments with a Disk of Stars. *Astrophysical Journal* **168**: 343–359 (266–268)

Hohl, F. 1976. Suppression of Bar Instability by a Massive Halo. *Astronomical Journal* **81**: 30–36 (271)

Hohl, F., and Hockney, R. W. 1969. A Computer Model of Disks of Stars. *Journal of Computational Physics* **4**: 306–324 (266)

Howell, T. F., and Shakeshaft, J. R. 1967. Spectrum of the 3° K Cosmic Microwave Radiation. *Nature* **216**: 753–754 (164)

Hoyle, F. 1948. A New Model for the Expanding Universe. *Monthly Notices of the Royal Astronomical Society* **108** 372–382 (50)

Hoyle, F. 1949. Stellar Evolution and the Expanding Universe. *Nature* **163**: 196–198 (140,154)

Hoyle, F. 1950. Nuclear Energy. *The Observatory* **70**: 194–195 (154)

Hoyle, F. 1951. The Origin of the Rotations of the Galaxies. In *Problems of Cosmical Aerodynamics; Proceedings of a Symposium on the Motion of Gaseous Masses of Cosmical Dimensions*, Paris, August 16–19, 1949. Dayton, OH: Central Air Document Offices, pp. 195–199 (216)

Hoyle, F. 1958. The Astrophysical Implications of Element Synthesis. *Ricerche Astronomiche* **5**: 279–284 (143)

Hoyle, F. 1959. The Ages of Type I and Type II Subgiants. *Monthly Notices of the Royal Astronomical Society* **119**: 124–133 (143)

Hoyle, F. 1980. *Steady-State Cosmology Revisited.* Cardiff: University College Cardiff Press (221)

Hoyle, F. 1981. The Big Bang in Astronomy. *New Scientist* **92** 521–524 (154)

Hoyle, F. 1988. Fifty Years in Cosmology. *Bulletin of the Astronomical Society of India* **16**: 1–9 (155,169)

Hoyle, F., Burbidge, G., and Narlikar, J. V. 1993. A Quasi-Steady State Cosmological Model with Creation of Matter. *Astrophysical Journal* **410**: 437–457 (55)

Hoyle, F., Burbidge, G., and Narlikar, J. V. 2000. A Different Approach to Cosmology: From a Static Universe through the Big Bang towards Reality. New York: Cambridge University Press, 2000 (55)

Hoyle, F., and Narlikar, J. V. 1966. A Radical Departure from the "Steady-State" Concept in Cosmology. *Proceedings of the Royal Society of London Series A* **290**: 162–176 (55,169)

Hoyle, F., and Tayler, R. J. 1964. The Mystery of the Cosmic Helium Abundance. *Nature* **203**: 1108–1110 (75,138,141,146–154,162)

Hoyle, F., and Wickramasinghe, N. C. 1988. Metallic Particles in Astronomy. *Astrophysics and Space Science* **147**: 245–256 (167–169)

Hu, W., Bunn, E. F., and Sugiyama, N. 1995. *COBE* Constraints on Baryon Isocurvature Models. *Astrophysical Journal Letters* **447**: L59–L63 (316)

Hu, W., Sugiyama, N., and Silk, J. 1997. The Physics of Microwave Background Anisotropies. *Nature* **386**: 37–43 (334)

Hu, W., and White, M. 1996. A New Test of Inflation. *Physical Review Letters* **77**: 1687–1690 (229)

Hubble, E. P. 1925. Cepheids in Spiral Nebulae. *The Observatory* **48**: 139–142 (41)

Hubble, E. P. 1926. Extragalactic Nebulae. *Astrophysical Journal* **64**: 321–369 (18,23, 42,51,78)

Hubble, E. 1929. A Relation between Distance and Radial Velocity among Extra-Galactic Nebulae. *Proceedings of the National Academy of Sciences* **15**: 168–173 (15,22,44,73,314)

Hubble, E. 1934. The Distribution of Extra-Galactic Nebulae. *Astrophysical Journal* **79**: 8–76 (18,33)

Hubble, E. 1936, *The Realm of the Nebulae*, New Haven, CT: Yale University Press (22,24,32,78,106,131,235,324)

Hubble, E. 1937, *The Observational Approach to Cosmology*, Oxford: Clarendon Press (131)

Hubble, E., and Humason, M. L. 1931. The Velocity-Distance Relation among Extra-Galactic Nebulae. *Astrophysical Journal* **74**: 43–80 (22,45)

Hubble, E., and Tolman, R. C. 1935. Two Methods of Investigating the Nature of the Nebular Redshift. *Astrophysical Journal* **82**: 302–337 (24)

Hudson, M. J., Dekel, A., Courteau, S., Faber, S. M., and Willick, J. A. 1995. Ω and Biasing from Optical Galaxies versus POTENT Mass. *Monthly Notices of the Royal Astronomical Society* **274**: 305–316 (103)

Hughes, D. J., Spatz, W. D., and Goldstein, N. 1949. Capture Cross Sections for Fast Neutrons. *Physical Review* **75**: 1781–1787 (133)

Humason, M. L., Mayall, N. U., and Sandage, A. R. 1956. Redshifts and Magnitudes of Extragalactic Nebulae. *Astronomical Journal* **61**: 97–162 (53,73)

Hut, P. 1977. Limits on Masses and Number of Neutral Weakly Interacting Particles. *Physics Letters B* **69**: 85–88 (291)

Hut, P., and White, S. D. M. 1984. Can a Neutrino-Dominated Universe Be Rejected? *Nature* **310**: 637–640 (288,314)

Huterer, D., Turner, M. S. 1999. Prospects for Probing the Dark Energy via Supernova Distance Measurements. *Physical Review D* **60**: 081301, 5 pp. (46)

Ibata, R. A., Lewis, G. F., Conn, A. R., et al. 2013. A Vast, Thin Plane of Corotating Dwarf Galaxies Orbiting the Andromeda Galaxy. *Nature* **493**: 62–65 (247)

Ikeuchi, S. 1981. Theory of Galaxy Formation Triggered by Quasar Explosions. *Publications of the Astronomical Society of Japan* **33**: 211–231 (222)

Ipser, J., and Sikivie, P. 1983. Can Galactic Halos Be Made of Axions? *Physical Review Letters* **50**: 925–927 (296–298)

Irvine, W. M. 1961. Local Irregularities in a Universe Satisfying the Cosmological Principle. PhD thesis, Harvard University (84)

Jacoby, G. H., Ciardullo, R., and Ford, H. C. 1990. Planetary Nebulae as Standard Candles. V. The Distance to the Virgo Cluster. *Astrophysical Journal* **356**: 332–349 (74)

Jaffe, A. H., Ade, P. A., Balbi, A., et al. 2001. Cosmology from MAXIMA-1, BOOMERANG, and COBE DMR Cosmic Microwave Background Observations. *Physical Review Letters* **86**: 3475–3479 (334)

Jeans, J. H. 1902. The Stability of a Spherical Nebula. *Philosophical Transactions of the Royal Society of London Series A* **199**: 1–53 (186,201,208)

Jedamzik, K., Fuller, G. M., Mathews, G. J., and Kajino, T. 1994. Enhanced Heavy-Element Formation in Baryon-Inhomogeneous Big Bang Models. *Astrophysical Journal* **422**: 423–429 (181)

Jones, B. T. J., and Peebles, P. J. E. 1972. Chaos in Cosmology. *Comments on Astrophysics and Space Physics* **4**: 121–128 (212,216,222)

Jordan, P. 1948. Fünfdimensionale Kosmologie. *Astronomische Nachrichten* **276**: 193–208 (49)

Kahn, F. D., and Woltjer, L. 1959. Intergalactic Matter and the Galaxy. *Astrophysical Journal* **130**: 705–717 (245)

Kahn, N. K. 1987. Desperately Seeking Supernovae. *Sky and Telescope* **73**: 594–597 (327)

Kaiser, N. 1984. On the Spatial Correlations of Abell Clusters. *Astrophysical Journal Letters* **284**: L9–L12 (30,67)

Kaiser, N. 1986. Statistics of Density Maxima and the Large-Scale Matter Distribution. In *Inner Space/Outer Space: The Interface between Cosmology and Particle Physics*. Chicago: University of Chicago Press, pp. 258–263 (64,67)

Kaiser, N. 1987. Clustering in Real Space and in Redshift Space. *Monthly Notices of the Royal Astronomical Society* **227**: 1–21 (88)

Kaiser, N., and Stebbins, A. 1984. Microwave Anisotropy Due to Cosmic Strings. *Nature* **310**: 391–393 (228)

Kalnajs, A. J. 1972. The Equilibria and Oscillations of a Family of Uniformly Rotating Stellar Disks. *Astrophysical Journal* **175**: 63–76 (268)

Kalnajs, A. J. 1983. In *Proceedings of IAU Symposium 100, Internal Kinematics and Dynamics of Galaxies*, Besançon, France, August 9–13, 1982. Dordrecht: Reidel, pp. 87–88 (261)

Kamionkowski, M., Ratra, B., Spergel, D. N., and Sugiyama, N. 1994. Cosmic Background Radiation Anisotropy in an Open Inflation, Cold Dark Matter Cosmogony. *Astrophysical Journal Letters* **434**: L1–L4 (63,315,320)

Kamionkowski, M., Spergel, D. N., and Sugiyama, N. 1994. Small-Scale Cosmic Microwave Background Anisotropies as a Probe of the Geometry of the Universe. *Astrophysical Journal Letters* **426**: L57–L60 (320)

Karachentsev, I. D. 1966. The Virial Mass–Luminosity Ratio and the Instability of Different Galactic Systems. *Astrophysics* **2**: 39–49 (243,273)

Katz, N., Hernquist, L., and Weinberg, D. H. 1992. Galaxies and Gas in a Cold Dark Matter Universe. *Astrophysical Journal Letters* **399**: L109–L112 (67)

Kauffmann, G., and White, S. D. M. 1992. The Observational Properties of an $\Omega = 0.2$ Cold Dark Matter Universe. *Monthly Notices of the Royal Astronomical Society* **258**: 511–520 (309)

Kelvin, L. 1901. Nineteenth Century Clouds over the Dynamical Theory of Heat and Light. *Philosophical Magazine* **Sixth Series** 1–40 (341)

Kent, S. M. 1981. Distances to the Galaxies Stephan's Quintet. *Publications of the Astronomical Society of the Pacific* **93**: 554–557 (246)

Kent, S. M. 1986. Dark Matter in Spiral Galaxies. I. Galaxies with Optical Rotation Curves. *Astronomical Journal* **91**: 1301–1327 (262)

Kent, S. M. 1987. Dark Matter in Spiral Galaxies. II. Galaxies with H I Rotation Curves. *Astronomical Journal* **93**: 816–832 (258,262,265,270)

Kibble, T. W. B. 1976. Topology of Cosmic Domains and Strings. *Journal of Physics A: Mathematical and General* **9**: 1387–1398 (225)

Kibble, T. W. B. 1980. Some Implications of a Cosmological Phase Transition. *Physics Reports* **67**: 183–199 (225,227)

King, I. R. 1977. Galaxies and Their Populations—The View on a Cloudy Day. In *Proceedings of a Conference on The Evolution of Galaxies and Stellar Populations, May 1977*. Eds B. M. Tinsley and R. B. Larson. New Haven: Yale University Observatory, pp. 1–17 (239,278)

Kirkman, D., Tytler, D., Suzuki, N., O'Meara, J. M., and Lubin, D. 2003. The Cosmological Baryon Density from the Deuterium-to-Hydrogen Ratio in QSO Absorption Systems: D/H toward Q1243+3047. *Astrophysical Journal Supplement* **149**; 1–28 (180,338)

Kirshner, R. P. 2010. Foundations of Supernova Cosmology. In *Dark Energy: Observational and Theoretical Approaches*, ed. Pilar Ruiz-Lapuente. Cambridge: Cambridge University Press, pp. 151–176 (329)

Kirshner, R. P., Oemler, A., Jr., and Schechter, P. L. 1978. A Study of Field Galaxies. I. Redshifts and Photometry of a Complete Sample of Galaxies. *Astronomical Journal* **83**: 1549–1563 (89)

Klein, O. 1956. On the Eddington Relations and Their Possible Bearing on an Early State of the System of Galaxies. *Helvetica Physica Acta Supplementum* **IV**: 147–149 (33,40)

Klein, O. 1966. Instead of Cosmology. *Nature* **211**: 1337–1341 (34)

Klypin, A., Holtzman, J., Primack, J., and Regős, E. 1993. Structure Formation with Cold Plus Hot Dark Matter. *Astrophysical Journal* **416**: 1–16 (318)

Knop, R. A., Aldering, G., Amanullah, R., et al. 2003. New Constraints on Ω_m, Ω_Λ, and w from an Independent Set of 11 High-Redshift Supernovae Observed with the Hubble Space Telescope. *Astrophysical Journal* **598**: 102–137 (330,337)

Kofman, L. A., Gnedin, N. Y., and Bahcall, N. A. 1993. Cosmological Constant, COBE Cosmic Microwave Background Anisotropy, and Large-Scale Clustering. *Astrophysical Journal* **413**: 1–9 (317,319)

Kofman, L. A., and Starobinsky, A. A. 1985. Effect of the Cosmological Constant on Large-Scale Anisotropies in the Microwave Background. *Pis'ma Astronomicheskii Zhurnal* **11**: 643–651. English translation in *Soviet Astronomy Letters* **11**: 271–274 (65, 314)

Kolb, E. W., and Turner, M. S. 1990. The Early Universe. *Frontiers of Physics* **69**: 547 pp. (291)

Komatsu, E., Smith, K. M., Dunkley, J., et al. 2011. Seven-Year Wilkinson Microwave Anisotropy Probe (WMAP) Observations: Cosmological Interpretation. *Astrophysical Journal Supplement* **192**: 18, 47 pp. (177,339)

Koo, D. C. 1981. Multi-Color Analysis of Galaxy Evolution and Cosmology. PhD thesis, University of California, Berkeley (94)

Kragh, H. 1996. *Cosmology and Controversy: The Historical Development of Two Theories of the Universe*. Princeton, NJ: Princeton University Press, 1996 (7,133,136)

Kragh, H. 2013. Cyclic Models of the Relativistic Universe: The Early History. arXiv:1308.0932, 29 pp. (161)

Krauss, L. M., Freese, K., Spergel, D. N., and Press, W. H. 1985. Cold Dark Matter Candidates and the Solar Neutrino Problem. *Astrophysical Journal* **299**: 1001–1006 (299)

Krauss, L. M., Srednicki, M., and Wilczek, F. 1986. Solar System Constraints and Signatures for Dark-Matter Candidates. *Physical Review D* **33**: 2079–2083 (299)

Krauss, L. M., and Turner, M. S. 1995. The Cosmological Constant Is Back. *General Relativity and Gravitation* **27**: 1137–1144 (320)

Kuhn, T. S. 1962. *The Structure of Scientific Revolutions*. Chicago: University of Chicago Press (3,354)

Lacey, C. G., and Ostriker, J. P. 1985. Massive Black Holes in Galactic Halos? *Astrophysical Journal* **299**: 633–652 (277)

Lanczos, C. 1922. Bemerkung zur de Sitterschen Welt. *Physikalische Zeitschrift* **23**: 539–543 (40)

Lanczos, K. 1923. Über die Rotverschiebung in der de Sitterschen Welt. *Zeitschrift für Physik* **17**: 168–188 (38)

Landau, L., and Lifshitz, E. 1951. *The Classical Theory of Fields*. Reading, MA: Addison-Wesley (17,57)

Larson, R. B. 1969. A Model for the Formation of a Spherical Galaxy. *Monthly Notices of the Royal Astronomical Society* **145**: 405–422 (308)

Larson, R. B. 1976. Models for the Formation of Disc Galaxies. *Monthly Notices of the Royal Astronomical Society* **176**: 31–52 (308)

Larson, R. B. 1983. Star Formation in Disks. *Highlights of Astronomy* **6**: 191–198 (308)

Layzer, D. 1954. Is the Origin of the Solar System Connected with the Overall Structure of the Universe? *Astronomical Journal* **59**: 170–172 (224,233)

Layzer, D. 1963. A Preface to Cosmogony. I. The Energy Equation and the Virial Theorem for Cosmic Distributions. *Astrophysical Journal* **138**: 174–184 (84)

Layzer, D. 1968. Black-Body Radiation in a Cold Universe. *Astrophysical Letters* **1**: 99–102 (167)

Lea, S. M. 1977. Hot Gas in Clusters of Galaxies. *Highlights of Astronomy* **4**: 329–339 (243)

Leavitt, H. S. 1912. Periods of 25 Variable Stars in the Small Magellanic Cloud. *Harvard College Observatory Circular* **173**: 1–3 (42,44)

Lee, B. W., and Weinberg, S. 1977a. Cosmological Lower Bound on Heavy-Neutrino Masses. *Physical Review Letters* **39**: 165–168 (290–295)

Lee, B. W., and Weinberg, S. 1977b. SU(3) ⊗ U(1) Gauge Theory of the Weak and Electromagnetic Interactions. *Physical Review Letters* **38**: 1237–1240 (293)

Lelli, F., McGaugh, S. S., and Schombert, J. M. 2017. Testing Verlinde's Emergent Gravity with the Radial Acceleration Relation. *Monthly Notices of the Royal Astronomical Society* **468**: L68–L71 (264)

Lemaître, G. 1925. Note on de Sitter's Universe. *MIT Journal of Mathematics and Physics* **4**: 188–192 (39)

Lemaître, G. 1927. Un Univers homogène de masse constante et de rayon croissant rendant compte de la vitesse radiale des nébuleuses extra-galactiques. *Annales de la Socité Scientifique de Bruxelles* **A47**: 49–59 (40–45,187)

Lemaître, G. 1929. La Grandeur de l'Espace. *Revue des Questions Scientifiques* **XV**: 189–216 (40)

Lemaître, G. 1931a. Expansion of the Universe, A Homogeneous Universe of Constant Mass and Increasing Radius Accounting for the Radial Velocity of Extra-Galactic Nebulae. *Monthly Notices of the Royal Astronomical Society* **91**: 483–490 (40,43,71)

Lemaître, G. 1931b. The Beginning of the World from the Point of View of Quantum Theory. *Nature* **127**: 706 (45,71)

Lemaître, G. 1931c. L'Expansion de l'Espace. *Revue des Questions Scientifiques* **XX**: 391–410 (45,71)

Lemaître, G. 1931d. The Expanding Universe. *Monthly Notices of the Royal Astronomical Society* **91**: 490–501 (54,69,190–193,316)

Lemaître, G. 1931e. Contribution to the British Association Discussion, The Evolution of the Universe. *Nature* **128**: 704–706 (114)

Lemaître, G. 1933a. L'Univers en expansion. *Annales de la société scientifique de Bruxelles* **53A**: 51–85 (71,160,193)

Lemaître, G. 1933b. Condensations sphériques dans l'universe en expansion. *Comptes rendus hebdomadaires des séances de l'Académie des sciences* **196**: 903–905 (188,191, 193)

Lemaître, G. 1933c. La formation des nébuleuses dans l'univers en expansion. *Comptes rendus hebdomadaires des séances de l'Académie des sciences* **196**: 1085–1087 (188,199)

Lemaître, G. 1934. Evolution of the Expanding Universe. *Proceedings of the National Academy of Science* **20**: 12–17 (46,59,69,190)

Lemaître, G. 1950. L'expansion de I'Univers, par Paul Couderc. *Annales d'Astrophysique* **13**: 344–345 (40)

Liddle, A. R., Lyth, D. H., and Sutherland, W. 1992. Structure Formation from Power Law (and Extended) Inflation. *Physics Letters B* **279**: 244–249 (313)

Liebes, S. 1964. Gravitational Lenses. *Physical Review* **133**: 835–844 (277)

Lifshitz, E. M. 1946. On the Gravitational Stability of the Expanding Universe. *Zhurnal Eksperimental'noi i Teoreticheskoi Fiziki* **16**: 587–602. English translation in *Journal of Physics of the USSR* **10**: 116–129, 1946 (119,130,188–193,198,215)

Lifshitz, E. M., and Khalatnikov, I. M. 1963. Investigations in Relativistic Cosmology. *Advances in Physics* **12**: 185–249 (191,211)

Lilje, P. B. 1992. Abundance of Rich Clusters of Galaxies: A Test for Cosmological Parameters. *Astrophysical Journal Letters* **386**: L33–L36 (100)

Lilje, P. B., Yahil, A., and Jones, B. J. T. 1986. The Tidal Velocity Field in the Local Supercluster. *Astrophysical Journal* **307**: 91–96 (222)

Limber, D. N. 1953. The Analysis of Counts of the Extragalactic Nebulae in Terms of a Fluctuating Density Field. *Astrophysical Journal* **117**: 134–144 (26)

Limber, D. N. 1954. The Analysis of Counts of the Extragalactic Nebulae in Terms of a Fluctuating Density Field. II. *Astrophysical Journal* **119**: 655–681 (26)

Lin, D. N. C., and Faber, S. M. 1983. Some Implications of Nonluminous Matter in Dwarf Spheroidal Galaxies. *Astrophysical Journal Letters* **266**: L21–L25 (285)

Linde, A. D. 1982. A New Inflationary Universe Scenario: A Possible Solution of the Horizon, Flatness, Homogeneity, Isotropy and Primordial Monopole Problems. *Physics Letters B* **108**: 389–393 (62)

Linde, A. D. 1986. Eternally Existing Self-Reproducing Chaotic Inflationary Universe. *Physics Letters B* **175**: 395–400 (62)

Linde, A. D. 1990. *Particle Physics and Inflationary Cosmology*. Switzerland: Harwood, Chur Publishers; available at https://arxiv.org/pdf/hep-th/0503203.pdf (161)

Lineweaver, C. H. 1998. The Cosmic Microwave Background and Observational Convergence in the $\Omega_m - \Omega_\Lambda$ Plane. *Astrophysical Journal Letters* **505**: L69–L73 (334)

Lineweaver, C. H., Barbosa, D., Blanchard, A., and Bartlett, J. G. 1997. Constraints on h, Ω_b and λ_0 from Cosmic Microwave Background Observations. *Astronomy and Astrophysics* **322**: 365–374 (335)

Livio, M. 2011. Lost in translation: Mystery of the missing text solved. *Nature* **479**: 171–173 (44)

Loh, E. D., and Spillar, E. J. 1986a. Photometric Redshifts of Galaxies. *Astrophysical Journal* **303**: 154–161 (94)

Loh, E. D., and Spillar, E. J. 1986b. A Measurement of the Mass Density of the Universe. *Astrophysical Journal Letters* **307**: L1–L4 (94,324)

Longair, M. S. 2006. The Cosmic Century. Cambridge: Cambridge University Press (38,52)

López Fune, E., Salucci, P., and Corbelli, E. 2017. Radial Dependence of the Dark Matter Distribution in M33. *Monthly Notices of the Royal Astronomical Society* **468**: 147–153 (262)

Lubimov, V. A., Novikov, E. G., Nozik, V. Z., Tretyakov, E. F., and Kosik, V. S. 1980. An Estimate of the ν_e Mass from the β-Spectrum of Tritium in the Valine Molecule. *Physics Letters B* **94**: 266–268 (285)

Lubin, L. M., and Bahcall, N. A. 1993. The Relation between Velocity Dispersion and Temperature in Clusters: Limiting the Velocity Bias. *Astrophysical Journal Letters* **415**: L17–L20 (244)

Lucchin, F., and Matarrese, S. 1985. Power-Law Inflation. *Physical Review D* **32**: 1316–1322 (313)

Luminet, J.-P. 2013. Editorial Note to: Georges Lemaître, A Homogeneous Universe of Constant Mass and Increasing Radius Accounting for the Radial Velocity of Extra-Galactic Nebulae. *General Relativity and Gravitation* **45**: 1619–1633 (42)

Lundmark, K. 1924. The Determination of the Curvature of Space-Time in de Sitter's World. *Monthly Notices of the Royal Astronomical Society* **84**: 747–770 (40)

Lundmark, K. 1925. The Motions and the Distances of Spiral Nebulæ. *Monthly Notices of the Royal Astronomical Society* **85**: 865–894 (41–44)

Lynden-Bell, D. 1962. On the Gravitational Collapse of a Cold Rotating Gas Cloud. *Mathematical Proceedings of the Cambridge Philosophical Society* **50**: 709–711 (234)

Lynden-Bell, D. 1969. Galactic Nuclei as Collapsed Old Quasars. *Nature* **223**: 690–694 (221)

Lynden-Bell, D., Faber, S. M., Burstein, D., et al. 1988. Spectroscopy and photometry of elliptical galaxies. V. Galaxy Streaming Toward the New Supergalactic Center. *Astrophysical Journal* **326**: 19–49 (85,97,222)

Lynden-Bell, D., Lahav, O., and Burstein, D. 1989. Cosmological Deductions from the Alignment of Local Gravity and Motion. *Monthly Notices of the Royal Astronomical Society* **241**: 325–345 (96)

Mach, E. 1960. The Science of Mechanics. Chicago: Open Court Publishing (14)

Mandelbrot, B. 1975. *Les Objects Fractals*. Paris: Flammarion Editeur (32)

Mandelbrot, B. 1989. *Les Objects Fractals*. Paris: Flammarion Editeur, third edition (32)

Marx, G., and Szalay, A. S. 1972. Cosmological Limit on Neutretto Mass. In *Neutrino '72, Balatonfüred Hungary, June 1972*. Eds. A. Frenkel and G. Marx. Budapest: OMKDT-Technoinform **I**, 191–195 (281)

Masjedi, M., Hogg, D. W., Cool, R. J., et al. 2006. Very Small Scale Clustering and Merger Rate of Luminous Red Galaxies. *Astrophysical Journal* **644**: 54–60 (30)

Mather, J. C. 1974. Far Infrared Spectrometry of the Cosmic Background Radiation. PhD thesis, University of California, Berkeley (171)

Mather, J. C., and Boslough, J. 1996. *The Very First Light: The True Inside Story of the Scientific Journey Back to the Dawn of the Universe*. New York: Basic Books (172)

Mather, J. C., Cheng, E. S., Eplee, R. E., Jr., et al. 1990. A Preliminary Measurement of the Cosmic Microwave Background Spectrum by the Cosmic Background Explorer (COBE) Satellite. *Astrophysical Journal Letters* **354**: L37–L40 (172)

Mathis, J. S. 1957. The Ratio of Helium to Hydrogen in the Orion Nebula. *Astrophysical Journal* **125**: 328–335 (142)

Matsumoto, T., Hayakawa, S., Matsuo, H., et al. 1988. The Submillimeter Spectrum of the Cosmic Background Radiation. *Astrophysical Journal* **329**: 567–571 (170)

Mayall, N. U. 1951. Comparison of Rotational Motions Observed in the Spirals M31 and M33 and in the Galaxy. *Publications of the Observatory of the University of Michigan* **10**: 19–24 (250–254)

Mayer, M. G., and Teller, E.1949. On the Origin of Elements. *Physical Review* **76**: 1226–1231 (182)

McCrea, W. H. 1971. The Cosmical Constant. *Quarterly Journal of the Royal Astronomical Society* **12**: 140–153 (59)

McCrea, W. H., and Milne, E. A. 1934. Newtonian Universes and the Curvature of Space. *Quarterly Journal of Mathematics* **5**: 73–80 (45,187)

McKellar, A. 1941. Molecular Lines from the Lowest States of Diatomic Molecules Composed of Atoms Probably Present in Interstellar Space. *Publications of the Dominion Astrophysical Observatory* **7**: 251–272 (153–155)

McVittie, G. C. 1967. *Quarterly Journal of the Royal Astronomical Society* **8**: 294–297 (43)

Melott, A. L., Einasto, J., Saar, E., et al. 1983. Cluster Analysis of the Nonlinear Evolution of Large-Scale Structure in an Axion/Gravitino/Photino-Dominated Universe. *Physical Review Letters* **51**: 935–938 (288,309)

Merton, R. K. 1962. Singletons and Multiples in Scientific Discovery: A Chapter in the Sociology of Science. *Proceedings of the American Philosophical Society* **105**: 470–486 (359)

Messina, A., Lucchin, F., Matarrese, S., and Moscardini, L. 1992. The Large-Scale Structure of the Universe in Skewed Cold Dark Matter Models. *Astroparticle Physics* **1**: 99–112 (316)

Mestel, L. 1963. On the Galactic Law of Rotation. *Monthly Notices of the Royal Astronomical Society* **126**: 553–575 (217)

Mészáros, P. 1974. The Behaviour of Point Masses in an Expanding Cosmological Substratum. *Astronomy and Astrophysics* **37**: 225–228 (201)

Michelson, A. 1903. *Light Waves and Their Uses* (340)

Michie, R. W. 1966. Galaxy Formation: Angular Momentum Problems in the Fragmentation Process. *Astronomical Journal* **71**: 171–172 (218)

Michie, R. W. 1969. On the Growth of Condensations in an Expanding Universe. Contribution Number 440 from the Kitt Peak National Observatory, 16 pp. (205)

Milgrom, M. 1983. A Modification of the Newtonian Dynamics as a Possible Alternative to the Hidden Mass Hypothesis. *Astrophysical Journal* **270**: 365–370 (263)

Miller, R. H. 1971. Numerical Experiments in Collisionless Systems. *Astrophysics and Space Science* **14**: 73–90 (267)

Miller, R. H. 1978a. Free Collapse of a Rotating Sphere of Stars. *Astrophysical Journal* **223**: 122–128 (271)

Miller, R. H. 1978b. Numerical Experiments on the Stability of Disklike Galaxies. *Astrophysical Journal* **223**: 811–823 (271)

Miller, R. H. 1983. Numerical Experiments on the Clustering of Galaxies. *Astrophysical Journal* **270**: 390–409 (309)

Miller, R. H., and Prendergast, K. H. 1968. Stellar Dynamics in a Discrete Phase Space. *Astrophysical Journal* **151**: 699–709 (266)

Miller, R. H., Prendergast, K. H., and Quirk, W. J. 1970. Numerical Experiments on Spiral Structure. *Astrophysical Journal* **161**: 903–916 (267)

Miller, R. H., and Smith, B. F. 1981. Numerical Experiments on Galaxy Formation: I. Introduction and First Results. *Astrophysical Journal* **244**: 467–475 (309)

Milne, E. A. 1933. World-Structure and the Expansion of the Universe. *Zeitschrift für Astrophysik* **6**: 1–95 (14)

Minnaert, M. 1957. The Determination of Cosmic Abundances. *Monthly Notices of the Royal Astronomical Society* **117**: 315–335 (145)

Misner, C. W. 1967. Transport Processes in the Primordial Fireball. *Nature* **214**: 40–41 (211)

Misner, C. W. 1969. Mixmaster Universe. *Physical Review Letters* **22**: 1071–1074 (211)

Misner, C. W., Thorne, K. S., and Wheeler, J. A. 1973. *Gravitation*. Princeton, NJ: Princeton University Press (57)

Mitchell, R. J., Culhane, J. L., Davison, P. J. N., and Ives, J. C. 1976. *Ariel 5* Observations of the X-Ray Spectrum of the Perseus Cluster. *Monthly Notices of the Royal Astronomical Society* **175**: 29P–34P (243)

Mitton, S. 2005. *Fred Hoyle: A Life in Science*. London: Aurum Press (7)

Muehlner, D., and Weiss, R. 1970. Measurement of the Isotropic Background Radiation in the Far infrared. *Physical Review Letters* **24**: 742–746 (166)

Mukhanov, V. F., and Chibisov, G. V. 1981. Quantum Fluctuations and a Nonsingular Universe. *Zhurnal Eksperimental'noi i Teoreticheskoi Fiziki* **33**: 549–553. English translation in *Soviet Physics—JETP Letters* **33**: 532–535 (62)

Muller, R. A., and Pennypacker, C. R. 1987. Giving Credit Where Due. *Sky and Telescope* **74**: 230 (327)

Mushotzky, R. F., Serlemitsos, P. J., Boldt, E. A., Holt, S. S., and Smith, B. W. 1978. OSO 8 X-Ray Spectra of Clusters of Galaxies. I. Observations of Twenty Clusters: Physical Correlations. *Astrophysical Journal* **225**: 21–39 (243)

Ne'eman, Y. 1965. Expansion as an Energy Source in Quasi-Stellar Radio Sources. *Astrophysical Journal* **141**: 1303 (221)

Netterfield, C. B., Ade, P. A. R., Bock, J. J., et al. 2002. A Measurement by BOOMERANG of Multiple Peaks in the Angular Power Spectrum of the Cosmic Microwave Background. *Astrophysical Journal* **571**: 604–614 (333)

Neyman, J. 1962. Alternative Stochastic Models of the Spatial Distribution of Galaxies. In *Proceedings of IAU Symposium 15, Problems of Extra-Galactic Research*, 294–314 (25)

Neyman, J., Page, T., and Scott, E. 1961. Summary of the Conference. *Astronomical Journal* **66**: 633–636 (243,274)

Neyman, J., Scott, E. L., and Shane, C. D. 1954. The Index of Clumpiness of the Distribution of Images of Galaxies. *Astronomical Journal Supplement* **1**: 269–293 (26)

Norberg, P., Baugh, C. M., Hawkins, E., et al. 2001. The 2dF Galaxy Redshift Survey: Luminosity Dependence of Galaxy Clustering. *Monthly Notices of the Royal Astronomical Society* **328**: 64–70 (69)

Norberg, P., Cole, S., Baugh, C. M., et al. 2002. The 2dF Galaxy Redshift Survey: The b_J-Band Galaxy Luminosity Function and Survey Selection Function. *Monthly Notices of the Royal Astronomical Society* **336**: 907–931 (84)

Nørgaard-Nielsen, H. U., Hansen, L., Jørgensen, H. E., Aragón Salamanca, A., Ellis, R. S., and Couch, W. J. 1989. The Discovery of a Type Ia Supernova at a Redshift of 0.31. *Nature* **339**: 523–525 (326)

Novikov, I. D. 1964a. On the Possibility of Appearance of Large Scale Inhomogeneities in the Expanding Universe. *Journal of Experimental and Theoretical Physics* **46**: 686–689; English translation in *Soviet Physics JETP* **19**: 467–469 (191)

Novikov, I. D. 1964b. Delayed Explosion of a Part of the Fridman Universe and Quasars. *Astronomicheskii Zhurnal* **41**: 1075–1083. English translation in *Soviet Astronomy—AJ* **8**, 857–863, 1965 (221)

Nusser, A., Davis, M., and Branchini, E. 2014. On the Recovery of the Local Group Motion from Galaxy Redshift Surveys. *Astrophysical Journal* **788**: 157, 12 pp. (223)

O'Dell, C. R. 1963. Photoelectric Spectrophotometry of Planetary Nebulae. *Astrophysical Journal* **138**: 1018–1034 (146,168)

O'Dell, C. R., Peimbert, M., and Kinman, T. D. 1964. The Planetary Nebula in the Globular Cluster M15. *Astrophysical Journal* **140**: 119–129 (146,168)

Ogburn, W. F., and Thomas, D. 1922. Are Inventions Inevitable? A Note on Social Evolution. *Political Science Quarterly* **37**: 83–98 (350)

Ohm, E. A. 1961. Project Echo Receiving System. *Bell System Technical Journal* **40**: 1065–1094 (156–162)

Olive, K. A., Seckel, D., and Vishniac, E. 1985. Recent Heavy-Particle Decay in a Matter-Dominated Universe. *Astrophysical Journal* **292**: 1–11 (314)

Oort, J. H. 1932. The Force Exerted by the Stellar System in the Direction Perpendicular to the Galactic Plane and Some Related Problems. *Bulletin of the Astronomical Institutes of the Netherlands* **6**: 249–287 (269)

Oort, J. H. 1940. Some Problems Concerning the Structure and Dynamics of the Galactic System and the Elliptical Nebulae NGC 3115 and 4494. *Astrophysical Journal* **91**: 273–306 (255)

Oort, J. H. 1958. Distribution of Galaxies and the Density of the Universe. In *Eleventh Solvay Conference*. Brussels: Editions Stoops, 21 pp. (18,22,32,75,78,83,214,245)

Oort, J. H. 1970. The Formation of Galaxies and the Origin of the High-Velocity Hydrogen. *Astronomy and Astrophysics* **7**: 381–404 (215,218)

Öpik, E. 1922. An Estimate of the Distance of the Andromeda Nebula. *Astrophysical Journal* **55**: 406–410 (41)

O'Raifeartaigh, C., O'Keeffe, M., Nahm, W., and Mitton, S. 2018. One Hundred Years of the Cosmological Constant: From "Superfluous Stunt" to Dark Energy. *European Physical Journal H* 43, 117 pp. (70)

Osterbrock, D. E. 2009. The Helium Content of the Universe. In *Finding the Big Bang*, Eds. P.J.E Peebles, Page, and Partridge, pp. 86–92. Cambridge: Cambridge University Press (142)

Osterbrock, D. E., and Rogerson, J. B., Jr. 1961. The Helium and Heavy-Element Content of Gaseous-Nebulae and the Sun. *Publications of the Astronomical Society of the Pacific* **73**: 129–134 (142–146,150,176)

Ostriker, J. P. 1982. Galaxy Formation. In *Astrophysical Cosmology, Proceedings of the Study Week on Cosmology and Fundamental Physics*, Vatican City State, September

28–October 2, 1981. Vatican City State: Pontifica Academia Scientiarum, pp. 473–493 (222)

Ostriker, J. P., and Bodenheimer, P. 1973. On the Oscillations and Stability of Rapidly Rotating Stellar Models. III. Zero-Viscosity Polytropic Sequences. *Astrophysical Journal* **180**: 171–180 (269)

Ostriker, J. P., and Cowie, L. L. 1981. Galaxy Formation in an Intergalactic Medium Dominated by Explosions. *Astrophysical Journal Letters* **243**: L127–L131 (222)

Ostriker, J. P., and Peebles, P. J. E. 1973. A Numerical Study of the Stability of Flattened Galaxies: or, Can Cold Galaxies Survive? *Astrophysical Journal* **186**: 467–480 (269–272)

Ostriker, J. P., Peebles, P. J. E., and Yahil, A. 1974. The Size and Mass of Galaxies, and the Mass of the Universe. *Astrophysical Journal Letters* **193**: L1–L4 (272–275,278,285,296)

Ostriker, J. P., and Steinhardt, P. J. 1995. The Observational Case for a Low-Density Universe with a Non-Zero Cosmological Constant. *Nature* **377**: 600–602 (320)

Ostriker, J. P., and Suto, Y. 1990. The Mach Number of the Cosmic Flow: A Critical Test for Current Theories. *Astrophysical Journal* **348**: 378–382 (97,318)

Ostriker, J. P., Thompson, C., and Witten, E. 1986. Cosmological Effects of Superconducting Strings. *Physics Letters B* **180**: 231–239 (222)

Ozernoi, L. M., and Chernin, A. D. 1967. The Fragmentation of Matter in a Turbulent Metagalactic Medium. I. *Astronomicheskii Zhurnal* **44**: 1131–1138. English translation in *Soviet Astronomy—AJ* **11**: 907–913, 1968 (215)

Paczyński, B. 1986. Gravitational Microlensing by the Galactic Halo. *Astrophysical Journal* **304**: 1–5 (277)

Pagel, B. E. J. 2000. Helium and Big Bang Nucleosynthesis. *Physics Reports* **333**: 433–447 (176,339)

Pagels, H., and Primack, J. R. 1982. Supersymmetry, Cosmology, and New Physics at Teraelectronvolt Energies. *Physical Review Letters* **48**: 223–226 (296,302,311)

Pais, A., 1982. *Subtle Is the Lord: The Science and the Life of Albert Einstein.* New York: Oxford University Press (56)

Palmer, P., Zuckerman, B., Buhl, D., and Snyder, L. E. 1969. Formaldehyde Absorption in Dark Nebulae. *Astrophysical Journal Letters* **56**: L147–L150 (154)

Pariiskii, Y. N. 1968. On the Origin of the Blackbody Radiation of the Universe. *Astronomicheskii Zhurnal* **45**: 279–285; English translation in *Soviet Astronomy—AJ* **12**: 219–224 (166)

Park, C., Gott, J. R., III, and da Costa, L. N. 1992. Large-Scale Structure in the Southern Sky Redshift Survey. *Astrophysical Journal Letters* **392**: L51–L54 (100)

Partridge, R. B., and Peebles, P. J. E. 1967. Are Young Galaxies Visible? *Astrophysical Journal* **147**: 868–886 (307)

Partridge, R. B., and Wilkinson, D. T. 1967. Isotropy and Homogeneity of the Universe from Measurements of the Cosmic Microwave Background. *Physical Review Letters* **18**: 557–559 (21)

Pauli, W. 1933. Die allgemeinen Prinzipien der Wellenmechanik. In H. Geiger, and K. Scheel Eds., Handbuch der Physik, Quantentheorie, XXIV, Part 1 (2nd ed.) (pp. 83–272) Berlin: Springer (59)

Pauli, W. 1958. Theory of Relativity. London: Pergamon Press (57)

Peacock, J. A., Cole, S., Norberg, P., et al. 2001. A Measurement of the Cosmological Mass Density from Clustering in the 2dF Galaxy Redshift Survey. *Nature* **410**: 169–173 (93)

Peacock, J. A., and Dodds, S. J. 1994. Reconstructing the Linear Power Spectrum of Cosmological Mass Fluctuations. *Monthly Notices of the Royal Astronomical Society* **267**: 1020–1034 (100)

Peebles, P. J. E. 1964. The Structure and Composition of Jupiter and Saturn. *Astrophysical Journal* **140**: 328–347 (145)

Peebles, P. J. E. 1965. The Black-Body Radiation Content of the Universe and the Formation of Galaxies. *Astrophysical Journal* **142**: 1317–1326 (119,182,205,210,233)

Peebles, P. J. E. 1966a. Primeval Helium Abundance and the Primeval Fireball. *Physical Review Letters* **16**: 410–413 (74,126,128,138,148,151,175,179,182)

Peebles, P. J. E. 1966b. Primordial Helium Abundance and the Primordial Fireball. II. *Astrophysical Journal* **146**: 542–552 (74,128,138,175,179)

Peebles, P. J. E. 1967a. The Gravitational Instability of the Universe. *Astrophysical Journal* **147**: 859–863 (86,189,191,212)

Peebles, P. J. E. 1967b. Primeval Galaxies. In *Proceedings of the Fourth Texas Symposium on Relativistic Astrophysics*, New York, 1967, January (not published) (205)

Peebles, P. J. E. 1968. Recombination of the Primeval Plasma. *Astrophysical Journal* **153**: 1–11 (120)

Peebles, P. J. E. 1969a. Origin of the Angular Momentum of Galaxies. *Astrophysical Journal* **155**: 393–401 (217)

Peebles, P. J. E. 1969b. Primeval Globular Clusters. II. *Astrophysical Journal* **157**: 1075–1093 (208)

Peebles, P. J. E. 1970. Structure of the Coma Cluster of Galaxies. *Astronomical Journal* **75**: 13–20 (269)

Peebles, P. J. E. 1971a. *Physical Cosmology*. Princeton, NJ: Princeton University Press (21,24,122,165,205,293)

Peebles, P. J. E. 1971b. Primeval Turbulence? *Astrophysics and Space Science* **11**: 443–450 (216)

Peebles, P. J. E. 1971c. Rotation of Galaxies and the Gravitational Instability Picture. *Astronomy and Astrophysics* **11**: 377–386 (218)

Peebles, P. J. E. 1972. Light out of Darkness vs Order out of Chaos. *Comments on Astrophysics and Space Physics* **4**: 53–58 (191,216)

Peebles, P. J. E. 1973a. Statistical Analysis of Catalogs of Extragalactic Objects. I. Theory. *Astronomical Journal* **185**: 413–440 (28)

Peebles, P. J. E. 1973b. Evolution of Irregularities in an Expanding Universe. In *Fundamental Interactions in Physics and Astrophysics, 9th Coral Gables Conference*, January 19–21, 1972. Eds. Geoffrey Iverson, Arnold Perlmutter, and Stephan Mintz. New York: Plenum Press, pp. 318–350 (309)

Peebles, P. J. E. 1973c. Comment on the Origin of Galactic Rotation. *Publications of the Astronomical Society of Japan* **25**: 291–294 (220)

Peebles, P. J. E. 1974a. The Effect of a Lumpy Matter Distribution on the Growth of Irregularities in an Expanding Universe. *Astronomy and Astrophysics* **32**: 391–397 (225)

Peebles, P. J. E. 1974b. The Gravitational-Instability Picture and the Nature of the Distribution of Galaxies. *Astrophysical Journal Letters* **189**: L51–L53 (233)

Peebles, P. J. E.1976a. The Peculiar Velocity Field in the Local Supercluster. *Astrophysical Journal* **205**: 318–328 (86,90)

Peebles, P. J. E. 1976b. A Cosmic Virial Theorem. *Astrophysics and Space Science* **45**: 3–19 (89)

Peebles, P. J. E. 1979. The Mean Mass Density Estimated from the Kirshner, Oemler, Schechter Galaxy Redshift Sample. *Astronomical Journal* **84**: 730–734 (89)

Peebles, P. J. E. 1980. *The Large-Scale Structure of the Universe.* Princeton, NJ: Princeton University Press (15,26,30,90,98,200,215,225,303)

Peebles, P. J. E. 1981a. Primeval Adiabatic Perturbations: Constraints from the Mass Distribution. *Astrophysical Journal* **248**: 885–897 (207,302)

Peebles, P. J. E. 1981b. Large-Scale Fluctuations in the Microwave Background and the Small-Scale Clustering of Galaxies. *Astrophysical Journal Letters* **243**: L119–L122 (300)

Peebles, P. J. E. 1982a. Primeval Adiabatic Perturbations: Effect of Massive Neutrinos. *Astrophysical Journal* **258**: 415–424 (288,302,306,311)

Peebles, P. J. E. 1982b. Large-Scale Background Temperature and Mass Fluctuations Due to Scale-Invariant Primeval Perturbations. *Astrophysical Journal Letters* **263**: L1–L5 (98,232,301–308,311)

Peebles, P. J. E. 1984a. Dark Matter and the Origin of Galaxies and Globular Star Clusters. *Astrophysical Journal* **277**: 470–477 (308)

Peebles, P. J. E. 1984b. Tests of Cosmological Models Constrained by Inflation. *Astrophysical Journal* **284**: 439–444 (65,314)

Peebles, P. J. E. 1986. The Mean Mass Density of the Universe. *Nature* **321**: 27–32 (63,69,81,86)

Peebles, P. J. E. 1987a. Origin of the Large-Scale Galaxy Peculiar Velocity Field: A Minimal Isocurvature Model. *Nature* **327**: 210–211 (232,315)

Peebles, P. J. E. 1987b. Cosmic Background Temperature Anisotropy in a Minimal Isocurvature Model for Galaxy Formation. *Astrophysical Journal Letters* **315**: L73–L76 (232,315)

Peebles, P. J. E. 1988. The Local Extragalactic Velocity Field as a Test of the Explosion and Gravitational Instability Pictures. *Astrophysical Journal Letters* **332**: 17–25 (223)

Peebles, P. J. E. 1989. Tracing Galaxy Orbits Back in Time. *Astrophysical Journal Letters* **344**: L53–L56 (87)

Peebles, P. J. E. 1993. *Principles of Physical Cosmology.* Princeton, NJ: Princeton University Press (203)

Peebles, P. J. E. 1999. An Isocurvature Cold Dark Matter Cosmogony. II. Observational Tests. *Astrophysical Journal* **510**: 531–540 (233,316)

Peebles, P. J. E. 2014. Discovery of the Hot Big Bang: What Happened in 1948. *European Physical Journal H* **39**: 205–223 (xv,8,115,133,136)

Peebles, P. J. E. 2017. Robert Dicke and the Naissance of Experimental Gravity Physics, 1957–1967. *European Physical Journal H* **42** 177–259 (xv,49,115,160,353)

Peebles, P. J. E., Daly, R. A., and Juszkiewicz, R. 1989. Masses of Rich Clusters of Galaxies as a Test of the Biased Cold Dark Matter Theory. *Astrophysical Journal* **347**: 563–574 (100)

Peebles, P. J. E., and Dicke, R. H. 1968. Origin of the Globular Star Clusters. *Astrophysical Journal* **154**: 891–908 (208)

Peebles, P. J. E., and Hauser, M. G. 1974. Statistical Analysis of Catalogs of Extragalactic Objects. III. The Shane-Wirtanen and Zwicky Catalogs. *Astrophysical Journal Supplement* **28**: 19–36 (30)

Peebles, P. J. E., Page, L. A., Jr., and Partridge, R. B. 2009. *Finding the Big Bang.* Cambridge: Cambridge University Press (xv,8,115,133,138,161,320,333)

Peebles, P. J. E., and Ratra, B. 1988. Cosmology with a Time-Variable Cosmological "Constant." *Astrophysical Journal* **325**: L17–L20 (60)

Peebles, P. J. E. and Ratra, B. 2003. The Cosmological Constant and Dark Energy. *Reviews of Modern Physics* **75**: 559–606 (59)

Peebles, P. J. E., and Yu, J. T. 1970. Primeval Adiabatic Perturbation in an Expanding Universe. *Astrophysical Journal* **162**: 815–836 (205,231,302,305)

Pen, U.-L., Seljak, U., and Turok, N. 1997. Power Spectra in Global Defect Theories of Cosmic Structure Formation. *Physical Review Letters* **79**: 1611–1614 (229)

Penzias, A. A., and Wilson, R. W. 1965. A Measurement of Excess Antenna Temperature at 4800 Mc/s. *Astrophysical Journal* **142**: 419–421 (21,128,149,155–164)

Penzias, A. A., Schraml, J., and Wilson, R. W. 1969. Observational Constraints on a Discrete-Source Model to Explain the Micro-Wave Background. *Astrophysical Journal Letters* **57**: L49–L51 (166)

Percival, W. J., Baugh, C. M., Bland-Hawthorn, J., et al. 2001. The 2dF Galaxy Redshift Survey: The Power Spectrum and the Matter Content of the Universe. *Monthly Notices of the Royal Astronomical Society* **327**: 1297–1306 (105,206,336)

Perl, M. L., Abrams, G. S., Boyarski, A. M., et al. 1975. Evidence for Anomalous Lepton Production in $e^+ - e^-$ Annihilation. *Physical Review Letters* **35**: 1489–1492 (293)

Perl, M. L., Feldman, G. J., Abrams, G. S., et al. 1976. Properties of Anomalous eμ Events Produced in e^+e^- Annihilation. *Physics Letters B* **63**: 466–470 (294)

Perl, M. L., Feldman, G. L., Abrams, G. S., et al. 1977. Properties of the Proposed τ Charged Lepton. *Physics Letters B* **70**: 487–490 (293)

Perlmutter, S. 2012. Nobel Lecture: Measuring the Acceleration of the Cosmic Expansion Using Supernovae. *Reviews of Modern Physics* **84**: 1127–1149 (327)

Perlmutter, S., Aldering, G., della Valle, M., et al. 1998. Discovery of a Supernova Explosion at Half the Age of the Universe. *Nature* **391**: 51–54 (328)

Perlmutter, S., Aldering, G., Goldhaber, G., et al. 1999. Measurements of Ω and Λ from 42 High-Redshift Supernovae. *Astrophysical Journal* **517**: 565–586 (328)

Perlmutter, S., Gabi, S., Goldhaber, G., et al. 1997. Measurements of the Cosmological Parameters Ω and Λ from the First Seven Supernovae at $z \geq 0.35$. *Astrophysical Journal* **483**: 565–581 (318,328)

Perlmutter, S., Goldhaber, G., Marvin, H. J., et al. 1990. The Program to Measure q_0 Using Supernovae at Cosmological Distances. *Bulletin of the American Astronomical Society* **22**: 1332 (327,330)

Perlmutter, S., and Schmidt, B. P. 2003. Measuring Cosmology with Supernovae. *Lecture Notes in Physics* **598**: 195–217 (329)

Petrosian, V., Salpeter, E., and Szekeres, P. 1967. Quasi-Stellar Objects in Universes with Non-Zero Cosmological Constant. *Astrophysical Journal* **147**:1222–1226 (69–72)

Phillips, M. M. 1993. The Absolute Magnitudes of Type Ia Supernovae. *Astrophysical Journal Letters* **413**: L105–L108 (325)

Pierce, M. J., and Tully, R. B. 1988. Distances to the Virgo and Ursa Major Clusters and a Determination of H_0. *Astrophysical Journal* **330**: 579–595 (74)

Pietronero, L., Gabrielli, A., and Sylos Labini, F. 2002. Statistical Physics for Cosmic Structures. *Physica A* **306**: 395–401 (32)

Pipher, J. L., Houck, J. R., Jones, B. W., and Harwit, M. 1971. Submillimetre Observations of the Night Sky Emission above 120 Kilometres. *Nature* **231**: 375–378 (166)

Planck Collaboration 2014a. Searches for Cosmic Strings and Other Topological Defects. *Astronomy and Astrophysics* **571**: A25, 21 pp. (228,237)

Planck Collaboration 2014b. Planck 2013 Results. I. Overview of Products and Scientific Results. *Astronomy and Astrophysics* **571**: A1, 48 pp. (336)

Planck Collaboration 2016. The Sunyaev-Zeldovich Signal from the Virgo Cluster. *Astronomy and Astrophysics* **596**: A101, 20 pp. (244)

Planck Collaboration 2018. Planck 2018 Results. VI. Cosmological Parameters. arXiv:1807.06209 (74,83,178,180,334)

Preskill, J., Wise, M. B., and Wilczek, F. 1983. Cosmology of the Invisible Axion. *Physics Letters B* **120**: 127–132 (296)

Press, W. H. 1976. Exact Evolution of Photons in an Anisotropic Cosmology with Scattering. *Astrophysical Journal* **205** 311–317 (211)

Press, W. H., and Gunn, J. E. 1973. Method for Detecting a Cosmological Density of Condensed Objects. *Astrophysical Journal* **185**: 397–412 (277)

Press, W. H., and Schechter, P. 1974. Formation of Galaxies and Clusters of Galaxies by Self-Similar Gravitational Condensation. *Astrophysical Journal* **187**: 425–438 (224,309)

Press, W. H., Teukolsky, S. A., Vetterling, W. T., and Flannery, B. P. 1992. *Numerical Recipes in FORTRAN. The Art of Scientific Computing.* Cambridge: Cambridge University Press, second ed. (327)

Primack, J. R. 1997. The Best Theory of Cosmic Structure Formation Is COLD+HOT Dark Matter (CHDM). In *Critical Dialogues in Cosmology, Proceedings of a Conference Held at Princeton in June 1996.* Ed. Neil Turok. Singapore: World Scientific, pp.535–554 (318)

Primack, J. R. and Blumenthal, G. R. 1983. Dark Matter, Galaxies, Superclusters and Voids. In *Proceedings of the Fourth Workshop on Grand Unification,* Philadelphia, April 21–23, 1983. Eds. H. A. Weldon, P. Langacker, and P. J. Steinhardt. Boston: Birkhäuser, pp. 256–288 (280)

Pskovskii, I. P. 1977. Light Curves, Color Curves, and Expansion Velocity of Type I Supernovae as Functions of the Rate of Brightness Decline. *Astronomicheskii Zhurnal* **54**: 1188–1201. English translation in *Soviet Astronomy* **21**: 675–682 (325)

Ratcliffe, A., Shanks, T., Parker, Q. A., et al. 1998. The Durham/UKST Galaxy Redshift Survey—V. The Catalogue. *Monthly Notices of the Royal Astronomical Society* **300**: 417–462 (93)

Ratra, B., and Peebles, P. J. E. 1988. Cosmological Consequences of a Rolling Homogeneous Scalar Field. *Physical Review D* **37**: 3406–3427 (50,60)

Ratra, B., and Peebles, P. J. E. 1995. Inflation in an Open Universe. *Physical Review D* **52**: 1837–1894 (63,315)

Raychaudhuri, A. 1952. Condensations in Expanding Cosmologic Models. *Physical Review* **86**: 90–92 (191)

Rees, M. J. 1972. Origin of the Cosmic Microwave Background Radiation in a Chaotic Universe. *Physical Review Letters* **28**: 1669–1671 (212)

Rees, M. J. 1977. Cosmology and Galaxy Formation. In *Proceedings of a Conference on The Evolution of Galaxies and Stellar Populations,* May, 1977. Eds B. M. Tinsley and R. B. Larson. New Haven: Yale University Observatory, pp. 339–368 (276)

Rees, M. J. 1978. Origin of Pregalactic Microwave Background. *Nature* **275**: 35–37 (168)

Rees, M. J. 1984. Is the Universe Flat? *Journal of Astrophysics and Astronomy* **5**: 331–348 (60–65,263,314)

Rees, M. J. 1985. Mechanisms for Biased Galaxy Formation. *Monthly Notices of the Royal Astronomical Society* **213**: 75p–81p (66)

Rees, M. J., and Ostriker, J. P. 1977. Cooling, Dynamics and Fragmentation of Massive Gas Clouds: Clues to the Masses and Radii of Galaxies and Clusters. *Monthly Notices of the Royal Astronomical Society* **179**: 541–559 (308)

Rees, M. J., and Sciama, D. W. 1969. The Evolution of Density Fluctuations in the Universe II. The Formation of Galaxies. *Comments on Astrophysics and Space Physics* **1**: 153–157 (210)

Reeves, H., Audouze, J., Fowler, W. A., and Schramm, D. N. 1973. On the Origin of Light Elements. *Astrophysical Journal* **179**: 909–930 (179)

Refsdal, S. 1964. The Gravitational Lens Effect. *Monthly Notices of the Royal Astronomical Society* **128**: 295–306 (277)

Regős, E., and Geller, M. J. 1989. Infall Patterns Around Rich Clusters of Galaxies. *Astronomical Journal* **98**: 755–765 (87,236)

Reines, F., Sobel, H. W., and Pasierb, E. 1980. Evidence for Neutrino Instability. *Physical Review Letters* **45**: 1307–1311 (285)

Reuter, M., and Wetterich, C. 1987. Time Evolution of the Cosmological "Constant." *Physics Letters B* **188**: 38–43 (60)

Riess, A. G. 2000. The Case for an Accelerating Universe from Supernovae. *Publications of the Astronomical Society of the Pacific* **112**: 1284–1299 (329)

Riess, A. G. 2012. Nobel Lecture: My Path to the Accelerating Universe. *Reviews of Modern Physics* **84**: 1165–1175 (328,332)

Riess, A. G., Casertano, S., Yuan, W., et al. 2018. New Parallaxes of Galactic Cepheids from Spatially Scanning the Hubble Space Telescope: Implications for the Hubble Constant. *Astrophysical Journal* **855**: 136,18 pp. (74)

Riess, A. G., Filippenko, A. V., Challis, P., et al. 1998. Observational Evidence from Supernovae for an Accelerating Universe and a Cosmological Constant. *Astronomical Journal* **116**: 1009–1038 (328)

Riess, A. G., Kirshner, R. P., Schmidt, B. P., et al. 1999. *BVRI* Light Curves for 22 Type Ia Supernovae. *Astronomical Journal* **117**: 707–724 (328)

Riess, A. G., Press, W. H., and Kirshner, R. P. 1995. Using Type Ia Supernova Light Curve Shapes to Measure the Hubble Constant. *Astrophysical Journal Letters* **438**: L17–L20 (327)

Riess, A. G., Press, W. H., and Kirshner, R. P. 1996. A Precise Distance Indicator: Type Ia Supernova Multicolor Light-Curve Shapes. *Astrophysical Journal* **473**: 88–109 (327)

Riess, A. G., Strolger, L.-G., Tonry, J., et al. 2004. Type Ia Supernova Discoveries at $z > 1$ from the Hubble Space Telescope: Evidence for Past Deceleration and Constraints on Dark Energy Evolution. *Astrophysical Journal* **607**: 665–687 (331)

Roberts, M. S. 1976. The Rotation Curves of Galaxies. *Comments on Astrophysics* **6**: 105–111 (263)

Roberts, M. S. 2008. M 31 and a Brief History of Dark Matter. A Celebration of NRAO's 50th Anniversary. *ASP Conference Series* **395**: 283–288 (275)

Roberts, M. S., and Rots, A. H. 1973. Comparison of Rotation Curves of Different Galaxy Types. *Astronomy and Astrophysics* **26**: 483–485 (261)

Roberts, M. S., and Whitehurst, R. N. 1975. The Rotation Curve and Geometry of M31 at Large Galactocentric Distances. *Astrophysical Journal* **201**: 327–346 (249,254,276)

Robertson, H. P. 1928. On Relativistic Cosmology. *Philosophical Magazine* **5**: 835–848 (39,44)

Robertson, H. P. 1929. On the Foundations of Relativistic Cosmology. *Proceedings of the National Academy of Science* **15**: 822–829 (37)

Robertson, H. P. 1955. The Theoretical Aspects of the Nebular Redshift. *Publications of the Astronomical Society of the Pacific* **67**: 82–98 (57)

Robertson, R. G. H., Bowles, T. J., Stephenson, G. J., Jr., et al. 1991. Limit on $\bar{\nu}_e$ Mass from Observation of the β Decay of Molecular Tritium. *Physical Review Letters* **67**: 957–960 (285)

Rogerson, J. B., and York, D. G. 1973. Interstellar Deuterium Abundance in the Direction of Beta Centauri. *Astrophysical Journal Letters* **186**: L95–L98 (76,180)

Rogstad, D. H. 1971. Aperture Synthesis Study of Neutral Hydrogen in the Galaxy M101: II. Discussion. *Astronomy and Astrophysics* **13**: 108–115 (259)

Rogstad, D. H., and Shostak, G. S. 1972. Gross Properties of Five Scd Galaxies as Determined from 21-Centimeter Observations. *Astrophysical Journal* **176**: 315–321 (260)

Roll, P. G., and Wilkinson, D. T. 1966. Cosmic Background Radiation at 3.2 cm–Support for Cosmic Black-Body Radiation. *Physical Review Letters* **16**: 405–407 (149)

Rots, A. H. 1974. Distribution and Kinematics of Neutral Hydrogen in the Spiral Galaxy M81. PhD thesis, University of Groningen, the Netherlands (261)

Rubin, V. C. 1954. Fluctuations in the Space Distribution of the Galaxies. *Proceedings of the National Academy of Science* **40**: 541–549 (26,214)

Rubin, V. C. 2011. An Interesting Voyage. *Annual Review of Astronomy and Astrophysics* **49**: 1–28 (253)

Rubin, V. C., and Ford, W. K., Jr. 1970. Rotation of the Andromeda Nebula from a Spectroscopic Survey of Emission Regions. *Astrophysical Journal* **159**: 379–403 (249,253, 268,269)

Rubin, V. C., Peterson, C. J., and Ford, W. K., Jr. 1976. Rotation Curve of the E7/S0 Galaxy NGC 3115. *Bulletin of the American Astronomical Society* **8**: 297 (255)

Rubin, V. C., Peterson, C. J., and Ford, W. K., Jr. 1980. Rotation and Mass of the Inner 5 kiloparsecs of the S0 Galaxy NGC 3115. *Astrophysical Journal* **239**: 50–53 (256)

Rudnicki, K., Dworak, T. Z., Flin, P., Baranowski, B., and Sendrakowski, A. 1973. A Catalogue of 15650 Galaxies in the Jagellonian Field. *Acta Cosmologica* **1**: 164 pp. (29)

Rugh, S. E., and Zinkernagel, H. 2002. The Quantum Vacuum and the Cosmological Constant Problem. *Studies in the History and Philosophy of Modern Physics* **33**: 663–705 (59)

Ryle, M. 1955. Radio Stars and Their Cosmological Significance. *The Observatory* **75**: 137–147 (52)

Ryle, M., and Scheuer, P. A. G. 1955. The Spatial Distribution and the Nature of Radio Stars. *Proceedings of the Royal Society of London Series A* **230**: 448–462 (52)

Sachs, R. K., and Wolfe, A. M. 1967. Perturbations of a Cosmological Model and Angular Variations of the Microwave Background. *Astrophysical Journal* **147**: 73–90 (304)

Sahni, V., Feldman, H., and Stebbins, A. 1992. Loitering Universe. *Astrophysical Journal* **385**: 1–8 (316)

Sakharov, A. D. 1967. Violation of CP Invariance, C Asymmetry, and Baryon Asymmetry of the Universe. *ZhETF Pis'ma* **5**: 32–35. English translation in *JETP Letters* **5**: 24–27 (63)

Salopek, D. S., Bond, J. R., and Bardeen, J. M. 1989. Designing Density Fluctuation Spectra in Inflation. *Physical Review D* **40**: 1753–1788 (313)

Sandage, A. 1958. Current Problems in the Extragalactic Distance Scale. *Astrophysical Journal* **127**: 513–526 (73)

Sandage, A. 1961a. The Ability of the 200-Inch Telescope to Discriminate between Selected World Models. *Astrophysical Journal* **133**, 355–392 (52,56,69,324)

Sandage, A. 1961b. The Light Travel Time and the Evolutionary Correction to Magnitudes of Distant Galaxies. *Astrophysical Journal* **134**: 916–926 (69,77)

Sandage, A. 1968. Observational Cosmology. *The Observatory* **88**: 91–106 (69)

Sandage, A. 1975. The Redshift-Distance Relation. VIII. Magnitudes and Redshifts of Southern Galaxies in Groups: A Further Mapping of the Local Velocity Field and an Estimate of q_0. *Astrophysical Journal* **202**: 563–582 (86)

Sandage, A., and Hardy, E. 1973. The Redshift-Distance Relation. VII. Absolute Magnitudes of the First Three Ranked Cluster Galaxies as Functions of Cluster Richness and Bautz-Morgan Cluster Type: The Effect on q_0. *Astrophysical Journal* **183**: 743–757 (324)

Sandage, A., and Tammann, G. A. 1975. Steps toward the Hubble Constant. V. The Hubble Constant from Nearby Galaxies and the Regularity of the Local Velocity Field. *Astrophysical Journal* **196**: 313–328 (86,97,222)

Sandage, A., and Tammann, G. A. 1981. *A Revised Shapley-Ames Catalog of Bright Galaxies.* Washington, DC: Carnegie Institution of Washington (86)

Sandage, A., and Tammann, G. A. 1984. The Hubble Constant as Derived from 21 cm Linewidths. *Nature* **307**: 326–329 (74)

Sargent, W. L. W. 1968. The Redshifts of Galaxies in the Remarkable Chain VV 172. *Astrophysical Journal Letters* **153**: L135–L137 (246)

Sarkar, S. 1996. Big Bang Nucleosynthesis and Physics beyond the Standard Model. *Reports on Progress in Physics* **59**: 1493–1609 (177,339)

Saslaw, W. C. 1972. The Kinetics of Gravitational Clustering. *Astrophysical Journal* **177**: 17–29 (233)

Saslaw, W. C., and Zipoy, D. 1967. Molecular Hydrogen in Pre-galactic Gas Clouds. *Nature* **216** 976–978 (121)

Sato, H., and Takahara, F. 1980. Clustering of the Relic Neutrinos in the Expanding Universe. *Progress of Theoretical Physics* **64**: 2029–2040 (286)

Sato, K. 1981. First-Order Phase Transition of a Vacuum and the Expansion of the Universe. *Monthly Notices of the Royal Astronomical Society* **195**: 467–479 (62)

Sato, K., and Kobayashi, M. 1977. Cosmological Constraints on the Mass and the Number of Heavy Lepton Neutrinos. *Progress of Theoretical Physics* **58**: 1775–1789 (291–293,313)

Saunders, W., Frenk, C., Rowan-Robinson, M., et al. 1991. The Density Field of the Local Universe. *Nature* **349**: 32–38 (99)

Saunders, W., Sutherland, W. J., Maddox, S. J., et al. 2000. The PSCz Catalogue. *Monthly Notices of the Royal Astronomical Society* **317**: 55–63 (93)

Schechter, P. 1976. An Analytic Expression for the Luminosity Function for Galaxies. *Astrophysical Journal* **203**: 297–306 (83)

Schmidt, B. P. 2012. Nobel Lecture: Accelerating Expansion of the Universe through Observations of Distant Supernovae. *Reviews of Modern Physics* **84**: 1151–1163 (327,332)

Schmidt, M. 1956. A Model of the Distribution of Mass in the Galactic System. *Bulletin of the Astronomical Institutes of the Netherlands* **13**: 15–41 (252)

Schmidt, M. 1957. The Distribution of Mass in M 31. *Bulletin of the Astronomical Institutes of the Netherlands* **14**: 17–19 (252)

Schmidt, M. 1959. The Rate of Star Formation. *Astrophysical Journal* **129**: 243–258 (141)

Schmoldt, I., Branchini, E., Teodoro, L., et al. 1999. Likelihood Analysis of the Local Group Acceleration. *Monthly Notices of the Royal Astronomical Society* **304**: 893–905 (96)

Schramm, D. N., and Steigman, G. 1981. Relic Neutrinos and the Density of the Universe. *Astrophysical Journal* **243**: 1–7 (284)

Schwartz, D. A. 1970. The Isotropy of the Diffuse Cosmic X-Rays Determined by OSO-III. *Astrophysical Journal* **162**: 439–444 (21)

Schwarzschild, M. 1946. On the Helium Content of the Sun. *Astrophysical Journal* **104**: 203–207 (126,139)

Schwarzschild, M. 1954. Mass Distribution and Mass-Luminosity Ratio in Galaxies. *Astronomical Journal* **59**: 273–284 (242,251–254)

Schwarzschild, M. 1970. Stellar Evolution in Globular Clusters. *Quarterly Journal of the Royal Astronomical Society* **11**: 12–22 (74)

Schwarzschild, M., Howard, R., and Härm, R. 1957. Inhomogeneous Stellar Models. V. A Solar Model with Convective Envelope and Inhomogeneous Interior. *Astrophysical Journal* **125**: 233–241 (127)

Sciama, D. W. 1955. On the Formation of Galaxies in a Steady State Universe. *Monthly Notices of the Royal Astronomical Society* **115**: 3–14 (185)

Sciama, D. W. 1966. On the Origin of the Microwave Background Radiation. *Nature* **211**: 277–279 (166)

Sciama, D. W. 1984. Massive Neutrinos and Photinos in Cosmology and Galactic Astronomy. In *The Big Bang and Georges Lemaître, Louvain-la-Neuve*, October 10–13, 1983. Ed. A. Berger. Dordrecht: Reidel. pp. 31–41 (298)

Seldner, M., and Peebles, P. J. E. 1977. A New Way to Estimate the Mean Mass Density Associated with Galaxies. *Astrophysical Journal Letters* **214**: L1–L4 (85)

Seldner, M., Siebers, B., Groth, E. J., and Peebles, P. J. E. 1977. New Reduction of the Lick Catalog of Galaxies. *Astronomical Journal* **82**: 249–256 (29)

Sellwood, J. A., and Evans, N. W. 2001. The Stability of Disks in Cusped Potentials. *Astrophysical Journal* **546**: 176–188 (270)

Shafi, Q., and Stecker, F. W. 1984. Implications of a Class of Grand-Unified Theories for Large-Scale Structure in the Universe. *Physical Review Letters* **53**: 1292–1295 (314)

Shakeshaft, J. R., Ryle, M., Baldwin, J. E., et al. 1955. A Survey of Radio Sources between Declinations −38° and +83°. *Memoirs of the Royal Astronomical Society* **67**: 106–154 (20)

Shane, C. D., and Wirtanen, C. A. 1954. The Distribution of Extragalactic Nebulae. *Astronomical Journal* **59**: 285–304 (26)

Shane, C. D., and Wirtanen, C. A. 1967. The Distribution of Galaxies. *Publications of the Lick Observatory* **XXII**: Part 1 (26,29,99)

Shanks, T. 1985. Arguments for an $\Omega = 1$, Low H_0 Baryon Dominated Universe. *Vistas in Astronomy* **28**: 595–609 (335)

Shapiro, P. R., Struck-Marcell, C., and Melott, A. L. 1983. Pancakes and the Formation of Galaxies in a Neutrino-Dominated Universe. *Astrophysical Journal* **275**: 413–429 (288)

Shapiro, S. L. 1971. The Density of Matter in the Form of Galaxies. *Astronomical Journal* **76**: 291–293 (179)

Shapley, H., and Ames, A. 1932. A Survey of the External Galaxies Brighter Than the Thirteenth Magnitude. *Annals of Harvard College Observatory* **88**: 43–75 (16–19)

Shaya, E. J., Peebles, P. J. E., and Tully, R. B. 1995. Action Principle Solutions for Galaxy Motions within 3000 Kilometers per Second. *Astrophysical Journal* **454**: 15–31 (87)

Shaya, E. J., Tully, R. B., and Pierce, M. J. 1992. Nearby Galaxy Flows Modeled by the Light Distribution. *Astrophysical Journal* **391**: 16–33 (103)

Sheldon, E. S., Johnston, D. E., Frieman, J. A., et al. 2004. The Galaxy-Mass Correlation Function Measured from Weak Lensing in the Sloan Digital Sky Survey. *Astronomical Journal* **127**: 2544–2564 (32)

Shklovsky, I. S., 1966. Relict Radiation in the Universe and Population of Rotational Levels of an Interstellar Molecule. Astronomical Circular 364, Soviet Academy of Science, 3 pp. (155)

Shklovsky, J. 1967. On the Nature of "Standard" Absorption Spectrum of the Quasi-Stellar Objects. *Astrophysical Journal Letters* **150**: L1–L3 (69)

Shobbrook, R. R., and Robinson, B. J. 1967. 21 cm Observations of NGC 300. *Australian Journal of Physics* **20**: 131–145 (257)

Shostak, G. S. 1972. Aperture Synthesis Observations of Neutral Hydrogen in Three Galaxies. PhD thesis, California Institute of Technology, Pasadena (258–261)

Shostak, G. S. 1973. Aperture Synthesis Study of Neutral Hydrogen in NGC 2403 and NGC 4236. II. Discussion. *Astronomy and Astrophysics* **24**: 411–419 (259)

Sikivie, P. 1983. Experimental Tests of the "Invisible" Axion. *Physical Review Letters* **51**: 1415–1417 (298)

Silk, J. 1967. Fluctuations in the Primordial Fireball. *Nature* **215**: 1155–1156 (205, 207,234)

Silk, J. 1968. Cosmic Black-Body Radiation and Galaxy Formation. *Astrophysical Journal* **151**: 459–471 (205,207,234)

Silk, J. 1974. Large-Scale Inhomogeneity of the Universe: Implications for the Deceleration Parameter. *Astrophysical Journal* **193**: 525–527 (86)

Silk, J. 1977. On the Fragmentation of Cosmic Gas Clouds. I. The Formation of Galaxies and the First Generation of Stars. *Astrophysical Journal* **211**: 638–648 (308)

Silk, J. 1982. Fundamental Tests of Galaxy Formation Theory. In *Astrophysical Cosmology, Proceedings of the Study Week on Cosmology and Fundamental Physics*, Vatican City State, September 28–October 2, 1981. Vatican City State: Pontifica Academia Scientiarum, pp. 427–472 (289)

Silk, J., Olive, K., and Srednicki, M. 1985. The Photino, the Sun, and High-Energy Neutrinos. *Physical Review Letters* **55**: 257–259 (299)

Silk, J., and Srednicki, M. 1984. Cosmic-Ray Antiprotons as a Probe of a Photino-Dominated Universe. *Physical Review Letters* **53**: 624–627 (298)

Silk, J., and Vilenkin, A. 1984. Cosmic Strings and Galaxy Formation. *Physical Review Letters* **53**: 1700–1703 (225)

Silk, J., and Wilson, M. L. 1981. Large-Scale Anisotropy of the Cosmic Microwave Background Radiation. *Astrophysical Journal Letters* **244**: L37–L41 (301)

Simha, V., and Steigman, G. 2008. Constraining the Universal Lepton Asymmetry. *Journal of Cosmology and Astroparticle Physics* **8**: 011, 12 pp. (181)

Simpson, J. J., and Hime, A. 1989. Evidence of the 17-keV Neutrino in the β Spectrum of ^{35}S. *Physical Review D* **39**: 1825–1836 (306)

Slipher, V. M. 1917. Radial Velocity Observations of Spiral Nebulae. *The Observatory* **40**: 304–306 (22,45,245)

Smirnov, Y. N. 1965. Hydrogen and He^4 Formation in the Prestellar Gamow Universe. *Astronomicheskii Zhurnal* **41**: 1084–1089; English translation in *Soviet Astronomy—AJ* **8**: 864–867 (146,182)

Smith, S. 1936. The Mass of the Virgo Cluster. *Astrophysical Journal* **83**: 23–30 (242)

Smoot, G. F., Bennett, C. L., Kogut, A., et al. 1992. Structure in the COBE Differential Microwave Radiometer First-Year Maps. *Astrophysical Journal Letters* **396**: L1–L5 (104,311)

Smoot, G. F., Gorenstein, M. V., and Muller, R. A. 1977. Detection of Anisotropy in the Cosmic Blackbody Radiation. *Physical Review Letters* **39**: 898–901 (95,222)

Soifer, B. T., Neugebauer, G., Wynn-Williams, C. G., et al. 1980. IR Observations of the Double Quasar 0957 + 561 A, B and the Intervening Galaxy. *Nature* **285**: 91–93 (228)

Sohn, J., Geller, M. J., Zahid, H. J., et al. 2017. The Velocity Dispersion Function of Very Massive Galaxy Clusters: Abell 2029 and Coma. *Astrophysical Journal Supplement* **229**, id. 20, 17 pp. (242)

Somerville, R. S., Behroozi, P., Pandya, V., et al. 2018. The Relationship between Galaxy and Dark Matter Halo Size from $z \sim 3$ to the Present. *Monthly Notices of the Royal Astronomical Society* **473**: 2714–2736 (220)

Soneira, R. M., and Peebles, P. J. E. 1978. A Computer Model Universe: Simulation of the Nature of the Galaxy Distribution in the Lick Catalog. *Astronomical Journal* **83**: 845–860 (30,99,233)

Songaila, A., Cowie, L. L., Hogan, C. J., and Rugers, M. 1994. Deuterium Abundance and Background Radiation Temperature in High-Redshift Primordial Clouds. *Nature* **368**: 599–604 (180)

Soucail, G., Mellier, Y., Fort, B., Mathez, G., and Cailloux, M. 1988. The Giant Arc in A 370: Spectroscopic Evidence for Gravitational Lensing from a Source at $z = 0.724$. *Astronomy and Astrophysics* **191**: L19–L21 (244)

Spergel, D. N., Bean, R., Doré, O., et al. 2007. Three-Year Wilkinson Microwave Anisotropy Probe (WMAP) Observations: Implications for Cosmology. *Astrophysical Journal Supplement* **170**: 377–408 (335)

Spergel, D. N., Verde, L., Peiris, H. V., et al. 2003. First-Year Wilkinson Microwave Anisotropy Probe (WMAP) Observations: Determination of Cosmological Parameters. *Astrophysical Journal Supplement* **148**: 175–193 (25,48,180,207,305,335,337)

Spinrad, H., Djorgovski, S., Marr, J., and Aguilar, L. 1985. A Third Update of the Status of the 3CR Sources: Further New Redshifts and New Identifications of Distant Galaxies. *Publications of the Astronomical Society of the Pacific* **97**: 932–961 (168)

Spitzer, L., Jr. 1956. On a Possible Interstellar Galactic Corona. *Astrophysical Journal* **124**: 20–34 (308)

Springob, C. M., Masters, K. L., Haynes, M. P., Giovanelli, R., and Marinoni, C. 2007. SFI++. II. A New I-Band Tully-Fisher Catalog, Derivation of Peculiar Velocities, and Data Set Properties. *Astrophysical Journal Supplement* **172**: 599–614 (96)

Starobinsky, A. A. 1980. A New Type of Isotropic Cosmological Models without Singularity. *Physics Letters B* **91**: 99–102 (62)

Stebbins, J., and Whitford, A. E. 1948. Six-Color Photometry of Stars. VI. The Colors of Extragalactic Nebulae. *Astrophysical Journal* **108** 413–428 (51)

Stecker, F. W. 1978. The Cosmic γ-Ray Background from the Annihilation of Primordial Stable Neutral Heavy Leptons. *Astrophysical Journal* **223**: 1032–1036 (298)

Steigman, G., Sarazin, C. L., Quintana, H., and Faulkner, J. 1978. Dynamical Interactions and Astrophysical Effects of Stable Heavy Neutrinos. *Astronomical Journal* **83**: 1050–1061 (295,298)

Steinhardt, P. J. 1983. Natural Inflation. In *The Very Early Universe, Proceedings of the Nuffield Workshop*, Cambridge, June 21–July 9, 1982. Eds. G. W. Gibbons and S. T. C. Siklos. Cambridge: Cambridge University Press, pp. 251–266 (62)

Steinhardt, P. J., and Turok, N. 2007. *Endless Universe: Beyond the Big Bang*. New York: Doubleday (161)

Strauss, M. A. 1989. A Redshift Survey of IRAS Galaxies. PhD thesis, University of California, Berkeley (92)

Strauss, M. A., and Davis, M. 1988. A Redshift Survey of IRAS Galaxies. In *IAU Symposium 130, Large Scale Structures of the Universe*. Balatonfured, Hungary, June 15–20, 1987. Eds. J. Audouze, M. C. Pelletan, and A. Szalay. Dordrecht: Reidel, pp. 191–201 (96)

Strauss, M. A., Yahil, A., Davis, M., Huchra, J. P., and Fisher, K. 1992. A Redshift Survey of *IRAS* Galaxies. V. The Acceleration on the Local Group. *Astrophysical Journal* **397**: 395–419 (91,96)

Strom, K. M., Strom, S. E., Jensen, E. B., et al. 1977. A Photometric Study of the S0 Galaxy NGC 3115. *Astrophysical Journal* **212**: 335–337 (256)

Strömberg, G. 1934. The Origin of the Galactic Rotation and of the Connection between Physical Properties of the Stars and Their Motions. *Astrophysical Journal* **79**: 460–474 (216–218)

Sunyaev, R. A., and Zel'dovich, Y. B. 1970. Small-Scale Fluctuations of Relic Radiation. *Astrophysics and Space Science* **7**: 3–19 (205,207)

Sunyaev, R. A., and Zel'dovich, Y. B. 1972. Formation of Clusters of Galaxies; Protocluster Fragmentation and Intergalactic Gas Heating. *Astronomy and Astrophysics* **20**: 189–200 (220,234,286)

Szalay, A. S. 1974. Finite Neutrino Rest Mass in Astrophysics. PhD thesis, Roland Eötvös University, Budapest (281–284)

Szalay, A. S., and Bond, J. R. 1983. Late Evolution of Adiabatic Fluctuations. In *Proceedings of the Fourth Workshop on Grand Unification*, Philadelphia, April 21–23, 1983, pp. 289–300 (280)

Szalay, A. S., and Marx, G. 1974. Limit on the Rest Masses from Big Bang Cosmology. *Acta Physica Academiae Scientiarum Hungaricae* **35**: 113–129 (77,282,284,291)

Szalay, A. S., and Marx, G. 1976. Neutrino Rest Mass from Cosmology. *Astronomy and Astrophysics* **49**: 437–441 (281,286,291)

Tadros, H., Ballinger, W. E., Taylor, A. N., et al. 1999. Spherical Harmonic Analysis of the PSCz Galaxy Catalogue: Redshift Distortions and the Real-Space Power Spectrum. *Monthly Notices of the Royal Astronomical Society* **305**: 527–546 (93)

Tammann, G. A. 1978. Some Statistical Properties of Supernovae. *Società Astronomica Italiana, Memorie* **49**: 315–329 (325)

Tayler, R. J. 1990. Neutrinos, Helium and the Early Universe—A Personal View. *Quarterly Journal of the Royal Astronomical Society* **31**: 371–375 (162)

Taylor, A. N., and Rowan-Robinson, M. 1992. The Spectrum of Cosmological Density Fluctuations and Nature of Dark Matter. *Nature* **359**: 396–399 (317)

Tegmark, M., and Peebles, P. J. E. 1998. The Time Evolution of Bias. *Astrophysical Journal Letters* **500**: L79–L82 (94)

Tegmark, M., and Zaldarriaga, M. 2000. Current Cosmological Constraints from a 10 Parameter Cosmic Microwave Background Analysis. *Astrophysical Journal* **544**: 30–42 (334)

Ter Haar, D. 1950. Cosmogonical Problems and Stellar Energy. *Reviews of Modern Physics* **22**: 119–152 (182)

Tinsley, B. M. 1967. Evolution of Galaxies and Its Significance for Cosmology. PhD thesis, University of Texas, Austin (75,77,94,324)

Tinsley, B. M. 1972. Stellar Evolution in Elliptical Galaxies. *Astrophysical Journal* **178**: 319–336 (77,94)

Tolman, R. C. 1931. On the Problem of the Entropy of the Universe as a Whole. *Physical Review* **37**: 1639–1660 (137)

Tolman, R. C. 1934a. *Relativity, Thermodynamics, and Cosmology*, Oxford: Clarendon Press. (94,115,122,135,160,323)

Tolman, R. C. 1934b. Effect of Inhomogeneity on Cosmological Models. *Proceedings of the National Academy of Sciences.* **20**: 169–176 (191,193)

Tomita, K. 1973. On the Origin of Galactic Rotation. *Publications of the Astronomical Society of Japan* **25**: 287–290 (220)

Tomita, K., and Hayashi, C. 1963. The Cosmical Constant and the Age of the Universe. *Progress of Theoretical Physics* **30**: 691–699 (75)

Tomita, K., Nariai, H., Satō, H., Matsuda, T., and Takeda, H. 1970. On the Dissipation of Primordial Turbulence in the Expanding Universe. *Progress of Theoretical Physics* **43**: 1511–1525 (215)

Tonry, J. L., and Davis, M. 1981. Velocity Dispersions of Elliptical and S0 Galaxies. II. Infall of the Local Group to Virgo. *Astrophysical Journal* **246**: 680–695 (85,87)

Tonry, J. L., Schmidt, B. P., Barris, B., et al. 2003. Cosmological Results from High-*z* Supernovae. *Astrophysical Journal* **594**: 1–24 (330,337)

Toomre, A. 1964. On the Gravitational Stability of a Disk of Stars. *Astrophysical Journal* **139**: 1217–1238 (265,269)

Toomre, A. 1977a. Theories of Spiral Structure. *Annual Review of Astronomy and Astrophysics* **15**: 437–478 (270)

Toomre, A. 1977b. Mergers and Some Consequences. In *Proceedings of the Conference Evolution of Galaxies and Stellar Populations*, Yale University, May 1977. New Haven, CT: Yale University, Observatory, pp. 401–426 (272)

Toomre, A., and Toomre, J. 1972. Galactic Bridges and Tails. *Astrophysical Journal* **178**: 623–666 (272)

Totsuji, H., and Kihara, T. 1969. The Correlation Function for the Distribution of Galaxies. *Publications of the Astronomical Society of Japan* **21**: 221–229 (26,29)

Tremaine, S., and Gunn, J. E. 1979. Dynamical Role of Light Neutral Leptons in Cosmology. *Physical Review Letters* **42**: 407–410 (283–285,295)

Truran, J. W., Hansen, C. J., and Cameron, A. G. W. 1965. The Helium Content of the Galaxy. *Canadian Journal of Physics* **43**: 1616–1635 (141)

Tully, R. B., and Fisher, J. R. 1977. A New Method of Determining Distances to Galaxies. *Astronomy and Astrophysics* **54**: 661–673 (87,264)

Tully, R. B., Libeskind, N. I., Karachentsev, I. D., et al. 2015. Two Planes of Satellites in the Centaurus A Group. *Astrophysical Journal Letters* **802**: L25, 5 pp. (247)

Tully, R. B., and Shaya, E. J. 1984. Infall of Galaxies into the Virgo Cluster and Some Cosmological Constraints. *Astrophysical Journal* **281**: 31–55 (87)

Turner, M. S., Steigman, G., and Krauss, L. M. 1984. Flatness of the Universe: Reconciling Theoretical Prejudices with Observational Data. *Physical Review Letters* **52**: 2090–2093 (65,314)

Turner, M. S., Wilczek, F., and Zee, A. 1983. Formation of Structure in an Axion-Dominated Universe. *Physics Letters B* **125**: 35–40 (304)

Turok, N. 1983. The Production of String Loops in an Expanding Universe. *Physics Letters B* **123**: 387–390 (225)

Tytler, D., Fan, X.-M., and Burles, S. 1996. Cosmological Baryon Density Derived from the Deuterium Abundance at Redshift $z = 3.57$. *Nature* **381**: 207–209 (180)

Umemura, M., and Ikeuchi, S. 1985. Formation of Subgalactic Objects within Two-Component Dark Matter. *Astrophysical Journal* **299**: 583–592 (314)

Underhill, A. B. 1958. Helium Abundance in Stellar Atmospheres. *The Observatory* **78**: 127–129 (145)

Vachaspati, T., and Vilenkin, A. 1984. Formation and Evolution of Cosmic Strings. *Physical Review D* **30**: 2036–2045 (228)

Valdarnini, R., and Bonometto, S. A. 1985. Fluctuation Evolution in a Two-Component Dark-Matter Model. *Astronomy and Astrophysics* **146**: 235–241 (314)

van Albada, T. S., and Sancisi, R. 1986. Dark Matter in Spiral Galaxies. *Philosophical Transactions of the Royal Society of London Series A* **320**: 447–464 (259,265)

van Dalen, A., and Schaefer, R. K. 1992. Structure Formation in a Universe with Cold Plus Hot Dark Matter. *Astrophysical Journal* **398**: 33–42 (318)

van de Hulst, H. C., Raimond, E., and van Woerden, H. 1957. Rotation and Density Distribution of the Andromeda Nebula Derived from Observations of the 21-cm Line. *Bulletin of the Astronomical Institutes of the Netherlands* **14**: 1–16 (249–254,269)

van den Bergh, S. 1961. The Luminosity Function of Galaxies. *Zeitschrift für Astrophysik* **53**: 219–222 (78,83,128,149)

van den Bergh, S. 1999. The Early History of Dark Matter. *Publications of the Astronomical Society of the Pacific* **111**: 657–660 (241)

van den Bergh, S. 2011. Discovery of the Expansion of the Universe. *Journal of the Royal Astronomical Society of Canada* **105**: 197–198 (44)

Viana, P. T. P., and Liddle, A. R. 1996. The Cluster Abundance in Flat and Open Cosmologies. *Monthly Notices of the Royal Astronomical Society* **281**: 323–332 (104,312)

Vilenkin, A. 1981a. Cosmological Density Fluctuations Produced by Vacuum Strings. *Physical Review Letters* **46**: 1169–1172 (225)

Vilenkin, A. 1981b. Gravitational Field of Vacuum Domain Walls and Strings. *Physical Review D* **23**: 852–857 (226–228)

Vilenkin, A. 1981c. Gravitational Radiation from Cosmic Strings. *Physics Letters B* **107**: 47–50 (237)

Vilenkin, A. 1983. Birth of Inflationary Universes. *Physical Review D* **27**: 2848–2855 (62)

Vittorio, N., and Silk, J. 1984. Fine-Scale Anisotropy of the Cosmic Microwave Background in a Universe Dominated by Cold Dark Matter. *Astrophysical Journal* **285**: L39–L43 (302)

Vogeley, M. S., Park, C., Geller, M. J., and Huchra, J. P. 1992. Large-Scale Clustering of Galaxies in the CfA Redshift Survey. *Astrophysical Journal Letters* **391**: L5–L8 (99)

von Hoerner, S. 1953. Beitrag zur Turbulenztheorie der Spiralnebel. *Zeitschrift für Astrophysik* **32**: 51–58 (213)

von Hoerner, S. 1960. Die numerische Integration des n-Körper-Problemes für Sternhaufen. I. *Zeitschrift für Astrophysik* **50**: 184–214 (191)

von Weizsäcker, C. F. 1951a. Turbulence in Interstellar Matter (Part 1). In *Proceedings of the Symposium on the Motions of Gaseous Masses of Cosmical Dimensions*, Paris, August 16–19, 1949, pp. 158–161 (213,236)

von Weizsäcker, C. F. 1951b. The Evolution of Galaxies and Stars. *Astrophysical Journal* **114**: 165–186 (213)

Vysotskii, M. I., Dolgov, A. D., and Zel'dovich, Y. B. 1977. Cosmological Limits on the Masses of Neutral Leptons. *ZhETF Pis'ma* **26** 200–202. English translation in *JETP Letters* **26**: 188–190 (291,293)

Wagoner, R. V. 1973. Big-Bang Nucleosynthesis Revisited. *Astrophysical Journal* **179**: 343–360 (175,179)

Wagoner, R. V., Fowler, W. A., and Hoyle, F. 1967. On the Synthesis of Elements at Very High Temperatures. *Astrophysical Journal* **148**: 3–49 (75,128,138,175–179)

Walker, A. G. 1935. On the Formal Comparison of Milne's Kinematical System with the Systems of General Relativity. *Monthly Notices of the Royal Astronomical Society* **95**: 263–269 (37)

Walsh, D., Carswell, R. F., and Weymann, R. J. 1979. 0957 + 561 A, B: Twin Quasistellar Objects or Gravitational Lens? *Nature* **279**: 381–384 (228)

Wasserman, I. 1981. On the Linear Theory of Density Perturbations in a Neutrino+Baryon Universe. *Astrophysical Journal* **248**: 1–12 (286)

Wasserman, I. 1986. Possibility of Detecting Heavy Neutral Fermions in the Galaxy. *Physical Review D* **33**: 2071–2078 (299)

Weinberg, S. 1971. Entropy Generation and the Survival of Protogalaxies in an Expanding Universe. *Astrophysical Journal* **168**: 175–194 (205)

Weinberg, S. 1972. *Gravitation and Cosmology: Principles and Applications of the General Theory of Relativity*. New York: Wiley (293)

Weinberg, S. 1974. Gauge and Global Symmetries at High Temperature. *Physical Review D* **9**: 3357–3378 (225)

Weinberg, S. 1977. *The First Three Minutes. A Modern View of the Origin of the Universe*. Basic Books (162)

Weinberg, S. 1987. Anthropic Bound on the Cosmological Constant. *Physical Review Letters* **59**: 2607–2610 (61)

Weinberg, S. 1989. The Cosmological Constant Problem. *Reviews of Modern Physics* **61**: 1–23 (60)

Wevers, B. M. H. R., van der Kruit, P. C., and Allen, R. J. 1986. The Palomar-Westerbork Survey of Northern Spiral Galaxies. *Astronomy and Astrophysics Supplement* **66**: 505–662 (259)

Weyl, H. 1923. Zur allgemeinen Relativitätstheorie. *Physikalische Zeitschrift* **24**: 230–232. (40)

Weymann, R. 1965. Diffusion Approximation for a Photon Gas Interacting with a Plasma via the Compton Effect. *Physics of Fluids* **8**: 2112–2114 (121)

Weymann, R. 1966. The Energy Spectrum of Radiation in the Expanding Universe. *Astrophysical Journal* **145**: 560–571 (121)

White, M., Scott, D., Silk, J., and Davis, M. 1995. Cold Dark Matter Resuscitated? *Monthly Notices of the Royal Astronomical Society* **276** L69–L75 (335)

White, S. D. M. 1984. Angular Momentum Growth in Protogalaxies. *Astrophysical Journal* **286**: 38–41 (218)

White, S. D. M. 1991. Dynamical Estimates of Ω_0 from Galaxy Clustering. In *Observational Tests of Cosmological Inflation*. Eds. T. Shanks, J. Banday, R. S. Ellis, and A. W. Wolfendale. Dordrecht: Lukwer, pp. 279–291. (101)

White, S. D. M., Efstathiou, G., and Frenk, C. S. 1993. The Amplitude of Mass Fluctuations in the Universe. *Monthly Notices of the Royal Astronomical Society* **262**: 1023–1028 (312)

White, S. D. M., and Frenk, C. S. 1991. Galaxy Formation through Hierarchical Clustering. *Astrophysical Journal* **379**: 52–79 (101)

White, S. D. M., Frenk, C. S., and Davis, M. 1983. Clustering in a Neutrino-Dominated Universe. *Astrophysical Journal Letters* **274**: L1–L5 (288)

White, S. D. M., Navarro, J. F., Evrard, A. E., and Frenk, C. S. 1993. The Baryon Content of Galaxy Clusters: A Challenge to Cosmological Orthodoxy. *Nature* **366**: 429–433 (102,319)

White, S. D. M., and Rees, M. J. 1978. Core Condensation in Heavy Halos: A Two-Stage Theory for Galaxy Formation and Clustering. *Monthly Notices of the Royal Astronomical Society* **183**: 341–358 (218–220,243,263,277,295,308)

Whitford, A. E. 1954. Observational Status of the Color-Excess Effect in Distant Galaxies. *Astrophysical Journal* **120** 599–602 (51)

Wilczek, F. 1983. Conference Summary and Concluding Remarks. In *The Very Early Universe, Proceedings of the Nuffield Workshop*. Cambridge: Cambridge University Press, pp. 475–480 (64)

Williams, J. G., Turyshev, S. G., and Boggs, D. H. 2012. Lunar Laser Ranging Tests of the Equivalence Principle. *Classical and Quantum Gravity* **29**: 184004, 11 pp. (49)

Williams, T. B. 1975. The Rotation Curve of NGC 3115. *Astrophysical Journal* **199**: 586–590 (255)

Willick, J. A., and Strauss, M. A. 1998. Maximum Likelihood Comparison of Tully-Fisher and Redshift Data. II. Results from an Expanded Sample. *Astrophysical Journal* **507**: 64–83 (104)

Willick, J. A., Strauss, M. A., Dekel, A., and Kolatt, T. 1997. Maximum Likelihood Comparisons of Tully-Fisher and Redshift Data: Constraints on Ω and Biasing. *Astrophysical Journal* **486** 629–644 (104,319)

Wilson, M. L. 1983. On the Anisotropy of the Cosmological Background Matter and Radiation Distribution. II. The Radiation Anisotropy in Models with Negative Spatial Curvature. *Astrophysical Journal* **273**: 2–15 (315)

Wilson, M. L., and Silk, J. 1981. On the Anisotropy of the Cosmological Background Matter and Radiation Distribution. I. The Radiation Anisotropy in a Spatially Flat Universe. *Astrophysical Journal* **243**: 14–25 (206)

Wilson, R. W., and Penzias, A. A. 1967. Isotropy of Cosmic Background Radiation at 4080 Megahertz. *Science* **156**: 1100–1101 (21)

Wolfe, A. M., and Burbidge, G. R. 1969. Discrete Source Models to Explain the Microwave Background Radiation. *Astrophysical Journal* **156**: 345–371 (166)

Wolfe, A. M., and Burbidge, G. R. 1970. Can the Lumpy Distribution of Galaxies Be Reconciled with the Smooth X-Ray Background? *Nature* **228**: 1170–1174 (21)

Woosley, S. E., and Weaver, T. A. 1986. The Physics of Supernova Explosions. *Annual Review of Astronomy and Astrophysics* **24**: 205–253 (325)

Wright, E. L. 2004. Theoretical Overview of Cosmic Microwave Background Anisotropy. In *Measuring and Modeling the Universe*. Ed. W. L. Freedman. Cambridge: Cambridge University Press, pp. 291–308 (335)

Wyse, A. B., and Mayall, N. U. 1941. Distribution of Mass in the Spiral Nebulae Messier 31 and 33. *Publications of the Astronomical Society of the Pacific* **53**: 269–276 (250)

Yahil, A., Sandage, A., and Tammann, G. A. 1980. The Determination of the Deceleration Parameter from Local Data. *Physica Scripta* **21**: 635–639 (86)

Yahil, A., Walker, D., and Rowan-Robinson, M. 1986. The Dipole Anisotropies of the IRAS Galaxies and the Microwave Background Radiation. *Astrophysical Journal Letters* **301**: L1–L5 (96)

Yang, J., Schramm, D. N., Steigman, G., and Rood, R. T. 1979. Constraints on Cosmology and Neutrino Physics from Big Bang Nucleosynthesis. *Astrophysical Journal* **227**: 697–704 (179,181)

Yoshimura, M. 1978. Unified Gauge Theories and the Baryon Number of the Universe. *Physical Review Letters* **41**: 281–284 (63)

Yu, J. T., and Peebles, P. J. E. 1969. Superclusters of Galaxies? *Astrophysical Journal* **158**: 103–113 (304)

Zaldarriaga, M., Spergel, D. N., and Seljak, U. 1997. Microwave Background Constraints on Cosmological Parameters. *Astrophysical Journal* **488**: 1–13 (334)

Zang, T. A. 1976. The Stability of a Model Galaxy. PhD thesis, Massachussetts Institute of Technology, Cambrigde, MA (270)

Zaritsky, D., and White, S. D. M. 1994. The Massive Halos of Spiral Galaxies. *Astrophysical Journal* **435**: 599–610 (247)

Zehavi, I., Zheng, Z., Weinberg, D. H., et al. 2011. Galaxy Clustering in the Completed SDSS Redshift Survey: The Dependence on Color and Luminosity. *Astrophysical Journal* **736**: 59, 30 pp. (30,69)

Zel'dovich, Y. B. 1962a. Prestellar State of Matter. *Journal of Experimental and Theoretical Physics U.S.S.R.* **43**: 1561–1562. English translation in *Soviet Journal of Experimental and Theoretical Physics* **16**: 1102–1103, 1963 (144,150)

Zel'dovich, Y. B. 1962b. Star Production in an Expanding Universe. *Journal of Experimental and Theoretical Physics U.S.S.R.* **43**: 1982–1984. English translation in *Soviet Journal of Experimental and Theoretical Physics* **16**: 1395–1396, 1963 (190,223)

Zel'dovich, Y. B. 1963a. The Initial Stages of the Evolution of the Universe. *Soviet Atomic Energy* **14**: 83–91. English translation in *Atomnoya Energiya* **14**: 92–99 (144,159)

Zel'dovich, Y. B. 1963b. The Theory of the Expanding Universe as Originated by A. A. Fridman. *Uspekhi Fizicheskikh Nauk* **80**: 357–390. English translation in *Soviet Physics Uspekhi* **6**: 475–494, 1964 (144,146)

Zel'dovich, Y. B. 1965. Survey of Modern Cosmology. *Advances in Astronomy and Astrophysics* **3**: 241–365 (190,224)

Zel'dovich, Y. B. 1967. The "Hot" Model of the Universe. *Uspekhi Fizicheskikh Nauk* **89** 647; English translation in *Soviet Physics-Uspekhi* **9**: 602–617, 1967 (150,229)

Zel'dovich, Y. B. 1970. Gravitational Instability: An Approximate Theory for Large Density Perturbations. *Astronomy and Astrophysics* **5**: 84–89 (201,234,286)

Zel'dovich, Y. B. 1972. A Hypothesis, Unifying the Structure and the Entropy of the Universe. *Monthly Notices of the Royal Astronomical Society* **160**: 1P–3P (231,302)

Zel'dovich, Y. B. 1978. The Theory of Large Scale Structure of the Universe. In *Proceedings of the Symposium on Large Scale Structure of the Universe*, Tallin, Estonia, September 12–16, 1977. Eds. M. S. Longair and J. Einasto. Dordrecht: Reidel, 409–421 (236)

Zel'dovich, Y. B. 1980. Cosmological Fluctuations Produced Near a Singularity. *Monthly Notices of the Royal Astronomical Society* **192**: 663–667 (225,227)

Zel'dovich, Y. B., Einasto, J., and Shandarin, S. F. 1982. Giant Voids in the Universe. *Nature* **300**: 407–413 (289)

Zel'dovich, Y. B., Kobzarev, I. Y., and Okun', L. B. 1975. Cosmological Consequences of a Spontaneous Breakdown of a Discrete Symmetry. *Zhurnal Eksperimental'noi i Teoreticheskoi Fiziki* **67**: 3–11. English translation in *Soviet Physics—JETP* **40**: 1–5 (226)

Zel'dovich, Y. B., Kurt, V. G., and Sunyaev, R. A. 1968. Recombination of Hydrogen in the Hot Model of the Universe. *Zhurnal Eksperimentalnoi i Teoreticheskoi Fiziki* **55**: 278–286. English translation in *Soviet Journal of Experimental and Theoretical Physics* **28**: 146–150. (120)

Zel'dovich, Y. B., and Novikov, I. D. 1966. Charge Asymmetry and Entropy of a Hot Universe. *ZhETF Pis'ma Redaktsiiu* **4**: 80–82 (160)

Zel'dovich, Y. B., and Sunyaev, R. A. 1969. The Interaction of Matter and Radiation in a Hot-Model Universe. *Astrophysics and Space Science* **4**: 301–316 (122)

Zel'dovich, Y. B., and Sunyaev, R. A. 1980. Astrophysical Implications of the Neutrino Rest Mass. I. The Universe. *Pis'ma Astronomicheskii Zhurnal* **6**: 451–456. English version *Soviet Astronomy Letters* **6**: 249–252 (286)

Zwicky, F. 1929. On the Red Shift of Spectral Lines through Interstellar Space. *Proceedings of the National Academy of Science* **15**: 773–779 (24)

Zwicky, F. 1933. Die Rotverschiebung von Extragalaktschen Nebeln. *Helvetica Physica Acta* **6**: 110–127 (240–242,276,279,292)

Zwicky, F. 1937a. On the Masses of Nebulae and of Clusters of Nebulae. *Astrophysical Journal* **86**: 217–246 (241–243,263,273,276,279,282)

Zwicky, F. 1937b. On a New Cluster of Nebulae in Pisces. *Proceedings of the National Academy of Sciences* **23**: 251 (279)

Zwicky, F., Herzog, E., Wild, P., Karpowicz, M., and Kowal, C. T. 1961–1968. *Catalogue of Galaxies and Clusters of Galaxies*, in 6 volumes. Pasadena: California Institute of Technology (29,88)

INDEX

A NOTE ON THE TYPE

THIS BOOK HAS been composed in Miller, a Scotch Roman typeface designed by Matthew Carter and first released by Font Bureau in 1997. It resembles Monticello, the typeface developed for The Papers of Thomas Jefferson in the 1940s by C. H. Griffith and P. J. Conkwright and reinterpreted in digital form by Carter in 2003.

Pleasant Jefferson ("P. J.") Conkwright (1905–1986) was Typographer at Princeton University Press from 1939 to 1970. He was an acclaimed book designer and aiga Medalist.

The ornament used throughout this book was designed by Pierre Simon Fournier (1712–1768) and was a favorite of Conkwright's, used in his design of the *Princeton University Library Chronicle*.